Imaging Sensors and Applications

Imaging Sensors and Applications

Editors

Changho Lee
Changhan Yoon

MDPI • Basel • Beijing • Wuhan • Barcelona • Belgrade • Manchester • Tokyo • Cluj • Tianjin

Editors
Changho Lee
Chonnam National University
Medical School
Korea

Changhan Yoon
Inje University
Korea

Editorial Office
MDPI
St. Alban-Anlage 66
4052 Basel, Switzerland

This is a reprint of articles from the Special Issue published online in the open access journal *Sensors* (ISSN 1424-8220) (available at: https://www.mdpi.com/journal/sensors/special_issues/sensorsapplication).

For citation purposes, cite each article independently as indicated on the article page online and as indicated below:

LastName, A.A.; LastName, B.B.; LastName, C.C. Article Title. *Journal Name* **Year**, *Volume Number*, Page Range.

ISBN 978-3-0365-2604-1 (Hbk)
ISBN 978-3-0365-2605-8 (PDF)

Contents

MDPI

Editorial

Recent Advances in Imaging Sensors and Applications

Changhan Yoon [1,2] and Changho Lee [3,4,*]

1 Department of Biomedical Engineering, Inje University, Gimhae 50834, Korea; cyoon@inje.ac.kr
2 Department of Nanoscience and Engineering, Inje University, Gimhae 50834, Korea
3 Department of Artificial Intelligence Convergence, Chonnam National University, 77 Yongbong-ro, Buk-gu, Gwangju 61186, Korea
4 Department of Nuclear Medicine, Chonnam National University Medical School & Hwasun Hospital, Hwasun 58128, Korea
* Correspondence: ch31037@jnu.ac.kr

Citation: Yoon, C.; Lee, C. Recent Advances in Imaging Sensors and Applications. *Sensors* **2021**, *21*, 3970. https://doi.org/10.3390/s21123970

Received: 29 May 2021
Accepted: 7 June 2021
Published: 9 June 2021

Publisher's Note: MDPI stays neutral with regard to jurisdictional claims in published maps and institutional affiliations.

Recent advances in sensor technology have allowed us to develop many interesting applications and enhance the quality of human life. In particular, imaging sensors have been regarded as critical elements in achieving high-level imaging methods such as laser-based imaging, ultrasound imaging, and X-ray imaging, as well as non-destructive inspection imaging, contributing to high sensitivity, real-time display, and compact implementation. These cutting-edge sensing technologies are a crucial player in the biomedical and industrial fields.

The purpose of this Special Issue is to cover some of the recent developments in imaging sensors and their applications. The Special Issue has been co-organized by Prof. Changho Lee, Department of Nuclear Medicine, Chonnam National University Medical School, Korea, and Prof. Changhan Yoon, Biomedical Engineering, Inje University, Korea. In this Special Issue, 17 original papers and two review papers have been published [1–19]. Most of the papers (15 papers) are in the field of biomedical engineering and four papers are related to the field of industrial applications.

Photoacoustic imaging is an emerging technology that combines optical contrast and ultrasonic resolution. This technology allows us to visualize functional information deep inside the body with high spatial resolution, which was not possible with a pure optical imaging modality. To further increase the depth-of-field, T. P Nguyen et al. proposed a multifocal point transducer for photoacoustic microscopy [1]. This work fabricated the multifocal point transducer with seven focal points by separated spherically focused surfaces. J. Jang et al. presented a transrectal ultrasound and photoacoustic probe for prostate cancer detection [2]. The goal of this work was to develop a transrectal hybrid probe, of which the size is similar to that of the currently used transrectal ultrasound transducer. T. T. Mai et al. performed a pilot study to monitor peripheral vascular dynamic to investigate the side effects of carfilzomib using quantitative photoacoustic imaging [3]. Additionally, new tracking and visualization using fast photoacoustic microscopy have been proposed to perform the safe and accurate navigation of balloon catheters for arterial stenosis dilatation, coronary artery disease, and gastrointestinal tracking applications [4]. R. Manwar et al. proposed the photoacoustic imaging approach to estimate the maximum thickness of the skull [5].

Many research papers have been published in the field of conventional medical imaging. High-resolution imaging techniques based on synthetic aperture and plane wave have been proposed for ophthalmic and abdominal applications and their performances were evaluated through ex vivo and in vivo studies [6,7]. C. Z.-H. Ma et al. proposed a new protocol of measuring bilateral back muscle stiffness along the thoracic and lumbar spine with ultrasound imaging [8]. In this work, they ascertained that ultrasound shearwave elastography and a tissue ultrasound palpation system produced reliable results for measuring back muscle stiffness. K. Kim et al. introduced an advanced bandwidth expander circuit composed of unique switching designs to support a wide range of the transducer with a single ultrasound imaging system [9]. In addition, two comprehensive review papers have been published in this Special Issue. One is about the recent development of super-resolution ultrasound imaging and another is about the limitation of clinical elastography diagnosis [10,11].

J. Kang et al. proposed a brain tumor classification method based on a combination of deep features and machine learning classifiers [12]. In this work, they tried to adopt the theory of transfer learning and utilized various pre-trained deep learning algorithms to acquire crucial deep features of brain magnetic resonance imaging data. For the remote sharing economy, two-photon laser scanning microscopy based on the internet of things was proposed as a remote research equipment sharing system [13]. Using the internet of things modules, they developed a web service system where data are transmitted to and received from remote users and installed in the two-photon laser scanning microscopy. S. A. Saleah et al. presented a new quad-scanner-based optical coherence tomography for visualizing the full-directional volumetric structure [14]. H. Wu et al. proposed a new approach for measuring the adjustable volumetric frequency and phase information of the human chest and abdomen surface regardless of motion artifacts [15].

For industrial applications, J. Lee et al. proposed a novel around view monitoring calibration method to avoid conventional exhaustive procedures which includes accurate positioning and estimating the calibration boards surrounding the vehicle [16]. This method only requires four pieces of random calibration information based on the correct position of individual calibrating boards. G. Lefever et al. investigated the effectiveness of elastic waves for a non-destructive testing method of cementitious samples and revealed their composites of the inner structure at the microscale [17]. K. A. Tiwari et al. presented a new analysis method of wave patterns from the macro-fiber composite transducer to overcome the limitation of accuracy issue of the previous analytical model [18]. They confirmed that the proposed model enhanced the analytical modeling for directivity pattern estimation. A multi-wavelength fluorescence LiDAR system was proposed for vegetation monitoring in forestry and agricultural applications [19]. The authors extended the system to the multi-channel fluorescence detection of laser-induced fluorescence based on the LiDAR scanning and ranging mechanism.

Funding: An NRF grant funded by the Korean government (MSIT) (NRF-2019R1F1A1062948 and NRF-2019R1A2C1089813) and Bio & Medical Technology Development Program of the NRF funded by the Korean government (MSIT) (NRF-2019M3E5D1A02067958).

Acknowledgments: The Guest Editors thank all the authors, reviewers, and members of MDPI's editorial team whose work has led to the publication of this Special Issue.

Conflicts of Interest: The authors declare no conflict of interest.

References

1. Nguyen, T.P.; Nguyen, V.T.; Mondal, S.; Pham, V.H.; Vu, D.D.; Kim, B.; Oh, J. Improved Depth-of-Field Photoacoustic Microscopy with a Multifocal Point Transducer for Biomedical Imaging. *Sensors* **2020**, *20*, 2020. [CrossRef] [PubMed]
2. Jang, J.; Kim, J.; Lee, H.J.; Chang, J.H. Transrectal Ultrasound and Photoacoustic Imaging Probe for Diagnosis of Prostate Cancer. *Sensors* **2021**, *21*, 1217. [CrossRef] [PubMed]
3. Mai, T.T.; Vo, M.; Chu, T.; Kim, J.Y.; Kim, C.; Lee, J.; Jung, S.; Lee, C. Pilot Study: Quantitative Photoacoustic Evaluation of Peripheral Vascular Dynamics Induced by Carfilzomib In Vivo. *Sensors* **2021**, *21*, 836. [CrossRef] [PubMed]
4. Kim, J.; Mai, T.T.; Kim, J.Y.; Min, J.; Kim, C.; Lee, C. Feasibility Study of Precise Balloon Catheter Tracking and Visualization with Fast Photoacoustic Microscopy. *Sensors* **2020**, *20*, 5585. [CrossRef] [PubMed]
5. Manwar, R.; Kratkiewicz, K.; Avanaki, K. Investigation of the Effect of the Skull in Transcranial Photoacoustic Imaging: A Preliminary Ex Vivo Study. *Sensors* **2020**, *20*, 4189. [CrossRef] [PubMed]
6. Lim, H.G.; Kim, H.H.; Yoon, C. Synthetic Aperture Imaging Using High-Frequency Convex Array for Ophthalmic Ultrasound Applications. *Sensors* **2021**, *21*, 2275. [CrossRef] [PubMed]
7. Bae, S.; Jang, J.; Choi, M.H.; Song, T.-K. In Vivo Evaluation of Plane Wave Imaging for Abdominal Ultrasonography. *Sensors* **2020**, *20*, 5675. [CrossRef] [PubMed]

8.   Ma, C.Z.; Ren, L.; Cheng, C.L.; Zheng, Y. Mapping of Back Muscle Stiffness along Spine during Standing and Lying in Young Adults: A Pilot Study on Spinal Stiffness Quantification with Ultrasound Imaging. *Sensors* **2020**, *20*, 7317. [CrossRef] [PubMed]

9.   Kim, K.; Choi, H. Novel Bandwidth Expander Supported Power Amplifier for Wideband Ultrasound Transducer Devices. *Sensors* **2021**, *21*, 2356. [CrossRef] [PubMed]

10.   Chen, Q.; Song, H.; Yu, J.; Kim, K. Current Development and Applications of Super-Resolution Ultrasound Imaging. *Sensors* **2021**, *21*, 2417. [CrossRef] [PubMed]

11.   Rus, G.; Faris, I.H.; Torres, J.; Callejas, A.; Melchor, J. Why Are Viscosity and Nonlinearity Bound to Make an Impact in Clinical Elastographic Diagnosis? *Sensors* **2020**, *20*, 2379. [CrossRef] [PubMed]

12.   Kang, J.; Ullah, Z.; Gwak, J. MRI-Based Brain Tumor Classification Using Ensemble of Deep Features and Machine Learning Classifiers. *Sensors* **2021**, *21*, 2222. [CrossRef] [PubMed]

13.   Park, E.; Lim, J.; Park, B.C.; Kim, D. IoT-Based Research Equipment Sharing System for Remotely Controlled Two-Photon Laser Scanning Microscopy. *Sensors* **2021**, *21*, 1533. [CrossRef] [PubMed]

14.   Saleah, S.A.; Seong, D.; Han, S.; Wijesinghe, R.E.; Ravichandran, N.K.; Jeon, M.; Kim, J. Integrated Quad-Scanner Strategy-Based Optical Coherence Tomography for the Whole-Directional Volumetric Imaging of a Sample. *Sensors* **2021**, *21*, 1305. [CrossRef] [PubMed]

15.   Wu, H.; Yu, S.; Yu, X. 3D Measurement of Human Chest and Abdomen Surface Based on 3D Fourier Transform and Time Phase Unwrapping. *Sensors* **2020**, *20*, 1091. [CrossRef] [PubMed]

16.   Lee, J.H.; Lee, D. A Novel AVM Calibration Method Using Unaligned Square Calibration Boards. *Sensors* **2021**, *21*, 2265. [CrossRef] [PubMed]

17.   Lefever, G.; Snoeck, D.; Belie, N.D.; Vlierberghe, S.V.; Hemelrijck, D.V.; Aggelis, D.G. The Contribution of Elastic Wave NDT to the Characterization of Modern Cementitious Media. *Sensors* **2020**, *20*, 2959. [CrossRef] [PubMed]

18.   Tiwari, K.A.; Raisutis, R. Mazeika, L. Analysis of Wave Patterns Under the Region of Macro-Fiber Composite Transducer to Improve the Analytical Modelling for Directivity Calculation in Isotropic Medium. *Sensors* **2021**, *21*, 836.

19.   Zhao, X.; Shi, S.; Yang, J.; Gong, W.; Sun, J.; Chen, B.; Guo, K.; Chen, B. Active 3D Imaging of Vegetation Based on Multi-Wavelength Fluorescence LiDAR. *Sensors* **2020**, *20*, 935. [CrossRef] [PubMed]

sensors

Article

Novel Bandwidth Expander Supported Power Amplifier for Wideband Ultrasound Transducer Devices

Kyeongjin Kim and Hojong Choi *

Department of Medical IT Convergence Engineering, Kumoh National Institute of Technology,
350-27 Gumi-daero, Gumi 39253, Korea; 20196092@kumoh.ac.kr
* Correspondence: hojongch@kumoh.ac.kr; Tel.: +82-054-478-7782

Abstract: Ultrasound transducer devices have their own frequency ranges, depending on the applications and specifications, due to penetration depth, sensitivity, and image resolution. For imaging applications, in particular, the transducer devices are preferable to have a wide bandwidth due to the specific information generated by the tissue or blood vessel structures. To support these ultrasound transducer devices, ultrasound power amplifier hardware with a wide bandwidth can improve the transducer performance. Therefore, we developed a new bandwidth expander circuit using specially designed switching architectures to increase the power amplifier bandwidth. The measured bandwidth of the power amplifier with the help of the bandwidth expander circuit increased by 56.9%. In addition, the measured echo bandwidths of the 15-, 20-, and 25-MHz transducer devices were increased by 8.1%, 6.0%, and 9.8%, respectively, with the help of the designed bandwidth expander circuit. Therefore, the designed architecture could help an ultrasound system hardware with a wider bandwidth, thus supporting the use of different frequency ultrasound transducer devices with a single developed ultrasound system.

Keywords: bandwidth expander; ultrasound transducer device; power amplifier

Citation: Kim, K.; Choi, H. Novel Bandwidth Expander Supported Power Amplifier for Wideband Ultrasound Transducer Devices. *Sensors* **2021**, *21*, 2356. https://doi.org/10.3390/s21072356

Academic Editor: Changho Lee

Received: 17 February 2021
Accepted: 25 March 2021
Published: 28 March 2021

Publisher's Note: MDPI stays neutral with regard to jurisdictional claims in published maps and institutional affiliations.

1. Introduction

Ultrasound transducers are the main sensor devices used in ultrasound systems [1]. Different types of ultrasound transducers are used, depending on their applications and specifications [2]. Engineers have manufactured ultrasound transducers with the required diagnostic applications. For diagnostic applications with different positions and locations, the imaging resolution and sensitivity performance are merits in the evaluation of ultrasound systems [3]. Typically, lower-frequency ultrasound transducer devices have a higher penetration depth and lower imaging resolution than higher-frequency ultrasound transducer devices [4]. Smaller-sized ultrasound transducer devices for intracardiac and intravascular applications need to be used such that the penetration depth of the devices is lower, but a higher imaging resolution could be preferable [5].

Figure 1 shows the penetration depth and imaging resolution for an ultrasound diagnostic analysis. Since the imaging resolution and penetration depth have a trade-off relationship, ultrasound transducers with smaller piezoelectric elements are used depending on the body parts [6]. For the heart and abdomen areas, 2–3 and 3–5 MHz transducer devices are preferable because of the penetration depth [5,7]. For the eyeballs and breast and thyroid areas, 7.5–12 MHz and 7.5–13 MHz transducer devices are used due to their relatively high imaging resolutions [5,7]. In particular, for peripheral blood vessels and digestive tract areas, small transducer devices with a high frequency, such as 15–30 MHz, are used because of the higher imaging resolution, which reduces the penetration depth [8–10].

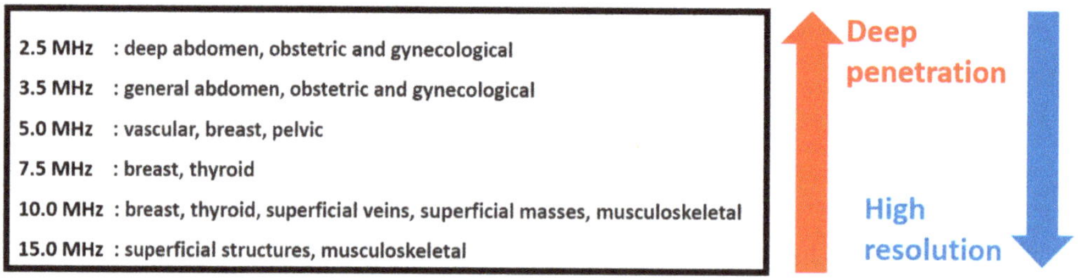

2.5 MHz : deep abdomen, obstetric and gynecological
3.5 MHz : general abdomen, obstetric and gynecological
5.0 MHz : vascular, breast, pelvic
7.5 MHz : breast, thyroid
10.0 MHz : breast, thyroid, superficial veins, superficial masses, musculoskeletal
15.0 MHz : superficial structures, musculoskeletal
Deep penetration
High resolution

Figure 1. Penetration depth and imaging resolution for ultrasound diagnostic analysis.

Figure 2 shows the ultrasound transducer probes (devices) used in ultrasound systems for specific applications [7,11]. Each ultrasound transducer probe has different frequency characteristics and shapes; thus, different anatomical cross-sectional images are obtained [11,12]. However, only one ultrasound system may not cover the transducer probes with different frequency ranges.

Abdomen
Obstetrics
Gynecology
Vascular
Musculoskeletal
Small parts
Urology
Obstetrics
Gynecology
Cardiac
Pediatric
TCD

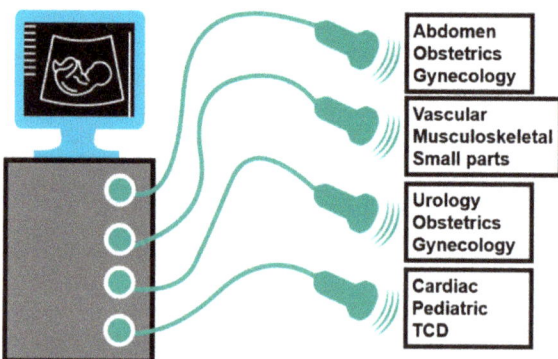

Figure 2. Transducer devices for various applications.

Various classes of power amplifiers have been developed for different types of ultrasound transducer devices. In power amplifiers, the input signals are amplified by active devices, which we call metal-oxide-semiconductor field-effect transistors (MOSFETs) or lateral-diffusion metal–oxide semiconductors (LDMOSs) [13]. Depending on the direct current (DC) bias voltages, the output currents generated by the active devices are fully or partially conducted during a single period of time [14]. This fundamental concept is used to categorize power amplifier classes, such as classes A, B, and C. In the Class A power amplifier, the operating bias voltage is located in the middle of the DC load line to minimize signal distortions [15]. Therefore, the active device in the power amplifier has a heavy load because the active device continues to operate regardless of the input signal [16,17]. Power is consumed continuously as long as an active device is operating. However, signal distortion by an active device is minimized, resulting in an output signal with high linearity [18]. In Class B power amplifiers, the bias voltage is located near the middle of the DC load line, and thus, the active device operates only for half the period of the signal [17]. Therefore, the active device operates for only half a period, which causes signal distortion, although the power consumption of the active device is reduced, compared to that of the Class A power amplifier. Class C power amplifiers operate for less than half a period of the signal because the bias voltage of the DC load line is lower than that of Class B amplifiers [13]. Therefore, the distortion of the signal is extremely high, whereas the power consumption of the active device is quite low. In addition, Class AB power amplifiers operate between Class A and B power amplifiers. In addition, there are other types of power amplifiers, such as Class D and Class E, which use harmonic components or modulated waveforms [19].

The Class B power amplifiers operate at only half of the pulse cycle, which has a much shorter time than Class A and Class AB power amplifiers. Therefore, Class B power amplifiers have much lower power consumption than those of Class A and Class AB power amplifiers, and thus, they have a higher efficiency [20]. The Class B power amplifier has lower signal distortions than the Class A power amplifier. Therefore, wire-type ultrasound machines have a Class A power amplifier because of the AC power cord [2]. Class B or Class C power amplifiers are preferable for mobile or portable ultrasound systems due to their limited battery modules [6]. Several studies related to power amplifiers have been conducted on ultrasound-transducer devices. For ultrasound signals, the burst or modulated waveforms with a limited time period and non-continuous signals are used. A typical Class A power amplifier was developed for ultrasound imaging applications [21]. A Class B power amplifier was developed to reduce the signal distortion [22].

There are a few studies of the power amplifiers used for ultrasound applications. There is a Class C power amplifier used for a 25 MHz transducer. This amplifier has high efficiency with high signal distortions. Therefore, the proposed circuit was developed to compensate for signal distortions [23]. Class D amplifier was developed for low-frequency piezoelectric transducers [24]. This amplifier was used for the high signal distortion and high-efficiency system. The Class E power amplifier was developed to improve the efficiency of low-frequency inductive piezoelectric converters [25]. As mentioned, various power amplifiers with various characteristics are used because all the amplifiers cannot satisfy some parameters of the signal distortion and the bandwidth. In ultrasound systems, transmitter and receiver architectures support ultrasound devices. The power amplifier in a transmitter is typically used in the last stage of electronics [2]. Thus, the bandwidth of the power amplifier only covers the specific transducer probes with their limited frequency ranges because the bandwidth of the power amplifier electronic devices decreases as the operating frequency of the power amplifier increases. This indicates that the output signals working at frequencies higher than the center frequency may deteriorate the image quality. Therefore, we first proposed a bandwidth-expander circuit for power amplifiers and various ultrasound devices.

There are several ways to increase the bandwidth of the power amplifier output. Impedance matching improves the amplitude and bandwidth of the power amplifiers and ultrasound transducers [26]. The output power can be increased through impedance matching, thus increasing the echo amplitude or the echo bandwidth of the ultrasound transducers [26]. However, the impedance of the ultrasound transducer varies significantly according to the frequency [27]. Furthermore, transducers are manufactured to have different central frequencies and resonance/anti-resonance frequency ranges [28]. Therefore, it can be difficult to match the impedance magnitudes within the desired frequency ranges for several ultrasound transducers.

Another method is to lower the gain and increase the bandwidth of the power amplifier by using a feedback loop circuit [14]. This method a fundamental approach for increasing the bandwidth of a power amplifier. This method can reduce the gain of the power amplifier and increase the bandwidth [16]. Thus, several-stage power amplifiers need to be utilized to increase the gain and bandwidth. As the number of stages increases, various problems such as a time delay and signal distortions due to line resistances and parasitic nonlinear components could affect the performance of the power amplifiers [29]. Moreover, there could be a bandwidth expansion limit because space and cost are finite in terms of manufacturing.

In this study, we developed a circuit that can enlarge the bandwidth and minimize the signal loss, thus applying it to a single-stage power amplifier. In our proposed circuit, the bandwidth can be improved by lowering the input poles of the power amplifiers. The designed circuit works as a switching mode that is simply turned on/off with DC power. This method can be useful to improve the bandwidth of the power amplifier and minimize the signal loss, thus supporting higher-frequency ultrasound devices. Section 2 describes the theoretical background and analysis of the proposed circuit and the power amplifier. Section 3 presents the measured performance and a discussion of the proposed circuit with several ultrasound transducers. Section 4 provides some concluding remarks regarding this research.

2. Materials and Methods

The designed bandwidth expander (BWE) is a type of switching circuit operated by different applied DCs. It is located before the lateral-diffusion metal-oxide-semiconductor field-effect transistor (PD57018-E, LDMOSFET, STMicroelec-tronics, Geneva, Switzerland), which is a type of high-voltage MOSFET (BSS123, active device in BWE) in the input port of the amplifier. When the active LDMOSFET device in the amplifier operates under higher-voltage amplitudes than a certain bias voltage, the amplifier is applied to the gate of the LDMOSFET [30]. Therefore, the drain and source of the MOSFET are applied such that the input impedance of the amplifier is changed accordingly. In addition, by adjusting the input impedance, the input poles of the amplifier can be tuned. The total transfer function of the amplifier is expected to increase by integrating the BWE circuit with the amplifier. Therefore, the amplifier with the help of the BWE has a higher output amplitude and wider bandwidth.

2.1. Designed Power Amplifier and BWE Schematic Diagram

Figure 3 shows an amplifier combined with a BWE circuit for the ultrasound transmitter. Table 1 shows the resistor, capacitor, and inductor elements of the amplifier, except for the LDMOSFET shown in Figure 3. The LDMOSFET (PD57018-E, STMicroelectronics, Geneva, Switzerland) was used as the core component of the amplifier. The simulation was conducted to make the impedance matching suitable for the 15 MHz transducer. In Figure 3, V_B is the main gate-source operating point of the LDMOSFET. In addition, LC1 and LC2 are choke coil inductors that prevent the inflow of the alternating current (AC) into the DC power supply [31]. The components used in the input line (LG1, LG2, CG1, CG2, and RG1) and the components used in the output line (LD1, LD2, CD1, CD2, and RD1) are tuned to be compatible with the 15 MHz transducer and to achieve the values below −10 dB of S (1,1) and S (2,2). In addition, they need to block the DC current from VGG and VDD using CG2 and CD2. If AC input voltages enter the DC power supply, oscillations may occur [32]. The polarizing electrolytic capacitors CG3 and CD2 are helpful in providing a constant direct voltage [33]. The ceramic capacitors CG4, CG5, CD4, and CD5 were used to reduce noise as bypass capacitors. The bias voltage was adjusted to operate through the voltage distributions of $RG2$ and $RG3$. The bias voltage equation at V_B is as follows [34,35]:

$$V_B = VGG \times \frac{RG3}{RG2 + RG3}. \tag{1}$$

Table 1. Numerical values of the circuit elements of Figure 3.

Components	Values	Components	Values
$RG1$	200 ohm	CD2	850 μF
$RG2$	1000 ohm	CD3	220 μF
$RG3$	Variable resistance	CD4	1000 pF
RD1	200 ohm	CD5	100 pF
CG1	550 pF	LG1	21 nH
CG2	340 pF	LG2	1000 nH
CG3	220 μF	LD1	130 nH
CG4	1000 pF	LD2	500 nH
CG5	100 pF	LC1	1 μH
CD1	340 pF	LC2	1 μH

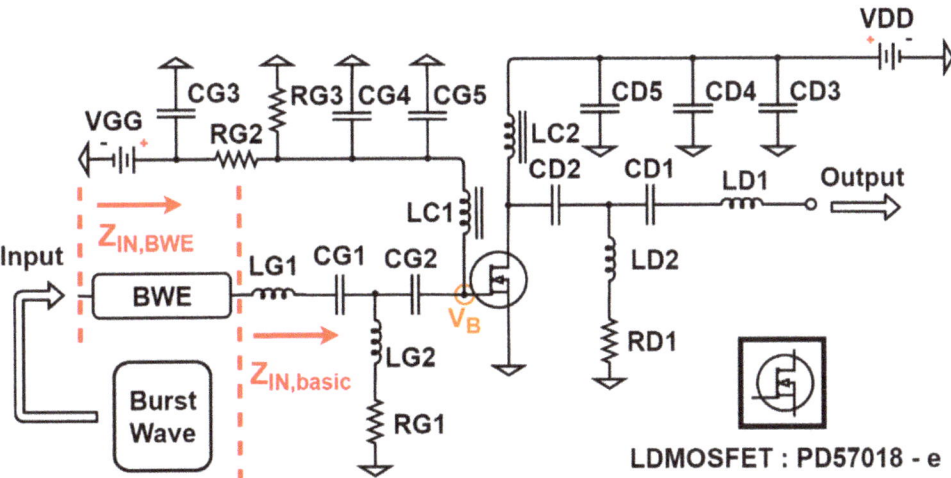

Figure 3. Designed amplifier schematic diagram.

In ultrasonic diagnostic equipment, the frequency of the input signal and the cycle of the burst wave were adjusted such that the Q factor used for providing an appropriate image quality was calculated [1]. Assuming that the LDMOSFET operates in the saturation region, the changes in the pole and transfer function can be estimated. As shown in Figure 3, the BWE circuit has little effect on the impedance. However, if a DC voltage is applied and operated, the input impedance changes. Therefore, the input pole changes. As the input pole is varied, the transfer function according to the frequency changes, and the output amplitude and bandwidth are changed. This concept is proved through several equations.

Figure 4 shows a schematic diagram of the designed BWE to show the circuit element values. Table 2 shows the numerical values of the resistors, capacitors, and inductor elements in Figure 4. R1 and R3 were used to be tuned to have proper voltage distribution and power consumption because M1 needs to be operated properly. The input signal was connected to the gate of M1 to form a feedback loop and can cause oscillation to be blocked through L1. However, the too high value of inductance may distort the input signal, so we properly selected the inductor value. R2 and C1 play a major role to lower the input impedance because the drain and source of M1 are shorted. If the impedance is too low, signal amplitude can be reduced. Thus, we properly selected those values for an input signal of 15 MHz. This circuit is added to the circuit, as shown in Figure 3, and thus the amplifier performance by applying different constant DC voltages is changed depending on the MOSFET (M1, BSS123) operation. When this circuit is assumed to be an ideal current source or ideal switch, the DC bias voltage V_{B1} in the BWE circuit operates the transistor, M1, as described below [36–38]:

$$V_{B1} = \left(V_{DC} - V_{D1(TH)} - V_{SD}\right) \times \frac{R_2||R_2}{R_1 + R_2||R_2 + R_3||R_3},\qquad(2)$$

where $V_{D1(TH)}$ and V_{SD} are the threshold voltage and drain-source diode forward voltage, respectively. Above a certain voltage, V_{B1} shortens the drain and source of M1. As the drain and source voltages at M1 are short-circuited, C1, $R2$, and L1, and the capacitance and on-state resistance inside M1 are connected in parallel to the input impedance.

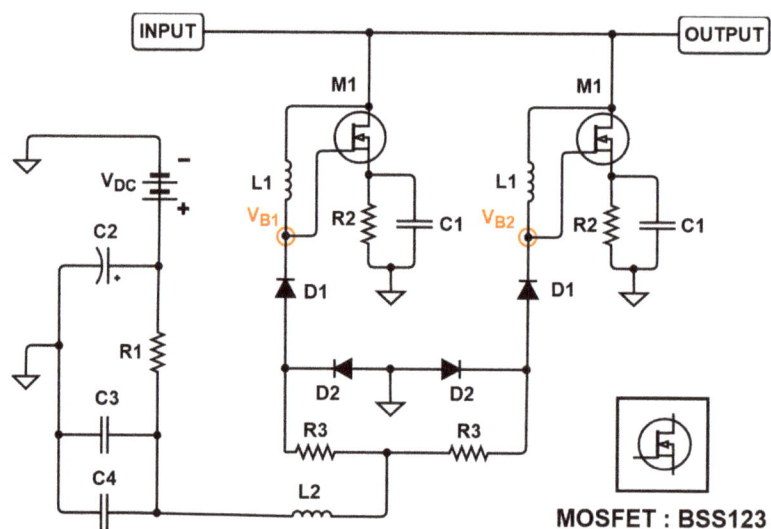

Figure 4. Schematic diagram of designed bandwidth expander (BWE) circuit.

Table 2. Numerical values of the circuit elements of Figure 4.

Components	Values	Components	Values
R1	150 ohm	C3	1000 pF
R2	750 ohm	C4	100 pF
R3	50 ohm	L1	560 nH
C1	47 pF	L2	1 µH
C2	220 µF		

2.2. Predicting Performance Results

Changes in the performance of the amplifier only and the amplifier with BWE can be expected and compared based on pole calculations. Therefore, the equivalent circuit models of the LDMOSFET in the amplifier and the MOSFET in the BWE were simplified to calculate the poles. First, the input impedance ($Z_{IN,basic}$) of the amplifier (Figure 5) was calculated as follows:

$$Z_{IN,basic} = X_{CG2}||(X_{LG2} + RG1) + X_{CG1} + X_{LG1}, \tag{3}$$

where X_{CG1} and X_{CG2} are the impedances of the capacitors, and X_{LG1} is the impedance of the inductor (see Figure 5). When the V_{DC} is applied to M1, the V_{PP} of the input signal is blocked at D1 because the DC level is increased by V_{B1}, as shown in Figure 4. In addition, the gate of M1 is short-circuited due to the applied V_{DC}. The on-state resistance R_D and internal capacitances C_{GD}, C_{GS}, and C_D exist. When M1 in the BWE circuit operates, it is expressed as an equivalent circuit for the AC analysis. Here, Z_{BWE} indicates the impedance of the equivalent circuit when the circuit operates and is expressed as follows:

$$Z_{BWE} = R_D||X_{C_D}||(X_{L1}||X_{C_{GD}} + X_{C_{GS}}) + R2||X_{C1}. \tag{4}$$

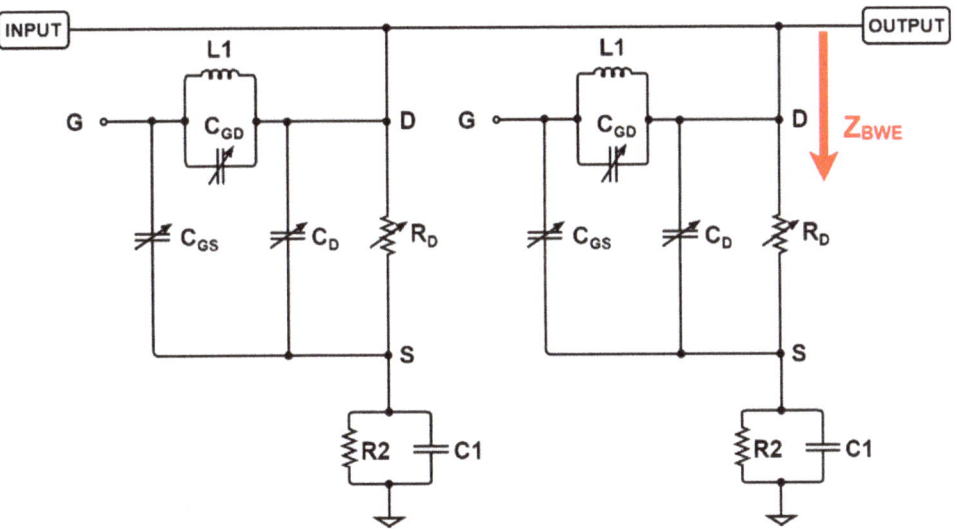

Figure 5. Equivalent circuit model when BWE circuit is operated.

The input impedance in the BWE circuit was operated with parallel circuits, as shown in Figure 5.

$$Z_{IN, \; BWE} = Z_{BWE} || Z_{BWE} || Z_{IN,basic}. \tag{5}$$

Figure 6 shows the equivalent circuit model of the amplifier [17]. Assuming that the bias voltage V_B of the amplifier only, and the amplifier with BWE are assumed to be the same, the internal capacitances $C_{L,GD}$, $C_{L,GS}$, $C_{L,DS}$, and $g_m V_{gs}$ are the same.

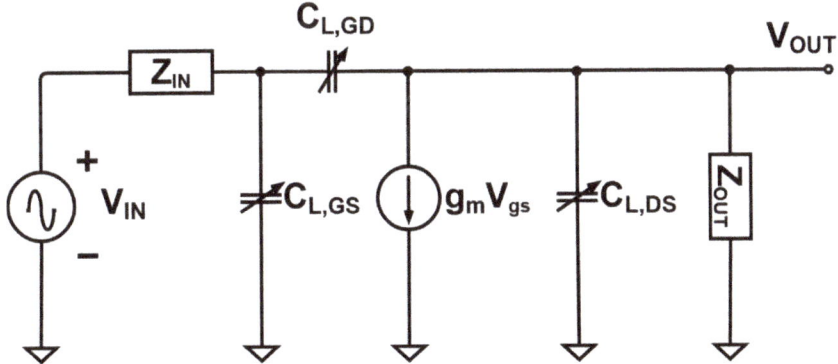

Figure 6. Equivalent circuit model of the amplifier.

As a result, the input and output poles and the transfer function can be predicted. The input pole of the amplifier only and the amplifier with the BWE circuit are given by Equation (6).

$$\omega_{IN,basic} = \frac{1}{Z_{IN,basic}[C_{L,GS} + (1 + g_m R_D)C_{GD}]} \tag{6}$$

$$\omega_{IN,BWE} = \frac{1}{Z_{IN,BWE}[C_{L,GS} + (1 + g_m R_D)C_{GD}]} \tag{7}$$

The output impedances of the amplifier only and the amplifier with the BWE circuit are assumed to be the same. The currents flowing from drain to source in the main transistor are the same if the same bias voltage is applied to both the amplifier and amplifier + BWE; the internal capacitance (C_{iss}, C_{oss}, and C_{rss}) is also the same. Thus, the internal capacitances ($C_{L,GD}$, $C_{L,GS}$, $C_{L,DS}$, and $g_m V_{gs}$) are supposed to be the same. Therefore, the output impedances of the amplifier and amplifier + BWE are the same as Z_{OUT}. Irrelevant to a transducer, the output signal does not change. Consequently, the output poles in the transfer functions of the amplifier only and the amplifier with the BWE circuit are expressed in Equations (8)–(10).

$$\omega_{OUT} = \frac{1}{Z_{OUT}\left[C_{DS} + \left(1 - A_v^{-1}\right)C_{GD}\right]} \cong \frac{1}{Z_{OUT}(C_{DS} + C_{GD})} \tag{8}$$

$$\frac{V_{OUT}}{V_{IN}}(s), \, basic = \frac{-g_m Z_{OUT}}{\left(1 + \frac{s}{\omega_{IN,basic}}\right)\left(1 + \frac{s}{\omega_{OUT}}\right)} \tag{9}$$

$$\frac{V_{OUT}}{V_{IN}}(s), \, BWE = \frac{-g_m Z_{OUT}}{\left(1 + \frac{s}{\omega_{IN, \, BWE}}\right)\left(1 + \frac{s}{\omega_{OUT}}\right)} \tag{10}$$

Equation (11) shows that $Z_{IN,BWE}$ has a relatively lower impedance than $Z_{IN,basic}$ because Z_{BWE} is connected in parallel to $Z_{IN,basic}$. Looking at Equations (6) and (7), the input pole has an inversely proportional relationship with the input impedance. Furthermore, from Equations (9) and (10), the transfer function has a proportional relationship with the input pole. As a result, the pole and transfer functions due to the different input impedances can be predicted as follows:

$$Z_{IN,basic} > Z_{IN,BWE} \tag{11}$$

$$\omega_{IN,basic} < \omega_{IN,BWE} \tag{12}$$

$$\frac{V_{OUT}}{V_{IN}}(s), \, basic < \frac{V_{OUT}}{V_{IN}}(s), \, BWE. \tag{13}$$

However, the predicted results should be operated in an ideal environment. The actual results are extremely different from the predicted results because the high-voltage amplifier operations are not accurately predictable due to several high-voltage environment variables such as parasitic impedances, high-voltage valuable variances, and unpredictable equivalent inductance models [39,40].

Figure 7 shows the predicted graph based on Equations (11)–(13). By adding a *BWE* circuit to the existing amplifier, the magnitude and input pole in the transfer function increase. In Figure 7, the entire line goes up, and the input pole moves to the right. As a result, we assume that the line decreases by 20, 40, and 20 dB/dec, respectively, in the interval between the input pole, the output pole, and the zero point. Comparing the two cases, the transfer function is expected to increase. As the pole location shifts, we can expect to achieve a wider bandwidth.

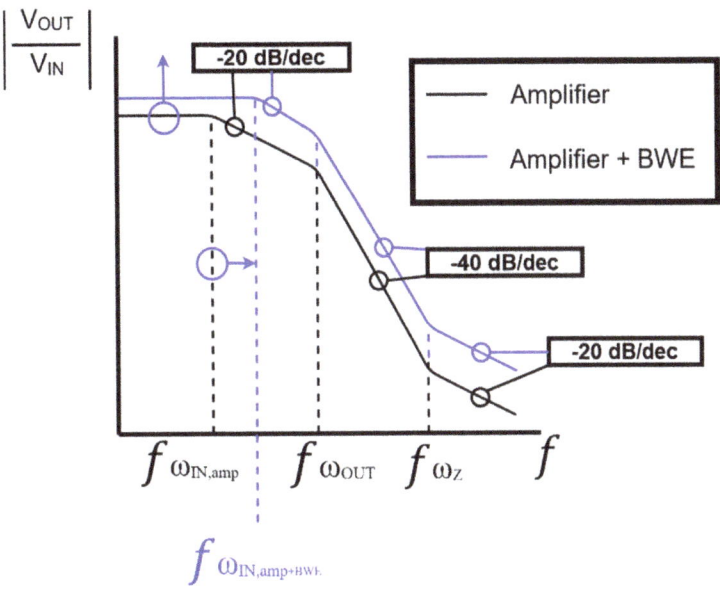

Figure 7. The predicted transfer function graph of the amplifier only and the amplifier with BWE.

2.3. Experimental Measurement Process

Figure 8 shows a block diagram showing the performance measurement of the amplifier only and the amplifier with *BWE* circuits. A function generator, DC power supply, attenuator, and oscilloscope were used to measure the performance of the designed circuits. A BWE circuit was installed between the function generator and amplifier. The BWE circuit operates when DC power is supplied by the power supply. The amplified signal can cause damage to the oscilloscope from a high voltage of 5 V_{P-P} or greater. Therefore, the output signals were attenuated by a 40 dB attenuator, and the performances were measured using an oscilloscope [41–43]. To measure the performance of the amplifiers, the frequency and input signal amplitudes were adjusted using a function generator. In addition, the voltage gain was obtained from the measured output signals. As a result, the outputs of the amplifier and BWE circuit-equipped amplifier were measured and compared. The voltage gain is a performance indicator for measuring the performance of an amplifier, and its high output helps to provide a clear ultrasound image [44–46]. In addition, the bandwidth of the gain over frequency of the amplifier only and the amplifier equipped with the BWE circuit can be compared. The output signal is an important performance indicator of an ultrasonic transmitter because it shows the sensitivity of the system.

Figure 9a shows the measurement procedures used to obtain the echo signal of the transducer with the designed amplifier with and without a BWE circuit. Various instruments have been used to measure the echo signal to determine its compatibility with ultrasonic transducer probes [47]. The amplified signal was passed through the expander [48,49]. The signal was transmitted through the transducer probe and reflected by the quartz to be received [50–52]. The expander was used to remove the noise and reduce the ringdown signal from amplified signals [53–55]. Since the received signal has an extremely low amplitude, it is amplified by an approximately 32 dB gain pre-amplifier and then displayed on the oscilloscope. During this process, because quartz reflects more than 99% of the signal, the data of the echo signal can be measured to estimate the amplifier performance [56,57]. The amplified signal, called a discharged signal, is required to vibrate the piezoelectric element of the transducer probe; however, it is not necessary to measure the echo signal, and the oscilloscope can be damaged with a voltage of higher than 5 V_{P-P}, and thus, it is minimized using a limiter [58–60]. Figure 9a,b shows the tested equipment components used in Figure 9a.

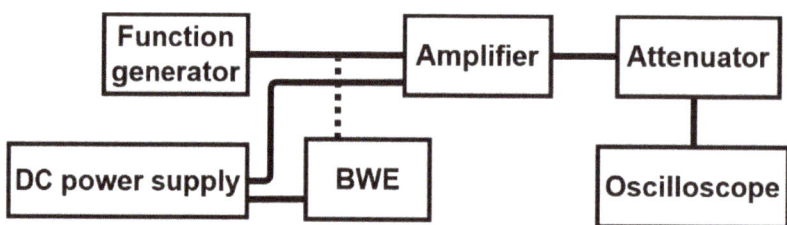

Figure 8. Block diagram showing the measured amplifier performances.

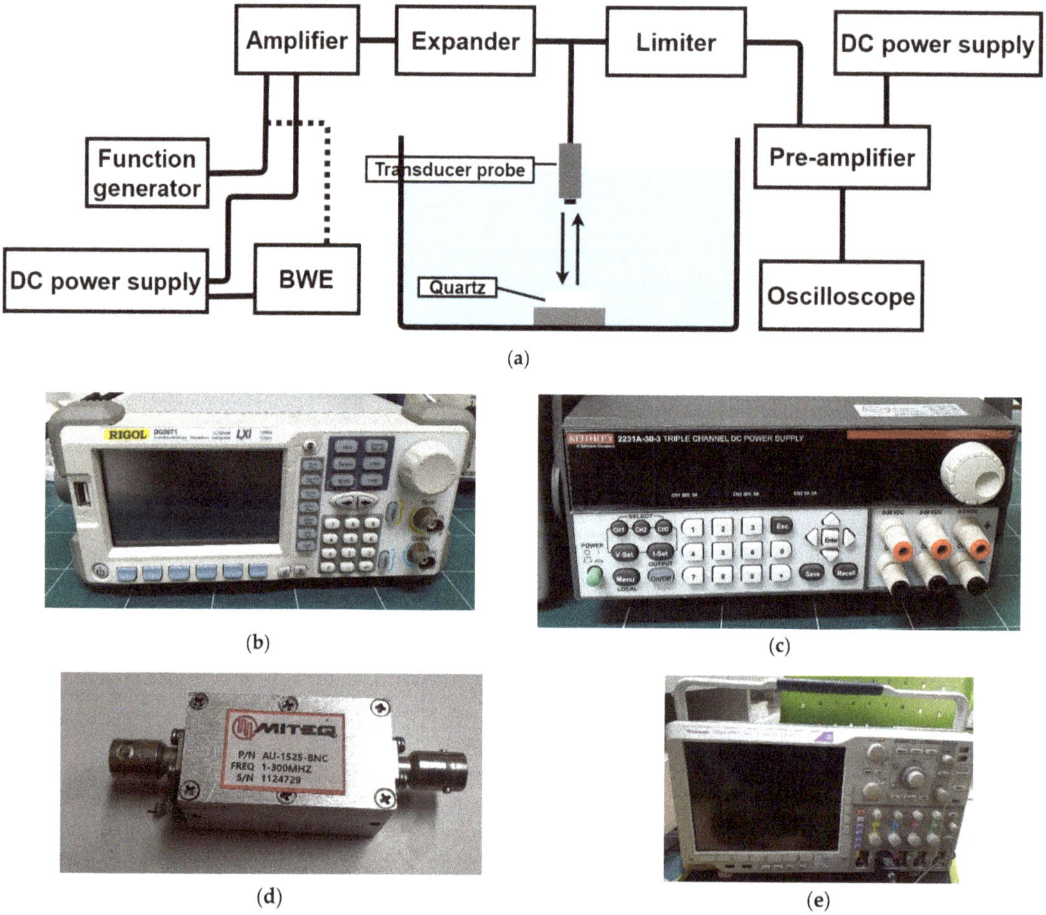

Figure 9. (**a**) Block diagram showing the measurement procedure of the echo signals using designed circuits and transducer probe, (**b**) function generator, (**c**) DC power supply, (**d**) pre-amplifier, and (**e**) oscilloscope.

In Figure 10, the transducer probe was used to measure the echo signal to estimate the amplifier equipped with and without the BWE circuit. The amplifier performance was measured by adjusting the input signal frequency according to the resonance frequency of each transducer [61–63]. All input parameters are the same when the amplifier is equipped with and without the BWE circuit. The measured performances are the amplitudes and pulse widths of the echo signals, −6 dB bandwidths, and harmonic components using a fast Fourier transform (FFT). The harmonic distortion characteristics were estimated using the total harmonic distortion (THD) equation [64–66]:

$$\mathrm{THD} = \frac{\sqrt{2ndharmonic^2 + 3rdharmonic^2}}{fundamental\ signal} \tag{14}$$

$$\mathrm{THD\ (dB)} = 20\log THD, \tag{15}$$

where the second and third harmonics are the amplitudes of the second and third harmonic distortion components, and the fundamental signal is the amplitude of the fundamental signal at the desired operating frequency.

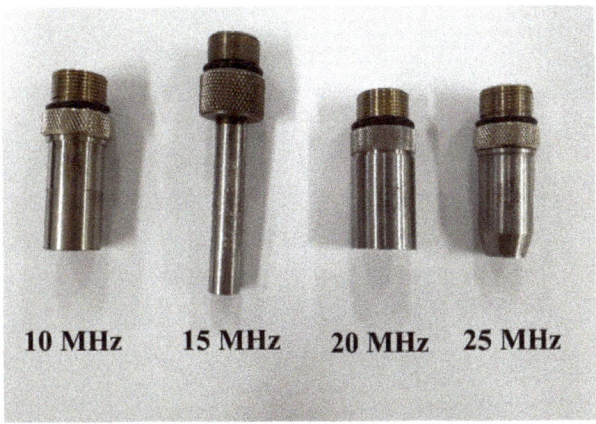

Figure 10. Transducer probes with each different frequency band used to measure the echo signals.

The higher the amplitude of the echo signal is, the higher the sensitivity of the transducer probe [47,67]. The narrower pulse width of the echo signal can result in a higher axial resolution of the transducer probe. The lateral resolution is related to the bandwidth [68]. The wider the bandwidth at the −6 dB point, the lower the Q factor, and thus, more image data can be realized [69]. However, the harmonic component generated unwanted image data [70,71]. Thus, these data need to be minimized. In this study, the measured performance factors were compared according to each transducer at different frequencies, as shown below. Therefore, the amplitudes, pulse widths, bandwidths, and THD were measured and compared by applying different transducer probes according to the frequency of each input signal with an amplifier equipped with and without the BWE circuit.

3. Results

Figure 11a,b shows the manufactured single-ended power amplifier and BWE circuits, respectively. The main transistor (LDMOSFET) with a heatsink was used to release heat more effectively [72–74]. The output port in Figure 11b is connected to the input port, as shown in Figure 11a.

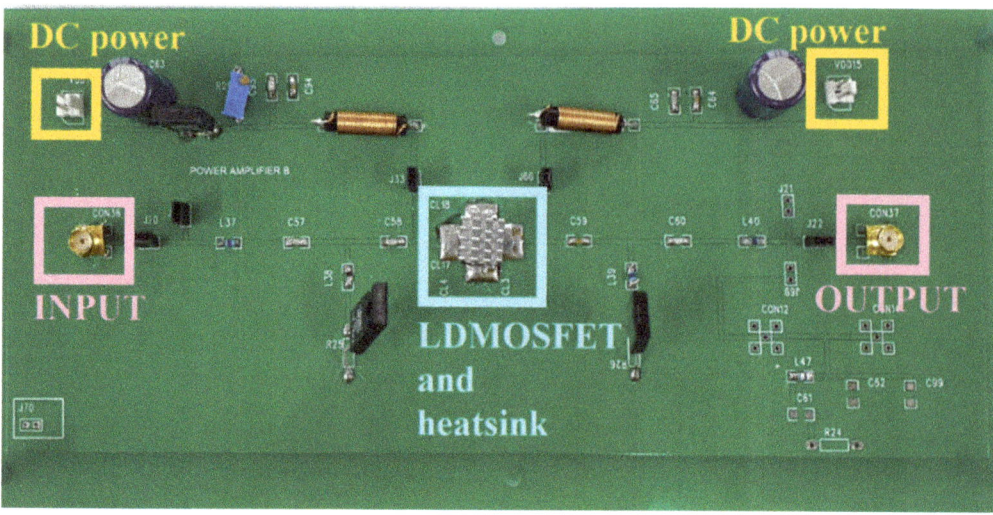

(**a**)

(**b**)

Figure 11. Manufactured (**a**) amplifier and (**b**) BWE circuits.

3.1. Performance Comparison and Analysis of the Amplifier Only and Amplifier + BWE Circuit

Figure 12a,b shows the P_{OUT} and gain variances as the input signal increases. The black line shows the performance of the power amplifier. The red, blue, and khaki lines show the performance of the amplifier with the BWE circuit using 0V, 1V, and 3V DC, respectively. In the case of amp + *BWE* (0 V) and amp + *BWE* (1 V), the active device of the BWE (see M1 in Figure 4) is not in operation, and thus, they have almost the same performance. Therefore, the red line is almost identical to the blue line. In Figure 12a,b, the performance of the power amplifier has a higher P_{OUT} and gain than the amplifier with a BWE (3 V) of between −10 dB$_m$ and 10 dB$_m$.

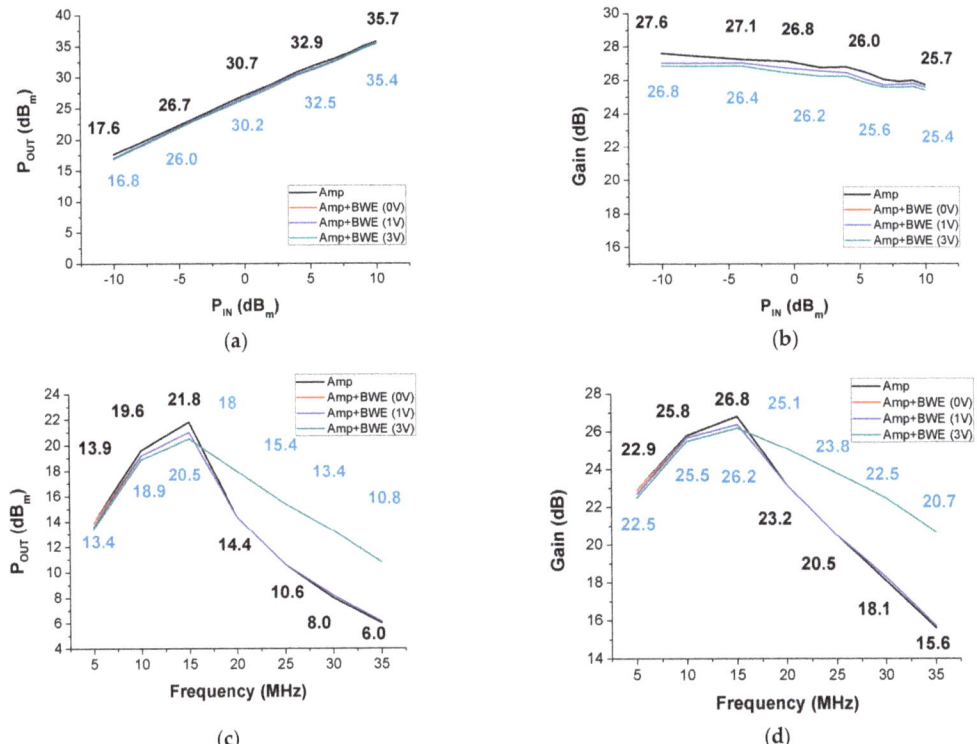

Figure 12. (a) P_{out} vs. P_{IN}, (b) gain vs. P_{IN}, (c) P_{out} vs. frequency, and (d) gain vs. frequency of the performance measurement results of the amplifier only and the amplifier with the addition of the BWE circuit.

Figure 12c,d shows the P_{OUT} and gain according to frequency variations at an input power of −6.5. In Figure 12c,d, the performances of the amplifier at 5–16 MHz have a higher P_{OUT} and gain than the power amplifier with BWE (3 V). The performance of the power amplifier with a BWE (3 V) has a higher P_{OUT} and gain of 17–35 MHz, compared to the power amplifier only. The P_{OUT} of the power amplifier with BWE (3V) outperformed that of the power amplifier after 16.132 MHz. In addition, the −3 dB P_{OUT} bandwidth of the power amplifier only and the power amplifier with BWE (3 V) were 51.5% and 81.5%, respectively. The −3 dB gain bandwidth of the power amplifier only and the power amplifier with BWE (3 V) are 84.1% and 141%, respectively. By incorporating the BWE circuit into the power amplifier, the P_{OUT} decreases to 0.8 dB_m, and the gain decreases by 0.8 dB when the P_{IN} is −10 dB_m. Theoretically, the magnitude of the transfer function increases with the addition of the BWE circuit to the power amplifier. Although the P_{OUT} and gain should be increased together, in practice, the final output signal can be slightly reduced because of the power loss of the passive components in the BWE circuit. However, in the graphs of the P_{OUT} and gain versus frequency, the P_{OUT} bandwidth increases by approximately 30%, and the gain bandwidth increases by approximately 56.9%. In this paper, the BWE circuit is used to widen the bandwidth by lowering the input impedance of the power amplifier. However, it does not decrease input impedance linearly at all frequencies. As shown in the experimental results, the bandwidth at the high-frequency range is wider than that of the low-frequency range because the BWE circuit has more impedance reduction at high frequency (See Figure 12c,d)

Figure 13 is the graph of the power added efficiency (PAE) of the amp and the amp + BWE (3 V) at 15 MHz. The PAE indicates how much DC power was used to amplify the input signal [15,75]. In Figure 13, the PAE versus P_{IN} of the amp (44.8%) is higher than the amp + BWE (3 V) (41.2%) when 10 dB$_m$ input power is applied. This is because the additional DC power is used for the designed BWE, and the amp has a higher P_{OUT} versus the P_{IN} than the amp + BWE (3 V). As a result, the PAEs of the amp and the amp + BWE (3V) do not show a big difference between them.

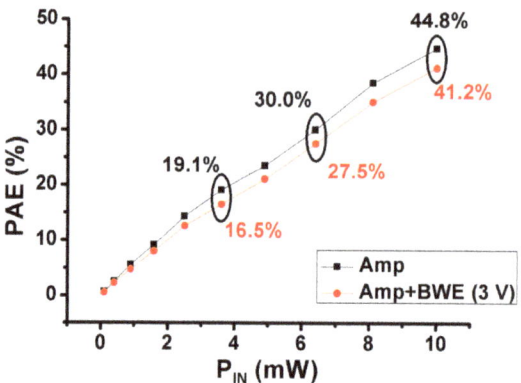

Figure 13. Power added efficiency (PAE) vs. P_{IN} measured results of the amplifier only and the amplifier with the addition of the BWE circuit.

3.2. Echo Signal Performance Comparison and Analysis

Figure 14 shows measured echo signal performances when using 10, 15, 20, and 25 MHz ultrasound transducers. When the ultrasound transducers with the same frequency were used, the distance to the target was exactly the same. The input signal was measured using a four-cycle burst wave with a suitable resonant frequency for the transducer probes. The measurement environment of each frequency is the same except for the presence of BWE with different DC voltages. As shown in Figure 14, the red, blue, and khaki lines show the performance of the power amplifier only and the power amplifier with BWE circuit using 0 V, 1 V, and 3V DC, respectively. In the case of the amplifier + *BWE* (0 V) and amplifier + *BWE* (1 V), the MOSFET (referring to M1 in Figure 4), which is the active device of the BWE circuit, is not operated and thus shows almost the same performances.

Figure 14a,b shows the measured results of the pulse widths and amplitudes over the time scale. Figure 14a shows the pulse width according to the frequency. The experimental results showed no significant difference in any of the measured frequency bands. Experimentally, the BWE circuit does not have a significant influence on the pulse width of the echo signals. Figure 14b shows the measured echo amplitudes of the peak-to-peak voltage according to the frequency using a 32 dB preamplifier. As shown in the graph, the amplitude of amp + *BWE* (3 V) was higher than that of the amp after 15 MHz. At 25 MHz, there is a difference of approximately 2.4 dB$_m$. Since the echo signal has an extremely low amplitude, a 2.4 dB$_m$ increment in the amplitude is an attractive result.

Figure 14c,d shows the calculated FFT data used to measure the harmonics and −6 dB bandwidths of the measured echo signals. Figure 14c shows the THD (%) according to the frequency. At 20 MHz, the THD of the amp was calculated as 7.15%, which is less than that of the amp + *BWE* (3 V). However, at 25 MHz, the THD of the amp was calculated as 15.63%, and that of the amp + *BWE* (3 V) was calculated as 5.74%. Figure 14d shows the echo bandwidth according to the frequency. By adding a BWE (3 V) circuit to the power amplifier, the −6 dB bandwidth of the echo signal was increased by 0.7%, 8.1%, 6.0%, and 9.8% at 10, 15, 20, and 25 MHz, respectively. The bandwidth of the echo signal is actually related to the axial resolution of the ultrasound image, and thus, a wider bandwidth can possibly improve the axial resolution [11]. By adding a *BWE* circuit to the power amplifier, the bandwidth of the echo signal is increased, which can help improve the quality of the echo signals.

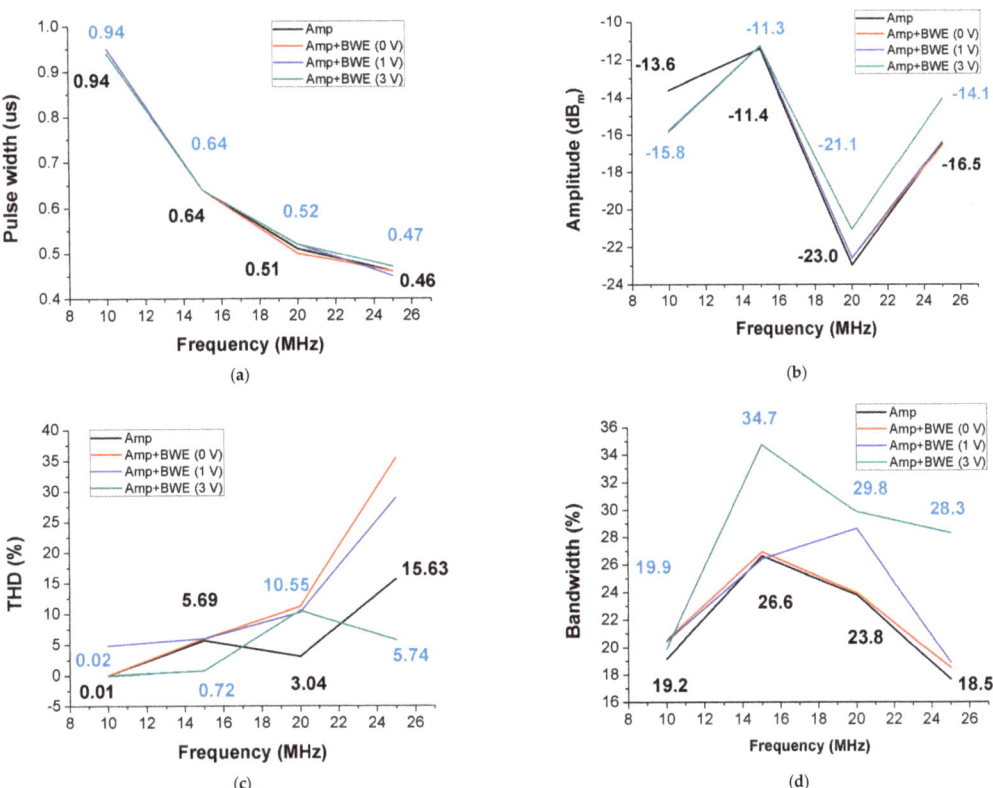

Figure 14. (**a**) Pulse width vs. frequency, (**b**) amplitude vs. frequency, (**c**) total harmonic distortion (THD) vs. frequency, and (**d**) bandwidth vs. frequency of the measured echo signal data using 10, 15, 20, and 25 MHz ultrasonic probes.

Both low and high-frequency transducers can be utilized by using the proposed BWE circuit with the amplifier. There are some ultrasound applications that utilize dual-band transducer applications [76]. For these applications, the signals of low- and high-frequency ranges from dual-band therapeutic/imaging transducer applications need to be obtained. From the paper, the ultrasound transducers enable therapeutic and imaging modes if needed. A treatment application requires to use many cycle sinusoidal waveforms [77]. The harmonic components generated when amplifying the input signal can affect the signal quality of the echo signals [78]. For low-frequency therapeutic applications, the harmonic components can distort the signal and attenuate the depth of penetration [79]. Therefore, the amplifiers used for therapeutic applications are preferred to have a narrow bandwidth in order to minimize harmonic components. For high-frequency imaging applications, the wider bandwidth, the higher axial resolution can be achieved [80]. Therefore, an amplifier with wide bandwidth is preferred. Hence, our proposed scheme could be useful for such dual-band transducer applications.

4. Conclusions

The transducers used in ultrasound systems have their own different frequency bands, depending on the particular purpose and testing area. Therefore, it is necessary to use an electrical circuit with a wide bandwidth such that the output signal of the transmitter can cover various ultrasound transducers. One way of expanding the bandwidth is impedance matching. Impedance matching is required to maximize the amplitudes or bandwidths of the output signals to the transducer. However, an impedance-matching job that can cover such wide transducers is extremely difficult because the impedance is different for each transducer. Although impedance matching is not taken into account in this document, it is clear that impedance matching can be used if we know the impedance values of the predetermined transducer. However, this method used to increase the bandwidth could possibly lower the output amplitude at the center frequency. In addition, the feedback loop circuit methodology can increase the bandwidth by reducing the output amplitude of the power amplifier. Therefore, we propose a switching mode transmit circuit that can widen the bandwidth and minimize the output amplitude as needed. The designed BWE circuit changes the performance of the power amplifier because the bandwidth in the transfer function is widened by moving the input pole of the power amplifier.

To verify our proposed concept and verify the performance results, we tested a power amplifier equipped with a BWE circuit under the same conditions. Comparing the performances of the manufactured amplifier only and the amplifier with the BWE circuit, the P_{OUT} and gain values of the amplifier with the BWE circuit were decreased slightly to 0.8 dB_m and 0.8 dB; however, the P_{OUT} bandwidth increased by approximately 30%, and the gain bandwidth increased by approximately 56.9% at -6.5 dB_m of input power. In addition, the echo bandwidths were expanded by 0.7%, 8.1%, 6.0%, and 9.8% at frequencies of 10, 15, 20, and 25 MHz, respectively.

In practice, the measured experimental data may be different from the theoretical data because there are various side effects caused by signal distortions of different frequency characteristics and parasitic components of the elements. From the experimental results, none of the measured performances were enhanced when adding a functional BWE circuit. Although the bandwidth is wider, there is a slight compromise, such as a decline in output power or an increase in THD (%). However, the manufactured BWE circuit improves the bandwidth and minimizes the amplitude of the power amplifier to support higher operating transducer probes, thus possibly helping improve the ultrasound system resolution.

Author Contributions: Conceptualization, K.K. and H.C.; methodology, K.K. and H.C.; formal analysis, K.K.; writing—original draft preparation, K.K. and H.C. All authors have read and agreed to the published version of the manuscript.

Funding: This work was supported by a National Research Foundation of Korea grant funded by the government (MSIT) (No. 2020R1A2C4001606). This work was supported by project for Industry-Academic Cooperation Based Platform R&D funded Korea Ministry of SMEs and Startups in 2020 (Project No. S3010583).

Institutional Review Board Statement: Not applicable.

Informed Consent Statement: Not applicable.

Data Availability Statement: The data presented in this study are available on request from the corresponding author.

Conflicts of Interest: The authors declare no conflict of interest.

References

1. Shung, K.K. *Diagnostic Ultrasound: Imaging and Blood Flow Measurements*; Taylor & Francis: Boca Raton, FL, USA, 2015.
2. Szabo, T.L. *Diagnostic Ultrasound Imaging: Inside Out*; Elsevier Academic Press: London, UK, 2013.
3. Zhou, Q.; Lau, S.; Wu, D.; Shung, K.K. Piezoelectric films for high frequency ultrasonic transducers in biomedical applications. *Prog. Mater. Sci.* **2011**, *56*, 139–174. [CrossRef] [PubMed]
4. Zhu, B.; Fei, C.; Wang, C.; Zhu, Y.; Yang, X.; Zheng, H.; Zhou, Q.; Shung, K.K. Self-focused AlScN film ultrasound transducer for individual cell manipulation. *ACS Sens.* **2017**, *2*, 172–177. [CrossRef] [PubMed]
5. Hoskins, P.R.; Martin, K.; Thrush, A. *Diagnostic Ultrasound: Physics and Equipment*; Cambridge University Press: Cambridge, UK, 2019.
6. Postema, M. *Fundamentals of Medical Ultrasound*; Taylor and Francis: New York, NY, USA, 2011.
7. Kremkau, F.W.; Forsberg, F. *Sonography Principles and Instruments*; Elsevier Health Sciences: Amsterdam, The Netherlands, 2015.

8. Li, X.; Wei, W.; Zhou, Q.; Shung, K.K.; Chen, Z. Intravascular photoacoustic imaging at 35 and 80 MHz. *J. Biomed. Opt.* **2012**, *17*, 106005. [CrossRef]
9. Ritter, T.A.; Shrout, T.R.; Tutwiler, R.; Shung, K.K. A 30-MHz piezo-composite ultrasound array for medical imaging applications. *IEEE Trans. Ultrason. Ferroelectr. Freq. Control.* **2002**, *49*, 217–230. [CrossRef]
10. Kim, J.; You, K.; Choi, H. Post-Voltage-Boost Circuit-Supported Single-Ended Class-B Amplifier for Piezoelectric Transducer Applications. *Sensors* **2020**, *20*, 5412. [CrossRef] [PubMed]
11. Bushberg, J.T.; Boone, J.M. *The Essential Physics of Medical Imaging*; Lippincott Williams & Wilkins: Philadelphia, PA, USA, 2011.
12. Zagzebski, J.A. *Essentials of Ultrasound Physics*; Mosby: Maryland Heights, MO, USA, 1996.
13. Lee, T.H. *The Design of CMOS Radio-Frequency Integrated Circuits*; Cambridge University Press: Cambridge, UK, 2006.
14. Razavi, B. *Design of Analog CMOS Integrated Circuits*; McGraw-Hill Science: New York, NY, USA, 2016.
15. Razavi, B. *RF Microelectronics*; Prentice Hall: Upper Saddel River, NJ, USA, 2011.
16. Albulet, M. *RF Power Amplifiers*; SciTech Publishing: London, UK, 2001.
17. Grebennikov, A. *RF and Microwave Power Amplifier Design*; McGraw-Hill: New York, NY, USA, 2005.
18. Katz, A. Linearization: Reducing distortion in power amplifiers. *IEEE Microw. Mag.* **2001**, *2*, 37–49. [CrossRef]
19. Cripps, S.C. *RF Power Amplifiers for Wireless Communications*; Artech House: Norwood, MA, USA, 2006.
20. Reynaert, P.; Steyaert, M. *RF Power Amplifiers for Mobile Communications*; Springer Science & Business Media: Berlin, Germany, 2006.
21. Park, J.; Hu, C.; Li, X.; Zhou, Q.; Shung, K.K. Wideband linear power amplifier for high-frequency ultrasonic coded excitation imaging. *IEEE Trans. Ultrason. Ferroelectr. Freq. Control.* **2012**, *59*, 825–832. [CrossRef]
22. Gao, Z.; Gui, P.; Jordanger, R. An integrated high-voltage low-distortion current-feedback linear power amplifier for ultrasound transmitters using digital predistortion and dynamic current biasing techniques. *IEEE Trans. Circuits Syst. II Express Briefs* **2014**, *61*, 373–377. [CrossRef]
23. Choi, H. Development of a Class-C Power Amplifier with Diode Expander Architecture for Point-of-Care Ultrasound Systems. *Micromachines* **2019**, *10*, 697. [CrossRef]
24. Agbossou, K.; Dion, J.-L.; Carignan, S.; Abdelkrim, M.; Cheriti, A. Class D Amplifier for a Power Piezoelectric Load. *IEEE Trans. Ultrason. Ferroelectr. Freq. Control.* **2000**, *47*, 1036–1041. [CrossRef]
25. Yuan, T.; Dong, X.; Shekhani, H.; Li, C.; Maida, Y.; Tou, T.; Uchino, K. Driving an inductive piezoelectric transducer with class E inverter. *Sens. Actuators A* **2017**, *261*, 219–227. [CrossRef]
26. Moon, J.-Y.; Lee, J.; Chang, J.H. Electrical impedance matching networks based on filter structures for high frequency ultrasound transducers. *Sens. Actuators A* **2016**, *251*, 225–233. [CrossRef]
27. Huang, H.; Paramo, D. Broadband electrical impedance matching for piezoelectric ultrasound transducers. *IEEE Trans. Ultrason. Ferroelectr. Freq. Control.* **2011**, *58*, 2699–2707. [CrossRef] [PubMed]
28. Zhou, Q.; Lam, K.H.; Zheng, H.; Qiu, W.; Shung, K.K. Piezoelectric single crystal ultrasonic transducers for biomedical applications. *Prog. Mater. Sci.* **2014**, *66*, 87–111. [CrossRef] [PubMed]
29. Vuolevi, J.; Rahkonen, T. *Distortion in RF Power Amplifiers*; Artech house: London, UK, 2003.
30. Chang, K. *Microwave Solid-State Circuits and Applications*; Wiley: New York, NY, USA, 1994.
31. Kazimierczuk, M.K. *RF Power Amplifier*; John Wiley & Sons: Hoboken, NJ, USA, 2014.
32. Malik, N.R. *Electronic Circuits: Analysis, Simulation, and Design*; Prentice Hall: Englewood Cliffs, NJ, USA, 1995.
33. Self, D. *Audio Power Amplifier Design*; Focal Press: Waltham, MA, 2013.
34. Larson, L.E. *RF and Microwave Circuit Design for Wireless Communications*; Artech House: Norwood, MA, USA, 1996.
35. Choe, S.-W.; Choi, H. Suppression Technique of HeLa Cell Proliferation Using Ultrasonic Power Amplifiers Integrated with a Series-Diode Linearizer. *Sensors* **2018**, *18*, 4248. [CrossRef]
36. You, K.; Choi, H. Wide Bandwidth Class-S Power Amplifiers for Ultrasonic Devices. *Sensors* **2020**, *20*, 290. [CrossRef] [PubMed]
37. Davidse, J. *Analog Electronic Circuit Design*; Prentice Hall: Upper Saddle River, NJ, USA, 1991.
38. Ullah, M.N.; Park, Y.; Kim, G.B.; Kim, C.; Park, C.; Choi, H.; Yeom, J.-Y. Simultaneous Acquisition of Ultrasound and Gamma Signals with a Single-Channel Readout. *Sensors* **2021**, *21*, 1048. [CrossRef]
39. Zhang, X.; Larson, L.E.; Asbeck, P. *Design of Linear RF Outphasing Power Amplifiers*; Artech House: Norwood, MA, USA, 2003.
40. Cripps, S.C. *Advanced Techniques in RF Power Amplifier Design*; Artech House: Norwood, MA, USA, 2002.
41. Kim, J.; You, K.; Choe, S.-H.; Choi, H. Wireless Ultrasound Surgical System with Enhanced Power and Amplitude Performances. *Sensors* **2020**, *20*, 4165. [CrossRef] [PubMed]
42. Zawawi, R.B.A.; Abbasi, W.H.; Kim, S.-H.; Choi, H.; Kim, J. Wide-Supply-Voltage-Range CMOS Bandgap Reference for In Vivo Wireless Power Telemetry. *Energies* **2020**, *13*, 2986. [CrossRef]
43. Chang-Hong, H.; Snook, K.A.; Poi-Jie, C.; Shung, K.K. High-frequency ultrasound annular array imaging. Part II: Digital beamformer design and imaging. *IEEE Trans. Ultrason. Ferroelectr. Freq. Control.* **2006**, *53*, 309–316. [CrossRef]
44. Kim, J.; Kim, K.; Choe, S.-H.; Choi, H. Development of an Accurate Resonant Frequency Controlled Wire Ultrasound Surgical Instrument. *Sensors* **2020**, *20*, 3059. [CrossRef]
45. You, K.; Kim, S.-H.; Choi, H. A Class-J Power Amplifier Implementation for Ultrasound Device Applications. *Sensors* **2020**, *20*, 2273. [CrossRef]
46. Li, G.; Qiu, W.; Zhang, Z.; Jiang, Q.; Su, M.; Cai, R.; Li, Y.; Cai, F.; Deng, Z.; Xu, D. Noninvasive Ultrasonic Neuromodulation in Freely Moving Mice. *IEEE Trans. Biomed. Eng.* **2018**, *66*, 217–224. [CrossRef] [PubMed]

47. Zhang, T.; Ou-Yang, J.; Yang, X.; Wei, W.; Zhu, B. High Performance KNN-Based Single Crystal Thick Film for Ultrasound Application. *Electron. Mater. Lett.* **2019**, *15*, 1–6. [CrossRef]

48. Choi, H.; Choe, S.-W. Acoustic Stimulation by Shunt-Diode Pre-Linearizer Using Very High Frequency Piezoelectric Transducer for Cancer Therapeutics. *Sensors* **2019**, *19*, 357. [CrossRef] [PubMed]

49. Jacobson, J.A. *Fundamentals of Musculoskeletal Ultrasound*; Elsevier Health Sciences: Amsterdam, The Netherlands, 2017.

50. Choi, H. Stacked Transistor Bias Circuit of Class-B Amplifier for Portable Ultrasound Systems. *Sensors* **2019**, *19*, 5252. [CrossRef]

51. Jeong, J.J.; Choi, H. An impedance measurement system for piezoelectric array element transducers. *Measurement* **2017**, *97*, 138–144. [CrossRef]

52. Oppelt, R.; Vester, M. Duplexer including a field-effect transistor for use in an ultrasound imaging system. U.S. Patent US 5,603,324, 18 February 1997.

53. Qiu, W.; Wang, X.; Chen, Y.; Fu, Q.; Su, M.; Zhang, L.; Xia, J.; Dai, J.; Zhang, Y.; Zheng, H. A Modulated Excitation Imaging System for Intravascular Ultrasound. *IEEE Trans. Biomed. Eng.* **2016**, *64*, 1935–1942. [CrossRef]

54. Choi, H. Development of negative-group-delay circuit for high-frequency ultrasonic transducer applications. *Sens. Actuators A* **2019**, *299*, 111616. [CrossRef]

55. Choi, H.; Choe, S.-W. Therapeutic Effect Enhancement by Dual-bias High-voltage Circuit of Transmit Amplifier for Immersion Ultrasound Transducer Applications. *Sensors* **2018**, *18*, 4210. [CrossRef]

56. Weibao, Q.; Yanyan, Y.; Fu Keung, T.; Lei, S. A multifunctional, reconfigurable pulse generator for high-frequency ultrasound imaging. *IEEE Trans. Ultrason. Ferroelectr. Freq. Control.* **2012**, *59*, 1558–1567. [CrossRef]

57. Li, J.; Xu, J.; Liu, X.; Zhang, T.; Lei, S.; Jiang, L.; Ou-Yang, J.; Yang, X.; Zhu, B. A novel CNTs array-PDMS composite with anisotropic thermal conductivity for optoacoustic transducer applications. *Compos. Part. B* **2020**, *196*, 108073. [CrossRef]

58. Snook, K.A.; Hu, C.-H.; Shrout, T.R.; Shung, K.K. High-frequency ultrasound annular-array imaging. Part I: Array design and fabrication. *IEEE Trans. Ultrason. Ferroelectr. Freq. Control.* **2006**, *53*, 300–308. [CrossRef] [PubMed]

59. Choi, H.; Yoon, C.; Yeom, J.-Y. A Wideband High-Voltage Power Amplifier Post-Linearizer for Medical Ultrasound Transducers. *Appl. Sci.* **2017**, *7*, 354. [CrossRef]

60. Choi, H. Prelinearized Class-B Power Amplifier for Piezoelectric Transducers and Portable Ultrasound Systems. *Sensors* **2019**, *19*, 287. [CrossRef] [PubMed]

61. Choi, H. Class-C Linearized Amplifier for Portable Ultrasound Instruments. *Sensors* **2019**, *19*, 898. [CrossRef]

62. Choi, H.; Park, C.; Kim, J.; Jung, H. Bias-Voltage Stabilizer for HVHF Amplifiers in VHF Pulse-Echo Measurement Systems. *Sensors* **2017**, *17*, 2425. [CrossRef]

63. Safari, A.; Akdogan, E.K. *Piezoelectric and Acoustic Materials for Transducer Applications*; Springer Science & Business Media: Berlin, Germany, 2008.

64. Eroglu, A. *Introduction to RF Power Amplifier Design and Simulation*; CRC Press: Boca Raton, FL, USA, 2018.

65. Leblebici, D.; Leblebici, Y. *Fundamentals of High.-Frequency CMOS Analog Integrated Circuits*; Cambridge University Press: New York, NY, USA; Cambridge, UK, 2009.

66. Wambacq, P.; Sansen, W.M. *Distortion Analysis of Analog Integrated Circuits*; Springer Science & Business Media: Berlin, Germany, 2013.

67. Choi, H.; Woo, P.C.; Yeom, J.-Y.; Yoon, C. Power MOSFET Linearizer of a High-Voltage Power Amplifier for High-Frequency Pulse-Echo Instrumentation. *Sensors* **2017**, *17*, 764. [CrossRef] [PubMed]

68. Choi, H.; Yeom, J.-Y.; Ryu, J.-M. Development of a Multiwavelength Visible-Range-Supported Opto–Ultrasound Instrument Using a Light-Emitting Diode and Ultrasound Transducer. *Sensors* **2018**, *18*, 3324. [CrossRef]

69. Flower, M.A. *The Physics of Medical Imaging*; CRC Press: Boca Raton, FL, USA, 2012.

70. Ullah, M.; Pratiwi, E.; Park, J.; Lee, K.; Choi, H.; Yeom, J. Wavelength discrimination (WLD) TOF-PET detector with DOI information. *Phys. Med. Biol.* **2019**, *65*, 55003. [CrossRef]

71. Zawawi, R.B.A.; Choi, H.; Kim, J. High-PSRR Wide-Range Supply-Independent CMOS Voltage Reference for Retinal Prosthetic Systems. *Electronics* **2020**, *9*, 2028. [CrossRef]

72. Kumar, N.; Grebennikov, A. *Distributed Power Amplifiers for RF and Microwave Communications*; Artech House: Norwood, MA, USA, 2015.

73. Gray, P.R. *Analysis and Design of Analog Integrated Circuits*; John Wiley & Sons: Hoboken, NJ, USA, 2009.

74. You, K.; Choi, H. Inter-Stage Output Voltage Amplitude Improvement Circuit Integrated with Class-B Transmit Voltage Amplifier for Mobile Ultrasound Machines. *Sensors* **2020**, *20*, 6244. [CrossRef] [PubMed]

75. Gilmore, R.; Besser, L. *Practical RF Circuit Design for Modern Wireless Systems Vol. 1 Passive Circuits and Systems*; Artech house: Norwood, MA, USA, 2003.

76. Lim, H.G.; Kim, H.; Kim, K.; Park, J.; Kim, Y.; Yoo, J.; Heo, D.; Baik, J.; Park, S.-M.; Kim, H.H. Thermal Ablation and High-Resolution Imaging Using a Back-to-Back (BTB) Dual-Mode Ultrasonic Transducer: In Vivo Results. *Sensors* **2021**, *21*, 1580. [CrossRef] [PubMed]

77. Mason, T.J.; Peters, D. *Practical Sonochemistry: Power Ultrasound Uses and Applications*; Woodhead Publishing: Cambridge, UK, 2002.

78. Miele, F.R. *Ultrasound Physics & Instrumentation*; Pegasus Lectures, Inc.: Forney, TX, USA, 2013.

79. Junru, W.; Wesley, N. *Emerging Therapeutic Ultrasound*; World Scientific Publishing: Hackensack, NJ, USA, 2006.

80. Suri, J.S.; Kathuria, C.; Chang, R.-F.; Molinar, F.; Fenster, A. *Advances in Diagnostic and Therapeutic Ultrasound Imaging*; Artech House: Norwood, MA, USA, 2008.

Communication

Synthetic Aperture Imaging Using High-Frequency Convex Array for Ophthalmic Ultrasound Applications

Hae Gyun Lim [1], Hyung Ham Kim [2],[*]and Changhan Yoon [3],[4],[*]

1 Department of Biomedical Engineering, Pukyong National University, Busan 48513, Korea; hglim@pknu.ac.kr
2 Department of Convergence IT Engineering, Pohang University of Science and Technology, Pohang 37673, Korea
3 Department of Biomedical Engineering, Inje University, Gimhae 50834, Korea
4 Department of Nanoscience and Engineering, Inje University, Gimhae 50834, Korea
* Correspondence: david.kim@postech.ac.kr (H.H.K.); cyoon@inje.ac.kr or yoonch80@gmail.com (C.Y.)

Abstract: High-frequency ultrasound (HFUS) imaging has emerged as an essential tool for pre-clinical studies and clinical applications such as ophthalmic and dermatologic imaging. HFUS imaging systems based on array transducers capable of dynamic receive focusing have considerably improved the image quality in terms of spatial resolution and signal-to-noise ratio (SNR) compared to those by the single-element transducer-based one. However, the array system still suffers from low spatial resolution and SNR in out-of-focus regions, resulting in a blurred image and a limited penetration depth. In this paper, we present synthetic aperture imaging with a virtual source (SA-VS) for an ophthalmic application using a high-frequency convex array transducer. The performances of the SA-VS were evaluated with phantom and ex vivo experiments in comparison with the conventional dynamic receive focusing method. Pre-beamformed radio-frequency (RF) data from phantoms and excised bovine eye were acquired using a custom-built 64-channel imaging system. In the phantom experiments, the SA-VS method showed improved lateral resolution (>10%) and sidelobe level (>4.4 dB) compared to those by the conventional method. The SNR was also improved, resulting in an increased penetration depth: 16 mm and 23 mm for the conventional and SA-VS methods, respectively. Ex vivo images with the SA-VS showed improved image quality at the entire depth and visualized structures that were obscured by noise in conventional imaging.

Keywords: high-frequency ultrasound; ophthalmic imaging; synthetic aperture; convex array transducer

Citation: Lim, H.G.; Kim, H.H.; Yoon, C. Synthetic Aperture Imaging Using High-Frequency Convex Array for Ophthalmic Ultrasound Applications. *Sensors* **2021**, *21*, 2275. https://doi.org/10.3390/s21072275

Academic Editors: Dipen N. Sinha, Changho Lee and Changhan Yoon

Received: 9 February 2021
Accepted: 22 March 2021
Published: 24 March 2021

Publisher's Note: MDPI stays neutral with regard to jurisdictional claims in published maps and institutional affiliations.

1. Introduction

High-frequency ultrasound (HFUS) imaging (>15 MHz) has evolved rapidly over the last decade and opened up new applications such as ophthalmic, dermatologic, intravascular, small animal, and molecular imaging [1–7]. It can provide sub-millimeter spatial resolution determined by $f_\# \cdot \lambda$ (where $f_\#$ is defined as a ratio of a focal distance to a length of the aperture used for transmission/reception, and λ is the wavelength) at the expense of a shallow penetration depth. Most custom-built or commercialized HFUS imaging systems have employed mechanically scanning single-element transducers to form an image [8–10]. While these single-element imaging systems have offered exciting potential for many applications, they suffered from low spatial resolution and signal-to-noise ratio (SNR) in the out-of-focus regions, thus deteriorating the image quality [11]. In addition, mechanical scanning limits the frame rate.

The adoption of array transducers in HFUS imaging has allowed for improving the spatial resolution and SNR [9,12,13]. The array transducer-based systems capable of dynamic receive focusing use electronic scanning to form an image; thus it provides a higher frame rate and image quality than those by the mechanical scanning systems. Although it can enhance the overall image quality of HFUS, two-way focusing is only achieved at the vicinity of the transmit focal depth. To mitigate this, multi-zone transmit

23

focusing, where transmit focusing is conducted at two or more depths for each scanline at the expense of frame rate (reduced by a factor of the number of transmitting foci), has been used [14]. In addition to the problem of one-way dynamic focusing, the HFUS imaging still suffers from low SNR due to diffraction and frequency-dependent attenuation that linearly increases with frequency [15]. Coded excitation can be a solution for improving the SNR [16,17]. However, the spatial resolution is still limited by the diffraction of the wave.

A viable solution to obtain a high spatial resolution, SNR, and frame rate is to employ synthetic aperture (SA) imaging techniques that are based on the superposition of unfocused transmit wave fields. Several different SA methods have been proposed and have shown their ability to enhance image quality at the expense of computational cost [18–21]. Among them, multi-element SA with a virtual source (SA-VS) that can achieve high SNR with full two-way dynamic focusing would be the most prominent method [20]. Clinical evaluations of the SA-VS on cancer diagnosis over conventional imaging were performed [22,23]. It was demonstrated that the improved image quality could be obtained using the SA-VS method and was perceived by radiologists. Recently, efficient architectures for SA-VS imaging have been proposed and implemented in prototype systems [21,24]. In addition, recent advances in graphic processor unit (GPU) computing in medical ultrasound imaging may facilitate more rapid commercialization of SA techniques [25,26].

The purpose of this study was to evaluate the feasibility of the SA-VS for ophthalmic imaging using a high-frequency convex array transducer by demonstrating its effectiveness in enhancing image quality compared to the conventional one-way dynamic focusing. Note that the high-frequency convex array transducer is the only transducer, and this is the first time we applied synthetic aperture imaging using this transducer for ophthalmic imaging. The main advantage of a convex array is that it can image the whole posterior segment at once. The performances of SA-VS were evaluated through phantoms and ex vivo experiments. Pre-beamformed radio-frequency data were acquired by using a custom-built research system. In the phantom experiments, spatial resolution and SNR were quantitatively assessed and compared with the conventional dynamic receive focusing method. In addition, an excised bovine eye was scanned, and the SA-VS image showed improved image quality.

2. Methods

2.1. Principle of Synthetic Aperture Imaging with a Virtual Source

Figure 1 shows the principle of synthesizing transmit fields in SA-VS imaging, which is capable of achieving two-way dynamic focusing at all imaging depths. A detailed description of SA-VS can be founded in [20]. Here, we briefly introduce the SA-VS. The SA-VS uses the same transmission (focused transmit) and reception procedures as in the conventional B-mode imaging. Thus, the frame rate of SA-VS is identical to that of the conventional method. In the SA-VS imaging method, a virtual source is regarded to be located at a transmit focal point where spherical waves assume to propagate from it. As can be seen, two transmit fields from different sub-apertures pass an imaging point, (x, z). Thus, the transmit focusing delay, τ_{tx}, for an imaging point can be obtained by computing the arrival time of wavefront, given by

$$\tau_{tx}(x, z) = \frac{z_{tx} \pm \sqrt{\left(x_f - x\right)^2 + \left(z_f - z\right)^2}}{c}, \tag{1}$$

where z_{tx} is the transmit focal depth, $\left(x_f, z_f\right)$ is the Cartesian coordinates of the transmit focal point, and c is the propagating speed of sound in soft tissue. The positive and negative signs in (1) are, respectively, applied in the areas after and before the transmit focal point. The receive focusing delay of nth element, (x_n, z_n), for the imaging point is identical to that in the conventional dynamic focusing method, which is computed by

$$\tau_{n,RX}(x,z) = \frac{\sqrt{(x-x_n)^2 + (z-z_n)^2}}{c}. \tag{2}$$

Based on these delays, the beamforming of the SA-VS can be achieved by

$$A(x,z) = \sum_{m=-M}^{M} \sum_{n=0}^{N-1} a_n \cdot r_{m,n}(t - (\tau_{TX}(x,z) + \tau_{n,RX}(x,z))), \tag{3}$$

where a_n is the apodization function, $r_{m,n}(t)$ is the radio-frequency (RF) data received by the nth element for the mth scanline, $2M+1$ is the number of sub-apertures used in synthesizing, and N is the number of channels at each sub-aperture.

As can be seen in Figure 1, the number of scanlines that can be used for synthesizing varies according to the imaging depth. At the transmit focal depth, there is no scanline that can be synthesized. However, the number of scanlines incorporated in the transmit field synthesis increases as the imaging point moves away from the transmit focal depth, resulting in an increment of signal strength after synthesis. Thus, it requires a compensation method in the SA-VS method to obtain uniform brightness similar to those in the conventional method. For this, the beamformed RF signal in the SA-VS is divided by $\sqrt{M_s}$ where M_s is the number of scanlines actually used for synthesis at a certain depth. This can be done by incorporating the values, $1/\sqrt{M_s}$, in the apodization function in (3).

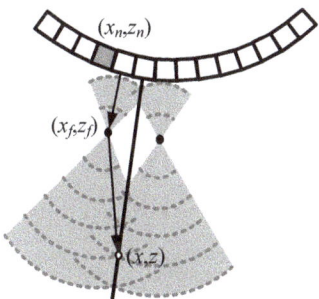

Figure 1. Principle of the transmit field synthesis in the synthetic aperture with a virtual source (SA-VS) imaging which represents the same sequences of transmit/receive as in the conventional imaging method.

2.2. Experimental Setup and Evaluation Metrics

To evaluate the performances, pre-beamformed RF data from phantoms and ex vivo bovine eye were acquired using a 64-channel research imaging system developed in our laboratory [13]. The system is composed of 256-channel of analog front-end pulser/receiver, 64-channel of time-gain compensation (TGC), and an analog-to-digital converter (ADC) with 12-bit resolution. A custom-built 20 MHz high-frequency transducer made with 1–3 composites was used in the experiments [27]. The array consists of 192 elements with an element pitch of 111 µm, and a −6 dB fractional bandwidth was 64%. The pre-beamformed RF data sampled at 100 MHz were stored in field-programmable gate arrays (FPGAs) that are embedded in the system and transferred to a PC. Off-line processing using MATLAB (MathWorks Inc., Natick, MA, USA) was carried out. The lateral resolutions were measured with 20 µm tungsten wire targets located at each depth. An agar phantom was made to estimate the SNR of both imaging methods, i.e., conventional and SA-VS. For ex vivo experiments, an excised bovine eye was purchased from Sierra Medical Inc. (Whittier, CA, USA). The eye was immersed in deionized water and fixed on a custom-made holder while scanning. In both beamformations (conventional and SA-VS), the optimal sound speed was estimated to minimize the effect from phase aberration artifacts, which is a first-order solution for phase aberration correction [28].

For quantitative comparison, −6 dB lateral resolutions were measured from the conventional and SA-VS images. In addition, the SNR as a function of depth was calculated by

$$SNR(z) = 10log10\left(\frac{P_{echo}(z)}{P_{noise}(z)}\right),\qquad(4)$$

where P_{echo} and P_{noise} are the mean power of echo and noise signals along with the imaging depth (z), respectively. The mean power at each depth was computed by summing the envelope signal laterally at the speckle region. The system noise was measured by acquiring pre-beamformed RF data without transmission. The acquired noise signal was processed in the same manner for each method, i.e., conventional and SA-VS methods. Based on the SNR, the penetration depths defined as the depth where SNR falls below 0 dB were estimated.

3. Results and Discussion

Figure 2 shows B-mode images of wire targets generated by the conventional and SA-VS methods, respectively. Two images were acquired for the conventional imaging with different transmit focal depths, 10 (Tx10) and 25 mm (Tx25), which are shown in Figure 2a,b, respectively. In the SA-VS imaging, the transmit focal depth was 10 mm to maximize the number of synthesizable sub-aperture at far depth. Note that the number of sub-aperture for synthetic aperture varies with different focal depth (see Figure 1). Since the main imaging target of the high-frequency convex array transducer is the posterior segment of the eye, the transmit focal depth of 10 mm was chosen. The maximum number of sub-aperture used in synthesis, in (3), was calculated based on the configuration of transducer (i.e., curvature and element pitch) and the transmit focal depth and was found to be 33. The optimal sound speed was estimated to be 1500 m/s, which was closed to the sound speed in water at room temperature [29]. All images were logarithmically compressed with a dynamic range of 40 dB. As shown in Figure 2, improved lateral resolution in the SA-VS image can be readily recognized under visual assessment.

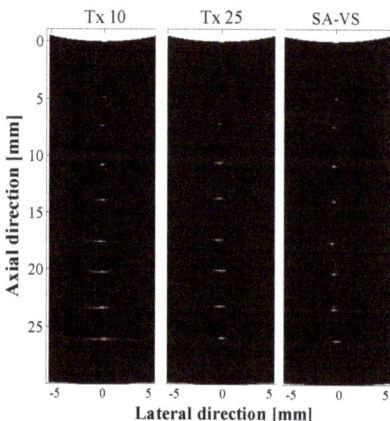

Figure 2. B-mode images of wire targets located at each depth obtained by the conventional method with transmitting focal depths of 10 mm (Tx 10), 25 mm (Tx 25), and the SA-VS method, respectively. The number of sub-aperture used in synthesis in the SA-VS was 33.

For quantitative comparison, the lateral beam profiles at depths of 4, 14, 18, and 24 mm were measured and plotted in Figure 3. As can be seen, the SA-VS method produced not only improved lateral resolutions but also decreased sidelobe levels compared to those from the conventional one-way dynamic focusing methods. At the depth of 24 mm, similar beam profiles were obtained from the SA-VS and the conventional method with the focal depth of 25 mm. The −6 dB lateral resolutions and first sidelobe levels are summarized in Table 1. As listed in Table 1, the −6 dB lateral resolutions and the sidelobe levels in the SA-VS method were improved at all imaging depths. The lateral resolutions were improved by more than 10% (maximally 65%) except at the depth of 24 mm where similar beam profiles were produced between the conventional (Tx25) and SA-VS methods. Considerable reductions in the sidelobe level were obtained by the SA-VS method; minimal and maximal enhancements were, respectively, 4.4 and 15.6 dB.

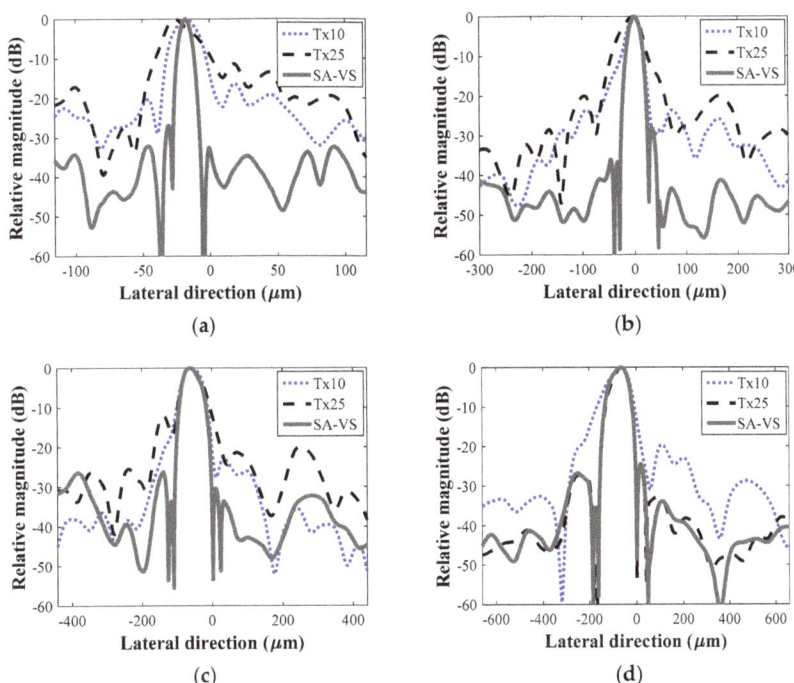

Figure 3. Lateral beam profiles from each method at depths of (**a**) 5 mm, (**b**) 14 mm, (**c**) 18 mm, and (**d**) 24 mm, respectively.

Table 1. The −6 dB lateral resolutions and first sidelobe levels for conventional Tx10, Tx25, and SA-VS.

| | −6 dB Lateral Resolution (μm)/First Sidelobe Level (dB) | | | |
	5 mm	14 mm	18 mm	24 mm
Conventional Tx10	24.0/−16.4	32.2/−23.7	71/−22.5	119.3/−19.6
Conventional Tx25	29.6/−11.4	51.3/−20.0	72.6/−12	106.0/−24.6
SA-VS	10.2/−27.0	26.9/−28.1	64.3/−26.2	105.5/−24.6

B-mode images of a custom-made agar phantom are shown in Figure 4. In conventional imaging, the transmit focal depths were 10 (Tx10) and 25 mm (Tx25), respectively. For the SA-VS imaging, a transmit focal depth was 10 mm, and the number of sub-aperture for synthesis was 33. As shown in Figure 4, the SA-VS produced speckle patterns with uniform brightness. The penetration depth was also increased in the SA-VS. The measured SNR curves for each method as a function of depth are shown in Figure 5. The mean power at each depth was computed from 20 scanlines at the center. Consistent with the visual assessment, the SA-VS method improved the SNR for the entire imaging depth compared to other methods. In the conventional method with the transmit focusing at 10 mm, the maximum SNR was achieved around 10 mm, and the SNR sharply decreased after the focal depth due to diffraction and attenuation. Due to the high attenuation, the SNR curve with the conventional method (Tx25) could not produce a peak at 25 mm. From the curves, the penetration depths were determined to be 16 and 23 mm for the conventional and SA-VS methods, respectively.

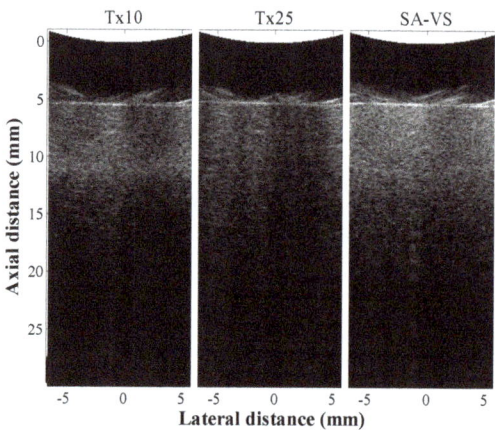

Figure 4. B-mode images of agar phantom obtained by the conventional method with transmit focal depths of 10 mm and 25 mm and the SA-VS method, respectively. The number of sub-aperture used in synthesis in the SA-VS was 33.

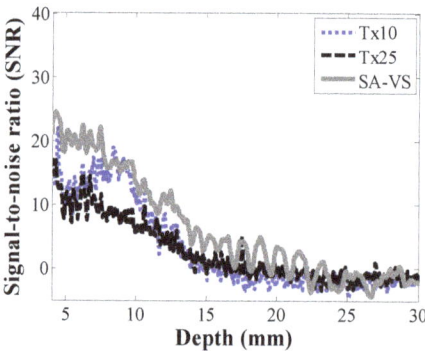

Figure 5. Measured SNRs for each method as a function of depth.

Figure 6 shows B-mode images of the bovine eye with the conventional and SA-VS methods. The image was acquired by avoiding the lens exhibiting high attenuation and propagation sound speed [30]. In this experiment, the optimal sound speed was estimated to be 1520 m/s and was well agreed with the previously reported value (i.e., 1513 m/s in the vitreous body) [30]. The transmit focal depth was 35 mm in the conventional method while it was 10 mm for the SA-VS method. The number of sub-aperture used in synthesis was 33. The images were rendered without any further processing such as filtering and post-image processing. Consistent with the results of phantom experiments, as shown in Figure 6, the SA-VS produced improved image quality compared to that by the conventional method. In the conventional method, the image was considerably blurred, especially in the anterior segment. On the other hand, the anterior and posterior segments were clearly visualized due to the enhancement of resolution. In addition, increased SNR in the SA-VS imaging allowed for visualizing structures that were ambiguous by noise in the conventional imaging; scattering from the vitreous body is clearly visualized proximal to the retina in the SA-VS imaging.

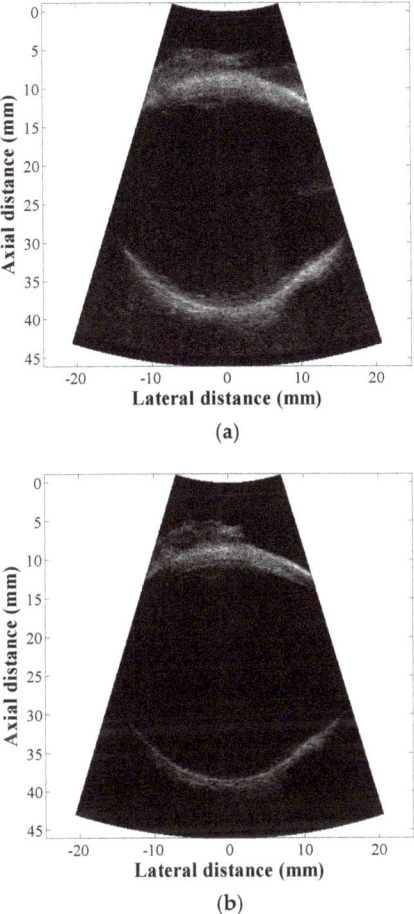

(a)

(b)

Figure 6. B-mode images of an excised bovine eye by (**a**) the conventional and (**b**) SA-VS methods. In the conventional method, a transmit focal depth was 35 mm while it was 10 mm for the SA-VS method. The number of sub-aperture used in synthesis in the SA-VS was 33.

In ophthalmic imaging, detailed information such as vitreous detachment, hemorrhage, and intraocular foreign body is important to diagnose and manage ocular emergencies, which can be achieved by increasing the center frequency of ultrasound imaging. Although the vitreous body is a gel-like substance and is known as acoustically transparent, ultrasound imaging with higher frequency (>20 MHz) still suffers from a high attenuation [31]. The method presented in the paper could resolve these problems (lower resolution at out-of-focus regions and SNR) and would be useful to diagnose and manage ocular emergencies.

Tissue motion and phase aberration are primary factors that limit the effectiveness of SA imaging [32]. Phase aberration correction based on correlation measurements would be the best solution. However, it requires a significant amount in computing correlation, which would be difficult to implement in real time. As a remedy, a method of estimating an optimal sound speed that can reduce the phase aberration artifacts has been proposed [28], which was used in the paper. Although it can partially resolve the problem, a previous study showed its potential in improving image quality in clinical practices [22]. Moreover, the effect of phase aberration in the ophthalmic SA imaging would be insignificant since the most of eye consists of a vitreous body that is a homogeneous medium.

4. Conclusions

Ultrasound imaging with the SA-VS was illustrated and showed its potential for ophthalmic imaging. The performances of the SA-VS method were evaluated through phantom and ex vivo experiments. The experimental results demonstrated that the SA-VS method can improve both lateral resolution and SNR. It was demonstrated that SA-VS imaging has the potential to deliver more additional significant information with diagnostic analytics compared to the conventional imaging method. Recent advances in electronics such as high-performance FPGA or GPU would support its high computational load and accelerate the commercialization of the SA-VS method. Further clinical evaluations of the SA-VS imaging need to be followed under various disease conditions to become an essential imaging tool for ophthalmic imaging.

Author Contributions: Conceptualization, H.G.L. and C.Y.; methodology, C.Y.; software, C.Y.; validation, H.G.L. and C.Y.; investigation, H.G.L.; resources, H.H.K.; writing—original draft preparation, C.Y.; writing—review and editing, H.G.L.; funding acquisition, H.H.K. and C.Y. All authors have read and agreed to the published version of the manuscript.

Funding: This work has been supported in part by the National Research Foundation of Korea (NRF) grant funded by the Korean government (MSIP; Ministry of Science, ICT and Future Planning) (No. 2019R1A2C1089813) and in part by the Korea Medical Device Development Fund grants funded by the Korea government (the Ministry of Science and ICT, the Ministry of Trade, Industry and Energy, the Ministry of Health and Welfare, the Ministry of Food and Drug Safety) (NTIS Number: 9991007146 and 202012E02).

Institutional Review Board Statement: Not applicable.

Informed Consent Statement: Not applicable.

Data Availability Statement: Not applicable.

Conflicts of Interest: The authors declare no conflict of interest.

References

1. Turnbull, D.; Starkoski, B.G.; Harasiewicz, K.A.; Semple, J.L.; Gupta, A.K.; Sauder, D.N.; Foster, F.S. A 40–100 MHz B-scan ultrasound backscatter microscope for skin imaging. *Ultrasound Med. Biol.* **1995**, *21*, 79–88. [CrossRef]
2. Lockwood, G.R.; Turnbull, D.H.; Christopher, D.A.; Foster, F.S. Beyond 30 MHz: Applications of high frequency ultrasonic imaging. *IEEE Eng. Med. Biol.* **1996**, *15*, 60–71. [CrossRef]
3. Passman, C.; Ermert, H. A 100 MHz ultrasound imaging system for dermatologic and ophthalmologic diagnostics. *IEEE Trans. Ultrason. Ferroelect. Freq. Control* **1996**, *43*, 545–552. [CrossRef]
4. Cannata, J.M.; Ritter, T.A.; Chen, W.-H.; Silverman, R.H.; Shung, K.K. Design of efficient, broadband single element (20–80 MHz) ultrasonic transducers for medical imaging applications. *IEEE Trans. Ultrason. Ferroelect. Freq. Control* **2003**, *50*, 1548–1557. [CrossRef]
5. Shung, K.K. High frequency ultrasonic imaging. *J. Med. Ultrasound* **2009**, *17*, 25–30. [CrossRef]
6. Li, X.; Wu, W.; Chung, Y.; Shih, W.Y.; Shih, W.-H.; Zhou, W.; Shung, K.K. 80-MHz Intravascular ultrasound transducer using PMN-PT free-standing film. *IEEE Trans. Ultrason. Ferroelect. Freq. Control* **2011**, *58*, 2281–2288.
7. Yeo, S.; Yoon, C.; Lien, C.-L.; Song, T.K.; Shung, K.K. Monitoring of Adult Zebrafish Heart Regeneration Using High-Frequency Ultrasound Spectral Doppler and Nakagami Imaging. *Sensors* **2019**, *19*, 4094. [CrossRef] [PubMed]
8. Liu, J.-H.; Jeng, G.-S.; Wu, T.-K.; Li, P.-C. ECG triggering and gating for ultrasound small animal imaging. *IEEE Trans. Ultrason. Ferroelect. Freq. Control* **2006**, *53*, 1590–1596.
9. Foster, F.S.; Hossack, J.; Adamson, S.L. Micro-ultrasound for preclinical imaging. *Interface Focus* **2011**, *1*, 576–601. [CrossRef]
10. Qiu, W.; Yu, Y.; Tsang, K.; Sun, L. An FPGA-based open platform for ultrasound biomicroscopy. *IEEE Trans. Ultrason. Ferroelectr. Freq. Control* **2012**, *59*, 1432–1442.
11. Li, M.-L.; Guan, W.-J.; Li, P.-C. Improved Synthetic Aperture Focusing Technique with Applications in High-Frequency Ultrasound Imaging. *IEEE Trans. Ultrason. Ferroelect. Freq. Contr.* **2004**, *51*, 63–70.

12. Hu, C.-H.; Snook, K.A.; Cao, P.-J.; Shung, K.K. High-frequency ultrasound annular array imaging. Part II: Digital beamformer design and imaging. *IEEE Trans. Ultrason. Ferroelectr. Freq. Control* **2006**, *53*, 309–316.

13. Yoon, C.; Kim, H.; Shung, K.K. Development of a low complexity, cost effective digital beamformer architecture for high-frequency ultrasound imaging. *IEEE Trans. Ultrason. Ferroelectr. Freq. Control* **2017**, *64*, 1002–1008. [CrossRef] [PubMed]

14. Yoon, C.; Yoo, Y.; Song, T.-K.; Chang, J.H. Orthogonal quadrature chirp signals for simultaneous multi-zone focusing in medical ultrasound imaging. *IEEE Trans. Ultrason. Ferroelectr. Freq. Control* **2012**, *59*, 1061–1069. [CrossRef] [PubMed]

15. Yoon, C.; Kim, G.-D.; Yoo, Y.; Song, T.-K.; Chang, J.H. Frequency equalized compounding for effective speckle reduction inmedical ultrasound imaging. *Biomed Signal Process. Control* **2013**, *8*, 876–887. [CrossRef]

16. Mamou, J.; Ketterling, J.A.; Silverman, R.H. Chirp-Coded Excitation Imaging With a High-Frequency Ultrasound Annular Array. *IEEE Trans. Ultrason. Ferroelectr. Freq. Control* **2008**, *55*, 508–513. [CrossRef] [PubMed]

17. Yoon, C.; Lee, W.; Chang, J.H.; Song, T.-K.; Yoo, Y. An Efficient Pulse Compression Method of Chirp-Coded Excitation in Medical Ultrasound Imaging. *IEEE Trans. Ultrason. Ferroelectr. Freq. Control* **2013**, *60*, 2225–2229. [CrossRef]

18. Karaman, M.; Li, P.-C.; O'Donnell, M. Synthetic aperture imaging for small scale systems. *IEEE Trans. Ultrason. Ferroelect. Freq. Control* **1995**, *42*, 429–442. [CrossRef]

19. Frazier, C.H.; O'Brien, W.D., Jr. Synthetic aperture techniques with a virtual source element. *IEEE Trans. Ultrason. Ferroelect. Freq. Control* **1998**, *45*, 196–207. [CrossRef]

20. Bae, M.-H.; Jeong, M.-K. A study of synthetic-aperture imaging with virtual source elements in B-mode ultrasound imaging systems. *IEEE Trans. Ultrason. Ferroelect. Freq. Control* **2000**, *47*, 1510–1519. [CrossRef] [PubMed]

21. Kortbek, J.; Jensen, J.A.; Gammelmark, K.L. Sequential beamforming for synthetic aperture imaging. *Ultrasonics* **2013**, *53*, 1–16. [CrossRef] [PubMed]

22. Kim, C.; Yoon, C.; Park, J.-H.; Lee, Y.; Kim, W.H.; Chang, J.M.; Choi, B.I.; Song, T.-K.; Yoo, Y. Evaluation of Ultrasound Synthetic Aperture Imaging Using Bidirectional Pixel-Based Focusing: Preliminary Phantom and In Vivo Breast Study. *IEEE Trans. Biomed. Eng.* **2013**, *60*, 2716–2724.

23. Hansen, P.M.; Hemmsen, M.; Brandt, A.; Rasmussen, J.; Lange, T.; Krohn, P.S.; Lonn, L.; Jensen, J.A.; Nielsen, M.B. Clinical evaluation of synthetic aperture sequential beamforming ultrasound in patients with liver tumors. *Ultrasound Med. Biol.* **2014**, *40*, 2805–2810. [CrossRef] [PubMed]

24. Park, J.-H.; Yoon, C.; Chang, J.H.; Yoo, Y.; Song, T.-K. A real-time synthetic aperture beamformer for medical ultrasound imaging. In Proceedings of the 2010 IEEE International Ultrasonics Symposium, San Diego, CA, USA, 11–14 October 2010.

25. So, H.K.H.; Junying, C.; Yu, A.C.H. Medical ultrasound imaging: To GPU or not to GPU? *IEEE Micro* **2011**, *31*, 54–65. [CrossRef]

26. Li, Y.-F.; Li, P.-C. Software beamforming: Comparison between a phased array and synthetic transmit aperture. *Ultrason. Imaging* **2011**, *33*, 109–118. [CrossRef] [PubMed]

27. Kim, H.H.; Hu, C.; Park, J.; Kang, B.J.; Wiliams, J.A.; Cannata, J.M.; Shung, K.K. Characterization and evaluation of high frequency convex array transducers. In Proceedings of the 2010 IEEE International Ultrasonics Symposium, San Diego, CA, USA, 11–14 October 2010.

28. Yoon, C.; Lee, Y.; Chang, J.H.; Song, T.-K.; Yoo, Y. In vitro estimation of mean sound speed based on minimum average phase variance in medical ultrasound imaging. *Ultrasonics* **2011**, *51*, 795–802. [CrossRef] [PubMed]

29. Carman, J.C. Classroom measurements of sound speed in fresh/saline water. *J. Acoust. Soc. Am.* **2012**, *131*, 2455. [CrossRef]

30. Dekorte, C.L.; Vandersteen, A.F.W.; Thijssen, J.M. Acoustic velocity and attenuation of eye tissues at 20 MHz. *Ultrasound Med. Biol.* **1994**, *20*, 471–480. [CrossRef]

31. Silverman, R.H.; Ketterling, J.A.; Mamou, J.; Lloyd, H.O.; Filoux, E.; Coleman, D.J. Pulse-Encoded Ultrasound Imaging of the Vitreous With an Annular Array. *Ophthalmic Surg. Lasers Imaging* **2012**, *43*, 82–86. [CrossRef]

32. Karaman, M.; Bilge, H.S.; O'Donnell, M. Adaptive Multi-element Synthetic Aperture Imaging with Motion and Phase Aberration Correction. *IEEE Trans. Ultrason. Ferroelect. Freq. Control* **1998**, *45*, 1077–1087. [CrossRef]

Review

Why Are Viscosity and Nonlinearity Bound to Make an Impact in Clinical Elastographic Diagnosis?

Guillermo Rus [1,2,3], **Inas H. Faris** [1,2], **Jorge Torres** [1,2,*], **Antonio Callejas** [1,2] and **Juan Melchor** [2,3,4]

1   Ultrasonics Group (TEP-959), Department of Structural Mechanics, University of Granada, 18071 Granada, Spain; grus@ugr.es (G.R.); inas@ugr.es (I.H.F.); acallejas@ugr.es (A.C.)
2   Biomechanics Group (TEC-12), Instituto de Investigación Biosanitaria, ibs.GRANADA, 18012 Granada, Spain; jmelchor@ugr.es
3   Excellence Research Unit "ModelingNature" MNat UCE.PP2017.03, University of Granada, 18071 Granada, Spain
4   Department of Statistics and Operations Research, University of Granada, 18071 Granada, Spain
*   Correspondence: geresez@ugr.es

Received: 25 March 2020; Accepted: 20 April 2020; Published: 22 April 2020

Abstract: The adoption of multiscale approaches by the biomechanical community has caused a major improvement in quality in the mechanical characterization of soft tissues. The recent developments in elastography techniques are enabling in vivo and non-invasive quantification of tissues' mechanical properties. Elastic changes in a tissue are associated with a broad spectrum of pathologies, which stems from the tissue microstructure, histology and biochemistry. This knowledge is combined with research evidence to provide a powerful diagnostic range of highly prevalent pathologies, from birth and labor disorders (prematurity, induction failures, etc.), to solid tumors (e.g., prostate, cervix, breast, melanoma) and liver fibrosis, just to name a few. This review aims to elucidate the potential of viscous and nonlinear elastic parameters as conceivable diagnostic mechanical biomarkers. First, by providing an insight into the classic role of soft tissue microstructure in linear elasticity; secondly, by understanding how viscosity and nonlinearity could enhance the current diagnosis in elastography; and finally, by compounding preliminary investigations of those elastography parameters within different technologies. In conclusion, evidence of the diagnostic capability of elastic parameters beyond linear stiffness is gaining momentum as a result of the technological and imaging developments in the field of biomechanics.

Keywords: elastography; soft tissue; nonlinearity; viscoelasticity

1. Introduction

Elastography is a medical imaging modality intended to map the elastic properties of soft tissues for diagnostic purposes that has recently been undergoing heavy development. It combines an imaging principle, that is, either ultrasonic or magnetic resonance imaging (MRI), with algorithms to reconstruct the stiffness maps from the raw shear wave propagation data [1–4]. The references are more detailed on ultrasound elastography, given the variety of techniques and the author's background, but the conclusions are fully applicable to magnetic resonance elastography (MRE). It follows that only dynamic or shear wave methods will be reviewed since strain elastography merely delivers relative deformability as the stress is unknown. However, in static ultrasonic methods, this current dependency of the stiffness on the probe pressure can become an opportunity instead of a drawback, since that dependency is caused by elastic nonlinearity, which is only quantifiable by dynamic techniques at this time. Further, the emerging field of elastography of viscous elastic parameters is finally gaining prominence. Therefore, beyond the current standard of elasticity maps, measuring the nonlinearity and viscosity might yield a more precise, pressure and operator-independent interpretation of the results, since for nonlinearity models, the dependence of the tangent stiffness modulus with deformation is correlated with operator-applied probe pressure, hence, decoupling operator dependency at the time a new biomarker is added. This proposed mechanical biomarkers, whose rationale is found in the tissue microstructure, and preliminary evidence, suggest a convincing diagnostic potential.

The purpose of this present work is not to address ultrasound elastography techniques in detail; there are several published works that deal with their differences and cut-off values, and the different systems available in the clinical market [2,5–8]. Instead, this article reviews the projected capabilities of viscous and nonlinear elastography parameters as clinical biomarkers from three perspectives: (1) the linear mechanics of soft tissue, focusing on the microstructure of the stroma, and therein mainly the fiber network organization; (2) how viscous and non-linear parameters are expected to be able to refine the diagnoses provided by classical elastography modalities; and (3) a spectrum of pathologies for which viscous and nonlinear elasticity quantification, conceived as mechanical biomarkers, has current or potential applications.

2. Mechanics of Soft Tissue

2.1. Soft Tissue Microstructure

The application of imaging techniques based on the propagation of shear acoustic waves aims to become a benchmark in terms of medical diagnosis. Pathologies such as tumors and fibrosis involve changes in consistency, since the structural properties of these anomalies imply a stiffer area that reflects histological differences in the microstructure of the tissue [9]. For current technologies to be effective and reliable, a sufficiently broad range of variation in the mechanical properties of the tissue must occur. This response can be addressed at the biological microscale, where the most relevant information can be gathered [10–12]. At this scale, there are two fundamental components, the extracellular matrix (ECM) and active cells, with fibroblasts and smooth muscle cells being the most prominent of this second group. The integrity of the tissue is ensured by the ECM, with a composition that provides support for the structural functionality through the formation of a fibrous scaffold. The components are organized hierarchically down to the macromolecular level, according to the morphology and function of the tissue they form. The primary constituent of the ECM is a crosslinked network of collagen and elastin, which is embedded within a gelatinous matrix of proteoglycans (PGs). This matrix is responsible for resisting and transmitting mechanical loads and regulating the hydrostatic pressure and fluid flow [13]. For illustrative purposes, the reader is referred to Figure 1, where the remodeling process of cervical ECM during pregnancy is graphically described.

All the elements that compose the ECM have a load-bearing role in the mechanical response, emphasizing the importance of the content and distribution of collagen fibers in the shear modulus of the tissue. During the synthesis of collagen there is a process of hydroxylation that determines the crosslink formation, setting the adhesion of the new fibrils [14,15]. This is a critical step in the development of pathologies related to collagen [16], such as fibrosis-associated pathologies, as shown on foreskin cell cultures [17]. When several fibrils are adhered, they increase the crosslink density, creating a stiffer fiber, 1–20 μm in diameter [18], and completing the fibrillogenesis process [19–23]. Apart from the diameter, the morphology of the collagen is defined by the interfibrillar space and the crimping. The origin of this wavy structure comes from the subfibril formation, very sensitive to different homeostasis levels affected by biochemical factors. It can resist very low compressions due to this crimping, which in turn is responsible for the existence of internal shear [24,25]. The collagenous matrix varies greatly depending on the organ and its state. For instance, breast tissue has a collagen content of 5–10% [26,27], similarly to the liver [28,29]. The collagen content is higher in the cervix and prostate tissues: around 60% for the cervix [30–32], and similar content for the prostate can be inferred from qualitative analysis [33].

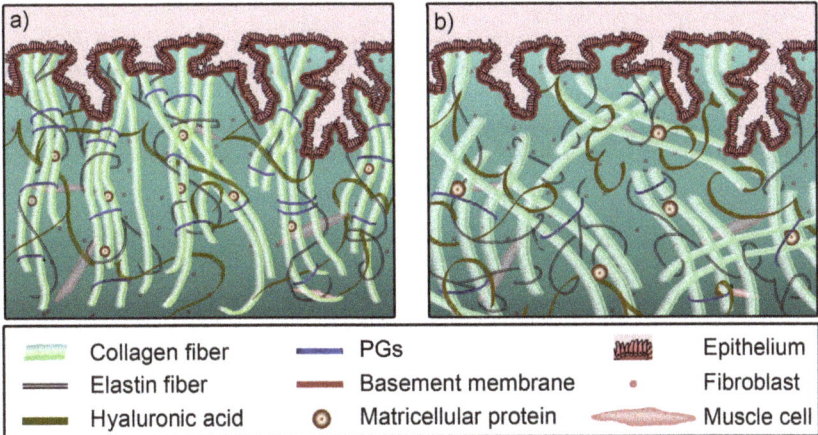

Figure 1. Histological illustration of the ECM remodeling as an example of cervical tissue during pregnancy. (**a**) The structure of the constituents in non-pregnant women. (**b**) The morphological evolution near the end of pregnancy. Quantitatively, there are increases in active cell and water contents, and crimping; and the diameter of collagen fibers increases, while PGs show a cyclic behavior. The legend at the bottom describes the symbol for each constituent; PGs: proteoglycans.

Elastin fibers are randomly distributed in the ECM, loosely interconnecting collagen fibers [34]. During their formation elastin fibers are prestressed, and once they are assembled they discharge stress, stretching and curling the attached collagen fibers [35,36]. They are used as a support in the mechanical response of collagen, operating as springs that recoil the structure to its initial state, allowing it to withstand repeated load cycles without reaching a plastic state [37]. They have a linear response up to 100% strain, with an average stiffness of 0.4 MPa depending on the tissue (two orders of magnitude lower than collagen) [38,39].

The gelatinous matrix is a ground material composed of water, proteins and PGs. PGs fill the spaces between the fibers in a perpendicular scattered network, conferring a supportive bending stiffness to collagen. At the same time, they contribute to resisting compression forces along with the interstitial fluid, balancing the fiber network [40]. PGs are composed of a core protein that covalently bonds with glycosaminoglycans (GAGs), thereby becoming a scaffold for the loose proteins of the ECM. Some of them can interlace their core with collagen, affecting the fibrillogenesis [41] and providing lateral stability [42,43]. GAGs are polysaccharide macromolecules with high electrical charges; among them, there is a particular GAG called hyaluronic acid, which is able to imbibe the surrounding elements. This is a hydrophilic process that attracts water, generating osmotic pressure, turning these components into a dampener against compression [44].

2.2. Linear Elasticity

The heterogeneous combination of the ECM components exhibits directional anisotropy, which is mainly attributed to variations in the morphology of the crosslinked fiber network [45]. Consequently, the stress at a point does not depend only on the gradient of deformation but also on the orientation, connection and distribution of its components. At the same time, the fiber network displays a nonlinear stress–strain relationship due to complex interactions that vary from point to point. The action of collagen and elastin can be lumped together, showing a stress–strain behavior divided into three regions (see Figure 2) [46]: (i) In the absence of load, collagen fibers are in their natural state of formation, wavy and loose. Due to its symmetrical organization, its behavior is frequently modeled as approximately isotropic. For strains lower than 2%, collagen offers little resistance, originated by fiber bending; thus, it is considered that elastin absorbs most of the energy, acting as a spring. This is the area with normal physiological activities called the toe region, showing nonlinear effects [47]. (ii) The progressive increase in deformation will disrupt the fibers that begin to line up in the direction of the load increasing the stiffness; this in turn, means that crosslinks are stressed and interfibrillar sliding is induced; the stress–strain relationship is approximately linear. (iii) At around 30% strain, depending on the tissue, crimping disappears and the fibers are arranged in parallel; the tissue reaches its highest stiffness [48–50]. Beyond these values, crosslinks and fibers begin to break, leaving severe damage to the tissue.

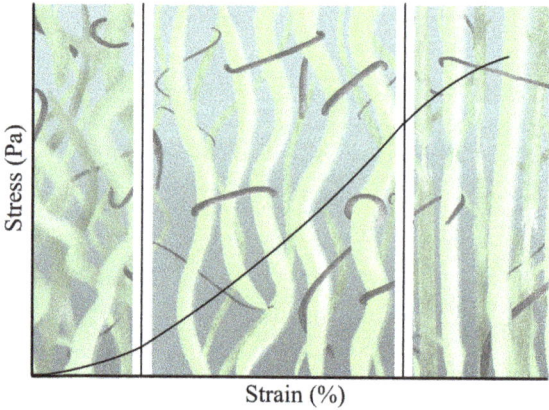

Figure 2. Stress–strain curve in soft tissues. The relationship is divided into three regions; namely, the toe, the nearly linear, and the failure regions: in the first, elastin fibers absorb most of the deformation and collagen forms a loose network that offers little resistance—primarily nonlinear behavior; in the second, collagen fibers line up and start to work under severe stress (nearly linear); and in the final region, the maximum capacity is reached. Color codes for the fibers are green for collagen and black for elastin.

Current imaging technologies are gaining prominence because they are based on the propagation of shear waves, which is directly proportional to the shear modulus, a very sensitive parameter to the microstructure of the material being examined [51]. Whether through biochemical modulations or the presence of a disease, the integrity of the tissue changes, which might be quantifiable with enough contrast for a clinical diagnosis; for that purpose a range of scores has been proposed [8]. However, it is difficult to maintain a standard, as the review article of Sigrist et al. notes [2], because the characterization of the shear modulus in the same tissue is variable. The variability in the commercial equipment methodologies and the existence of mechanisms of contrast make achieving standardization unfeasible. Another factor comes from the physical nature of shear waves; the displacement generated is characterized by being usually oriented perpendicularly to their propagation. However, the waves do not propagate with the fibrous matrix direction necessarily. This is dependent on the interaction of the wavelength relative to the interrogated fibrous matrix; therefore, its inherent anisotropy defines the examined direction. The dependence on tissue anisotropy, albeit interesting, is outside the scope of this work; for further information the reader is referred to [45]. Additionally, when the viscoelastic nature of tissues is considered, their mechanical response is time and frequency-dependent. Finally, the microscopic dimensions of the ECM components concerning the exciter wavelength must be taken into account. The key is to find a trade-off in the excitation frequency between a small enough wavelength that is able to interrogate internal components of the target tissue and a distance to the source of excitation that reduces wave attenuation [52–54].

The next step is to introduce the mathematical basis for soft tissue biomechanics, in this case, from the perspective of the continuum making two simplifications. The first is incompressibility, which stems from the high water content that does not allow the tissue to alter its volume under deformation; thus, the Poisson's ratio is considered close to 0.5. The second simplification is isotropy, since the anisotropy of the tissue increases the difficulty of formulating robust constitutive relationships. In most soft tissues, these simplifications have enabled researchers to work with more manageable problems, enabling progress in the understanding of the mechanics of soft tissues.

The total stress (σ_{ij}) and strain (ε_{ij}) can be deconstructed into two linear parts that naturally decompose the basic constituents of soft tissue [55,56]. On one hand, the volumetric, spherical or hydrostatic part is associated with the ground substance, mainly fluid, which provides no significant stiffness against shear deformations but is highly incompressible. On the other hand, shear stiffness is provided by the stroma, which governs the deviatoric components, the fiber and protein structure of the ECM.

$$\sigma_{ij} = -p\delta_{ij} + \tau_{ij} \qquad p = -1/3\sigma_{kk} \tag{1}$$

$$\varepsilon_{ij} = -v\delta_{ij} + d_{ij} \qquad v = -1/3\varepsilon_{kk} \tag{2}$$

where δ_{ij} is the delta of Kronecker, p is the hydrostatic pressure, v is the volumetric strain, τ_{ij} is the deviatoric stress tensor and d_{ij} is the deviatoric strain tensor. The previous relations can be combined to derive a constitutive relation, which is linear at first approximation and is similarly divided into volumetric and deviatoric components.

$$\sigma_{ij} = \lambda\delta_{ij}\varepsilon_{kk} + 2\mu\varepsilon_{ij} \tag{3}$$

where λ and μ are known as the Lamé constants, which characterize the elastic behavior of the material and must be obtained experimentally. The constant λ has no direct physical meaning; nevertheless, it is often associated with the bulk modulus $K = \lambda + 2/3\mu$, which describes the response in volume change under volumetric pressure. Since the compressibility of soft tissues tends toward that of water, which is orders of magnitudes higher than shear stiffness provided by the stroma, a good approximation is $K \approx \lambda$. As for the constant μ, it is usually called shear modulus and represents the resistance to shear deformation and can be written in terms of elasticity modulus and Poisson's ratio $\mu = E/(2(1+v))$ [57]. The volumetric and deviatoric decomposition naturally splits the former linear constitutive relationship into

$$p = 3Kv \qquad \tau_{ij} = 2\mu d_{ij} \tag{4}$$

Nevertheless, these parameters and assumptions do not provide a full representation of the behavior of soft tissues; they are limited to low levels of strain, such as image-guided interventions. New methodologies to interrogate other mechanical properties, such as shear viscosity and shear nonlinearity, are now appearing. The viscoelasticity of soft tissues implies the search for high order models that characterize the dispersion associated with shear wave propagation. As input, some studies have used the shear wave group velocity, which approximates as a series of derivative orders [58,59]. Likewise, taking advantage of the acoustoelasticity phenomenon, wherein the shear wave velocity is altered when a stress is applied due to wave propagation, new parameters become measurable [60–62]. For the specific case of nonlinear values, several theoretical methodologies have been proposed, and some experimental results have been obtained through acoustoelasticity, high amplitude shear wave propagation and nonlinear shear wave interaction [63]. The extracted information refers to the structure and functionality of the tissue, allowing one to identify conditions that elasticity alone is not able to capture so that the diagnosis is refined.

2.3. Viscoelasticity

From the mechanical viewpoint, two phenomena contribute to the time-dependent or rheological behavior of soft tissues: viscoelasticity and poroelasticity [64]. Although both viscosity and porosity contribute additively to the same phase lag between stress and strain dynamics, they are commonly quantified as an unique value called viscosity within the elastography community. However, viscoelasticity and poroelasticity stem from fundamentally different origins and are only separable playing with space and time scales. In other words, at large-size scales, tissues are viscoelastic in the short-time period and poroelastic in the long-time period, whereas the small-size scales, tissues are poroelastic in the short-time period and viscoelastic in the long-time period [65], which is clinically intractable given the limited region and frequency ranges. For this reason, it might be appropriate to rename viscoelastic elastography to rheological or dynamic elastography.

Soft tissues are generally assumed to be decomposed into their porous solid phases and their fluid phases [66]. The high fluid content in tissues is combined with the poroelastic structure of the ECM to allow motion between components under load, creating a time delay in the strain and triggering the viscoelastic response [67]. This biphasic nature implies a phase lag between the stress and strain associated with a relaxation time, or in the case of oscillatory mechanical tests, a phase angle. Then it would be advisable to start considering time-dependent effects, since the strain response to load and unload conditions is a function of time, often called the velocity of deformation. During the loading cycle there is dissipation of energy, reflecting the existence of hysteretic effects. At the same time, the strain evolution is slowed to allow the viscous flow to settle. Thus, the duration and rate of loading define the dynamics of the tissue strain. Without this characteristic, the stress during physiological activities would be harmful to the active structure [68].

One of the key features of viscoelastic tissues comes from the physics of wave propagation, where the dispersion is defined as a compound expression of the poroelastic and microstructural media governed by the complex fibrous multiscale microstructure of the stroma [69–72]. It is also known that the amplitude and intensity of waves decays proportionally to the distance traveled. Additionally, in a highly viscous environment, where the microvasculature and hemodynamics play an important role, it is observed that wave phase velocity changes with frequency, and wave amplitude is affected by geometric factors, such as boundary conditions and the sizes of scattering particles, similar or smaller than the wavelength [73]. Another important point is that the frequency-dependent behavior complicates the comparison of different technologies, since each author chooses a suitable range [6]. Neglecting the viscous part introduces bias for the estimation of elasticity, since the effect of wave dispersion is ignored.

The possibility of explaining these mechanical parameters by the internal structure and function of the tissue seems to be the key to improving the specificity of a pathology diagnosis. Collagen by itself exhibits viscoelastic behavior, attributed to fiber and fibril sliding and the crosslinking density; however, due to its short time of relaxation, it seems that the global response is dominated by non-collagenous components [74]. Elastin has been found to contribute to stress relaxation, since when it was removed in arteries, the relaxation time dropped significantly [75]. Nonetheless, PGs are considered to be the main viscous constituents, via embedding the collagen fibers and creating a lubricating effect. Their hydrophilia generates hydrostatic pressure, which, coupled with HA [76] and its large molecular size, entails water attraction, filling the porous matrix [77]. The roles of PGs and HA have been reviewed in tumor biology [78] and in inflammatory processes [79]. They are capable of acting as signaling pathways, interacting with diverse receptors, which affect the ultrastructure of the ECM that is transformed during inflammatory and neoplastic diseases [80]. In the case of pregnancy, as the time of delivery approaches, an inflammatory process is triggered, during which the proportion of PGs to collagen increases; therefore, higher viscosity is expected [81–83]. As for fibrosis disorders, there is an increased deposition of ECM constituents, especially collagen, accompanied by PGs and HA that help in cell signaling and proliferation [84]. A better understanding of these proteins and their relationship with viscosity might allow for the development of concrete diagnostic and therapeutic strategies.

Similarly, higher smooth muscle cell (SMC) tone in the carotid wall has been linked to higher viscosity [85]. For its part, it has been seen that there is an increase in SMC in the internal walls of the cervix as delivery approaches, and at the time of induction it became the most sensitive part [86]. In the liver, the development of fibrosis has been accompanied by an increase of SMC actin [87]. Investigations about the arterial viscoelasticity linked it to wall pressure [88].

From the perspective of tumors, there are changes at the cellular level which promote different reactions of the stroma. In breasts, the viscosity of lesions has been studied in order to discriminate the nature of the masses [89–92]. Higher viscosity was registered compared to healthy tissue and different ranges allowed researchers to distinguish between benign and malignant lesions.

Thus far, most studies have ignored this behavior, relying only on approaches based on linear elasticity simplifications. Although this has enabled progress to be made in quantitative imaging techniques, diagnoses sometimes fail because they do not deal with all the information. [93]. To reduce false-negative and false-positive results and to better understand pathological changes in soft tissues, extended dynamic mechanical parameters such as viscosity need to be investigated [94] and eventually be used as new diagnostic biomarkers. Ex vivo studies evidence the predictive relationship between viscosity and pathology; for instance, the marked Ex vivo neuronal demyelination with development of apparent vacuoles associated with a loss of interneuronal connections and thus with a reduction of matrix dimensionality, causing an observed alteration of viscosity [72,95–97]. The collected data from either traditional testing methods (creep and relaxation tests) or state-of-the-art imaging combined with the current computational power are allowing for the retrieval of viscous parameters from empirical or computational models. Table 1 presents a preview of the experimental evidence from which viscosity parameters have been estimated with different methods, along with applications to soft tissues whose results are described later in the manuscript.

Table 1. Qualitative overview of the work done on the description of the viscoelastic nature of selected soft tissues. The techniques that have achieved remarkable results are: shear wave dispersion ultrasound vibrometry (SDUV), dynamic mechanical analysis (DMA), magnetic resonance elastography (MRE), shear wave elastography (SWE) and torsional wave elastography (TWE). KVFD: Kelvin–Voigt fractional derivate; KV: Kelvin–Voigt.

Technique	Soft Tissue	Study Objective	Method	Reference
SDUV	Liver in vivo porcine	Regular characterization	Dispersion curve Voigt model	Chen et al. [98]
	Liver in vivo	Regular characterization	Dispersion curve Voigt model	Chen et al. [99]
	Liver in vitro rat	Fibrosis staging	Dispersion curve Voigt model	Lin et al. [100]
	Prostate in vitro	Regular characterization	Dispersion curve Voigt model	Mitri et al. [101]
	Breast in vivo	Malignant vs. Benign vs. Healthy state	Dispersion curve Voigt model	Kumar et al. [89]
DMA	Prostate in vitro	Healthy vs. Cancerous state	Dispersion curve KVFD model	Zhang et al. [102]
MRE	Breast in vivo	Malignant vs. Benign vs. Healthy state	Phase offset imaging reconstruction	Sinkus et al. [103]
	Breast in vivo	Malignant vs. Benign vs. Healthy state	Transversely isotropic model	Sinkus et al. [104]
	Liver in vivo	Transplant rejection	Attenuation Measuring Ultrasound Shearwave Elastography (AMUSE)	Nenadic et al. [105]
	Liver in vivo	Regular characterization	Dispersion curve Zener model	Klatt et al. [106]
	Liver in vivo	Fibrosis staging	Dispersion curve Zener model	Asbach et al. [107]
	Prostate in vivo	Prostate cancer vs. Benign prostatitis	Phase offset imaging reconstruction	Li et al. [108]
SWE	Liver in vivo	Fibrosis	Shear Wave Dispersion Slope	Sugimoto et al. [109]
	Liver in vivo	Healthy vs. Fibrosis staging	Shear Wave Spectroscopy	Deffieux et al. [110]
TWE	Cervix Ex vivo	Regular characterization	Dispersion curve KV and KVFD model	Callejas et al. [111]

It is important to note that if tissues are precompressed when they are examined, the estimation of parameters will be biased, as the time-dependency of the response is relevant. Changes over time due to mechanical stimulation are attributed to rapid alterations in cellular activity, mainly the synthesis and modification of components of the ECM (collagen and proteinases) [112]. To avoid this situation, preconditioning protocols should be proposed whenever the specimen studied allows it, so that a stabilization in the response is achieved [113]. With the aim of capturing this material behavior, the most popular approach considers soft tissues as uniphase solids and their response to external loads or deformation is represented as a lumped relationship. This method uses linear viscoelastic models that generally include a solid-related characteristic (e.g., spring) and a viscous fluid element (e.g., dashpot). To name a few, Maxwell, Kelvin–Voigt (KV) and Zener viscoelastic models provide information on how the different scales are linked to each other [98,111]. However, in order to fit a model when the soft tissue shows several characteristic times, generalized linear viscoelastic models are used, such as generalized Maxwell or KV models [114,115]. When large strains are expected, these linear models are not suitable; thus, the proposed Fung's quasilinear viscoelastic model is frequently adopted [116].

One of the models in the literature most used to fit the parameters is the KV model, due to its simplicity [117]. Other models have been explored, such as Maxwell; fractional derivative versions of the above; and combined models, such as the springpot model [118]. The KV formulation in terms of the stress tensor (Equation (1)), assuming constitutive and viscous linearity have been derived with the aim of simplifying equations [119]. Following the references found in the literature [120–122],

$$p = 3Kv + 3\eta^v \dot{v}$$

$$\tau_{ij} = 2\mu d_{ij} + 2\eta \dot{d}_{ij}$$

$$(5)$$

where K is the compressional modulus; η and η^v are the shear and volumetric viscosities, respectively; and $\dot{v}$ and $\dot{d}_{ij}$ are the derivate of the volumetric and deviatoric strains, respectively.

Assuming incompressibility, only deviatoric components (τ_{ij}, $p = v = 0$) are considered. According to the schematic representation of the KV model, the total stress is the sum of the elastic and viscous terms,

$$\sigma_{ij} = \tau_{ij} = 2\mu d_{ij} + 2\eta \dot{d}_{ij} = 2\mu \epsilon_{ij} + 2\eta \dot{\epsilon}_{ij} \tag{6}$$

Following the same steps as in the Kelvin–Voigt model, the implementation of the Maxwell model stems from the strain tensor of Equation (2). For the same reasons stated for the KV case (d_{ij}, $p = v = 0$), exclusively deviatoric components are considered. Only elastic and viscous components of the deviatoric term of the strain tensor are adopted,

$$d_{ij} = \tau_{ij}/2\mu, \quad \dot{d}_{ij} = \tau_{ij}/2\eta \tag{7}$$

The constitutive equation for this model is obtained by adding the elastic and viscous terms by,

$$\dot{d}_{ij} = \dot{\tau}_{ij}/2\mu + \tau_{ij}/2\eta \tag{8}$$

All this evidence suggests that the viscous phase may become a biomarker for the characterization of microstructural changes [123–127]. Table 2 shows some indications of the current status of this parameter in terms of limitations and characteristics that have been specified for some ultrasound elastography methods. Phase-sensitive imaging techniques might become a monitoring tool for early diagnose, able to keep track of quick dynamic changes in the tissue, before significant or unclear changes in elasticity and also reducing the number of unnecessary biopsies [1,128].

Table 2. Comparison of the current methods that have been able to successfully estimate viscosity parameters using ultrasound elastography.

Method	Advantages	Disadvantages
Shear wave speed dispersion curve: estimation of vicosity parameters by fitting a rheological model	Most relevant and extended technique Considerable amount of previous work for different types of organs to compare with Depends on shear wave methods: noninvasive both internally and externally in contact with the soft tissue	No consensus on the most appropriate rheological model for soft tissue characterization Studies report values of viscosity for a specific rheological model (not comparable)
Shear Wave Dispersion Imaging	Dispersion slope value: physical quantity not based on a rheological model (model-free)	Integrated into commercial ultrasound systems not accessible for researchers (black box software)
Shear Wave Spectroscopy: new signal processing of the SSI data (Supersonic Shear Imaging)	Frequency-dependent measurement of the shear wave speed, quantitative and noninvasive	Limits its use to scans via SSI

2.4. Nonlinearity

One of the main hypotheses about the pathology-mediated origin of nonlinearity changes is based on the nonlinear character of the strain response. The organization of collagen fibers and elastin, as well as their amounts, combined with the synthesis and degradation processes that are experienced due to growth and remodeling enhance the nonlinear behavior [129,130]. Additionally, the stress–strain behavior of the stroma is nonlinear between tension and compression, with a stiffer response and reduced extensibility in tension, and a more compliant response in compression [131,132].

Several experimental studies, including the recent study of Aristizabal et al. [94], estimated the nonlinear shear modulus in Ex vivo samples. Particularly, that paper was about Ex vivo kidneys diagnosing end-stage renal disease, for which a better contrast in the diagnosis was shown. Based on the principle of acoustoelasticity, the feasibility of obtaining nonlinear parameters through changes in the deformation and its consequent interaction with the propagated wave is proven. The application of a deformation and the use of radio frequency ultrasonic signals to quantify it, was the work of Goenezen et al. [133]; they obtained spatial maps of nonlinear elastic parameters in patients with malignant and benign tumors. Their conclusions highlight a greater magnitude in the case of malignant tumors. In the context of preterm birth assessment, Myers et al. [131] investigated the interaction between mechanical and chemical properties of several cervical samples from different human hysterectomy specimens: non-pregnant patients with previous vaginal deliveries; non-pregnant patients with no previous vaginal deliveries; and pregnant patients at the time of cesarean section. The samples were tested under confined compression, unconfined compression and tension. Results indicated that the cervical stroma has a nonlinear behavior that could be explained with an accurate multi-scale model.

The significant hyperelasticity that soft tissues exhibit can manifest itself as quantifiable shear wave harmonic generation (via ultrasonic shear elastography); the stored strain energy is variable with the fiber orientation. Taking this opportunity, an efficient application of nonlinear or hyperelastic constitutive equations for either finite element analysis or experimental analysis requires the derivation of a strain energy function to consider an adequate stress–strain relationship. A diversity of approaches to nonlinear mechanics have been developed since Landau and Murnaghan [134,135], which are particularly well-suited for nonlinear wave modeling; then came the recent proposals lead by Ogden, Mooney-Rivlin, Yeoh and Fung [136,137] which cover adjustment theories based on modeling of physiological mechanics [138].

These behaviors can be modeled from the perspective of the continuum, making the assumption of Landau third and fourth-order elastic constants (TOEC and FOEC),

$$S_{ij} = \lambda \delta_{ij} \varepsilon_{kk} + 2\mu \varepsilon_{ij} + \mathcal{A} \varepsilon_{ij}^2 + 2\mathcal{B} \varepsilon_{ij} \varepsilon_{kk} + \mathcal{C}(\varepsilon_{kk})^2 + h.o.t. \tag{9}$$

limited to the third order, where $\mathcal{A}, \mathcal{B}$ and $\mathcal{C}$ are the TOEC, and δ_{ij} is the delta of Kronecker, and where S_{ij} is the second Piola–Kirchoff Stress tensor [139]. The simplification of nonlinear strain energy function in the case of incompressible tissues, and extended to fourth order, was derived by Hamilton [140], and it is considered as the most representative.

$$S_{ij} = 2\mu \varepsilon_{ij} + \mathcal{A} \varepsilon_{ij}^2 + 4\mathcal{D}(\text{tr}\varepsilon_{ij}^2)\varepsilon_{ij} \tag{10}$$

Experimentally, nonlinear parameters can either be estimated by measuring the change of apparent speed of shear wave propagation after a precompression [141], or by quantifying the cumulative harmonic generation during the propagation of shear waves across nonlinear tissue [142]. The nonlinear shear wave equation depending on TOEC and FOEC in the soft solid isotropic state was derived by Hamilton and Zabolotskaya [140]. Then, through a strain energy function they were able to separate the compressional and shear parts. In that approach, nonlinear propagation depends only on three elastic constants of the first (linear), the third and the fourth-order (nonlinear). Therefore, the generation of harmonics in soft tissue and biomaterials is likely to be studied under this prism. However, it is also possible to describe a theoretical model of shear waves propagating in soft biological tissue induced remotely by the nonlinear radiation force of the focused ultrasound. The spatial and temporal profiles of the shear displacement confirm the results of the mathematical modeling previously described. The experimental procedures based on acustoelasticity techniques are also performed to obtain the nonlinear coefficients of the Burgers Equation by describing the behavior of tissue [63]. For example, another experience in this line of research is the use of MRE by visualizing the nonlinear propagation of shear waves providing valuable information about the nonlinear mechanical behavior of the soft tissue [143]. From this procedure, it is shown that both odd and even higher harmonics are processed, with their amplitudes depending on the actuator details, the image geometry and the nonlinear properties of the tissue. With an adequate analysis of the displacement, it is possible to derive the harmonics that arise from the nonlinear soft tissue response. They have been extracted, for example, in phantoms at 600 and 750 Hz. Thus, if strain energy is modulated, it is feasible to determine the nonlinear biomechanical properties of the tissue [51]. The second approach has been proposed in combination with torsional wave elastography, described later [144,145] following Landau's theory [134] and its adaptation for quasi-incompressible media coupled with multiscale hyperelastic models [146,147]. The formulations of the nonlinear torsional wave propagation on a hyperelastic material should be taken into account in cylindrical coordinates characterized by strain energy functions [148,149].

Analogously, it is also possible to accurately and quantitatively recover the local Landau $\mathcal{A}$ parameter. The characterization of the shear nonlinearity of soft tissues by applying the acustoelasticity techniques in quasi-fluids could be correlated to the ultrasonic shear wave speed [150]. But these theories should be tackled by more profound studies due to the dispersion and variability of the outputs. It is also possible to deduce the nonlinear coefficients in the modified Burgers model using the numerical simulation from the quadratic wave equation rewritten in its nondimensional form [63]. It has been introduced to calculate nonlinear parameters of hydrogels and in Ex vivo porcine kidneys [94], but the cubic orders are valid under a relation that should be verified in some cases depending on acoustic nonlinearity [140].

In summary, since shear waves are believed to be far more sensitive to tissue classification than standard compressional waves, but they are complicated to quantify, some experimental observations may tangentially suggest that nonlinear mechanical properties may be a key signature withh which to quantify and classify soft tissue behavior [151–156]. The advantages and disadvantages of the current scene of nonlinearity in biomechanics are summarized in the Table 3. Therefore, the focus on developing nonlinear models in the clinical field will provide a better understanding of soft tissue biomechanics alongside new diagnostic biomarkers. Techniques such as shear wave elastography and torsional waves are postulated to be crucial tools, sensitive to the measurements of these nonlinear parameters, provided a consistent and efficient complete formulation is established.

Table 3. Summary of the current state of implementation of nonlinearity in the quantification of soft tissue mechanical properties.

Advantages	Disadvantages
New set of parameters to interpret biological and physiological disorders	Several proposed models to be chosen depending on the problem, pathology or tissue considered
Characterization of tissue microscale in terms of harmonics	Inhomogeneus measurements due to the nature of propagation in the tissue
Open questions that add a new branch in biomedical engineering	Mathematically intractable in exact terms

3. Clinical Applications

Since the 80s, elastography has gradually become a widely applied medical imaging technique [157]. The different techniques of elastography are based on the assumption that soft tissues are deformed more than rigid tissues, and that these differences can be quantified [158]. However, this conventional perspective is undergoing a change of scenery; recently, emphasis has been placed on the complex structures that soft tissues exhibit, deeming not only elastic but strongly nonlinear hyperelastic, viscoelastic and poroelastic behavior important. Linear elastic models have been used extensively to characterize soft tissues by the biomechanics community, though it is known that this simplification in the characterization provides incomplete information in their results. Additional biomarkers, such as viscosity and nonlinearity, are herein proposed as hypotheses to enable new diagnostic standards in a broad spectrum of pathologies. In the following subsections, because of the prevalence of the diseases from which they suffer, the focus is on prostate, breast, liver and labor disorders, not to mention that the conclusions could be extended to solid tumors, atherosclerosis and osteoarticular syndromes, to name a few.

3.1. Prostate

Prostate cancer is the second most common cancer in men worldwide (almost 1.3 million diagnoses) and the fifth leading cause of cancer death among men (350,000 deaths worldwide) [159]. Furthermore, the increase in longevity and awareness of the disease is leading to more men requesting screening, which in turn will dramatically increase the number of patients diagnosed [160]. Barr et al. [161] provided an extensive study of guidelines and recommendations on the clinical use of ultrasound elastography on the prostate.

Ex vivo and in vivo results have demonstrated that acoustic radiation force impulse (ARFI) can be applied to visualize internal structures and to detect suspicious lesions in the prostate [162,163]. Among all the elastography techniques for prostate cancer detection that provide quantitative elasticity results at present, transrectal SWE (TR-SWE) by Aixplorer® (SuperSonic Imagine, Aix-en-Provence, France) is the most prolific in terms of the number of publications. Recent in vivo studies on prostate cancer diagnosis using TR-SWE presented auspicious results [151,164]. However, TR-SWE has some drawbacks [165]: the pressure artifacts induced by the transducer, as the end-fire design of the probe requires bending to image mid prostate and apex; the slow frame rate, i.e., one image per second; the limited size of the ROI, since only half of the prostate is covered; the delay in stabilizing the signals at each acquisition plane; and the signal attenuation in large prostates was making the evaluation of the anterior transitional zone of said prostates difficult or impossible [166]. Most of the quantitative elastography results of tissue elasticity of the prostate have been achieved by using TR-SWE in different states of in vivo prostatic tissue [151,166–170]. The frequency range is expected to be between 50 and 450 Hz according to other TR-SWE publications [171]. By analyzing these results, differentiation between benign and malignant tissue in terms of stiffness is not a trivial matter, since ranges of values overlap. In order to discriminate in vivo malignant tissues from benign tissues using TR-SWE, Correas et al. [164] and Barr et al. [151] proposed Young's modulus thresholds of 35 and 37 kPa respectively. According to their conclusions, these thresholds provided additional criteria for prostate cancer detection and biopsy guidance and enabled a substantial reduction in the number of biopsies.

The application of point shear wave elastography (pSWE) allowed Zhai et al. [172] to reconstruct the shear modulus values from excised human prostates with different pathologies. The limitation of the work was the low spatial resolution, which may cause variations in the reconstructed shear modulus. Another in vivo study by Zheng et al. stated that pSWE could effectively measure the stiffness of prostate nodular lesions between prostate cancer and benign prostatic hyperplasia [173]. Even so, the authors specified that the limited detected depth and the fixed box dimensions of the target region of interest (ROI) could hamper the broader application of pSWE technology.

As for the viscoelastic characterization of human prostatic tissue, few studies based on ultrasound elastography have addressed the issue. Shear wave dispersion ultrasound vibrometry (SDUV), one of the few techniques that has been used in the prostate, consists of monitoring the propagation of the shear wave by a separate ultrasound detector and reconstruction of the wave speed from two different phases [98] (refer to Figure 3 for an illustrative example of the principle). The in vitro study of Mitri et al. [101] used a KV model aimed at the characterization of the mechanical shear parameter for frequencies between 50 and 400 kHz. They obtained shear elastic modulus values of 1.31–12.81 kPa and viscosity values between 1.10 and 6.82 Pa.s. These data proved the viscoelastic nature of the properties of prostatic tissue.

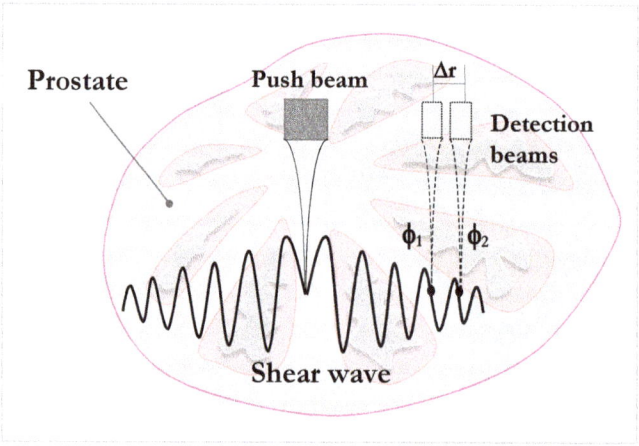

Figure 3. Illustration of shearwave dispersion ultrasound vibrometry (SDUV) principle. A harmonic shear wave is produced by a push beam; the propagation is monitored by separated detection beams at two positions. The shear wave speed is reconstructed from its phase ϕ_1, ϕ_2, separated a distance Δr.

Two other studies used a Kelvin–Voigt fractional derivative (KVFD) constitutive law, a more generalized case of the KV model, for measuring the variation of the complex Young's modulus E^* between normal and cancerous prostatic tissue [102,174]. In the first in vitro study, Zhang et al. [102] extracted the complex Young's modulus by fitting data from a dynamic mechanical analysis (DMA) test to a KVFD model. In Table 4 the viscosity parameter and the order of the fractional derivative associated with the KVFD Young's modulus is presented. In the second Ex vivo study, Hoyt et al. [174], made a comparative study between crawling wave spectroscopy and the same DMA test used in the first study for two samples of human excised prostate. Results showed relative similarities between techniques with errors below 12%. In any case, the sample sizes were too small to be statistically significant in both studies.

Table 4. Viscosity parameters derived from different methods, including a Kelvin–Voigt fractional derivative (KVFD) fitting using dynamic mechanical analysis (DMA), KV fitting on shear wave dispersion ultrasonic vibrometry (SDUV) and magnetic resonance elastography (MRE) results of prostatic tissue. Values are reported as means and standard deviations.

Tissue State	Viscosity Parameter (Pa.s)	Fractional Derivate Order	Method	Reference
Healthy	3.61 ± 1.25	0.215 ± 0.042	DMA	Zhang et al. [102]
Cancerous	8.65 ± 3.40	0.225 ± 0.03		
Healthy	1.10–6.82 (range)	-	SDUV	Mitri et al. [101]
Benign prostatitis	2.13 ± 0.21	-	MRE	Li et al. [108]
Cancerous	6.56 ± 0.99	-		

In the field of MRE some studies have addressed the generation of shear waves using transurethral devices. Chopra et al. [175] designed a transurethral actuator to produce shear waves in the prostate with adequate propagation at a reasonable frequency. A canine experiment demonstrated the feasibility of transurethral MRE in vivo. Shear waves have a penetration depth of 3–5 cm, as opposed to 15 cm for an external driver, allowing high spatial resolution. An alternative to the invasive transurethral driver was subsequently proposed by Arani et al. [176]. The driver was tested in prostate-mimicking gelatin phantoms to explore the imaging parameters of transurethral MRE and to determine whether they encompass the requirements for prostate cancer localization. A more recent study carried out by Reiter et al. [177] investigated the limitations present in MRI, such as interobserver variability and low specificity. For this purpose, fourteen fresh prostate specimens from men were examined. A piezoelectric actuator induced radially converging shear waves in the sample. The results of the work suggested that prostate MRE has the potential to improve the diagnostic performance of multiparametric MRI. An in vivo study carried out by Li et al. [108] showed that MRE could be used to distinguish between prostate cancer and benign prostatic disease in terms of shear viscosity. The study included 18 patients (eight with prostate cancer, 10 with prostatitis). The mean shear viscosity was significantly higher in prostate cancer (6.56 ± 0.99 Pa.s) than in benign prostatitis (2.13 ± 0.21 Pa.s).

Further experimental characterization studies of prostatic tissue are required to accurately model the real viscoelastic behavior of the prostatic tissue in all its conditions. As far as we know, no clinical studies taking into account the effect of the nonlinearity of prostate tissues have been reported.

3.2. Breast

The International Agency for Research on Cancer concluded in 2018 [159] that breast cancer is the most commonly diagnosed (over 2 million cases) and leading cause of cancer death (over 600,000 cases worldwide) among females. In the last few decades, several studies have compared the efficacy of the diagnosis of mammary elastography versus conventional ultrasound in the evaluation of different breast lesions. Ultrasound evaluation is established through the BIRADS classification [178], while the elastographic assessment is based on building a pattern between stress and size relationships [179,180]. Despite these efforts, it remains a significant healthcare problem, and what is more, countries in transition are experiencing a rise in their rates [181].

The representative clinical cases whose applications are relevant to include are benign lesions, malignant lesions and lymphatic and metastatic lesions [182]. The anatomy of the breast has allowed several elastography-based studies to be performed for the characterization and detection of masses. An extensive work of the World Federation of Ultrasound in Medicine and Biology (WFUMB) dealing with the guidelines and recommendations for clinical use of ultrasound elastography on the breast could be consulted for further information about elastography systems and their cut off values [183]. However, no clear consensus has been reached as to what measure of the shear modulus should be used and what ROI is the most appropriate for the estimation of elasticity. What is clear is that malignant lesions show a larger shear modulus than benign [92,184–190]. Still, several forms of misdiagnosis have been considered. The size of the lesion combined with a high density of the tissue could complicate the detection [191]. If benign and malignant lesions overlap, the power of the elasticity estimation is reduced [171,192]. Another issue comes from the effect of calcification: the surrounding zones are hardened, making the elasticity estimation higher. If the ROI selected matches this area, a misdiagnosis may be expected [193].

In contrast, viscosity is emerging as a better indicator, especially for tumor differentiation [194]. The first studies in using this parameter for in vivo diagnosis were attempted by Qiu et al. [195]. They compared the retardation times of benign and malignant lesions. The time for the benign state was clearly larger than the malignant. This was justified because malignant tumors increase their collagen and crosslinking densities, while there is a reduction of proteoglycans that declines the lubricating effect. Benign lesions are dominated by the fluid viscous phase of the tissue, hypothesized in part to be the lubricated motion of collagen. Those results are opposed to the studies on the quantification of the shear viscosity summarized in Table 5. Sinkus et al. performed two in vivo studies using MRE [103] and transversely isotropic models [104]. The idea behind the use of models with transverse waves is to remove the contribution of compressional spurious waves in order to reconstruct viscoelastic parameters. The SDUV technique has also been used in combination with viscoelastic models [89] (refer to Figure 4 for the reconstruction process). Another recent study on in vivo tissue applied the data from the creep test to a first order KV model fit, where the retardation time allowed them to distinguish between benign, malignant and healthy tissue [196]. These techniques are not feasible to compare, since, as previously stated, soft tissues are frequency-dependent and each author uses a different range of frequencies. The common finding that emerged was that shear viscosity was higher in all malignant states, and despite the great dispersion showed, these masses were heterogeneous in terms of their viscosity values. Additionally, the studies inferred that the maximum values were well correlated with malignant diagnosis in MR mammographies, encouraging further exploration.

Table 5. Viscosity parameters are calculated for the malignant, benign and healthy states in the breast tissue. The methods applied were magnetic resonance elastography (MRE), transverse acoustic waves and shear wave dispersion ultrasound vibromerty (SDUV). Values are reported as means and standard deviations.

Tissue State	Viscosity (Pa.s)	Parameter	Method	Reference
Malignant	2.40 ± 1.70		MRE	Sinkus et al. [103]
Benign	2.10 ± 1.40			
Healthy	0.55 ± 0.12			
Malignant	3.00 ± 0.80		Transverse Acoustic Waves	Sinkus et al. [104]
Benign	2.40 ± 1.90			
Healthy	0.70 ± 0.55			
Malignant	8.22 ± 3.36		SDUV + Kelvin-Voigt	Kumar et al. [89]
Benign	2.83 ± 1.47			
Healthy	1.41 ± 0.67			

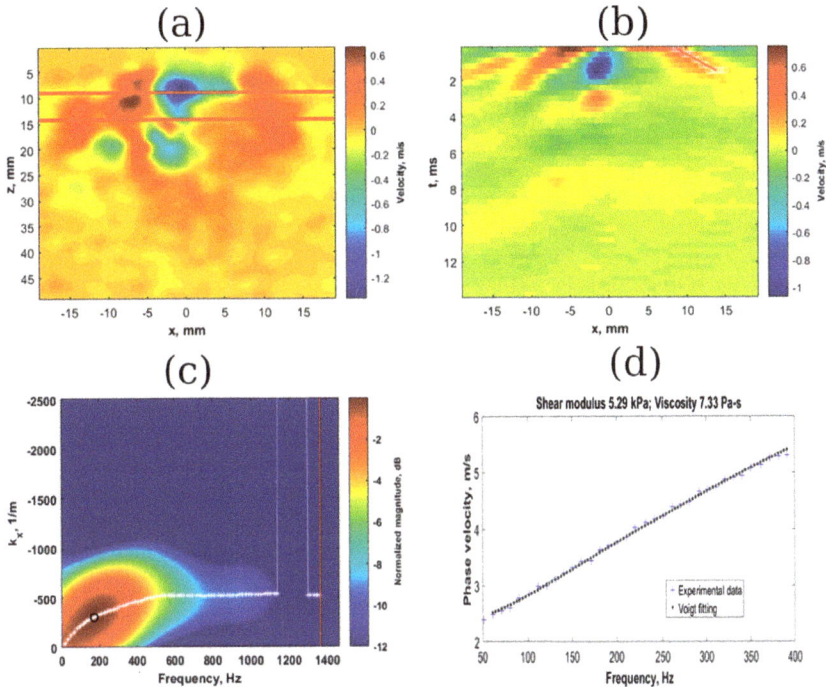

Figure 4. An illustrative process for the estimation of the viscosity parameter of a malignant mass. (**a,b**) Maps of particle velocity; (**c**) a k-space map displaying the phase velocity with the energy of each frequency; (**d**) the final result as a dispersion curve, based on the phase velocity, which is fitted using a Voigt model for estimation of viscoelastic parameters. Source: PLoS ONE, modified from 2018 Kumar et al. [89].

Bernal et al. [141] focused on the detection of early breast cancer in vivo by nonlinear quantification. In their study they implemented a technique that combines shear wave elastography with a prestress that modifies the shear wave speed due to the Landau-type elastic nonlinearity, to measure the nonlinear shear modulus. The mean values of the nonlinear parameter A were -95 kPa for healthy tissue, -619 kPa for benign lesions, and -806 kPa for malignant lesions, a considerable variability that show signs of its utility.

These techniques suggest a promising scenario, but the recent expansion of elastography among all device designers and manufacturers has led to a dizzying increase in the number of tests whose results call for consistency improvements [197]. Despite this, it has been exhibited that both linear and nonlinear elastography, possibly together, promise better sensitivity and specificity with which to characterize benign and malignant mammary lesions [198].

3.3. Liver

Over two million people are estimated to die every year due to chronic liver diseases: one million due to complications of cirrhosis and the rest due to viral hepatitis and hepatocellular carcinoma [199]. These diseases remain a burdening health problem [200] that demands better mechanisms for prevention, correct detection and treatment [201]. Different organizations have published a quite number of reviews of utlrasound elastography and clinical guidelines, and they can be consulted to deepen knowledge in the technical and clinical domains [2,8,202–206].

There are several tests available in the clinical protocols to assess the extent of fibrosis and cirrhosis. The most common is a percutaneous liver biopsy, a procedure performed without hospital admission that consists of introducing a biopsy needle through the ribs to the liver [207]. Although it is a standardized method to determine the state of the liver, its limitations stem from being an invasive method, which can cause minor or severe complications [208]. Additionally, the liver is a large organ and the biopsy represents only 1 of 50,000 of its total volume, whereby it can provide false negatives or misinterpretations of the real state of the disease [209,210]. The METAVIR scale and the Scheuer classification [211] divides fibrosis into five stages. Stage 0: there is no fibrosis. Stage 1: mild fibrosis. Stage 2: fibrosis extends to areas near the portal vein. Stage 3: fibrosis extends out from the areas of the portal vein. In this stage, many bridges of fibrosis connect the portal vein with the central areas of the liver. Stage 4: fibrosis has evolved to cirrhosis, which is an advanced pathological stage with distortion of the hepatic vasculature and architecture [212].

The most important advance for fibrosis staging has been obtained with the appearance of transient elastography (TE) using Fibroscan® (Echosens, Paris, France), which has pioneered efforts since its first commercialization in 2003. Fibroscan® generates images corresponding to the propagated elastic wave associated with values of hepatic rigidity measured in kilopascals (kPa). In vivo results of Ziol et al. [213] and Castera et al. [214] indicate that TE allows differentiating significant states of fibrosis. Chon et al. in [215] confirmed in a meta-analysis that TE is more accurate for detecting F4 fibrosis than mild fibrosis. Similar results were obtained by Afdhal et al. [216]. Transient elastography has been shown to be effective in diagnosing cirrhosis (stage F4 fibrosis) and generally in distinguishing significant fibrosis ($\geq$F2) from non-significant fibrosis (F0 and F1). Cassinoto et al. [217] made a comparison study between TE and 2D-SWE and pSWE using biopsy as a gold standard. Results demonstrate that shear wave elastography (SWE) is more accurate in the diagnosis of severe fibrosis than TE. Similar results can be found in [218–220]. However, the distinction between individual fibrosis stages is still not well validated. These studies did not change the frequency of vibration, thereby disregarding the viscoelastic properties of the liver, and this presumably could lead to errors in the early detection of liver fibrosis because the elasticity can be kept within normal values in those stages [110,220].

The highly viscoelastic structure of the liver suggests a strong diagnostic potential of viscosity, since shear wave velocity is frequency-dependent; this means that it is possible to in vivo quantify the tissue viscosity from the dispersion curves [99,122,221–223]. The elasticity of the liver depends mainly upon the fibrosis stage, but additionally on factors such as edema, inflammation, extrahepatic cholestasis and congestion [224]. For these cases of hepatic diseases, having an additional biomarker to quantify the stage of the disease may yield a significant advantage. Viscosity also plays a vital role in cases where the contrast of the elastography is not good enough [128].

In terms of attenuation of shear waves, viscosity has been used to propose a technique to separate transplanted livers with severe rejection from livers with no rejection by Nenadic et al. [105]. The study computed the attenuation of shear wave elastography (AMUSE), which allows the characterization of viscoelastic parameters without using rheological models. Shear wave velocity and attenuation of 15 transplanted livers in patients with severe rejection were measured; the results were correlated with biopsy findings, confirming a high ratio of concordance.

SSI was also used to staging liver fibrosis, with several studies reporting that shear wave imaging was more accurate than TE [225,226], but again, SWE can not reliably differentiate between mild stages of fibrosis. The importance of this potential biomarker has led to supersonic shear imaging (SSI) to recently release AIXPLORER MACH30® (SuperSonic Imagine, Aix-en-Provence, France) with new liver tools as the viscosity imaging feature. Figure 5 shows and imaging of a healthy liver with real-time viscosity values.

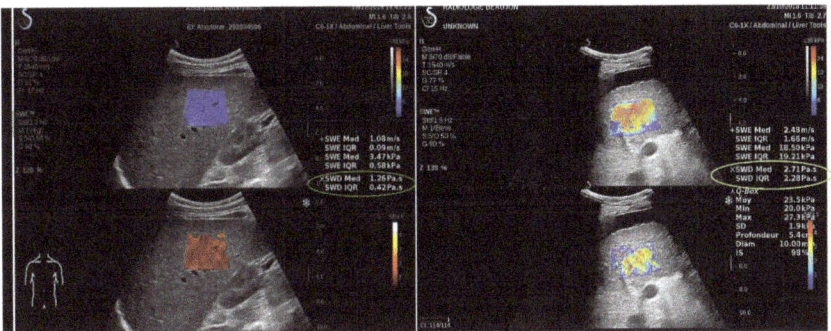

Figure 5. Measuring real-time viscosity of a volunteer patient using supersonic imagine (SSI) AIXPLORER MACH30®. The image on the left shows a healthy patient, while the right subfigure clearly distinguishes differences in the viscosity of a cirrhotic liver. Courtesy of Pr V.Vilgrain—Hopital Beaujon.

Conversely, several authors obtained results that have shown that viscosity does not notably improve liver fibrosis staging [99,110]. The works of Chen et al. [98] and Lin et al. [100] used SDUV, reporting values of 1.96 ± 0.34 Pa·s for the in vivo healthy porcine liver and 1.07 ± 0.12 (F0), 1.22 ± 0.25 (F1), 1.61 ± 0.17 (F2), 1.64 ± 0.11 (F3) and 1.61 ± 0.21 Pa·s for the in vivo fibrotic rat liver, respectively. But the common understanding is that viscosity in the human liver increases with higher fibrosis stages, as summarized in Table 6. Likewise, a recent study of Sugimoto et al. [109] has tried to overcome the limitations of the former studies by enrolling subjects with a single etiology, and using the dispersion slope value instead of a simple Voigt model since there is no consensus in the clinical/elastography community with the most appropriate rheological model for soft tissue characterization. They put the focus of dispersion slope measurements on the lack of practical guidance. Furthermore, the work has confirmed that shear wave speed (SWS) is superior to shear wave dispersion slope in delimiting the degree of fibrosis. On the other hand, they found that the dispersion slope is superior to SWS in the prognostics of the degree of necroinflammation.

Table 6. Human liver range of viscoelastic biomarkers for healthy state and different grades of fibrosis. Results were obtained using magnetic resonance elastography (MRE) and shear wave pectroscopy (SW spectroscopy). Values are reported as means and standard deviations.

Tissue State	Viscosity Parameter (Pa.s)	Method	Reference
Healthy	6.7 ± 1.3	MRE + Zener model	Klatt et al. [106]
Healthy	7.3 ± 2.3	MRE + Zener model	Asbach et al. [107]
Healthy	2.0 ± 0.8 (F0) 2.3 ± 0.7 (F1)	SW spectroscopy	Deffieux et al. [110]
Fibrosis	2.6 ± 0.5 (F2) 2.7 ± 1.9 (F3) 3.7 ± 2.5 (F4)	SW spectroscopy	Deffieux et al [110]
Fibrosis	14.4 ± 6.6 (F3–4)	MRE + Zener model	Asbach et al. [107]

Moreover, it has been found that shear wave dispersion is strongly correlated with the degree of steatosis in non-alcoholic fatty liver (NAFLD). In the most severe cases NAFLD could progress to cirrhosis, requiring liver transplant [227]. Preliminary Ex vivo and in vivo studies in mouse, porcine, duck and goose livers manifest that viscosity may become a key biomarker in distinguishing fatty liver [128].

Recent publications have highlighted the interest MRE causes as a method for detection and staging of liver fibrosis. Sherman et al. [228] examined performance characteristics of the enhanced liver fibrosis (ELF) index compared to MRE. The conclusions stated that the ELF index was a highly sensitive and specific marker of cirrhosis when compared with MRE. A posterior study evaluated the relationship between an increase in liver stiffness on MRE and fibrosis progression in nonalcoholic fatty liver disease (NAFLD) [229]. The prospective cohort study included 102 patients who underwent contemporaneous MRE and liver biopsy. The study concluded that a 15% increase in liver stiffness on MRE may be associated with histological fibrosis progression. Although high mortality is associated with significant hepatic fibrosis, data on the estimated prevalence of liver fibrosis in the general population is scarce. Kang et al. [230] carried out a study with 2170 participants. The prevalence values of significant and advanced liver fibrosis were 5.1% and 1.3% in the overall health-clinic cohort.

Viscosity imaging seems to be an essential non-invasive biomarker, providing additional information to diffuse liver pathology. Even so, it is believed that suboptimal shear wave signal quality measured in vivo could be one of the causes of worse performance of viscosity over elasticity. The precise quantification of the viscosity is a challenging inquiry; besides, the selected viscoelastic model determines the accuracy of the results. Exploring the nonlinear parameters to evaluate the degree of fibrosis has not yet been achieved at any level.

3.4. Labor Disorders

The World Health Organization (WHO) estimated in 2017 that approximately 15 million babies would be born preterm (<37 weeks of gestation); this is a rate above 1 in 10 newborns [86]. The problem of cervical insufficiency is intimately related to the mechanical properties of the cervix, and hence any approach must involve means to quantify the biomechanical state of the cervix. The mechanical parameters are sensitive to the collagen remodeling that progresses throughout cervical ripening, and which ultimately controls the cervix's mechanical ability to dilate [231].

Cervical tissue elasticity has been studied extensively. The first investigations were carried out by using static elastography (SE) [232]. However, researchers have claimed since then that we should not depend on SE to capture the changes that the cervical tissue undergoes during gestation because it highly depends on the pressure applied by the operator. Standardization of the measurement method is a call in many in vivo studies [233,234]. Molina et al. [235] came up with the idea of restricting the induced probe displacement. Controlling the pressure was an objective of Hernandez et al. [236], using a reference elastomer material [237]. Thus far it seems that there is no way to bypass the limitation of strain elastography [234,238,239].

Research moved towards looking for solutions, adopting the dynamic technique named shear wave elasticity imaging (SWEI) [51,240]. It has been widely used for the assessment of cervical changes [4,241–245]. Carlson et al. [4] measured SWS in human Ex vivo samples. Results showed that SWS was able to distinguish between ripened and unripened cervical tissue. Feltovich et al. [233] proposed the elasticity as an interesting biomarker for physicians, since the elastic modulus varies more than 80 kPa while SWS varies from approximately 1.2 to 5.5 m/s over the cervix. Carlson et al. [246] found in a longitudinal study that stiffness decreased over the course of pregnancy, and the same group explored the feasibility of SWS in capture the cervical softness in pre and post ripening in women experiencing induction for labor [241]. Peralta et al. [247] used the commercial SSI to quantify the cervical stiffness at four ROIs, which evidenced that microstructural changes generate a measurable shear stiffness reduction that gradually undergoes throughout gestation. This remodeling has been further investigated in the regions of the external os that have been proven to be softer than the internal os [235,248]. If pregnant women score small strain values at the internal os, it is unusual to experience spontaneous preterm birth [248]. SWS was found to decrease versus gestational age at the internal os [243]. Related results were obtained by Muller [242] in pregnant women compared to a control group. SWS before and after prostaglandin application were measured prior to term induction of labor in 20 women. Significant results were obtained (2.53 ± 0.75 m/s before and 1.54 ± 0.31 m/s 4 h after prostaglandin application) [241]. Authors also compared SWS between pregnant women in the first trimester and the third one; results of 4.42 ± 0.32 m/s and 2.13 ± 0.66 m/s were reported respectively.

Although the SWEI technique has been effective in the cervical tissue description, it presents some limitations: first, shear waves are highly attenuated due to the microstructural complexity of the cervix, and secondly, the complexity of producing adequate shear waves in its boundaries. The use of torsional waves (shear elastic waves that propagate radially and in-depth in a curved geometry to sense soft tissue architecture) has been demonstrated to enable a new class of characterization to quantify the mechanical functionality of any soft tissue [249–252].

Given these limitations, Melchor et al. [253] and Callejas et al. [111] introduced a novel technique, torsional wave elastography (TWE). The method is based on the transmission of shear waves by a rotational electromechanical actuator and received by a sensing ring. One of the advantages of this technique when compared with SWEI, is that it is highly adequate for cylindrical, small organs, such as the uterine cervix, since TWE generates low energy that does not generate rebounds as SWEI. Torsional wave elastography was used to quantify the stiffness of cervix in pregnant women in vivo by Masso et al. [254]. Preliminary results reveal that TWE could become an advantageous technique capable of quantifying the decrease of cervical stiffness during gestation.

Up to this point, the studies presented earlier have ignored the viscosity and nonlinearity of the uterine cervical tissue. Substantial hydration changes and inflammatory processes are well known to occur during maturation, as is collagen decrimping, which suggests that viscous and nonlinear parameters may be of significant importance. TWE explored viscosity in Ex vivo cervix tissue—results are shown in Table 7 and Figure 6 [111]—and nonlinear parameters by the harmonic generation of torsional shear waves [145].

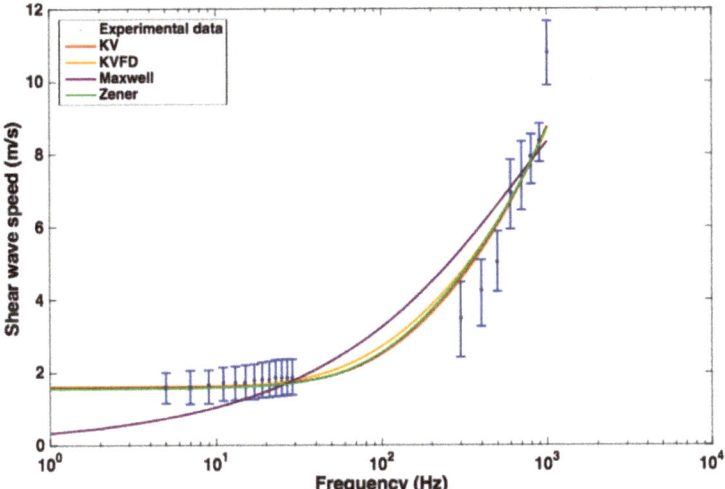

Figure 6. Fitting of the most popular rheological models to the experimental results obtained by rheometry (the lowest frequencies) and TWE (the highest frequencies) in the cervix Ex vivo. The Kelvin–Voigt (KV), Kelvin–Voigt fractional derivative (KVFD) and Zener models are successfully adjusted while the Maxwell model is not able to represent the full frequency range satisfactorily. Source: Sensors, reproduced from 2017 Callejas et al. [111].

Table 7. Viscoelastic parameters of Ex vivo cervical tissue using data from rheometry (R), torsional wave elastography (TWE) and a combination of both techniques (R + TWE) for Kelvin–Voigt (KV) and Kelvin–Voigt fractional derivative (KVFD) models. Values are reported as means and standard deviations.

Models	Rheometry (R)	TWE	R + TWE
Elasticity μ (kPa)			
KV	1.79 ± 0.08	2.43 ± 0.26	1.92 ± 0.15
KVFD	0.92 ± 0.15	2.06 ± 0.11	2.01 ± 0.24
Viscosity η (Pa.s)			
KV	6.34 ± 0.95	4.59 ± 0.29	4.5 ± 0.25
KVFD	23 ± 9.84	4.23 ± 0.22	4.64 ± 0.09
Fractional Derivative Power α			
KVFD	0.25 ± 0.15	0.97 ± 0.02	0.98 ± 0.01

McFarlin et al. [255] suggested that cervical ultrasonic attenuation, which is theoretically linked to compressional viscosity (independent from shear viscosity), could identify women at risk of spontaneous preterm birth (SPTB). It seemed that low attenuation may be an additional biomarker with which to identify SPTB. SWEI was conducted in vivo on the pregnant cervix of Rhesus macaque, divided into two groups; ripened and unripened specimens [81]. Authors found dispersion (the slope of dispersion curve of SWS versus frequency) in both groups (median 5.5 m/s/kHz, interquartile range: 1.5–12.0 m/s/kHz). Peralta et al. [71] proposed Maxwell's model as the best model to use in preliminary estimations of cervical viscoelastic properties. Myers et al. [256] suggested that since the cervical tissue is mechanically anisotropic, the uniaxial response of Ex vivo human cervix samples would depend on the load direction.

Jiang et al. [257] employed 3D multifrequency MRE to the uterus and analyzed the viscoelasticity of the uterine tissue in healthy volunteers. They observed that the uterine corpus has higher elasticity, but similar viscosity compared with the cervix, in terms of complex shear modulus (uterine corpus = 2.58 ± 0.52 kPa vs. cervix = 2.00 ± 0.34 kPa). They concluded that the proposed technique shows sensitivity to structural and functional changes of the endometrium and myometrium during the menstrual cycle. Shi et al. [258] measured the compressive viscoelastic mechanical properties of Ex vivo human cervical tissue using indentation and an inverse finite element analysis, to conclude that the human cervix is nonlinear and the area of the internal os is stiffer than the external os.

No human in vivo measurements of cervical viscosity changes during gestation have yet been reported in the literature, and no measurements of nonlinear biomarkers have been published as far as we know.

4. Discussion

In perspective, the purpose of this review was to present ground and clinical evidence that goes a step beyond linear elasticity. Abnormalities in the viscosity and nonlinearity of soft tissues are intimately linked to a broad range of pathologies, including labor disorders, solid tumors, atherosclerosis, liver fibrosis and osteoarticular syndromes, just to name but a few. This suggests that it is crucial to rethink where we are in terms of soft tissue mechanics and how pathologies affect them, opening a timely opportunity of moving forward defining new mechanical biomarkers, enabling earlier, more specific and precise diagnostic and therapeutic decision making.

On the one hand, viscoelasticity, or more generally, tissue rheology or dynamic dispersion, is recognized from the physics of wave propagation as a compound expression of the rheological, poroelastic and microstructural scattering phenomena governed by the complex fibrous multiscale microstructure of the stroma, which mainly stems from the interaction of collagen and elastin with the viscous proteoglycans, which undergo characteristic changes during pathologies.

Sensors **2020**, *20*, 2379

On the other hand, the significant hyperelasticity that soft tissues exhibit can manifest itself as quantifiable shear wave harmonic generation, and one of the main hypotheses about the pathology-mediated origin of nonlinearity changes is based on the crimping and crosslinking of tissue fibers. In the same manner that shear waves have recently been believed far more sensitive to tissue classification than standard compressional waves but are troublesome to quantify; some experimental observations may tangentially suggest nonlinear mechanical properties may be a key signature with which to quantify and classify and diagnose a range of soft tissue pathologies.

Only scarce clinical elastography measurements of viscous or nonlinear parameters have been reported for diagnostic purposes, despite the promising perspectives that both unveil from the underlying rationale and from Ex vivo or animal testings. For instance, within the field of labor disorders, despite the decrimping of fibers along gestation as well as the inflammatory process, it is suggested to be a strong diagnostic potential of those biomarkers. Nonetheless, no attempts to measure viscoelastic and nonlinear parameters using elastography as biomarkers have been reported in the literature, which opens a promising research field. Similarly, Ex vivo measurements together with non-elastographic data evidence strong correlations between viscosity and pathology in the liver and prostate, supporting promising clinical potential and opening future research prospects. Within the field of breast cancer, only one attempt using shear wave elastography for nonlinear measurements in vivo has been reported to date, combining elastography with a prestress that modifies the shear wave speed due to the Landau-type elastic nonlinearity, though it exhibited limited repeatability. Still, MRE delivers more extensive results with a clearly discriminant potential. Despite these preliminary experiences, linear and nonlinear elastography, possibly together, promise an improved sensitivity and specificity to characterize benign and malignant mammary lesions.

Regarding the limitations of these recent methodologies, it is difficult to describe them objectively, since it is not possible to compare studies and draw conclusions. Viscosity measurement with ultrasonic techniques is currently less extended than by MR techniques, but this shortcoming is only attributable to the immaturity of the ultrasonic technique; thus, barriers to its future potential are foreseen. The two origins of dispersion: viscosity and poroelasticity will probably remain indistinguishable in vivo, since their separation would require measurements at timescales too far away. Hence, a single biomarker will probably describe both. Evidence towards the potential of elastic nonlinearity biomarkers has been provided, whilst the technology is still too immature to state any potential limitations towards nonlinearity quantification.

The key open research questions involve a detailed formulation for the nonlinear and viscous components of the microstructure as the ideal procedure to understand the changes and functions in tissues that exhibit these behaviors. However, the diverse interactions between fluid components and fibers do not allow the validation of complete models, where the stored energy is considered individually for each component, ignoring physiological processes of mixed nature that should not be underestimated. In the specific case of viscosity, the industry has already taken its firsts steps to address it at the clinical level, and the challenge now is that commercial elastography techniques must converge on a common framework for the estimation of viscosity and accurate differentiation of disease states, not only regarding whether there is a pathological condition, but whether it is of malignant or benign nature. In particular, ways to enhance the dispersion biomarker applicability, by widening the interrogation frequency range, promise to enable not only storage and loss moduli, but also poroelastic and a range of viscoelastic models simultaneously. This would yield more profound understanding of tissue rheological ultrastructure and histology parameters, eventually allowing prediction of how disease processes change mechanical properties. As regards nonlinearity, it is a yet pending biomarker, an emerging concept where the still-modest clinical experiences such as breast cancer A parameter promise strong diagnostic potential once the technical issues are solved. Nonlinearity is, to our knowledge, still not available on commercial systems.

In conclusion, several front lines have been exposed, yet many other questions call for a response. How do soft tissue properties change in the case of anisotropy tissues? How about on a cellular scale in the presence of tumors? How will the ultrasound elastography industry develop techniques considering these biomarkers to adapt them to a real application? Quantitative answers to these questions would definitely improve many clinical protocols.

Author Contributions: Conceptualization, G.R.; writing—original draft preparation, G.R. and J.M.; writing—review and editing, I.H.F., J.T. and A.C.; supervision, G.R. and A.C. All authors have read and agreed to the published version of the manuscript.

Funding: This research was funded by Ministerio de Educación, Cultura y Deporte grant numbers DPI2017-83859-R, DPI2014-51870-R, UNGR15-CE-3664 and EQC2018-004508-P; Ministerio de Sanidad, Servicios Sociales e Igualdad grant numbers DTS15/00093 and PI16/00339; Instituto de Salud Carlos III y Fondos Feder; Junta de Andalucía grant numbers PI-0107-2017, PIN-0030-2017 and IE2017-5537; Juan de la Cierva Incorporación IJC2018-037167-I, Ministerio de Ciencia, Innovación y Universidades grant number PRE2018-086085.

Acknowledgments: The contributions by Antonio Gomez, Monica Contreras and Francisca S. Molina are gratefully acknowledged.

Conflicts of Interest: The authors declare no conflict of interest.

Abbreviations

The following abbreviations are used in this manuscript:

MRI	Magnetic Resonance Imaging
ECM	Extracellular Matrix
PGs	Proteoglycans
GAGs	Glycosaminoglycans
SMC	Smooth Muscle Cell
KV	Kelvin–Voigt
TOEC	Third-Order Elastic Constant
FOEC	Fourth-Order Elastic Constant
ARFI	Acoustic Radiation Force Impulse
pSWE	Point Shear Wave Elastography
TR-SWE	Transrectal Shear Wave Elastography
ROI	Region Of Interest
SDUV	Shear Wave Dispersion Ultrasound Vibrometry
KVFD	Kelvin-Voigt Fractional Derivative
DMA	Dynamic Mechanical Analysis
MRE	Magnetic Resonance Elastography
ELF	Enhanced Liver Fibrosis
WFUMB	World Federation of Ultrasound in Medicine and Biology
TE	Transient Elastography
SWE	Shear Wave Elastography
NAFLD	Non-alcoholic Fatty Liver Disease
SSI	Supersonic Shear Imaging
WHO	World Health Organization
SE	Static Elastography
SWEI	Shear Wave Elasticity Imaging
SWS	Shear Wave Speed
TWE	Torsional Wave Elastography
SPTB	Sponteneous Preterm Birth

References

1. Urban, M.W.; Chen, S.; Fatemi, M. A review of shearwave dispersion ultrasound vibrometry (SDUV) and its applications. *Curr. Med. Imaging Rev.* **2012**, *8*, 27–36. [CrossRef] [PubMed]
2. Sigrist, R.M.; Liau, J.; El Kaffas, A.; Chammas, M.C.; Willmann, J.K. Ultrasound elastography: Review of techniques and clinical applications. *Theranostics* **2017**, *7*, 1303. [CrossRef] [PubMed]

3. Fruscalzo, A.; Londero, A.P.; Fröhlich, C.; Möllmann, U.; Schmitz, R. Quantitative Elastography for Cervical Stiffness Assessment during Pregnancy. *BioMed Res. Int.* **2014**, *2014*, 826535. [CrossRef] [PubMed]

4. Carlson, L.C.; Feltovich, H.; Palmeri, M.L.; Dahl, J.J.; del Rio, A.M.; Hall, T.J. Estimation of shear wave speed in the human uterine cervix. *Ultrasound Obstet. Gynecol.* **2014**, *43*, 452–458. [CrossRef]

5. Bamber, J.; Cosgrove, D.; Dietrich, C.; Fromageau, J.; Bojunga, J.; Calliada, F.; Cantisani, V.; Correas, J.M.; D'onofrio, M.; Drakonaki, E.; et al. EFSUMB guidelines and recommendations on the clinical use of ultrasound elastography. Part 1: Basic principles and technology. *Ultraschall Med. Eur. J. Ultrasound* **2013**, *34*, 169–184. [CrossRef]

6. Shiina, T.; Nightingale, K.R.; Palmeri, M.L.; Hall, T.J.; Bamber, J.C.; Barr, R.G.; Castera, L.; Choi, B.I.; Chou, Y.H.; Cosgrove, D.; et al. WFUMB guidelines and recommendations for clinical use of ultrasound elastography: Part 1: Basic principles and terminology. *Ultrasound Med. Biol.* **2015**, *41*, 1126–1147. [CrossRef]

7. Săftoiu, A.; Gilja, O.H.; Sidhu, P.S.; Dietrich, C.F.; Cantisani, V.; Amy, D.; Bachmann-Nielsen, M.; Bob, F.; Bojunga, J.; Brock, M.; et al. The EFSUMB guidelines and recommendations for the clinical practice of elastography in non-hepatic applications: Update 2018. *Ultraschall Med. Eur. J. Ultrasound* **2019**, *40*, 425–453. [CrossRef]

8. Cosgrove, D.; Piscaglia, F.; Bamber, J.; Bojunga, J.; Correas, J.M.; Gilja, O.; Klauser, A.; Sporea, I.; Calliada, F.; Cantisani, V.; et al. EFSUMB guidelines and recommendations on the clinical use of ultrasound elastography. Part 2: Clinical applications. *Ultraschall Med. Eur. J. Ultrasound* **2013**, *34*, 238–253.

9. Riegler, J.; Labyed, Y.; Rosenzweig, S.; Javinal, V.; Castiglioni, A.; Dominguez, C.X.; Long, J.E.; Li, Q.; Sandoval, W.; Junttila, M.R.; et al. Tumor elastography and its association with collagen and the tumor microenvironment. *Clin. Cancer Res.* **2018**, *24*, 4455–4467. [CrossRef]

10. Peralta, L.; Rus, G.; Bochud, N.; Molina, F. Mechanical assessment of cervical remodelling in pregnancy: Insight from a synthetic model. *J. Biomech.* **2015**, *48*, 1557–1565. [CrossRef]

11. Carleton, J.B.; D'Amore, A.; Feaver, K.R.; Rodin, G.J.; Sacks, M.S. Geometric characterization and simulation of planar layered elastomeric fibrous biomaterials. *Acta Biomater.* **2015**, *12*, 93–101. [CrossRef]

12. Torres, J.; Faris, I.; Callejas, A. Histobiomechanical Remodeling of the Cervix during Pregnancy: Proposed Framework. *Math. Probl. Eng.* **2019**, *2019*, 5957432. [CrossRef]

13. Holzapfel, G.A. Biomechanics of Soft Tissue. *Handb. Mater. Behav. Model.* **2001**, *3*, 1049–1063.

14. Buehler, M.J. Nanomechanics of collagen fibrils under varying cross-link densities: Atomistic and continuum studies. *J. Mech. Behav. Biomed. Mater.* **2008**, *1*, 59–67. [CrossRef] [PubMed]

15. Canty, E.G.; Kadler, K.E. Procollagen trafficking, processing and fibrillogenesis. *J. Cell Sci.* **2005**, *118*, 1341–1353. [CrossRef] [PubMed]

16. Rosini, S.; Pugh, N.; Bonna, A.M.; Hulmes, D.J.; Farndale, R.W.; Adams, J.C. Thrombospondin-1 promotes matrix homeostasis by interacting with collagen and lysyl oxidase precursors and collagen cross-linking sites. *Sci. Signal.* **2018**, *11*, eaar2566. [CrossRef]

17. Elbjeirami, W.M.; Yonter, E.O.; Starcher, B.C.; West, J.L. Enhancing mechanical properties of tissue-engineered constructs via lysyl oxidase crosslinking activity. *J. Biomed. Mater. Res. Part A Off. J. Soc. Biomater. Jpn. Soc. Biomater. Aust. Soc. Biomater. Korean Soc. Biomater.* **2003**, *66*, 513–521. [CrossRef]

18. Ushiki, T. Collagen fibers, reticular fibers and elastic fibers. A comprehensive understanding from a morphological viewpoint. *Arch. Histol. Cytol.* **2002**, *65*, 109–126. [CrossRef]

19. Kadler, K.E.; Holmes, D.F.; Trotter, J.A.; Chapman, J.A. Collagen fibril formation. *Biochem. J.* **1996**, *316*, 1–11. [CrossRef]

20. Herchenhan, A.; Uhlenbrock, F.; Eliasson, P.; Weis, M.; Eyre, D.; Kadler, K.E.; Magnusson, S.P.; Kjaer, M. Lysyl oxidase activity is required for ordered collagen fibrillogenesis by tendon cells. *J. Biol. Chem.* **2015**, *290*, 16440–16450. [CrossRef]

21. Mäki, J.M.; Sormunen, R.; Lippo, S.; Kaarteenaho-Wiik, R.; Soininen, R.; Myllyharju, J. Lysyl oxidase is essential for normal development and function of the respiratory system and for the integrity of elastic and collagen fibers in various tissues. *Am. J. Pathol.* **2005**, *167*, 927–936. [CrossRef]

22. Siegel, R.C.; Pinnell, S.R.; Martin, G.R. Cross-linking of collagen and elastin. Properties of lysyl oxidase. *Biochemistry* **1970**, *9*, 4486–4492. [CrossRef] [PubMed]

23. Cronlund, A.L.; Smith, B.D.; Kagan, H.M. Binding of lysyl oxidase to fibrils of type I collagen. *Connect. Tissue Res.* **1985**, *14*, 109–119. [CrossRef] [PubMed]

24. Lanir, Y. A structural theory for the homogeneous biaxial stress-strain relationships in flat collagenous tissues. *J. Biomech.* **1979**, *12*, 423–436. [CrossRef]

25. Chernoff, E.A.; Chernoff, D.A. Atomic force microscope images of collagen fibers. *J. Vac. Sci. Technol. A Vac. Surfaces Films* **1992**, *10*, 596–599. [CrossRef]

26. Taroni, P.; Comelli, D.; Pifferi, A.; Torricelli, A.; Cubeddu, R. Absorption of collagen: Effects on the estimate of breast composition and related diagnostic implications. *J. Biomed. Opt.* **2007**, *12*, 014021. [CrossRef]

27. Ambekar, R.; Lau, T.Y.; Walsh, M.; Bhargava, R.; Toussaint, K.C. Quantifying collagen structure in breast biopsies using second-harmonic generation imaging. *Biomed. Opt. Express* **2012**, *3*, 2021–2035. [CrossRef]

28. Rojkind, M.; Ponce-Noyola, P. The extracellular matrix of the liver. *Collagen Relat. Res.* **1982**, *2*, 151–175. [CrossRef]

29. Aycock, R.S.; Seyer, J.M. Collagens of normal and cirrhotic human liver. *Connect. Tissue Res.* **1989**, *23*, 19–31. [CrossRef]

30. Oxlund, B.S.; Ørtoft, G.; Brüel, A.; Danielsen, C.C.; Bor, P.; Oxlund, H.; Uldbjerg, N. Collagen concentration and biomechanical properties of samples from the lower uterine cervix in relation to age and parity in non-pregnant women. *Reprod. Biol. Endocrinol.* **2010**, *8*, 82. [CrossRef]

31. Sundtoft, I.; Langhoff-Roos, J.; Sandager, P.; Sommer, S.; Uldbjerg, N. Cervical collagen is reduced in non-pregnant women with a history of cervical insufficiency and a short cervix. *Acta Obstet. Gynecol. Scand.* **2017**, *96*, 984–990. [CrossRef] [PubMed]

32. Myers, K.M.; Paskaleva, A.; House, M.; Socrate, S. Mechanical and biochemical properties of human cervical tissue. *Acta Biomater.* **2008**, *4*, 104–116. [CrossRef] [PubMed]

33. Tang, J.; Zhang, Y.; Zhang, M.B.; Li, Y.M.; Fei, X.; Song, Z.G. Tissue elasticity displayed by elastography and its correlation with the characteristics of collagen type I and type III in prostatic stroma. *Asian J. Androl.* **2014**, *16*, 305. [CrossRef] [PubMed]

34. Ross, R.; Fialkow, P.J.; Altman, L.K. The morphogenesis of elastic fibers. In *Elastin and Elastic Tissue*; Springer: Boston, MA, USA, 1977; pp. 7–17.

35. Ferruzzi, J.; Collins, M.J.; Yeh, A.T.; Humphrey, J.D. Mechanical assessment of elastin integrity in fibrillin-1-deficient carotid arteries: Implications for Marfan syndrome. *Cardiovasc. Res.* **2011**, *92*, 287–295. [CrossRef]

36. Higuita Castro, N.; Hansford, D.J. Mechanical characterization of cells and tissues at the micro scale. *Rev. Ing. Biomédica* **2008**, *2*, 56–64.

37. Rauscher, S.; Pomès, R. Structural disorder and protein elasticity. In *Fuzziness*; Springer: New York, NY, USA, 2012; pp. 159–183.

38. Gosline, J.; Lillie, M.; Carrington, E.; Guerette, P.; Ortlepp, C.; Savage, K. Elastic proteins: Biological roles and mechanical properties. *Philos. Trans. R. Soc. Lond. Ser. B Biol. Sci.* **2002**, *357*, 121–132. [CrossRef]

39. Lokshin, O.; Lanir, Y. Micro and macro rheology of planar tissues. *Biomaterials* **2009**, *30*, 3118–3127. [CrossRef]

40. Rozario, T.; DeSimone, D.W. The extracellular matrix in development and morphogenesis: A dynamic view. *Dev. Biol.* **2010**, *341*, 126–140. [CrossRef]

41. Scott, J.E. Proteoglycan-fibrillar collagen interactions. *Biochem. J.* **1988**, *252*, 313. [CrossRef]

42. Cöster, L. Structure and properties of dermatan sulphate proteoglycans. *Biochem. Soc. Trans.* **1991**, *19*, 866–868. [CrossRef]

43. Ruoslahti, E.; Engvall, E. Complexing of fibronectin glycosaminoglycans and collagen. *Biochim. Biophys. Acta (BBA) Gen. Subj.* **1980**, *631*, 350–358. [CrossRef]

44. Cowin, S.C.; Doty, S.B. *Tissue Mechanics*; Springer Science & Business Media: New York, USA, 2007.

45. Aristizabal, S. Anisotropic Shear Wave Elastography. In *Ultrasound Elastography for Biomedical Applications and Medicine*; John Wiley & Sons: West Sussex, UK, 2018; pp. 399–421.

46. Depalle, B.; Qin, Z.; Shefelbine, S.J.; Buehler, M.J. Influence of cross-link structure, density and mechanical properties in the mesoscale deformation mechanisms of collagen fibrils. *J. Mech. Behav. Biomed. Mater.* **2015**, *52*, 1–13. [CrossRef] [PubMed]

47. Svensson, R.B.; Mulder, H.; Kovanen, V.; Magnusson, S.P. Fracture mechanics of collagen fibrils: Influence of natural cross-links. *Biophys. J.* **2013**, *104*, 2476–2484. [CrossRef] [PubMed]

48. Peacock, C.J.; Kreplak, L. Nanomechanical mapping of single collagen fibrils under tension. *Nanoscale* **2019**, *11*, 14417–14425. [CrossRef] [PubMed]

49. Schriefl, A.J.; Schmidt, T.; Balzani, D.; Sommer, G.; Holzapfel, G.A. Selective enzymatic removal of elastin and collagen from human abdominal aortas: Uniaxial mechanical response and constitutive modeling. *Acta Biomater.* **2015**, *17*, 125–136. [CrossRef]

50. Zou, Y.; Zhang, Y. The orthotropic viscoelastic behavior of aortic elastin. *Biomech. Model. Mechanobiol.* **2011**, *10*, 613–625. [CrossRef]

51. Sarvazyan, A.P.; Rudenko, O.V.; Swanson, S.D.; Fowlkes, J.; Emelianov, S.Y. Shear wave elasticity imaging: A new ultrasonic technology of medical diagnostics. *Ultrasound Med. Biol.* **1998**, *24*, 1419–1435. [CrossRef]

52. Sarvazyan, A.P.; Urban, M.W.; Greenleaf, J.F. Acoustic waves in medical imaging and diagnostics. *Ultrasound Med. Biol.* **2013**, *39*, 1133–1146. [CrossRef]

53. Xu, B.; Li, H.; Zhang, Y. Understanding the viscoelastic behavior of collagen matrices through relaxation time distribution spectrum. *Biomatter* **2013**, *3*, e24651. [CrossRef] [PubMed]

54. Shen, Z.L.; Kahn, H.; Ballarini, R.; Eppell, S.J. Viscoelastic properties of isolated collagen fibrils. *Biophys. J.* **2011**, *100*, 3008–3015. [CrossRef] [PubMed]

55. Cowin, S.C. Deviatoric and hydrostatic mode interaction in hard and soft tissue. *J. Biomech.* **1990**, *23*, 11–14. [CrossRef]

56. Holzapfel, G.A.; Gasser, T.C.; Ogden, R.W. A new constitutive framework for arterial wall mechanics and a comparative study of material models. *J. Elast. Phys. Sci. Solids* **2000**, *61*, 1–48.

57. Duck, F.A. *Physical Properties of Tissues: A Comprehensive Reference Book*; Academic Press: London, UK, 2013.

58. Nightingale, K.R.; Rouze, N.C.; Rosenzweig, S.J.; Wang, M.H.; Abdelmalek, M.F.; Guy, C.D.; Palmeri, M.L. Derivation and analysis of viscoelastic properties in human liver: Impact of frequency on fibrosis and steatosis staging. *IEEE Trans. Ultrason. Ferroelectr. Freq. Control* **2015**, *62*, 165–175. [CrossRef] [PubMed]

59. Trutna, C.A.; Rouze, N.C.; Palmeri, M.L.; Nightingale, K.R. Measurement of Viscoelastic Material Model Parameters using Fractional Derivative Group Shear Wave Speeds in Simulation and Phantom Data. *IEEE Trans. Ultrason. Ferroelectr. Freq. Control.* **2019**, *67*, 286–295. [CrossRef] [PubMed]

60. Yang, Y.; Song, P.; Chen, S.; Urban, M.W.; McGough, R.J. Shear elasticity and shear viscosity imaging in viscoelastic phantoms. *J. Acoust. Soc. Am.* **2017**, *141*, 3720–3721. [CrossRef]

61. Amador, C.; Kinnick, R.R.; Urban, M.W.; Fatemi, M.; Greenleaf, J.F. Viscoelastic tissue mimicking phantom validation study with shear wave elasticity imaging and viscoelastic spectroscopy. In Proceedings of the 2015 IEEE International Ultrasonics Symposium (IUS), Taipei, Taiwan, 21–24 October 2015; pp. 1–4.

62. Qiang, B.; Brigham, J.C.; Aristizabal, S.; Greenleaf, J.F.; Zhang, X.; Urban, M.W. Modeling transversely isotropic, viscoelastic, incompressible tissue-like materials with application in ultrasound shear wave elastography. *Phys. Med. Biol.* **2015**, *60*, 1289. [CrossRef]

63. Gennisson, J.L.; Aristizabal, S. Nonlinear Shear Elasticity. In *Ultrasound Elastography for Biomedical Applications and Medicine*; John Wiley & Sons: West Sussex, UK, 2018; pp. 451–469.

64. Hu, Y.; Suo, Z. Viscoelasticity and poroelasticity in elastomeric gels. *Acta Mech. Solida Sin.* **2012**, *25*, 441–458. [CrossRef]

65. Wang, Q.M.; Mohan, A.C.; Oyen, M.L.; Zhao, X.H. Separating viscoelasticity and poroelasticity of gels with different length and time scales. *Acta Mech. Sin.* **2014**, *30*, 20–27. [CrossRef]

66. Al Mayah, A. *Biomechanics of Soft Tissues: Principles and Applications*; CRC Press: Boca Raton, FL, USA, 2018.

67. Muir, H. Proteoglycans as organizers of the intercellular matrix. *Biochem. Soc. Trans.* **1983**, *11*, 613–622. [CrossRef]

68. Soltz, M.A.; Ateshian, G.A. Experimental verification and theoretical prediction of cartilage interstitial fluid pressurization at an impermeable contact interface in confined compression. *J. Biomech.* **1998**, *31*, 927–934. [CrossRef]

69. Cardoso, L.; Cowin, S.C. Role of structural anisotropy of biological tissues in poroelastic wave propagation. *Mech. Mater.* **2012**, *44*, 174–188. [CrossRef] [PubMed]

70. Brum, J.; Bernal, M.; Gennisson, J.; Tanter, M. In vivo evaluation of the elastic anisotropy of the human Achilles tendon using shear wave dispersion analysis. *Phys. Med. Biol.* **2014**, *59*, 505. [CrossRef] [PubMed]

71. Peralta, L.; Rus, G.; Bochud, N.; Molina, F. Assessing viscoelasticity of shear wave propagation in cervical tissue by multiscale computational simulation. *J. Biomech.* **2015**, *48*, 1549–1556. [CrossRef] [PubMed]

72. Nitta, N.; Shiina, T.; Ueno, E. Quantitative assessment and imaging of viscoelastic properties of soft tissue. In Proceedings of the 2002 IEEE Ultrasonics Symposium, Munich, Germany, 8–11 October 2002; Volume 2, pp. 1885–1889.

73. Sack, I.; Jöhrens, K.; Würfel, J.; Braun, J. Structure-sensitive elastography: On the viscoelastic powerlaw behavior of in vivo human tissue in health and disease. *Soft Matter* **2013**, *9*, 5672–5680. [CrossRef]

74. Gautieri, A.; Vesentini, S.; Redaelli, A.; Ballarini, R. Modeling and measuring visco-elastic properties: From collagen molecules to collagen fibrils. *Int. J. Non-Linear Mech.* **2013**, *56*, 25–33. [CrossRef]

75. García, A.; Martínez, M.; Pena, E. Viscoelastic properties of the passive mechanical behavior of the porcine carotid artery: Influence of proximal and distal positions. *Biorheology* **2012**, *49*, 271–288. [CrossRef]

76. Soby, L.; Jamieson, A.; Blackwell, J.; Choi, H.; Rosenberg, L. Viscoelastic and rheological properties of concentrated solutions of proteoglycan subunit and proteoglycan aggregate. *Biopolym. Orig. Res. Biomol.* **1990**, *29*, 1587–1592. [CrossRef]

77. Sridhar, M.; Liu, J.; Insana, M. Viscoelasticity imaging using ultrasound: Parameters and error analysis. *Phys. Med. Biol.* **2007**, *52*, 2425. [CrossRef]

78. Schaefer, L.; Tredup, C.; Gubbiotti, M.A.; Iozzo, R.V. Proteoglycan neofunctions: Regulation of inflammation and autophagy in cancer biology. *FEBS J.* **2017**, *284*, 10–26. [CrossRef]

79. Frey, H.; Schroeder, N.; Manon-Jensen, T.; Iozzo, R.V.; Schaefer, L. Biological interplay between proteoglycans and their innate immune receptors in inflammation. *FEBS J.* **2013**, *280*, 2165–2179. [CrossRef] [PubMed]

80. Losa, G.A.; Alini, M. Sulfated proteoglycans in the extracellular matrix of human breast tissues with infiltrating carcinoma. *Int. J. Cancer* **1993**, *54*, 552–557. [CrossRef] [PubMed]

81. Rosado-Mendez, I.M.; Palmeri, M.L.; Drehfal, L.C.; Guerrero, Q.W.; Simmons, H.; Feltovich, H.; Hall, T.J. Assessment of structural heterogeneity and viscosity in the cervix using shear wave elasticity imaging: Initial results from a rhesus macaque model. *Ultrasound Med. Biol.* **2017**, *43*, 790–803. [CrossRef] [PubMed]

82. Petersen, L.K.; Oxlund, H.; Uldbjerg, N.; Forman, A. In vitro analysis of muscular contractile ability and passive biomechanical properties of uterine cervical samples from nonpregnant women. *Obstet. Gynecol.* **1991**, *77*, 772–776. [PubMed]

83. Yao, W.; Yoshida, K.; Fernandez, M.; Vink, J.; Wapner, R.J.; Ananth, C.V.; Oyen, M.L.; Myers, K.M. Measuring the compressive viscoelastic mechanical properties of human cervical tissue using indentation. *J. Mech. Behav. Biomed. Mater.* **2014**, *34*, 18–26. [CrossRef]

84. Ghatak, S.; Maytin, E.V.; Mack, J.A.; Hascall, V.C.; Atanelishvili, I.; Moreno Rodriguez, R.; Markwald, R.R.; Misra, S. Roles of proteoglycans and glycosaminoglycans in wound healing and fibrosis. *Int. J. Cell Biol.* **2015**, *2015*, 834893. [CrossRef]

85. Armentano, R.L.; Barra, J.G.; Santana, D.B.; Pessana, F.M.; Graf, S.; Craiem, D.; Brandani, L.M.; Baglivo, H.P.; Sanchez, R.A. Smart damping modulation of carotid wall energetics in human hypertension: Effects of angiotensin-converting enzyme inhibition. *Hypertension* **2006**, *47*, 384–390. [CrossRef]

86. Vink, J.Y.; Qin, S.; Brock, C.O.; Zork, N.M.; Feltovich, H.M.; Chen, X.; Urie, P.; Myers, K.M.; Hall, T.J.; Wapner, R.; et al. A new paradigm for the role of smooth muscle cells in the human cervix. *Am. J. Obstet. Gynecol.* **2016**, *215*, 478.e1. [CrossRef]

87. Yamaoka, K.; Nouchi, T.; Marumo, F.; Sato, C. α-Smooth-muscle actin expression in normal and fibrotic human livers. *Dig. Dis. Sci.* **1993**, *38*, 1473–1479. [CrossRef]

88. Langewouters, G.; Wesseling, K.; Goedhard, W. The pressure dependent dynamic elasticity of 35 thoracic and 16 abdominal human aortas in vitro described by a five component model. *J. Biomech.* **1985**, *18*, 613–620. [CrossRef]

89. Kumar, V.; Denis, M.; Gregory, A.; Bayat, M.; Mehrmohammadi, M.; Fazzio, R.; Fatemi, M.; Alizad, A. Viscoelastic parameters as discriminators of breast masses: Initial human study results. *PLoS ONE* **2018**, *13*, e0205717. [CrossRef]

90. Nabavizadeh, A.; Bayat, M.; Kumar, V.; Gregory, A.; Webb, J.; Alizad, A.; Fatemi, M. Viscoelastic biomarker for differentiation of benign and malignant breast lesion in ultra-low frequency range. *Sci. Rep.* **2019**, *9*, 5737. [CrossRef]

91. Sinkus, R.; Siegmann, K.; Xydeas, T.; Tanter, M.; Claussen, C.; Fink, M. MR elastography of breast lesions: Understanding the solid/liquid duality can improve the specificity of contrast-enhanced MR mammography. *Magn. Reson. Med. Off. J. Int. Soc. Magn. Reson. Med.* **2007**, *58*, 1135–1144. [CrossRef]

92. Balleyguier, C.; Canale, S.; Hassen, W.B.; Vielh, P.; Bayou, E.; Mathieu, M.; Uzan, C.; Bourgier, C.; Dromain, C. Breast elasticity: Principles, technique, results: An update and overview of commercially available software. *Eur. J. Radiol.* **2013**, *82*, 427–434. [CrossRef] [PubMed]

93. Palmeri, M.L.; Nightingale, K.R. What challenges must be overcome before ultrasound elasticity imaging is ready for the clinic? *Imaging Med.* **2011**, *3*, 433. [CrossRef] [PubMed]

94. Aristizabal, S.; Carrascal, C.A.; Nenadic, I.Z.; Greenleaf, J.F.; Urban, M.W. Application of Acoustoelasticity to Evaluate Nonlinear Modulus inEx VivoKidneys. *IEEE Trans. Ultrason. Ferroelectr. Freq. Control* **2018**, *65*, 188–200. [CrossRef]

95. Schregel, K.; née Tysiak, E.W.; Garteiser, P.; Gemeinhardt, I.; Prozorovski, T.; Aktas, O.; Merz, H.; Petersen, D.; Wuerfel, J.; Sinkus, R. Demyelination reduces brain parenchymal stiffness quantified in vivo by magnetic resonance elastography. *Proc. Natl. Acad. Sci. USA* **2012**, *109*, 6650–6655. [CrossRef] [PubMed]

96. Streitberger, K.J.; Sack, I.; Krefting, D.; Pfüller, C.; Braun, J.; Paul, F.; Wuerfel, J. Brain viscoelasticity alteration in chronic-progressive multiple sclerosis. *PLoS ONE* **2012**, *7*, e29888. [CrossRef] [PubMed]

97. Sack, I.; Beierbach, B.; Wuerfel, J.; Klatt, D.; Hamhaber, U.; Papazoglou, S.; Martus, P.; Braun, J. The impact of aging and gender on brain viscoelasticity. *Neuroimage* **2009**, *46*, 652–657. [CrossRef]

98. Chen, S.; Urban, M.W.; Pislaru, C.; Kinnick, R.; Zheng, Y.; Yao, A.; Greenleaf, J.F. Shearwave dispersion ultrasound vibrometry (SDUV) for measuring tissue elasticity and viscosity. *IEEE Trans. Ultrason. Ferroelectr. Freq. Control* **2009**, *56*, 55–62. [CrossRef]

99. Chen, S.; Sanchez, W.; Callstrom, M.R.; Gorman, B.; Lewis, J.T.; Sanderson, S.O.; Greenleaf, J.F.; Xie, H.; Shi, Y.; Pashley, M.; et al. Assessment of liver viscoelasticity by using shear waves induced by ultrasound radiation force. *Radiology* **2013**, *266*, 964–970. [CrossRef]

100. Lin, H.; Zhang, X.; Shen, Y.; Zheng, Y.; Guo, Y.; Zhu, Y.; Diao, X.; Wang, T.; Chen, S.; Chen, X. Model-dependent and model-independent approaches for evaluating hepatic fibrosis in rat liver using shearwave dispersion ultrasound vibrometry. *Med. Eng. Phys.* **2017**, *39*, 66–72. [CrossRef] [PubMed]

101. Mitri, F.G.; Urban, M.W.; Fatemi, M.; Member, S.; Greenleaf, J.F.; Fellow, L. Shear Wave Dispersion Ultrasonic Vibrometry for Measuring Prostate Shear Stiffness and Viscosity: An In Vitro Pilot Study. *IEEE Trans. Biomed. Eng.* **2011**, *58*, 235–242. [CrossRef] [PubMed]

102. Zhang, M.; Nigwekar, P.; Castaneda, B.; Hoyt, K.; Joseph, J.V.; di Sant'Agnese, A.; Messing, E.M.; Strang, J.G.; Rubens, D.J.; Parker, K.J. Quantitative characterization of viscoelastic properties of human prostate correlated with histology. *Ultrasound Med. Biol.* **2008**, *34*, 1033–1042. [CrossRef] [PubMed]

103. Sinkus, R.; Tanter, M.; Xydeas, T.; Catheline, S.; Bercoff, J.; Fink, M. Viscoelastic shear properties of in vivo breast lesions measured by MR elastography. *Magn. Reson. Imaging* **2005**, *23*, 159–165. [CrossRef] [PubMed]

104. Sinkus, R.; Tanter, M.; Catheline, S.; Lorenzen, J.; Kuhl, C.; Sondermann, E.; Fink, M. Imaging anisotropic and viscous properties of breast tissue by magnetic resonance-elastography. *Magn. Reson. Med. Off. J. Int. Soc. Magn. Reson. Med.* **2005**, *53*, 372–387. [CrossRef]

105. Nenadic, I.Z.; Qiang, B.; Urban, M.W.; Zhao, H.; Sanchez, W.; Greenleaf, J.F.; Chen, S. Attenuation measuring ultrasound shearwave elastography and in vivo application in post-transplant liver patients. *Phys. Med. Biol.* **2016**, *62*, 484. [CrossRef]

106. Klatt, D.; Hamhaber, U.; Asbach, P.; Braun, J.; Sack, I. Noninvasive assessment of the rheological behavior of human organs using multifrequency MR elastography: A study of brain and liver viscoelasticity. *Phys. Med. Biol.* **2007**, *52*, 7281–7294. [CrossRef]

107. Asbach, P.; Klatt, D.; Hamhaber, U.; Braun, J.; Somasundaram, R.; Hamm, B.; Sack, I. Assessment of liver viscoelasticity using multifrequency MR elastography. *Magn. Reson. Med.* **2008**, *60*, 373–379. [CrossRef]

108. Li, S.; Chen, M.; Wang, W.; Zhao, W.; Wang, J.; Zhao, X.; Zhou, C. A feasibility study of MR elastography in the diagnosis of prostate cancer at 3.0 T. *Acta Radiol.* **2011**, *52*, 354–358. [CrossRef]

109. Sugimoto, K.; Moriyasu, F.; Oshiro, H.; Takeuchi, H.; Yoshimasu, Y.; Kasai, Y.; Itoi, T. Clinical utilization of shear wave dispersion imaging in the diffuse liver disease. *Ultrasonography* **2019**, *39*, 3. [CrossRef]

110. Deffieux, T.; Gennisson, J.L.; Bousquet, L.; Corouge, M.; Cosconea, S.; Amroun, D.; Tripon, S.; Terris, B.; Mallet, V.; Sogni, P.; et al. Investigating liver stiffness and viscosity for fibrosis, steatosis and activity staging using shear wave elastography. *J. Hepatol.* **2015**, *62*, 317–324. [CrossRef] [PubMed]

111. Callejas, A.; Gomez, A.; Melchor, J.; Riveiro, M.; Massó, P.; Torres, J.; López-López, M.; Rus, G. Performance study of a torsional wave sensor and cervical tissue characterization. *Sensors* **2017**, *17*, 2078. [CrossRef] [PubMed]

112. Susilo, M.E.; Paten, J.A.; Sander, E.A.; Nguyen, T.D.; Ruberti, J.W. Collagen network strengthening following cyclic tensile loading. *Interface Focus* **2016**, *6*, 20150088. [CrossRef] [PubMed]

113. Han, L.; Burcher, M.; Noble, J.A. Non-invasive measurement of biomechanical properties of in vivo soft tissues. In Proceedings of the International Conference on Medical Image Computing and Computer-Assisted Intervention, Tokyo, Japan, 25–28 September 2002; pp. 208–215.

114. Serra-Aguila, A.; Puigoriol-Forcada, J.; Reyes, G.; Menacho, J. Viscoelastic models revisited: Characteristics and interconversion formulas for generalized Kelvin–Voigt and Maxwell models. *Acta Mech. Sin.* **2019**, *35*, 1911–1209. [CrossRef]

115. Schmitt, C.; Henni, A.H.; Cloutier, G. Characterization of blood clot viscoelasticity by dynamic ultrasound elastography and modeling of the rheological behavior. *J. Biomech.* **2011**, *44*, 622–629. [CrossRef]

116. Fung, Y.C. *Biomechanics: Mechanical Properties of Living Tissues*; Springer Science & Business Media: New York, NY, USA, 2013.

117. Parker, K.; Szabo, T.; Holm, S. Towards a consensus on rheological models for elastography in soft tissues. *Phys. Med. Biol.* **2019**, *64*, 215012. [CrossRef]

118. Koeller, R. Applications of fractional calculus to the theory of viscoelasticity. *J. Appl. Mech.* **1984**, *51*, 299–307. [CrossRef]

119. Rus, G. Nature of acoustic nonlinear radiation stress. *Appl. Phys. Lett.* **2014**, *105*, 121904. [CrossRef]

120. Gomez, A. Transurethral Shear Wave Elastography for Prostate Cancer. Ph.D. Thesis, University College London, London, UK, 2018.

121. Orescanin, M.; Wang, Y.; Insana, M.F. 3-D FDTD simulation of shear waves for evaluation of complex modulus imaging. *IEEE Trans. Ultrason. Ferroelectr. Freq. Control* **2011**, *58*, 389–398. [CrossRef]

122. Bercoff, J.; Tanter, M.; Muller, M.; Fink, M. The role of viscosity in the impulse diffraction field of elastic waves induced by the acoustic radiation force. *IEEE Trans. Ultrason. Ferroelectr. Freq. Control* **2004**, *51*, 1523–1536. [CrossRef]

123. Chen, L.H.; Ng, S.P.; Yu, W.; Zhou, J.; Wan, K.F. A study of breast motion using non-linear dynamic FE analysis. *Ergonomics* **2013**, *56*, 868–878. [CrossRef]

124. Salameh, N.; Peeters, F.; Sinkus, R.; Abarca-Quinones, J.; Annet, L.; Ter Beek, L.C.; Leclercq, I.; Van Beers, B.E. Hepatic viscoelastic parameters measured with MR elastography: correlations with quantitative analysis of liver fibrosis in the rat. *J. Magn. Reson. Imaging Off. J. Int. Soc. Magn. Reson. Med.* **2007**, *26*, 956–962. [CrossRef] [PubMed]

125. Palacio-Torralba, J.; Hammer, S.; Good, D.W.; McNeill, S.A.; Stewart, G.D.; Reuben, R.L.; Chen, Y. Quantitative diagnostics of soft tissue through viscoelastic characterization using time-based instrumented palpation. *J. Mech. Behav. Biomed. Mater.* **2015**, *41*, 149–160. [CrossRef] [PubMed]

126. Vappou, J.; Maleke, C.; Konofagou, E.E. Quantitative viscoelastic parameters measured by harmonic motion imaging. *Phys. Med. Biol.* **2009**, *54*, 3579. [CrossRef] [PubMed]

127. Zhang, M.; Castaneda, B.; Wu, Z.; Nigwekar, P.; Joseph, J.V.; Rubens, D.J.; Parker, K.J. Congruence of imaging estimators and mechanical measurements of viscoelastic properties of soft tissues. *Ultrasound Med. Biol.* **2007**, *33*, 1617–1631. [CrossRef] [PubMed]

128. Bhatt, M.; Moussu, M.A.; Chayer, B.; Destrempes, F.; Gesnik, M.; Allard, L.; Tang, A.; Cloutier, G. Reconstruction of Viscosity Maps in Ultrasound Shear Wave Elastography. *IEEE Trans. Ultrason. Ferroelectr. Freq. Control* **2019**, *66*, 1065–1078. [CrossRef] [PubMed]

129. Lanir, Y. Multi-scale structural modeling of soft tissues mechanics and mechanobiology. *J. Elast.* **2017**, *129*, 7–48. [CrossRef]

130. Pritchard, R.H.; Huang, Y.Y.S.; Terentjev, E.M. Mechanics of biological networks: From the cell cytoskeleton to connective tissue. *Soft Matter* **2014**, *10*, 1864–1884. [CrossRef]

131. Myers, K.M.; Feltovich, H.; Mazza, E.; Vink, J.; Bajka, M.; Wapner, R.J.; Hall, T.J.; House, M. The mechanical role of the cervix in pregnancy. *J. Biomech.* **2015**, *48*, 1511–1523. [CrossRef]

132. Muñoz, R.; Melchor, J. Nonlinear Classical Elasticity Model for Materials with Fluid and Matrix Phases. *Math. Probl. Eng.* **2018**, *2018*, 5049104. [CrossRef]

133. Goenezen, S.; Dord, J.F.; Sink, Z.; Barbone, P.E.; Jiang, J.; Hall, T.J.; Oberai, A.A. Linear and nonlinear elastic modulus imaging: An application to breast cancer diagnosis. *IEEE Trans. Med. Imaging* **2012**, *31*, 1628–1637. [CrossRef] [PubMed]

134. Landau, L.D.; Lifshitz, E.M. *Course of Theoretical Physics Volume 7: Theory and Elasticity*; Pergamon Press: Oxford, UK, 1959.

135. Murnaghan, F. The compressibility of media under extreme pressures. *Proc. Natl. Acad. Sci. USA* **1944**, *30*, 244. [CrossRef] [PubMed]

136. Gasser, T.C.; Ogden, R.W.; Holzapfel, G.A. Hyperelastic modelling of arterial layers with distributed collagen fibre orientations. *J. R. Soc. Interface* **2005**, *3*, 15–35. [CrossRef] [PubMed]

137. Ogden, R.W.; Holzapfel, G.A. *Mechanics of Biological Tissue*; Springer: Berlin, Germany, 2006.

138. Ovenden, N.; Walsh, C. Fundamentals of Physiological Solid Mechanics. In *Fluid and Solid Mechanics*; World Scientific: London, UK, 2016; pp. 169–217.

139. Truesdell, C.; Noll, W. The non-linear field theories of mechanics. In *The Non-Linear Field Theories of Mechanics*; Springer: Berlin, Germany, 2004; pp. 1–579.

140. Zabolotskaya, E.A.; Hamilton, M.F.; Ilinskii, Y.A.; Meegan, G.D. Modeling of nonlinear shear waves in soft solids. *J. Acoust. Soc. Am.* **2004**, *116*, 2807–2813. [CrossRef]

141. Bernal, M.M.; Chammings, F.; Couade, M.; Bercoff, J.; Tanter, M.; luc Gennisson, J. In Vivo Quantification of the Nonlinear Shear Modulus in Breast Lesions: Feasibility Study. *IEEE Trans. Ultrason. Ferroelectr. Freq. Control* **2016**, *63*, 101–109. [CrossRef] [PubMed]

142. Melchor, J.; Parnell, W.; Bochud, N.; Peralta, L.; Rus, G. Damage prediction via nonlinear ultrasound: A micro-mechanical approach. *Ultrasonics* **2019**, *93*, 145–155. [CrossRef] [PubMed]

143. Sack, I.; Mcgowan, C.K.; Samani, A.; Luginbuhl, C.; Oakden, W.; Plewes, D.B. Observation of nonlinear shear wave propagation using magnetic resonance elastography. *Magn. Reson. Med. Off. J. Int. Soc. Magn. Reson. Med.* **2004**, *52*, 842–850. [CrossRef] [PubMed]

144. Melchor, J. Mechanics of Nonlinear Ultrasound in Tissue. Ph.D. Thesis, Universidad de Granada, Granada, Spain, 2016.

145. Naranjo-Pérez, J.; Riveiro, M.; Callejas, A.; Rus, G.; Melchor, J. Nonlinear torsional wave propagation in cylindrical coordinates to assess biomechanical parameters. *J. Sound Vib.* **2019**, *445*, 103–116. [CrossRef]

146. Zahedmanesh, H.; Lally, C. A multiscale mechanobiological modelling framework using agent-based models and finite element analysis: Application to vascular tissue engineering. *Biomech. Model. Mechanobiol.* **2012**, *11*, 363–377. [CrossRef]

147. Maceri, F.; Marino, M.; Vairo, G. A unified multiscale mechanical model for soft collagenous tissues with regular fiber arrangement. *J. Biomech.* **2010**, *43*, 355–363. [CrossRef]

148. Hamilton, M.F.; Blackstock, D.T. (Eds.) *Nonlinear Acoustics*; Academic Press: San Diego, CA, USA, 1998; Volume 237.

149. Rushchitsky, J.; Simchuk, Y.V. Quadratic nonlinear torsional hyperelastic waves in isotropic cylinders: Primary analysis of evolution. *Int. Appl. Mech.* **2008**, *44*, 304–312. [CrossRef]

150. Latorre-Ossa, H.; Gennisson, J.L.; De Brosses, E.; Tanter, M. Quantitative imaging of nonlinear shear modulus by combining static elastography and shear wave elastography. *IEEE Trans. Ultrason. Ferroelectr. Freq. Control* **2012**, *59*, 833–839. [CrossRef] [PubMed]

151. Barr, R.G.; Memo, R.; Schaub, C.R. Shear wave ultrasound elastography of the prostate: Initial results. *Ultrasound Q.* **2012**, *28*, 13–20. [CrossRef] [PubMed]

152. Sandrin, L.; Tanter, M.; Catheline, S.; Fink, M. Shear modulus imaging with 2-D transient elastography. *IEEE Trans. Ultrason. Ferroelectr. Freq. Control* **2002**, *49*, 426–435. [CrossRef]

153. Gennisson, J.L.; Deffieux, T.; Fink, M.; Tanter, M. Ultrasound elastography: Principles and techniques. *Diagn. Interv. Imaging* **2013**, *94*, 487–495. [CrossRef]

154. Jiang, Y.; Li, G.; Qian, L.X.; Liang, S.; Destrade, M.; Cao, Y. Measuring the linear and nonlinear elastic properties of brain tissue with shear waves and inverse analysis. *Biomech. Model. Mechanobiol.* **2015**, *14*, 1119–1128. [CrossRef]

155. Jacob, X.; Catheline, S.; Gennisson, J.L.; Barrière, C.; Royer, D.; Fink, M. Nonlinear shear wave interaction in soft solids. *J. Acoust. Soc. Am.* **2007**, *122*, 1917–1926. [CrossRef]

156. Jiang, Y.; Li, G.Y.; Qian, L.X.; Hu, X.D.; Liu, D.; Liang, S.; Cao, Y. Characterization of the nonlinear elastic properties of soft tissues using the supersonic shear imaging (SSI) technique: Inverse method, ex vivo and in vivo experiments. *Med. Image Anal.* **2015**, *20*, 97–111. [CrossRef]

157. Parker, K.J.; Doyley, M.M.; Rubens, D.J. Imaging the elastic properties of tissue: The 20 year perspective. *Phys. Med. Biol.* **2011**, *56 1*, R1–R29. [CrossRef]

158. Ozturk, A.; Grajo, J.R.; Dhyani, M.; Anthony, B.W.; Samir, A.E. Principles of ultrasound elastography. *Abdom. Radiol.* **2018**, *43*, 773–785. [CrossRef]

159. Bray, F.; Ferlay, J.; Soerjomataram, I.; Siegel, R.L.; Torre, L.A.; Jemal, A. Global cancer statistics 2018: GLOBOCAN estimates of incidence and mortality worldwide for 36 cancers in 185 countries. *CA Cancer J. Clin.* **2018**, *68*, 394–424. [CrossRef]

160. Abdellaoui, A.; Iyengar, S.; Freeman, S. Imaging in prostate cancer. *Futur. Oncol.* **2011**, *7*, 679–691. [CrossRef] [PubMed]

161. Barr, R.G.; Cosgrove, D.; Brock, M.; Cantisani, V.; Correas, J.M.; Postema, A.W.; Salomon, G.; Tsutsumi, M.; Xu, H.X.; Dietrich, C.F. WFUMB guidelines and recommendations on the clinical use of ultrasound elastography: Part 5. Prostate. *Ultrasound Med. Biol.* **2017**, *43*, 27–48. [CrossRef]

162. Zhai, L.; Madden, J.; Foo, W.C.; Palmeri, M.L.; Mouraviev, V.; Polascik, T.J.; Nightingale, K.R. Acoustic radiation force impulse imaging of human prostates ex vivo. *Ultrasound Med. Biol.* **2010**, *36*, 576–588. [CrossRef]

163. Zhai, L.; Polascik, T.J.; Foo, W.C.; Rosenzweig, S.; Palmeri, M.L.; Madden, J.; Nightingale, K.R. Acoustic Radiation Force Impulse Imaging of Human Prostates: Initial In Vivo Demonstration. *Ultrasound Med. Biol.* **2012**, *38*, 50–61. [CrossRef] [PubMed]

164. Correas, J.M.; Tissier, A.M.; Khairoune, A.; Vassiliu, V.; Méjean, A.; Hélénon, O.; Memo, R.; Barr, R.G. Prostate Cancer: Diagnostic Performance of Real-time Shear-Wave Elastography. *Radiology* **2015**, *275*, 280–289. [CrossRef] [PubMed]

165. Correas, J.; Drakonakis, E.; Isidori, A.M.; Hélénon, O.; Pozza, C.; Cantisani, V.; Di Leo, N.; Maghella, F.; Rubini, a.; Drudi, F.; D'ambrosio, F. Update on ultrasound elastography: Miscellanea. Prostate, testicle, musculo-skeletal. *Eur. J. Radiol.* **2013**, *82*, 1904–1912. [CrossRef] [PubMed]

166. Zhang, M.; Fu, S.; Zhang, Y.; Tang, J.; Zhou, Y. Elastic modulus of the prostate: A new non-invasive feature to diagnose bladder outlet obstruction in patients with benign prostatic hyperplasia. *Ultrasound Med. Biol.* **2013**, *40*, 1408–1413. [CrossRef] [PubMed]

167. Woo, S.; Kim, S.Y.; Cho, J.Y.; Kim, S.H. Shear wave elastography for detection of prostate cancer: A preliminary study. *Korean J. Radiol.* **2014**, *15*, 346–355. [CrossRef] [PubMed]

168. Correas, J.M.; Khairoune, A.; Tissier, A.M.; Vassiliu, V. Trans-rectal quantitative Shear Wave Elastrography: Application to prostate cancer—A feasibility study. In Proceedings of the European Congress of Radiology, Vienna, Austria, 3–7 March 2011.

169. Boehm, K.; Salomon, G.; Beyer, B.; Schiffmann, J.; Simonis, K.; Graefen, M.; Budaeus, L. Shear Wave Elastography for Localization of Prostate Cancer Lesions and Assessment of Elasticity Thresholds: Implications for Targeted Biopsies and Active Surveillance Protocols. *J. Urol.* **2015**, *193*, 794–800. [CrossRef] [PubMed]

170. Woo, S.; Kim, S.Y.; Lee, M.S.; Cho, J.Y.; Kim, S.H. Shear wave elastography assessment in the prostate: An intraobserver reproducibility study. *Clin. Imaging* **2015**, *39*, 484–487. [CrossRef] [PubMed]

171. Tanter, M.; Bercoff, J.; Athanasiou, A.; Deffieux, T.; Gennisson, J.L.; Montaldo, G.; Muller, M.; Tardivon, A.; Fink, M. Quantitative assessment of breast lesion viscoelasticity: Initial clinical results using supersonic shear imaging. *Ultrasound Med. Biol.* **2008**, *34*, 1373–1386. [CrossRef] [PubMed]

172. Zhai, L.; Madden, J.; Foo, W.C.; Mouraviev, V.; Polascik, T.J.; Palmeri, M.L.; Nightingale, K.R. Characterizing Stiffness of Human Prostates Using Acoustic Radiation Force. *Ultrason. Imaging* **2010**, *32*, 201–213. [CrossRef]

173. Zheng, X.; Ji, P.; Mao, H.; Hu, J. A comparison of virtual touch tissue quantification and digital rectal examination for discrimination between prostate cancer and benign prostatic hyperplasia. *Radiol. Oncol.* **2012**, *46*, 69–74. [CrossRef] [PubMed]

174. Hoyt, K.; Castaneda, B.; Zhang, M.; Nigwekar, P.; Anthony, P.; Agnese, S.; Joseph, J.V.; Strang, J.; Rubens, D.J.; Parker, K.J. Tissue elasticity properties as biomarkers for prostate cancer. *Cancer Biomark.* **2008**, *4*, 213–225. [CrossRef]

175. Chopra, R.; Arani, A.; Huang, Y.; Musquera, M.; Wachsmuth, J.; Bronskill, M.; Plewes, D. In vivo MR elastography of the prostate gland using a transurethral actuator. *Magn. Reson. Med. Off. J. Int. Soc. Magn. Reson. Med.* **2009**, *62*, 665–671. [CrossRef]

176. Arani, A.; Plewes, D.; Chopra, R. Transurethral prostate magnetic resonance elastography: Prospective imaging requirements. *Magn. Reson. Med.* **2011**, *65*, 340–349. [CrossRef]

177. Reiter, R.; Majumdar, S.; Kearney, S.; Kajdacsy-Balla, A.; Macias, V.; Crivellaro, S.; Caldwell, B.; Abern, M.; Royston, T.J.; Klatt, D. Prostate cancer assessment using MR elastography of fresh prostatectomy specimens at 9.4 T. *Magn. Reson. Med.* **2019**, *84*, 396–404. [CrossRef]

178. Orel, S.G.; Kay, N.; Reynolds, C.; Sullivan, D.C. BI-RADS categorization as a predictor of malignancy. *Radiology* **1999**, *211*, 845–850. [CrossRef]

179. Leong, L.; Sim, L.; Lee, Y.; Ng, F.; Wan, C.; Fook-Chong, S.; Jara-Lazaro, A.; Tan, P. A prospective study to compare the diagnostic performance of breast elastography versus conventional breast ultrasound. *Clin. Radiol.* **2010**, *65*, 887–894. [CrossRef]

180. Goddi, A.; Bonardi, M.; Alessi, S. Breast elastography: A literature review. *J. Ultrasound* **2012**, *15*, 192–198. [CrossRef] [PubMed]

181. Bray, F.; McCarron, P.; Parkin, D.M. The changing global patterns of female breast cancer incidence and mortality. *Breast Cancer Res.* **2004**, *6*, 229. [CrossRef] [PubMed]

182. Culpan, A.M. Breast Elastography. Imaging & Therapy Practice. 2016, 30. Available online: https://search.proquest.com/openview/a5377f86d950c0ca1132920c07c00430/1.pdf?pq-origsite=gscholar&cbl=46803 (accessed on 25 March 2020).

183. Barr, R.G.; Nakashima, K.; Amy, D.; Cosgrove, D.; Farrokh, A.; Schafer, F.; Bamber, J.C.; Castera, L.; Choi, B.I.; Chou, Y.H.; et al. WFUMB guidelines and recommendations for clinical use of ultrasound elastography: Part 2: Breast. *Ultrasound Med. Biol.* **2015**, *41*, 1148–1160. [CrossRef] [PubMed]

184. Barr, R.G.; Destounis, S.; Lackey, L.B.; Svensson, W.E.; Balleyguier, C.; Smith, C. Evaluation of breast lesions using sonographic elasticity imaging: A multicenter trial. *J. Ultrasound Med.* **2012**, *31*, 281–287. [CrossRef] [PubMed]

185. Li, G.; Li, D.W.; Fang, Y.X.; Song, Y.J.; Deng, Z.J.; Gao, J.; Xie, Y.; Yin, T.S.; Ying, L.; Tang, K.F. Performance of shear wave elastography for differentiation of benign and malignant solid breast masses. *PLoS ONE* **2013**, *8*, e76322. [CrossRef]

186. D'Orsi, C.; Bassett, L.; Feig, S. Breast imaging reporting and data system (BI-RADS). In *Breast Imaging Atlas*, 4th ed.; American College of Radiology: Reston, VA, USA, 1998.

187. Chen, L.; He, J.; Liu, G.; Shao, K.; Zhou, M.; Li, B.; Chen, X. Diagnostic performances of shear-wave elastography for identification of malignant breast lesions: A meta-analysis. *Jpn. J. Radiol.* **2014**, *32*, 592–599. [CrossRef]

188. Evans, A.; Whelehan, P.; Thomson, K.; McLean, D.; Brauer, K.; Purdie, C.; Jordan, L.; Baker, L.; Thompson, A. Quantitative shear wave ultrasound elastography: Initial experience in solid breast masses. *Breast Cancer Res.* **2010**, *12*, R104. [CrossRef]

189. Denis, M.; Mehrmohammadi, M.; Song, P.; Meixner, D.D.; Fazzio, R.T.; Pruthi, S.; Whaley, D.H.; Chen, S.; Fatemi, M.; Alizad, A. Comb-push ultrasound shear elastography of breast masses: Initial results show promise. *PLoS ONE* **2015**, *10*, e0119398. [CrossRef] [PubMed]

190. Raza, S.; Odulate, A.; Ong, E.M.; Chikarmane, S.; Harston, C.W. Using real-time tissue elastography for breast lesion evaluation: Our initial experience. *J. Ultrasound Med.* **2010**, *29*, 551–563. [CrossRef]

191. Yoon, J.H.; Jung, H.K.; Lee, J.T.; Ko, K.H. Shear-wave elastography in the diagnosis of solid breast masses: What leads to false-negative or false-positive results? *Eur. Radiol.* **2013**, *23*, 2432–2440. [CrossRef]

192. Youk, J.H.; Gweon, H.M.; Son, E.J.; Han, K.H.; Kim, J.A. Diagnostic value of commercially available shear-wave elastography for breast cancers: Integration into BI-RADS classification with subcategories of category 4. *Eur. Radiol.* **2013**, *23*, 2695–2704. [CrossRef] [PubMed]

193. Gregory, A.; Mehrmohammadi, M.; Denis, M.; Bayat, M.; Stan, D.L.; Fatemi, M.; Alizad, A. Effect of calcifications on breast ultrasound shear wave elastography: An investigational study. *PLoS ONE* **2015**, *10*, e0137898. [CrossRef] [PubMed]

194. Madani, N.; Mojra, A. Quantitative diagnosis of breast tumors by characterization of viscoelastic behavior of healthy breast tissue. *J. Mech. Behav. Biomed. Mater.* **2017**, *68*, 180–187. [CrossRef]

195. Qiu, Y.; Sridhar, M.; Tsou, J.K.; Lindfors, K.K.; Insana, M.F. Ultrasonic viscoelasticity imaging of nonpalpable breast tumors: preliminary results. *Acad. Radiol.* **2008**, *15*, 1526–1533. [CrossRef] [PubMed]

196. Bayat, M.; Nabavizadeh, A.; Kumar, V.; Gregory, A.; Insana, M.; Alizad, A.; Fatemi, M. AutomatedIn VivoSub-Hertz Analysis of Viscoelasticity (SAVE) for Evaluation of Breast Lesions. *IEEE Trans. Biomed. Eng.* **2017**, *65*, 2237–2247. [CrossRef]

197. Amy, D.; Bercoff, J.; Bibby, E. Breast Elastography. In *Lobar Approach to Breast Ultrasound*; Springer: New York, NY, USA, 2018; pp. 85–106.
198. Barr, R.G. Future of breast elastography. *Ultrasonography* **2019**, 38, 93. [CrossRef]
199. Asrani, S.K.; Devarbhavi, H.; Eaton, J.; Kamath, P.S. Burden of liver diseases in the world. *J. Hepatol.* **2018**, *70*, 151–171. [CrossRef]
200. Byass, P. The global burden of liver disease: A challenge for methods and for public health. *BMC Med.* **2014**, *12*, 159. [CrossRef]
201. Marcellin, P.; Kutala, B.K. Liver diseases: A major, neglected global public health problem requiring urgent actions and large-scale screening. *Liver Int.* **2018**, *38*, 2–6. [CrossRef]
202. Ferraioli, G.; Wong, V.W.S.; Castera, L.; Berzigotti, A.; Sporea, I.; Dietrich, C.F.; Choi, B.I.; Wilson, S.R.; Kudo, M.; Barr, R.G. Liver ultrasound elastography: An update to the world federation for ultrasound in medicine and biology guidelines and recommendations. *Ultrasound Med. Biol.* **2018**, *44*, 2419–2440. [CrossRef]
203. Ferraioli, G.; Filice, C.; Castera, L.; Choi, B.I.; Sporea, I.; Wilson, S.R.; Cosgrove, D.; Dietrich, C.F.; Amy, D.; Bamber, J.C.; et al. WFUMB guidelines and recommendations for clinical use of ultrasound elastography: Part 3: Liver. *Ultrasound Med. Biol.* **2015**, *41*, 1161–1179. [CrossRef] [PubMed]
204. Barr, R.G.; Ferraioli, G.; Palmeri, M.L.; Goodman, Z.D.; Garcia-Tsao, G.; Rubin, J.; Garra, B.; Myers, R.P.; Wilson, S.R.; Rubens, D.; et al. Elastography assessment of liver fibrosis: Society of radiologists in ultrasound consensus conference statement. *Radiology* **2015**, *276*, 845–861. [CrossRef] [PubMed]
205. European Association for Study of Liver. EASL Clinical Practice Guidelines: Management of hepatitis C virus infection. *J. Hepatol.* **2014**, *60*, 392. [CrossRef] [PubMed]
206. Dietrich, C.F.; Bamber, J.; Berzigotti, A.; Bota, S.; Cantisani, V.; Castera, L.; Cosgrove, D.; Ferraioli, G.; Friedrich-Rust, M.; Gilja, O.H.; et al. EFSUMB guidelines and recommendations on the clinical use of liver ultrasound elastography, update 2017 (long version). *Ultraschall Med. Eur. J. Ultrasound* **2017**, *38*, e16–e47.
207. Gebo, K.A.; Herlong, H.F.; Torbenson, M.S.; Jenckes, M.W.; Chander, G.; Ghanem, K.G.; El-Kamary, S.S.; Sulkowski, M.; Bass, E.B. Role of liver biopsy in management of chronic hepatitis C: A systematic review. *Hepatology* **2002**, *36*, s161–s172. [CrossRef]
208. Seeff, L.B.; Everson, G.T.; Morgan, T.R.; Curto, T.M.; Lee, W.M.; Ghany, M.G.; Shiffman, M.L.; Fontana, R.J.; Di Bisceglie, A.M.; Bonkovsky, H.L.; et al. Complication rate of percutaneous liver biopsies among persons with advanced chronic liver disease in the HALT-C trial. *Clin. Gastroenterol. Hepatol.* **2010**, *8*, 877–883. [CrossRef]
209. Regev, A.; Berho, M.; Jeffers, L.J.; Milikowski, C.; Molina, E.G.; Pyrsopoulos, N.T.; Feng, Z.Z.; Reddy, K.R.; Schiff, E.R. Sampling error and intraobserver variation in liver biopsy in patients with chronic HCV infection. *Am. J. Gastroenterol.* **2002**, *97*, 2614. [CrossRef]
210. Stotland, B.; Lichtenstein, G. Liver biopsy complications and routine ultrasound. *Am. J. Gastroenterol.* **1996**, *91*, 1295.
211. Bedossa, P.; Poynard, T. An algorithm for the grading of activity in chronic hepatitis C. *Hepatology* **1996**, *24*, 289–293. [CrossRef]
212. Goodman, Z.D. Grading and staging systems for inflammation and fibrosis in chronic liver diseases. *J. Hepatol.* **2007**, *47*, 598–607. [CrossRef]
213. Ziol, M.; Handra-Luca, A.; Kettaneh, A.; Christidis, C.; Mal, F.; Kazemi, F.; de Lédinghen, V.; Marcellin, P.; Dhumeaux, D.; Trinchet, J.C.; et al. Noninvasive assessment of liver fibrosis by measurement of stiffness in patients with chronic hepatitis C. *Hepatology* **2005**, *41*, 48–54. [CrossRef] [PubMed]
214. Castéra, L.; Vergniol, J.; Foucher, J.; Le Bail, B.; Chanteloup, E.; Haaser, M.; Darriet, M.; Couzigou, P.; de Lédinghen, V. Prospective comparison of transient elastography, Fibrotest, APRI, and liver biopsy for the assessment of fibrosis in chronic hepatitis C. *Gastroenterology* **2005**, *128*, 343–350. [CrossRef]
215. Chon, Y.E.; Choi, E.H.; Song, K.J.; Park, J.Y.; Han, K.H.; Chon, C.Y.; Ahn, S.H.; Kim, S.U.; et al. Performance of transient elastography for the staging of liver fibrosis in patients with chronic hepatitis B: A meta-analysis. *PLoS ONE* **2012**, *7*, e44930. [CrossRef] [PubMed]
216. Afdhal, N.H.; Bacon, B.R.; Patel, K.; Lawitz, E.J.; Gordon, S.C.; Nelson, D.R.; Challies, T.L.; Nasser, I.; Garg, J.; Wei, L.J.; et al. Accuracy of fibroscan, compared with histology, in analysis of liver fibrosis in patients with hepatitis B or C: A United States multicenter study. *Clin. Gastroenterol. Hepatol.* **2015**, *13*, 772–779. [CrossRef] [PubMed]

217. Cassinotto, C.; Lapuyade, B.; Mouries, A.; Hiriart, J.B.; Vergniol, J.; Gaye, D.; Castain, C.; Le Bail, B.; Chermak, F.; Foucher, J.; et al. Non-invasive assessment of liver fibrosis with impulse elastography: comparison of Supersonic Shear Imaging with ARFI and FibroScan®. *J. Hepatol.* **2014**, *61*, 550–557. [CrossRef] [PubMed]

218. Boursier, J.; Isselin, G.; Fouchard-Hubert, I.; Oberti, F.; Dib, N.; Lebigot, J.; Bertrais, S.; Gallois, Y.; Calès, P.; Aubé, C. Acoustic radiation force impulse: A new ultrasonographic technology for the widespread noninvasive diagnosis of liver fibrosis. *Eur. J. Gastroenterol. Hepatol.* **2010**, *22*, 1074–1084. [CrossRef]

219. Palmeri, M.L.; Wang, M.H.; Dahl, J.J.; Frinkley, K.D.; Nightingale, K.R. Quantifying hepatic shear modulus in vivo using acoustic radiation force. *Ultrasound Med. Biol.* **2008**, *34*, 546–558. [CrossRef]

220. Muller, M.; Gennisson, J.L.; Deffieux, T.; Tanter, M.; Fink, M. Quantitative viscoelasticity mapping of human liver using supersonic shear imaging: Preliminary in vivo feasability study. *Ultrasound Med. Biol.* **2009**, *35*, 219–229. [CrossRef]

221. Deffieux, T.; Montaldo, G.; Tanter, M.; Fink, M. Shear wave spectroscopy for in vivo quantification of human soft tissues visco-elasticity. *IEEE Trans. Med. Imaging* **2008**, *28*, 313–322. [CrossRef]

222. Barry, C.T.; Hah, Z.; Partin, A.; Mooney, R.A.; Chuang, K.H.; Augustine, A.; Almudevar, A.; Cao, W.; Rubens, D.J.; Parker, K.J. Mouse liver dispersion for the diagnosis of early-stage fatty liver disease: A 70-sample study. *Ultrasound Med. Biol.* **2014**, *40*, 704–713. [CrossRef]

223. Huwart, L.; Sempoux, C.; Salameh, N.; Jamart, J.; Annet, L.; Sinkus, R.; Peeters, F.; ter Beek, L.C.; Horsmans, Y.; Van Beers, B.E. Liver fibrosis: Noninvasive assessment with MR elastography versus aspartate aminotransferase–to-platelet ratio index. *Radiology* **2007**, *245*, 458–466. [CrossRef] [PubMed]

224. Millonig, G.; Friedrich, S.; Adolf, S.; Fonouni, H.; Golriz, M.; Mehrabi, A.; Stiefel, P.; Pöschl, G.; Büchler, M.W.; Seitz, H.K.; Mueller, S. Liver stiffness is directly influenced by central venous pressure. *J. Hepatol.* **2010**, *52*, 206–210. [CrossRef] [PubMed]

225. Bavu, É.; Gennisson, J.L.; Couade, M.; Bercoff, J.; Mallet, V.; Fink, M.; Badel, A.; Vallet-Pichard, A.; Nalpas, B.; Tanter, M.; et al. Noninvasive in vivo liver fibrosis evaluation using supersonic shear imaging: A clinical study on 113 hepatitis C virus patients. *Ultrasound Med. Biol.* **2011**, *37*, 1361–1373. [CrossRef] [PubMed]

226. Sporea, I.; Bota, S.; Gradinaru-Taşcău, O.; Şirli, R.; Popescu, A.; Jurchiş, A. Which are the cut-off values of 2D-Shear Wave Elastography (2D-SWE) liver stiffness measurements predicting different stages of liver fibrosis, considering Transient Elastography (TE) as the reference method? *Eur. J. Radiol.* **2014**, *83*, e118–e122. [CrossRef]

227. Benedict, M.; Zhang, X. Non-alcoholic fatty liver disease: An expanded review. *World J. Hepatol.* **2017**, *9*, 715. [CrossRef]

228. Sherman, K.E.; Abdel-Hameed, E.A.; Ehman, R.L.; Rouster, S.D.; Campa, A.; Martinez, S.S.; Huang, Y.; Zarini, G.G.; Hernandez, J.; Teeman, C.; et al. Validation and refinement of noninvasive methods to assess hepatic fibrosis: Magnetic resonance elastography versus enhanced liver fibrosis index. *Dig. Dis. Sci.* **2019**, *65*, 1252–1257. [CrossRef]

229. Ajmera, V.; Liu, A.; Singh, S.; Yachoa, G.; Ramey, M.; Bhargava, M.; Zamani, A.; Lopez, S.; Mangla, N.; Bettencourt, R.; et al. Clinical Utility of an Increase in Magnetic Resonance Elastography in Predicting Fibrosis Progression in Nonalcoholic Fatty Liver Disease. *Hepatology* **2020**, *71*, 849–860. [CrossRef]

230. Kang, K.A.; Jun, D.W.; Kim, M.S.; Kwon, H.J.; Nguyen, M.H. Prevalence of significant hepatic fibrosis using magnetic resonance elastography in a health check-up clinic population. *Aliment. Pharmacol. Ther.* **2020**, *51*, 388–396. [CrossRef]

231. O'Hara, S.; Zelesco, M.; Sun, Z. Can shear wave elastography of the cervix be of use in predicting imminent cervical insufficiency and preterm birth?—Preliminary results. *Ultrasound Med. Biol.* **2019**, *45*, S111–S112. [CrossRef]

232. Thomas, A. Imaging of the cervix using sonoelastography. *Ultrasound Obstet. Gynecol. Off. J. Int. Soc. Ultrasound Obstet. Gynecol.* **2006**, *28*, 356–357. [CrossRef]

233. Feltovich, H.; Hall, T. Quantitative imaging of the cervix: Setting the bar. *Ultrasound Obstet. Gynecol.* **2013**, *41*, 121–128. [CrossRef] [PubMed]

234. Maurer, M.; Badir, S.; Pensalfini, M.; Bajka, M.; Abitabile, P.; Zimmermann, R.; Mazza, E. Challenging the in-vivo assessment of biomechanical properties of the uterine cervix: A critical analysis of ultrasound based quasi-static procedures. *J. Biomech.* **2015**, *48*, 1541–1548. [CrossRef] [PubMed]

235. Molina, F.; Gómez, L.; Florido, J.; Padilla, M.; Nicolaides, K. Quantification of cervical elastography: A reproducibility study. *Ultrasound Obstet. Gynecol.* **2012**, *39*, 685–689. [CrossRef] [PubMed]

236. Hernandez-Andrade, E.; Hassan, S.S.; Ahn, H.; Korzeniewski, S.J.; Yeo, L.; Chaiworapongsa, T.; Romero, R. Evaluation of cervical stiffness during pregnancy using semiquantitative ultrasound elastography. *Ultrasound Obstet. Gynecol.* **2013**, *41*, 152–161. [CrossRef] [PubMed]

237. Hee, L.; Sandager, P.; Petersen, O.; Uldbjerg, N. Quantitative sonoelastography of the uterine cervix by interposition of a synthetic reference material. *Acta Obstet. Gynecol. Scand.* **2013**, *92*, 1244–1249. [CrossRef] [PubMed]

238. Mazza, E.; Parra-Saavedra, M.; Bajka, M.; Gratacos, E.; Nicolaides, K.; Deprest, J. In vivo assessment of the biomechanical properties of the uterine cervix in pregnancy. *Prenat. Diagn.* **2014**, *34*, 33–41. [CrossRef]

239. Fruscalzo, A.; Mazza, E.; Feltovich, H.; Schmitz, R. Cervical elastography during pregnancy: A critical review of current approaches with a focus on controversies and limitations. *J. Med. Ultrason.* **2016**, *43*, 493–504. [CrossRef]

240. Nightingale, K.; Palmeri, M.; Trahey, G. Analysis of contrast in images generated with transient acoustic radiation force. *Ultrasound Med. Biol.* **2006**, *32*, 61–72. [CrossRef]

241. Carlson, L.C.; Romero, S.T.; Palmeri, M.L.; del Rio, A.M.; Esplin, S.M.; Rotemberg, V.M.; Hall, T.; Feltovich, H.M. Changes in shear wave speed pre- and post-induction of labor: A feasibility study. *Ultrasound Obstet. Gynecol. Off. J. Int. Soc. Ultrasound Obstet. Gynecol.* **2015**, *46*, 93–98. [CrossRef]

242. Muller, M.; Aït-Belkacem, D.; Hessabi, M.; luc Gennisson, J.; Grange, G.; Goffinet, F.; Lecarpentier, E.R.; Cabrol, D.; Tanter, M.; Tsatsaris, V. Assessment of the Cervix in Pregnant Women Using Shear Wave Elastography: A Feasibility Study. *Ultrasound Med. Biol.* **2015**, *41*, 2789–2797. [CrossRef]

243. Hernández-Andrade, E.; Aurioles-Garibay, A.; Garcia, M.; Korzeniewski, S.J.; Schwartz, A.G.; Ahn, H.; Martínez-Varea, A.; Yeo, L.; Chaiworapongsa, T.; Hassan, S.S.; et al. Effect of depth on shear-wave elastography estimated in the internal and external cervical os during pregnancy. *J. Perinat. Med.* **2014**, *42*, 549–557. [CrossRef] [PubMed]

244. Carlson, L.C.; Hall, T.J.; Rosado-Mendez, I.M.; Palmeri, M.L.; Feltovich, H.M. Detection of Changes in Cervical Softness Using Shear Wave Speed in Early versus Late Pregnancy: An in Vivo Cross-Sectional Study. *Ultrasound Med. Biol.* **2017**, *44*, 515–521. [CrossRef] [PubMed]

245. Rosado-Mendez, I.M.; Carlson, L.C.; Woo, K.M.; Santoso, A.P.; Guerrero, Q.W.; Palmeri, M.L.; Feltovich, H.; Hall, T.J. Quantitative assessment of cervical softening during pregnancy in the Rhesus macaque with shear wave elasticity imaging. *Phys. Med. Biol.* **2018**, *63*, 085016. [CrossRef] [PubMed]

246. Carlson, L.C.; Hall, T.J.; Rosado-Mendez, I.M.; Mao, L.; Feltovich, H. Quantitative assessment of cervical softening during pregnancy with shear wave elasticity imaging: An in vivo longitudinal study. *Interface Focus* **2019**, *9*, 20190030. [CrossRef]

247. Peralta, L.; Molina, F.; Melchor, J.; Gomez, L.; Masso, P.; Florido, J.; Rus, G. Transient Elastography to Assess the Cervical Ripening during Pregnancy: A Preliminary Study. *Ultrasound Obstet. Gynecol.* **2015**, *38*, 395–402. [CrossRef]

248. Hernández-Andrade, E.; Romero, R.; Korzeniewski, S.J.; Ahn, H.; Aurioles-Garibay, A.; Garcia, M.; Schwartz, A.G.; Yeo, L.; Chaiworapongsa, T.; Hassan, S.S. Cervical strain determined by ultrasound elastography and its association with spontaneous preterm delivery. *J. Perinat. Med.* **2014**, *42*, 159–169. [CrossRef]

249. Reissner, E.; Sagoci, H. Forced torsional oscillations of an elastic half-space. I. *J. Appl. Phys.* **1944**, *15*, 652–654. [CrossRef]

250. Hadj Henni, A.; Schmitt, C.; Trop, I.; Cloutier, G. Shear wave induced resonance elastography of spherical masses with polarized torsional waves. *Appl. Phys. Lett.* **2012**, *100*, 133702. [CrossRef]

251. Ouared, A.; Montagnon, E.; Cloutier, G. Generation of remote adaptive torsional shear waves with an octagonal phased array to enhance displacements and reduce variability of shear wave speeds: Comparison with quasi-plane shear wavefronts. *Phys. Med. Biol.* **2015**, *60*, 8161. [CrossRef]

252. Yengul, S.S.; Barbone, P.E.; Madore, B. Dispersion in Tissue-Mimicking Gels Measured with Shear Wave Elastography and Torsional Vibration Rheometry. *Ultrasound Med. Biol.* **2019**, *45*, 586–604. [CrossRef]

253. Melchor, J.; Rus, G. Torsional ultrasonic transducer computational design optimization. *Ultrasonics* **2014**, *54*, 1950–1962. [CrossRef] [PubMed]

254. Massó, P.; Callejas, A.; Melchor, J.; Molina, F.S.; Rus, G. In Vivo Measurement of Cervical Elasticity on Pregnant Women by Torsional Wave Technique: A Preliminary Study. *Sensors* **2019**, *19*, 3249. [CrossRef]

255. McFarlin, B.L.; Kumar, V.; Bigelow, T.A.; Simpson, D.G.; White-Traut, R.C.; Abramowicz, J.S.; O'Brien, W.D. Beyond Cervical Length: A Pilot Study of Ultrasonic Attenuation for Early Detection of Preterm Birth Risk. *Ultrasound Med. Biol.* **2015**, *41*, 3023–3029. [CrossRef] [PubMed]

256. Myers, K.M.; Socrate, S.; Paskaleva, A.; House, M. A study of the anisotropy and tension/compression behavior of human cervical tissue. *J. Biomech. Eng.* **2010**, *132*, 021003. [CrossRef] [PubMed]

257. Jiang, X.; Asbach, P.; Streitberger, K.; Thomas, A.Z.; Hamm, B.; Braun, J.; Sack, I.; Guo, J. In vivo high-resolution magnetic resonance elastography of the uterine corpus and cervix. *Eur. Radiol.* **2014**, *24*, 3025–3033. [CrossRef] [PubMed]

258. Shi, L.; Yao, W.; Gan, Y.; Zhao, L.; McKee, W.E.; Vink, J.; Wapner, R.; Hendon, C.; Myers, K. Anisotropic Material Characterization of Human Cervix Tissue based on Indentation. *J. Biomech. Eng.* **2019**, *141*, 091017. [CrossRef]

sensors

MDPI

Article

A Novel AVM Calibration Method Using Unaligned Square Calibration Boards

Jung Hyun Lee and Dong-Wook Lee *

Department of Electronics and Electrical Engineering, Dongguk University, Seoul 04620, Korea;
jhlee36@dongguk.edu
* Correspondence: dlee@dongguk.edu; Tel.: +82-2-2260-3350

Abstract: An around view monitoring (AVM) system acquires the front, rear, left, and right-side information of a vehicle using four cameras and transforms the four images into one image coordinate system to monitor around the vehicle with one image. Conventional AVM calibration utilizes the maximum likelihood estimation (MLE) to determine the parameters that can transform the captured four images into one AVM image. The MLE requires reference data of the image coordinate system and the world coordinate system to estimate these parameters. In conventional AVM calibration, many aligned calibration boards are placed around the vehicle and are measured to extract the reference sample data. However, accurately placing and measuring the calibration boards around a vehicle is an exhaustive procedure. To remediate this problem, we propose a novel AVM calibration method that requires only four randomly placed calibration boards by estimating the location of each calibration board. First, we define the AVM errors and determine the parameters that minimize the error in estimating the location. We then evaluate the accuracy of the proposed method through experiments using a real-sized vehicle and an electric vehicle for children to show that the proposed method can generate an AVM image similar to the conventional AVM calibration method regardless of a vehicle's size.

Keywords: around view monitoring system; automatic camera calibration; vision-based advanced driver assistance systems

Citation: Lee, J.H.; Lee, D.-W. A Novel AVM Calibration Method Using Unaligned Square Calibration Boards. *Sensors* **2021**, *21*, 2265. https://doi.org/10.3390/s21072265

Academic Editor: Changho Lee

Received: 22 February 2021
Accepted: 22 March 2021
Published: 24 March 2021

Publisher's Note: MDPI stays neutral with regard to jurisdictional claims in published maps and institutional affiliations.

1. Introduction

Around view monitoring (AVM) systems eliminate blind spots around the vehicle to prevent car accidents [1]. Because AVM systems create images that show the surrounding view of the vehicle, various vision-based advanced driver assistance systems (ADAS) utilize these AVM-produced images. For example, the parking space detection system detects the parking lines in the AVM images to determine the parking space area [2–4], the automated driving system detects the road lanes in the AVM images to track the position of the vehicle [5], and the downward view generation operation transforms an AVM image to generate a downward view image [6]. Therefore, these systems all require well-calibrated AVM images.

The AVM system transforms four captured images to generate an AVM image, as shown in Figure 1. In AVM calibration, image transformation parameters that are required to generate the AVM images are estimated. These parameters describe the geometrical relationship between the captured image coordinate system and the world coordinate system. In conventional AVM calibration, the maximum likelihood estimation (MLE) is used to estimate this relationship.

The MLE assumes that the location of the calibration board on the surface of the road represents the world coordinate system. Figure 1c shows the reconstructed world coordinate system using the calibration board location. The MLE computes the Euclidean distance, which is the re-projection error between the calibration boards in the reconstructed world coordinate system and the calibration boards in the captured image coordinate system, and minimizes this error to determine the image transformation parameters. Therefore, accurately measuring the calibration board location is a significant and operative procedure of conventional AVM calibration methods to generate well-calibrated AVM images.

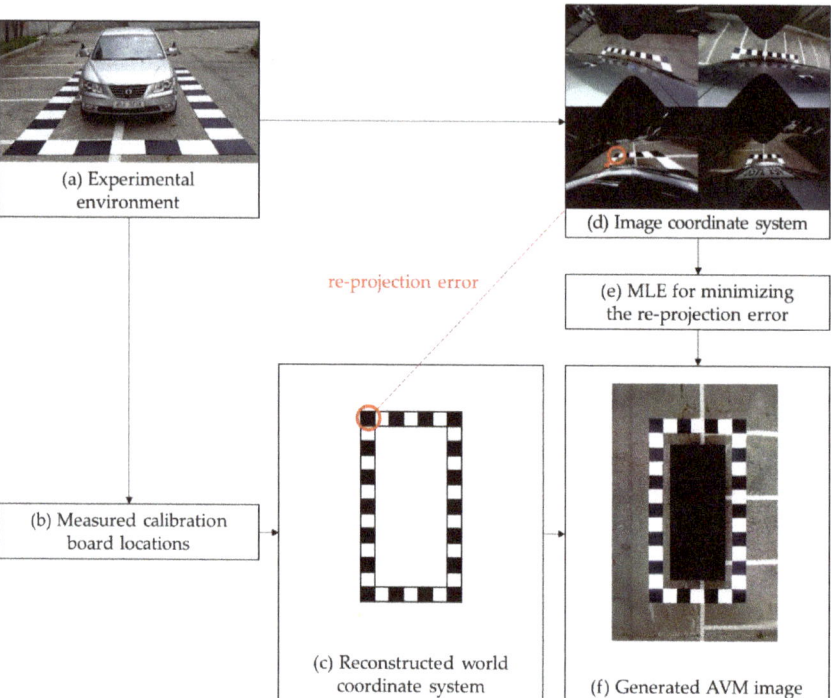

Figure 1. The procedure for conventional around view monitoring (AVM) calibration.

Conventional AVM calibration requires the alignment of the calibration boards for measurements, as shown in Figure 1a. Because calibration boards are spread over a large area, accurately measuring calibration boards is an exhaustive procedure. Vehicle manufacturers use AVM calibration facilities to measure the calibration board locations accurately, as shown in Figure 2 [7]. Various AVM calibration studies are also based on well-aligned calibration boards [8–19]. The details of these studies are provided in Section 2.1.

Figure 2. The AVM calibration facility.

Some researchers have utilized alternative devices to facilitate camera calibration [20–23]. They used odometry or an inertial measurement unit. However, adjacent images utilizing these approach methods cannot be aligned because these methods focus on the calibration of only one camera.

Other approaches detect road lanes or the host vehicle instead of utilizing additional devices [24–28]. These approaches also focus on the calibration of only one camera. Choi et al. [29] calibrated four AVM cameras to align adjacent images using detected road lanes. These calibration methods must repeat the road lane detection process until the integrity of the detected lanes is verified. The methods we have surveyed indicate that camera calibration without the use of calibration boards can face various challenges.

Lee et al. [30] calibrated AVM cameras using only two circle-shaped calibration boards. This method takes multiple photos while the vehicle passes between the two calibration boards to achieve the effect of having more calibration boards placed. However, driving perfectly straight ahead is as exhausting as accurately measuring the calibration board locations. Furthermore, only one calibration board per image with the smallest mean square error is selected from among the multiple images taken while driving. Therefore, this approach is not suitable for the MLE because only one calibration board is used to represent the world coordinate system.

We propose an MLE-based AVM calibration method that uses minimal calibration boards, as shown in Figure 3. This method estimates the location of the four calibration boards instead of measuring them. To this end, we divide the AVM image into two areas, as shown in Figure 4. The first area is the overlapping region of interest (ROI) where the fields of view of adjacent cameras overlap. The other area is the nonoverlapping ROI.

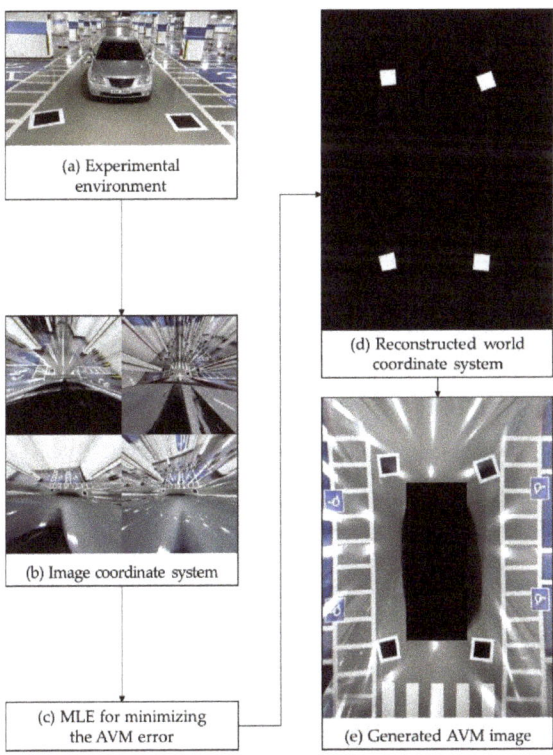

Figure 3. The procedure for the proposed AVM calibration.

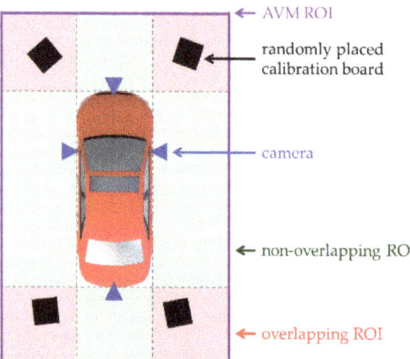

Figure 4. The proposed AVM calibration environment. The region of interest (ROI) of the AVM image (purple rectangle) can be divided into the overlapping ROI (red area) and the nonoverlapping ROI (green area).

At least one calibration board must be placed in each overlapping ROI. If we place additional calibration boards in the nonoverlapping ROI, the accuracy of the MLE will increase. However, the human eye can hardly distinguish between the AVM image results with and without calibration boards placed in the nonoverlapping ROI because the nonoverlapping ROI errors are distributed equally for each pixel and are, therefore, not significant. In contrast, the human eye can easily recognize the overlapping ROI errors because the overlapping ROI is where adjacent images are stitched. Therefore, it is possible to generate an AVM image even if the calibration boards are placed only in the overlapping ROI.

We define two errors to calibrate the AVM cameras using square-shaped calibration boards: a square-shaped error (SSE) and an alignment error (AME). An SSE indicates the difference between the square shape and the quadrilateral shape. A square-shaped calibration board can become a quadrilateral-shaped calibration board in the captured images based on the camera orientation. Therefore, we can estimate the camera orientation by minimizing the SSE. An AME indicates the Euclidean distance between the same calibration boards in the adjacent images. By minimizing the AME, we can estimate the camera position and align the adjacent images. Therefore, we use the sum of the two errors as the loss function of the proposed method.

The proposed AVM calibration offers the following various advantages:

- A measuring procedure is not required.
- Only four calibration boards are used to minimize the placing procedure.
- The proposed method can still generate an AVM image similar to that generated by the conventional method.
- In a small repair shop, the four calibration boards need to be in place only when AVM calibration is being done.

2. Related Works

Camera calibration has been extensively researched in a wide range of fields. Therefore, this literature review focuses on two types of AVM calibration-related studies: AVM calibration and vehicle-mounted camera calibration. AVM calibration methods consider the geometric relationship of adjacent AVM cameras. Vehicle-mounted camera calibration methods cannot estimate the adjacent AVM camera relationships, but they can estimate the orientation and position of a mono camera so that these methods can be utilized for AVM calibration.

2.1. AVM Calibration

Most of the AVM calibration methods we surveyed use well-aligned calibration boards. Chang et al. [8] proposed a method to determine accurate vertexes of calibration boards when the edges of the calibration boards were blurred and jagged. Zhao et al. [9] reduced the brightness difference among fisheye images and achieved a smooth transition around stitching seam. Two methods [8,9] utilized direct linear transform (DLT) to estimate the image transformation matrix required to generate an AVM image and focused on increasing the accuracy of the AVM calibration.

Gao et al. [10] projected a 2D AVM image generated by the DLT on a 3D ship model. The 3D AVM image helps drivers to be aware of the driving environment and eliminates visual blind spots. Yang et al. [11] proposed a flexible central-around coordinate mapping (CACM) model for vehicle surround view synthesis. The CACM model calculates the geometric relationship between the world coordinate system and the virtual AVM camera coordinate system. These studies focused on mapping models for AVM systems.

Jeon et al. [12] and Lo [13] focused on improving the performance of the embedded system. They also generated an AVM image using the DLT and upload a lookup table including image transformation parameters for generating an AVM image.

No matter how well-aligned calibration boards are used, errors will occur if the coordinates of the calibration boards are not accurately detected in an image. Some researchers proposed a method that can determine the coordinates of calibration boards in an image more accurately. Kim [14] patented a technology for a robot that revises the coordinates of calibration boards in an image. Pyo et al. [15] drew straight lines between calibration boards and detected the vanishing points using the drawn lines. The detected vanishing points help calibration board detection accurately detect the coordinates of calibration boards.

Natroshvili et al. [16] utilized MLE to estimate the orientation and location of cameras. The DLT-based method can only estimate a homography matrix used to transform an image, but the MLE-based method can estimate parameters indicating the orientation and location of cameras. When an AVM image requires revision, adjusting the orientation and location parameters is more intuitive and convenient than adjusting the homography matrix.

Zeng et al. [17] patented an AVM calibration method that paints calibration boards on all grounds, including under the vehicle, to determine the vehicle coordinates accurately. Since the calibration boards under the vehicle are obscured by the vehicle, the coordinates of the vehicle can be estimated.

Ko et al. [18] and Li [19] used a hyperbolic reflector and a spherical image sensor instead of a fisheye lens, respectively. The hyperbolic reflector is a mirror that increases the field of view of a camera by more than 180 degrees. The spherical image sensor can see all 360-degree surroundings by combining two cameras having a field of view of 180 degrees or more.

2.2. Vehicle-Mounted Camera Calibration

Camera calibration methods for vehicle-mounted cameras focus on estimating the orientation and location of the camera. The estimated parameters can be used to inverse perspective mapping (IPM). IPM is a method that transforms a captured image into a top view image that removes perspective distortion using the orientation and location of the camera.

Some researchers used additional devices instead of calibration boards. Wang et al. [20] proposed a camera-encoder fusion system to estimate extrinsic parameters. The extracted and tracked natural features provide the Euclidean distance information of the image coordinate system, and the encoder measures the camera travel distance. This method estimates the extrinsic parameters by comparing the Euclidean distance of the natural features with the camera travel distance. Schneider et al. [21] and Chien et al. [22] also measured the camera travel distance using odometry and visual-odometry, respectively. Li et al. [23] used an inertial measurement unit to measure the orientation of the camera.

Other researchers detected road lanes instead of using additional devices or calibration boards. Xu et al. [24] and Prakash et al. [25] detected road lanes and used them for estimating the orientation and location of the front camera. The estimated parameters are used for IPM. A top view image generated by IPM provides the distance between obstacles and the host vehicle. Wang et al. [26] and de Paula et al. [27] also detected road lanes to estimate the orientation and location of a front camera. They estimated the distance between obstacles and the host vehicle without IPM.

Lee et al. [28] proposed a camera calibration method detecting the host vehicle instead of detecting the road lanes. More specifically, this method detects the host vehicle surface to avoid the problems of utilizing detected road lanes, but it can only estimate the orientation of the camera.

3. AVM Calibration Using Four Randomly Placed Calibration Boards

The proposed method can generate an AVM image without the location information of the calibration boards. To this end, we estimate the calibration board locations by minimizing the AVM error, which consists of the SSE and AME. In the following sections, we first describe the difference between conventional AVM calibration and the proposed AVM calibration and then define the SSE, AME, and AVM error used to generate an AVM image.

3.1. Conventional AVM Calibration

The MLE-based conventional AVM calibration estimates the geometrical relationship between the calibration board locations in the world coordinate system and the image coordinate system. Because lens distortion parameters do not change even if the camera orientation and location are changed, we assume that the source images of the AVM calibration are lens distortion-corrected images. The relationship between the world coordinate system and the source image coordinate system can be expressed as follows:

$$\tilde{\mathbf{u}}_s = \mathbf{K}_s[\mathbf{R}_s|\mathbf{T}_s]\tilde{\mathbf{u}}_w \tag{1}$$

where $\tilde{\mathbf{u}}_s$ is the homogeneous source image coordinate system, $\tilde{\mathbf{u}}_w$ is the homogeneous world coordinate system, $\mathbf{R}_s$ is the rotation matrix describing the camera orientation, $\mathbf{T}_s$ is the translation matrix describing the camera location, and $\mathbf{K}_s$ is the intrinsic matrix describing the optical properties of the camera.

$$\mathbf{K}_s = \begin{bmatrix} f_s\mathbf{I}_{2\times2} & \mathbf{p}_s \\ \mathbf{0}_{2\times1} & 1 \end{bmatrix} \tag{2}$$

where f_s is the focal length, $\mathbf{I}_{2\times2}$ is a 2×2 identity matrix, and $\mathbf{p}_s$ is a 2D principal point. We assume that a virtual AVM camera is over the vehicle and looks at the vehicle vertically downward to generate an AVM image, as shown in Figure 5.

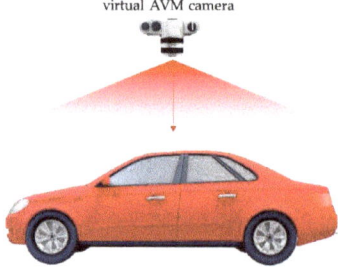

Figure 5. Visualization of the orientation and location of the virtual AVM camera.

The relationship between the world coordinate system and the AVM image coordinate system can be expressed in the same way as in Equation (1).

$$\tilde{\mathbf{u}}_v = \mathbf{K}_v[\mathbf{R}_v|\mathbf{T}_v]\tilde{\mathbf{u}}_w \tag{3}$$

where $\tilde{\mathbf{u}}_v$ is the homogeneous coordinate system of the virtual AVM image, $\mathbf{R}_v$ is the rotation matrix describing the virtual AVM camera orientation, $\mathbf{T}_v$ is the translation matrix describing the virtual AVM camera location, and $\mathbf{K}_v$ is the intrinsic matrix describing the optical properties of the virtual AVM camera. From Equations (1) and (3), we can express the relationship between the source image coordinate system and the AVM image coordinate system as

$$\tilde{\mathbf{u}}_v = (\mathbf{K}_v[\mathbf{R}_v|\mathbf{T}_v])(\mathbf{K}_s[\mathbf{R}_s|\mathbf{T}_s])^{-1}\tilde{\mathbf{u}}_s = \mathbf{H}_{AVM}\tilde{\mathbf{u}}_s \tag{4}$$

where $\mathbf{H}_{AVM}$ is a 3×3 homography matrix describing the relationship between the source image coordinate system and the AVM image coordinate system. The matrix $\mathbf{K}_v[\mathbf{R}_v|\mathbf{T}_v]$ consists of known parameters because the properties of the virtual AVM camera are determined by the drivers or manufacturers, as shown in Figure 6. Furthermore, because the camera optical properties do not change even if the camera orientation and location are changed, we can assume that the intrinsic matrix $\mathbf{K}_s$ is known. Therefore, conventional AVM calibration focuses only on estimating the extrinsic matrix $[\mathbf{R}_s|\mathbf{T}_s]$ to compute $\mathbf{H}_{AVM}$. To estimate the extrinsic matrix $[\mathbf{R}_s|\mathbf{T}_s]$, conventional AVM calibration defines a re-projection error e_{rp} and determines the extrinsic matrix that minimizes the re-projection error.

$$e_{rp} = \|\bar{\mathbf{u}}_v - \mathbf{H}_{AVM}\bar{\mathbf{u}}_s\| \tag{5}$$

$$[\bar{\mathbf{R}}_s|\bar{\mathbf{T}}_s] = \underset{[\mathbf{R}_s|\mathbf{T}_s]}{\mathrm{argmin}}(e_{rp}) \tag{6}$$

where e_{rp} is the re-projection error, $\bar{\mathbf{u}}_v$ represents the measured calibration board coordinates for the virtual AVM image coordinate system, $\bar{\mathbf{u}}_s$ represents the measured calibration board coordinates for the source image coordinate system, and $[\bar{\mathbf{R}}_s|\bar{\mathbf{T}}_s]$ is the estimated extrinsic matrix. Equation (5) is the loss function of the conventional AVM calibration method. Because the calibration board locations are not measured, the measured calibration board coordinates representing the virtual AVM image coordinate system, $\bar{\mathbf{u}}_v$, in Equation (5) is unknown. Therefore, we estimate the calibration board coordinates in the virtual AVM image, $\bar{\mathbf{u}}_v$, to generate an AVM image.

Figure 6. Reconstructed AVM images according to the properties of the virtual AVM camera.

3.2. Calibration Board Detection

Calibration board detection occurs in the preprocessing phase of the proposed method. We detect the calibration boards in the source images and utilize them to compute the SSE and AME. Because one calibration board is placed in each overlapping ROI, two calibration boards are photographed in one source image (one source image has two overlapping ROIs). The photographed square-shaped calibration boards become quadrilateral shapes in the source images due to camera tilting. Therefore, we detect two quadrilateral shapes in the source images using simple and commonly used image processing techniques, as shown in Figure 7.

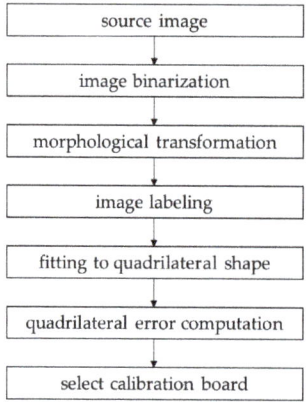

Figure 7. The procedure for calibration board detection.

We utilize the adaptive thresholding image binarization method to binarize the source images [31]. This method computes the local threshold values instead of the global threshold value to accurately binarize an image. The morphological transformation can remove noise [32], and the labeling algorithm assigns the pixels to the same group if the values between the neighboring pixels are identical [33]. Next, we detect the edge points of the labeled object and fit the edge points to four straight lines using K-mean clustering [34].

If the labeled object is a quadrilateral, the fitted four straight lines indicate four sides of the quadrilateral. To find the two calibration boards among the labeled objects, we compute the quadrilateral error. The quadrilateral error is the sum of the Euclidean distance between the edge points and the fitted four straight lines. If the labeled object is a quadrilateral, the quadrilateral error is close to zero. Because there are two calibration boards in one source image, we divide the source image into left and right areas and select the labeled object with the least quadrilateral error in each area as the calibration board.

3.3. Square-Shaped Error

We can estimate the geometrical relationship between the quadrilateral shape and the square shape because a square-shaped calibration board has a quadrilateral shape in the source image. The square-shaped calibration board can be transformed into a parallelogram shape by an affine transformation matrix, and the parallelogram shape can be transformed into a quadrilateral shape by a perspective transformation matrix.

$$\tilde{\mathbf{u}}_{\text{parall}} = \mathbf{H}_A \tilde{\mathbf{u}}_{\text{square}} = \begin{bmatrix} a_{11} & a_{12} & 0 \\ 0 & 1 & 0 \\ 0 & 0 & 1 \end{bmatrix} \tilde{\mathbf{u}}_{\text{square}}$$

$$\tilde{\mathbf{u}}_{\text{quad}} = \mathbf{H}_P \mathbf{H}_A \tilde{\mathbf{u}}_{\text{square}} = \mathbf{H}_P \tilde{\mathbf{u}}_{\text{parall}} = \begin{bmatrix} 1 & 0 & 0 \\ 0 & 1 & 0 \\ p_{31} & p_{32} & 1 \end{bmatrix} \tilde{\mathbf{u}}_{\text{parall}} \tag{7}$$

where $\tilde{\mathbf{u}}_{\text{parall}}$ represents the homogeneous coordinates of the parallelogram-shaped calibration board, $\tilde{\mathbf{u}}_{\text{square}}$ represents the homogeneous coordinates of the square-shaped calibration board, $\tilde{\mathbf{u}}_{\text{quad}}$ represents the homogeneous coordinates of the quadrilateral-shaped calibration board, $\mathbf{H}_P$ is a perspective transformation matrix, and $\mathbf{H}_A$ is an affine transformation matrix. The parameter a_{11} of the affine transformation matrix $\mathbf{H}_A$ transforms a square into a rectangle, the parameter a_{12} transforms a rectangle into a parallelogram, the parameter p_{31} of the perspective transformation matrix $\mathbf{H}_P$ transforms a square into a trapezoid with a parallel pair of opposite sidelines in the horizontal direction, and the parameter p_{32} of the perspective transformation matrix $\mathbf{H}_P$ transforms a square into a trapezoid with a parallel pair of opposite sidelines in the vertical direction. We can transform the quadrilateral-shaped calibration boards into square-shaped calibration boards with the perspective and affine matrices:

$$\tilde{\mathbf{u}}_{\text{square}} = (\mathbf{H}_P\mathbf{H}_A)^{-1}\tilde{\mathbf{u}}_{\text{quad}} \tag{8}$$

To estimate the matrix $(\mathbf{H}_P\mathbf{H}_A)^{-1}$ in Equation (8), we define a SSE to indicate the difference between the coordinates $\tilde{\mathbf{u}}_{\text{quad}}$ and $\tilde{\mathbf{u}}_{\text{square}}$ using the characteristics of a square shape. The characteristics of a square is that the four angles and the intersection angle of two diagonals are 90 degrees, the length of the four sidelines are equal, and the two diagonals are $\sqrt{2}$ times longer than the sidelines. We define two types of errors based on these characteristics: angle-based SSE (ASSE) and length-based SSE (LSSE). The reason for classifying the SSE into two types is to simultaneously minimize the SSE and AME, details of which are described in Section 3.5.

3.3.1. Angle-Based SSE

An angle-based SSE (ASSE) refers to the difference between an internal angle of a square and the corresponding quadrilateral angle. Let a line vector $\overline{\mathbf{l}}_{\text{quad},i}$ represent an i-th sideline of a detected quadrilateral-shaped calibration board. By the matrix $(\mathbf{H}_P\mathbf{H}_A)^{-1}$ in Equation (8), the detected quadrilateral-shaped calibration board can be transformed into a square-shaped calibration board $\overline{\mathbf{l}}_{\text{square},i} = (\mathbf{H}_P\mathbf{H}_A)^{-1}\overline{\mathbf{l}}_{\text{quad},i}$. The included angle of the square-shaped calibration board can be determined by the dot product of i-th and the j-th line vectors where $\overline{\mathbf{l}}_{\text{square},i} = \begin{bmatrix} \overline{l}_{1,i} & \overline{l}_{2,i} & \overline{l}_{3,i} \end{bmatrix}^{\text{T}}$.

$$\phi = \cos^{-1}\left(\frac{\overline{l}_{1,i}\overline{l}_{1,j} + \overline{l}_{2,i}\overline{l}_{2,j}}{\sqrt{\left(\overline{l}_{1,i}\right)^2 + \left(\overline{l}_{2,i}\right)^2} \cdot \sqrt{\left(\overline{l}_{1,j}\right)^2 + \left(\overline{l}_{2,j}\right)^2}} \right) \tag{9}$$

Therefore, we can define the ASSE as follows:

$$e_{\text{ASSE}} = \left| \frac{\pi}{2} - \phi \right| \tag{10}$$

Equation (10) can be simplified by the cosine function as:

$$
\begin{aligned}
e_{\text{ASSE}} &= \left|\cos\left(\tfrac{\pi}{2}\right) - \cos(\phi)\right| = \left|-\cos(\phi)\right| \\
&= \frac{\overline{l}_{1,i}\overline{l}_{1,j} + \overline{l}_{2,i}\overline{l}_{2,j}}{\sqrt{\left(\overline{l}_{1,i}\right)^2 + \left(\overline{l}_{2,i}\right)^2} \cdot \sqrt{\left(\overline{l}_{1,j}\right)^2 + \left(\overline{l}_{2,j}\right)^2}}
\end{aligned} \tag{11}
$$

where $0 \leq \phi \leq \pi$. We then determine the parameters that minimize the ASSE and the calibration boards in the source image can be transformed into square shapes:

$$\left(\overline{\mathbf{H}}_P, \overline{\mathbf{H}}_A \right) = \underset{\mathbf{H}_P, \mathbf{H}_A}{\arg\min} \left(\sum_{n=1}^{2} \sum_{k=1}^{5} e_{\text{ASSE}}(n,k) \right) \tag{12}$$

where $e_{ASSE}(n, k)$ is the ASSE of the k-th angle of the n-th calibration board, $\overline{\mathbf{H}}_P$ is the estimated perspective transformation matrix, and $\overline{\mathbf{H}}_A$ is the estimated affine transformation matrix. There are two calibration boards in the source image and five intersection points in the square (four vertices and one center of the square); thus, n is from 1 to 2 and k is from 1 to 5, respectively.

3.3.2. Length-Based SSE

A length-based SSE (LSSE) refers to the sideline length difference between the quadrilateral and square shapes. Let homogeneous coordinates $\overline{\mathbf{v}}_{quad,i}$ represent the i-th vertex of a detected quadrilateral-shaped calibration board, then the transformed homogeneous coordinates by the matrix variable is $\overline{\mathbf{v}}_{square,i} = (\mathbf{H}_P \mathbf{H}_A)^{-1} \overline{\mathbf{v}}_{quad,i} = \begin{bmatrix} \overline{v}_{1,i} & \overline{v}_{2,i} & 1 \end{bmatrix}^T$. We can calculate the length of one side using the transformed coordinates $\overline{\mathbf{v}}_{square,i}$ as:

$$\overline{m}_i = \sqrt{\left(\overline{v}_{1,i} - \overline{v}_{1,j}\right)^2 + \left(\overline{v}_{2,i} - \overline{v}_{2,j}\right)^2} \tag{13}$$

where $\overline{m}_i$ is the length of the i-th side of the transformed calibration board. The LSSE can be defined as Equation (14), where the length of one side of the calibration board is m:

$$e_{LSSE} = \sum_{i=1}^{4} |m - \overline{m}_i| + \sum_{j=1}^{2} \left| \sqrt{2}m - \overline{d}_j \right| \tag{14}$$

where $\overline{d}_j$ is the length of the j-th diagonal of the transformed calibration board. We then find the parameters that minimize the LSSE, and the calibration boards in the source image can be transformed into square shapes with:

$$\left(\overline{\mathbf{H}}_P, \overline{\mathbf{H}}_A \right) = \underset{\mathbf{H}_P, \mathbf{H}_A}{\text{argmin}} \left(\sum_{n=1}^{2} e_{LSSE}(n) \right) \tag{15}$$

where $e_{LSSE}(n)$ is the ASSE of the n-th calibration board, $\overline{\mathbf{H}}_P$ is the estimated perspective transformation matrix, and $\overline{\mathbf{H}}_A$ is the estimated affine transformation matrix.

3.4. Alignment Error

An alignment error (AME) is defined as the Euclidean distance between the same square-shaped calibration boards in adjacent images. Because the quadrilateral-shaped calibration board can be transformed into square-shaped calibration boards by minimizing the SSE, we focus only on estimating the similarity transformation matrix $\mathbf{H}_S$ consisting of a scale parameter s, an image rotation parameter θ, and image translation parameters t_x and t_y to align the square-shaped calibration boards in adjacent images.

$$\mathbf{H}_S = \begin{bmatrix} s \cdot \cos(\theta) & s \cdot \sin(\theta) & t_x \\ -s \cdot \sin(\theta) & s \cdot \cos(\theta) & t_y \\ 0 & 0 & 1 \end{bmatrix} \tag{16}$$

Square-shaped calibration boards of a front image and a left image can be aligned using Equation (17).

$$\mathbf{H}_S^{front} \tilde{\mathbf{v}}_{square}^{front} = \mathbf{H}_S^{left} \tilde{\mathbf{v}}_{square}^{left} \tag{17}$$

where $\mathbf{H}_S^{front}$ is the similarity transformation matrix of a front image, and $\tilde{\mathbf{v}}_{square}^{front}$ represents the homogeneous coordinates of the vertex of the square-shaped calibration board of the front image. Therefore, we can define the AME as follows:

$$
\begin{aligned}
e_{AME} = &\|\mathbf{H}_S^{front}\bar{\mathbf{v}}_{square}^{front} - \mathbf{H}_S^{left}\bar{\mathbf{v}}_{square}^{left}\| + \|\mathbf{H}_S^{left}\bar{\mathbf{v}}_{square}^{left} - \mathbf{H}_S^{rear}\bar{\mathbf{v}}_{square}^{rear}\| \\
&+\|\mathbf{H}_S^{rear}\bar{\mathbf{v}}_{square}^{rear} - \mathbf{H}_S^{right}\bar{\mathbf{v}}_{square}^{right}\| + \|\mathbf{H}_S^{right}\bar{\mathbf{v}}_{square}^{right} - \mathbf{H}_S^{front}\bar{\mathbf{v}}_{square}^{front}\|
\end{aligned}
\tag{18}
$$

where $\bar{\mathbf{v}}_{square}$ represents the homogeneous coordinates of the vertex of the transformed calibration boards by the perspective and affine transformation matrices. We can estimate the similarity transformation matrix by minimizing the AME.

$$
\left(\bar{\mathbf{H}}_S^{front},\bar{\mathbf{H}}_S^{left},\bar{\mathbf{H}}_S^{rear},\bar{\mathbf{H}}_S^{right}\right) = \underset{\substack{\mathbf{H}_S^{front}\\\mathbf{H}_S^{left}\\\mathbf{H}_S^{rear}\\\mathbf{H}_S^{right}}}{\mathrm{argmin}}\left(e_{AME}\right)
\tag{19}
$$

where $\bar{\mathbf{H}}_S$ is the estimated similarity transformation matrix.

3.5. AVM Error

We can estimate the image transformation parameters for generating the AVM image by minimizing the AVM error, which consists of an SSE and AME. Because there are two types of SSEs, the ASSE and LSSE, the AVM error can be expressed as a combination of the two types: the ASSE–AME and the LSSE–AME.

The problem with the ASSE–AME combination is that the units of the two measurements are not consistent. The ASSE is in radians whereas the AME is in pixels. In contrast, the units for the LSSE and AME are both in pixels. Therefore, we focus on using the LSSE–AME combination. However, the LSSE–AME combination is not without limitations. The LSSE–AME combination suffers from the local minimum problem because the range of the parameters searched by the MLE changes according to the size of the calibration board.

To solve this problem, we find the appropriate initial parameters by minimizing the ASSE. To minimize the ASSE, we utilize the Levenberg–Marquardt algorithm, which is most widely used to solve the maximum likelihood problems of camera calibration. The estimated matrices $\bar{\mathbf{H}}_P$ and $\bar{\mathbf{H}}_A$, by minimizing the ASSE, are used as initial values to minimize the LSSE–AME combination, as shown in Figure 8. Since the matrices $\bar{\mathbf{H}}_P$ and $\bar{\mathbf{H}}_A$ are already optimized, the local minimum problem caused by the size of the calibration board can be solved. The LSSE–AME combination is also minimized by utilizing the Levenberg–Marquardt algorithm.

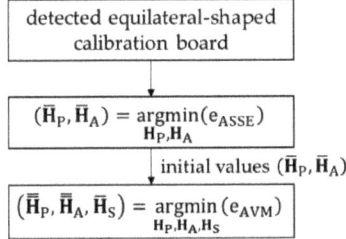

Figure 8. The procedure for the AVM error minimization, where $e_{AVM} = e_{LSSE} + e_{AME}$.

4. Experiments

We performed several experiments to evaluate the proposed method. We used Kodak's PIXPRO SP360 cameras with a 235° field of view and a 2880 $px \times$ 2880 px resolution [35]. The cameras were installed on a Hyundai SONATA vehicle, as shown in Figure 9 [36]. The installation heights of the front, rear, left, and right cameras are approximately 57, 84, 92, and 92 cm, respectively. Each camera is tilted approximately 30°. The overall length of the vehicle is 480 cm, the overall width is 183 cm, and the overall height is 147.5 cm. The dimensions of the calibration boards are 50 cm× 50 cm and we set the calibration board dimensions in the AVM image to 100 $px \times$ 100 px.

(**a**) Front camera   (**b**) Rear camera   (**c**) Left camera   (**d**) Right camera

Figure 9. Four installed cameras for the field experiments.

The size of the calibration board must be experimentally determined based on the size of the vehicle and the field of view of the cameras. More specifically, the calibration board size must increase with the increase in the size of the vehicle or the range of the camera field of view. However, the larger the calibration boards, the more inefficient it is to carry and place them. When we used calibration boards with dimensions smaller than 50 cm× 50 cm, sometimes the calibration board detection algorithm failed. When we used calibration boards with dimensions larger than 100 cm× 100 cm, it was difficult to place the calibration boards in the overlapping ROI. Therefore, for the purpose of our experiment, we set the dimensions of the calibration board as 50 cm× 50 cm.

4.1. Performance Evaluation Using a Real-Sized Vehicle

We placed four calibration boards around the vehicle to evaluate the performance of the proposed method, as shown in Figure 10. Because the camera manufacturer provides the lens distortion parameters and intrinsic parameters, we can easily correct the lens distortion, as shown in Figure 11. In the lens distortion-corrected images, the shape of the calibration boards is quadrilateral. The calibration board detection detects two quadrilaterals per image, as shown in Figure 11c. The proposed method transforms the source images such that the detected quadrilateral calibration boards become squares. Figure 12 shows the generated AVM image using the proposed method. We can observe that all the calibration boards are similar to squares and the adjacent images are well aligned. Furthermore, even though there are no calibration boards in the nonoverlapping ROI, the source image in the nonoverlapping ROI can also be transformed into a well-calibrated AVM image.

Figure 10. The experimental environment with four randomly placed calibration boards.

(**a**) Captured images

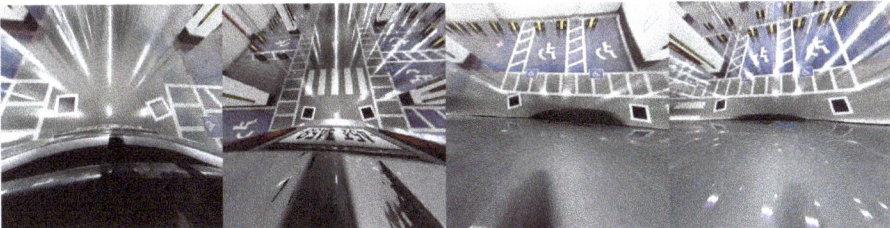

(**b**) Lens distortion corrected images

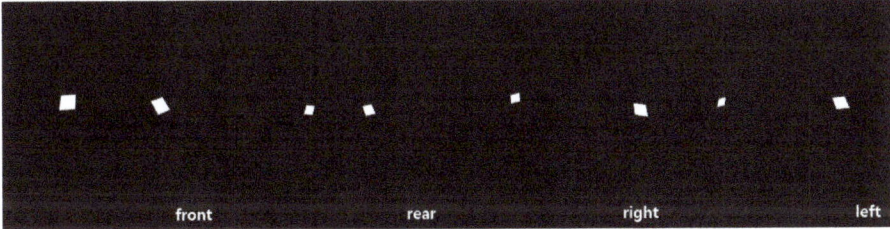

(**c**) Detected calibration boards

Figure 11. Example images for the performance evaluation of the proposed method.

Table 1 shows the estimated image transformation parameters corresponding to the AVM image in Figure 12a. Because the parameters a_{11}, a_{12}, p_{31}, and p_{32} are normalized, the affine and perspective distortion-corrected images are scaled and rotated, as shown in Figure 13. For example, the front image in Figure 13a is rotated 0.2524π clockwise and the average of the side lengths is $1.9278\ px$ when the affine and perspective distortions are corrected. Therefore, the product of s and γ is close to $100\ px$ and the sum of θ and ϕ of the front, left, rear, and right images are close to 0π, -0.5π, $-\pi$, and -1.5π, respectively, as shown in Table 2.

Table 1. Estimated image transformation parameters.

Parameters	Front	Left	Rear	Right
s	51.8717	3.5862	8.7625	37.9317
$\theta\ (rad)$	-0.2611π	0.0857π	-0.3056	-0.8748
$t_x\ (px)$	-7428.2071	2825.3564	-3613.53	-7673.6563
$t_y\ (px)$	-7496.9527	-2347.6187	-2979.6379	10,336.4311
a_{11}	0.0843	0.5115	0.2502	0.1971
a_{12}	-0.1002	-0.0410	0.0879	-0.1692
p_{31}	-0.0009	-0.0007	-0.0003	-0.001
p_{32}	0.0044	-0.0012	-0.0019	-0.0034

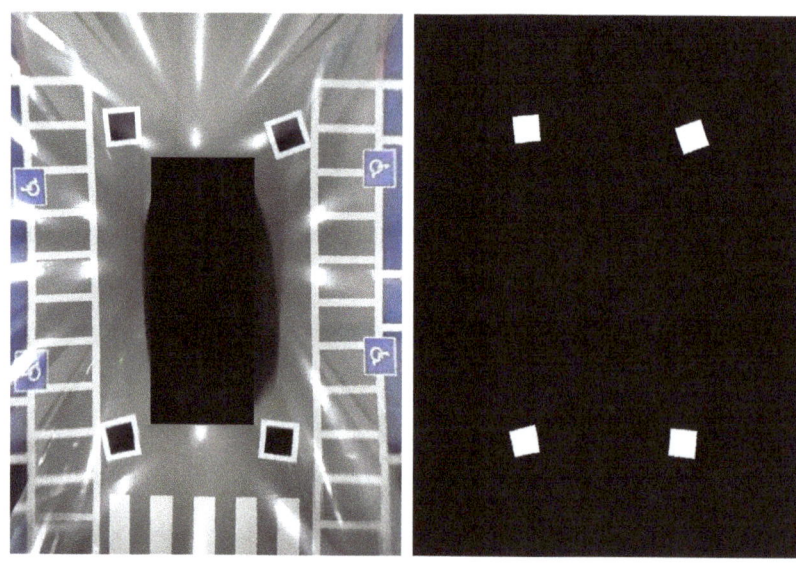

(**a**) Generated AVM image   (**b**) Estimated calibration boards

Figure 12. The results of the proposed method.

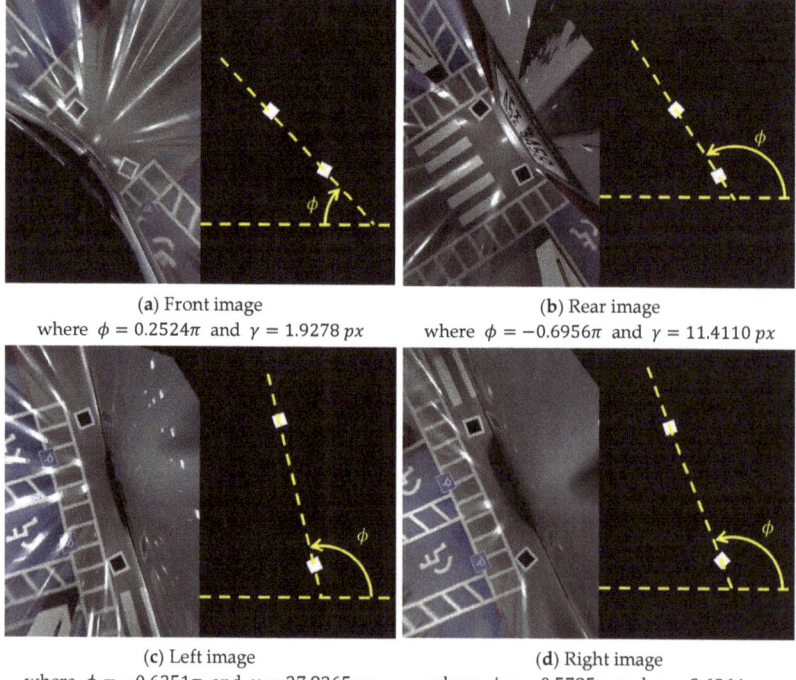

(**a**) Front image
where $\phi = 0.2524\pi$ and $\gamma = 1.9278\,px$

(**b**) Rear image
where $\phi = -0.6956\pi$ and $\gamma = 11.4110\,px$

(**c**) Left image
where $\phi = -0.6251\pi$ and $\gamma = 27.8365\,px$

(**d**) Right image
where $\phi = -0.5785\pi$ and $\gamma = 2.6364\,px$

Figure 13. Affine and perspective corrected images, where ϕ is the rotation angle and γ is the scale value by the normalized parameters.

Table 2. The relationship between the normalized coefficients and estimated parameters.

Parameters	Front	Left	Rear	Right
s	51.8717	3.5862	8.7625	37.9317
$\gamma\ (px)$	1.9278	27.8365	11.4110	2.6364
$s \times \gamma\ (px)$	99.9983	99.8273	99.9889	100.0031
$\theta\ (rad)$	-0.2611π	0.0857π	-0.3056π	-0.8748π
$\phi\ (rad)$	0.2524π	-0.6251π	-0.6956π	-0.5785π
$\theta + \phi\ (rad)$	-0.0087π	-0.5394π	-1.0012π	-1.4533π

For quantitative evaluation, we calculated the AVM errors, as shown in Table 3. Because there are two boards in one image, the LSSE per calibration board is approximately $17.6571/2 = 8.8285\ px$. The LSSE is the sum of the errors of the four sides and two diagonal lines; thus, the error for each sideline is approximately $8.8285/6 \approx 1.4714\ px$. That is, the length of one side of the calibration board is approximately $100 \pm 1.4714\ px$ in the generated AVM image. The AME indicates the offset of the adjacent images when two images are stitched. Because one calibration board has four vertexes, the offset of the calibration board is approximately $10.1691/4 \approx 2.5423\ px$. These values are significantly small enough to be difficult for the human eye to recognize.

Table 3. AVM errors of the proposed method.

Calibration Board	e_{AVM}	e_{LSSE}	e_{AME}
front-left	25.1992	16.3466	8.8526
left-rear	37.231	25.1189	12.1121
rear-right	24.2864	14.6869	9.5994
right-front	24.5884	14.4762	10.1122
average	27.8262	17.6571	10.1691

4.2. Performance Evaluation Using an Electric Vehicle for Children

The orientation and location of the camera can change depending on the type and size of a vehicle. Because the proposed method should be able to generate an AVM image regardless of vehicle type, we experimented using an electric vehicle for children to verify this aspect, as shown in Figure 14. The installation height of each camera is approximately 40 cm and each camera is tilted approximately 30°. The overall length of the miniature vehicle is 126 cm, the overall width is 73 cm, and the overall height is 64.5 cm. The dimensions of calibration boards are 20 cm× 20 cm and we set the calibration board dimensions in the AVM image to 100 px× 100 px.

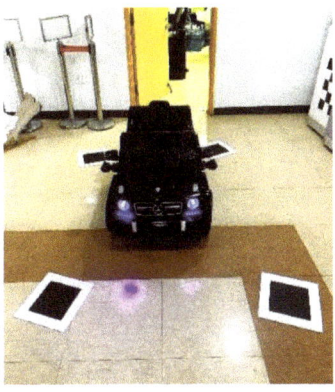

Figure 14. The experimental environment using an electric vehicle for children.

Figure 15 shows a generated AVM image using the proposed method for an electric vehicle for children. We can observe that the proposed method can generate a well-calibrated AVM image even though the size of the vehicle is small.

Figure 15. The results of the proposed method using an electric vehicle for children.

Table 4 shows the calculated AVM errors corresponding to the AVM image in Figure 15. The error for each sideline is approximately $39.5697/12 \approx 3.2974 \ px$ and the offset of the calibration board is approximately $6.6367/4 \approx 1.6591 \ px$. These resulting values are similar to those of the experimental environment using a real-sized vehicle because the calibration board dimensions in the AVM image are the same in both experiments. From the results of the experiments using real-sized and miniature vehicles, it can be verified that the proposed method can generate an AVM image regardless of the size of the vehicles.

Table 4. AVM errors of the proposed method using an electric vehicle for children.

Calibration Board	e_{AVM}	e_{LSSE}	e_{AME}
front-left	37.9732	31.5650	6.4082
left-rear	16.0779	8.3447	7.7332
rear-right	34.7484	25.7785	8.9699
right-front	96.0263	92.5907	3.4357
average	46.2065	39.5697	6.6367

4.3. Comparison Experiments with the Conventional Method

The proposed method can generate an AVM image using only four randomly placed calibration boards. In contrast, the conventional methods require calibration boards with known locations. Therefore, to compare the proposed method with the conventional method, we aligned and measured the calibration board locations, as shown in Figure 16, and provided the measured data as input to the conventional method.

Figure 17 shows the AVM images generated by the proposed method and the conventional method. We can observe that the results of the two methods are very similar, even though we did not input information regarding calibration board location to the proposed method. To compare the two methods in more detail, we calculated the root mean square error (RMSE), optical flow, and AVM errors for the two AVM images. The RMSE can be expressed as follows:

$$e_{RMSE} = \sqrt{\frac{1}{mn} \sum_{i=0}^{m-1} \sum_{j=0}^{n-1} \left[I_c(i,j) - I_p(i,j) \right]^2} \tag{20}$$

where $I_c(i, j)$ is the grayscale value of the AVM image from the conventional method at the (i, j) point, I_p is the grayscale value of the AVM image from the proposed method, m is the width of the AVM images, and n is the height of the AVM images. The calculated RMSE of the two AVM images in Figure 17a,b is 0.0457 when the range of the grayscale is 0–1.

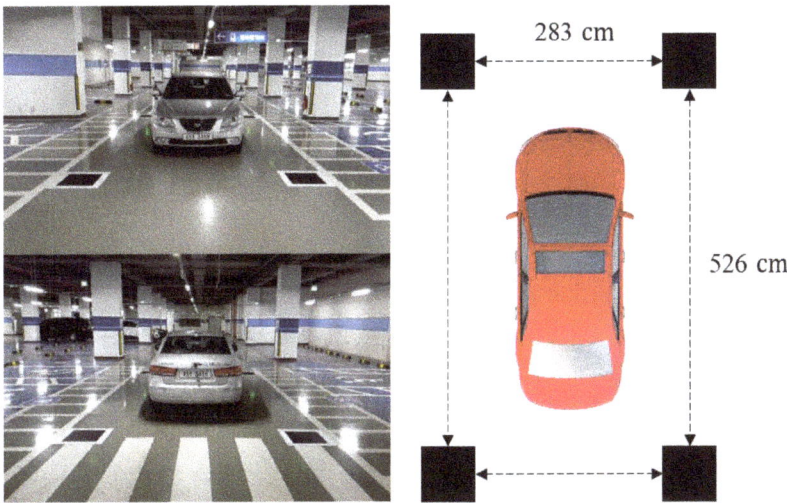

(**a**) Experimental environment   (**b**) Location of calibration boards

Figure 16. Aligned calibration boards in the conventional method.

(**a**) With the proposed method   (**b**) With the conventional method   (**c**) Difference of images between the two results

Figure 17. Experimental results of the proposed method and the conventional method. The magenta and green regions show where the grayscale intensities differ.

Since the RMSE can depend on the content of the source images, we additionally compute optical flow to measure the displacement. We utilize a method of Farneback [37] to compute optical flow. Figure 18 shows the optical flow between the AVM images of the proposed method and the conventional method. The average of the optical flow is 7.1239 px where the resolution of the AVM image is 1170 $px \times$ 1000 px. The RMSE value and the average of the optical flow indicate that the two AVM images are very similar.

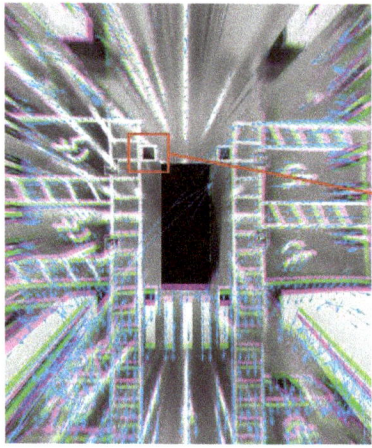

Figure 18. The optical flow between the AVM images of the proposed method and the conventional method, where the blue arrows indicate magnitudes and orientations of the optical flow.

Table 5 shows the AVM errors of the proposed method and the conventional method. We can observe that the results of the proposed method are analogous to those of the conventional method. The AVM error in the conventional method is caused by the measurement data error and the calibration board detection error. The AVM error in the proposed method is caused only by the calibration board detection error, not the measurement data error. Therefore, the AVM error in the conventional method is bound to be larger than that of the proposed method.

Table 5. AVM errors in the proposed method and the conventional method.

Calibration Board	e_{AVM}		e_{LSSE}		e_{AME}	
	Proposed	Conventional	Proposed	Conventional	Proposed	Conventional
front-left	20.3643	44.3193	16.5563	26.0181	3.8080	18.3012
left-rear	24.4981	62.4852	20.4678	40.2469	4.0303	22.2383
rear-right	41.5034	57.8872	27.8327	36.4387	13.6708	21.4485
right-front	67.4672	64.4724	64.0037	46.3019	3.4635	18.1706
average	38.4583	57.291	32.2151	37.2514	6.2431	20.0396

If we used the AVM calibration facility, the measurement data error would be very small, so the AVM error of the conventional method would have been less or similar to those of the proposed method. However, since we experimented in the same environment without the calibration facility, the AVM error of the conventional method is larger than the proposed method.

These evaluations along with the comparison experiments verify that the proposed method is able to generate an AVM image similar to that of the conventional method without requiring the calibration board location.

5. Conclusions

We propose an AVM calibration method using four randomly placed calibration boards and define a novel loss function to utilize the MLE for AVM calibration without the need for information regarding the calibration board locations. The proposed method offers more advantages than the conventional method. The most important advantage is that the proposed method does not require the procedure of measuring the calibration board locations. With this advantage, we can save time and costs that would otherwise be spent on accurately measuring the calibration board locations over a large area. Additionally, as the size of the vehicle increases, the time and cost in using the conventional method also increase, but this is not the case when using the proposed method.

The second advantage of the proposed method is the ability to use the MLE. The most recent AVM calibration method using only two circle-shaped calibration boards cannot utilize the MLE because the MLE requires multiple calibration boards. In contrast, the AVM errors of the proposed method are evenly distributed in all images because we are able to utilize the MLE. The human eye cannot detect the evenly distributed errors.

Flexibility regarding the vehicle size and board size is the third advantage offered by the proposed method. We verify through various experiments that the proposed method can generate AVM images for both real-sized vehicles with large-sized calibration boards and electric vehicles for children with small-sized calibration boards.

Lastly, it is simpler to calibrate AVM systems in the proposed method because there is no need for expert handling facilities for AVM calibration. These advantages were verified through experiments with the vehicle in a parking lot. Based on these advantages, we expect that AVM calibration will be possible in a small repair shop or even in parking lots, resolving the inconvenience of having to visit a large repair shop with AVM facilities for AVM calibration.

Author Contributions: J.H.L. developed the algorithm and performed the experiments. D.-W.L. developed the system architecture and analyzed the experimental results. Both authors together wrote the paper and have read and agreed to the published version of the manuscript.

Funding: This research received no external funding.

Institutional Review Board Statement: Not applicable.

Informed Consent Statement: Not applicable.

Data Availability Statement: The data presented in this study are available on request from the corresponding author. The data are not publicly available due to ethics.

Conflicts of Interest: The authors declare no conflict of interest.

References

1. Ehlgen, T.; Pajdla, T.; Ammon, D. Eliminating blind spots for assisted driving. *IEEE Trans. Intell. Transp. Syst.* **2008**, *9*, 657–665. [CrossRef]
2. Hsu, C.-M.; Chen, J.-Y. Around view monitoring-based vacant parking space detection and analysis. *Appl. Sci.* **2019**, *9*, 3403. [CrossRef]
3. Kim, C.; Cho, S.; Jang, C.; Sunwoo, M.; Jo, K. Evidence filter of semantic segmented image from around view monitor in automated parking system. *IEEE Access* **2019**, *7*, 92791–92804. [CrossRef]
4. Jang, C.; Sunwoo, M. Semantic segmentation-based parking space detection with standalone around view monitoring system. *Mach. Vis. Appl.* **2019**, *30*, 309–319. [CrossRef]
5. Kim, D.; Kim, B.; Chung, T.; Yi, K. Lane-level localization using an AVM camera for an automated driving vehicle in urban environments. *IEEE/ASME Trans. Mechatron.* **2016**, *22*, 280–290. [CrossRef]
6. Lee, J.; Kim, M.; Lee, S.; Hwang, S. Real-time downward view generation of a vehicle using around view monitor system. *IEEE Trans. Intell. Transp. Syst.* **2019**, *21*, 3447–3456. [CrossRef]
7. HYUNDAI MOTOR GROUP. Available online: https://www.hyundai.co.kr/ (accessed on 3 February 2021).
8. Chang, Y.-L.; Hsu, L.-Y.; Chen, O.T.-C. Auto-calibration around-view monitoring system. In Proceedings of the IEEE Vehicular Technology Conference, Las Vegas, NV, USA, 2–5 September 2013.
9. Zhao, J.; Gao, H.; Zhang, X.; Zhang, Y.; Liu, Y. Ring fusion of fisheye images based on corner detection algorithm for around view monitoring system of intelligent driving. *J. Robot.* **2018**, *2018*, 9143290. [CrossRef]
10. Gao, Y.; Lin, C.; Zhao, Y.; Wang, X.; Wei, S.; Huang, Q. 3-D surround view for advanced driver assistance systems. *IEEE Trans. Intell. Transp. Syst.* **2017**, *19*, 320–328. [CrossRef]
11. Yang, Z.; Zhao, Y.; Hu, X.; Yin, Y.; Zhou, L.; Tao, D. A flexible vehicle surround view camera system by central-around coordinate mapping model. *Multimed. Tools Appl.* **2019**, *78*, 11983–12006. [CrossRef]
12. Jeon, B.; Park, G.; Lee, J.; Yoo, S.; Jeong, H. A memory-efficient architecture of full HD around view monitor systems. *IEEE Trans. Intell. Transp. Syst.* **2014**, *15*, 2683–2695. [CrossRef]
13. Lo, W.-J.; Lin, D.-T. Embedded system implementation for vehicle around view monitoring. In Proceedings of the International Conference on Advanced Concepts for Intelligent Vision Systems, Catania, Italy, 26–29 October 2015.
14. Kim, D.M. Auto Revising System for around View Monitoring and Method Thereof. U.S. Patent No. 9,517,725, 13 December 2016.

15. Pyo, J.; Hyun, S.; Jeong, Y. Auto-image calibration for AVM system. In Proceedings of the International SoC Design Conference, Gyungju, Korea, 2–5 November 2015.
16. Natroshvili, K.; Scholl, K.-U. Automatic extrinsic calibration methods for surround view systems. In Proceedings of the IEEE Intelligent Vehicles Symposium, Los Angeles, CA, USA, 11–14 June 2017.
17. Zeng, S.; Zhang, W.; Wang, J. Automatic Calibration of Extrinsic and Intrinsic Camera Parameters for Surround-View Camera System. U.S. Patent No. 9,386,302, 5 July 2016.
18. Ko, Y.-J.; Yi, S.-Y. Catadioptric imaging system with a hybrid hyperbolic reflector for vehicle around-view monitoring. *J. Math. Imaging Vis.* **2018**, *60*, 503–511. [CrossRef]
19. Li, S. Monitoring around a vehicle by a spherical image sensor. *IEEE Trans. Intell. Transp. Syst.* **2006**, *7*, 541–550. [CrossRef]
20. Wang, X.; Chen, H.; Li, Y.; Huang, H. Online extrinsic parameter calibration for robotic camera-encoder system. *IEEE Trans. Ind. Inform.* **2019**, *15*, 4646–4655. [CrossRef]
21. Schneider, S.; Luettel, T.; Wuensche, H.-J. Odometry-based online extrinsic sensor calibration. In Proceedings of the IEEE/RSJ International Conference on Intelligent Robots and Systems, Tokyo, Japan, 3–7 November 2013.
22. Chien, H.-J.; Klette, R.; Schneider, N.; Franke, U. Visual odometry driven online calibration for monocular lidar-camera systems. In Proceedings of the IEEE International Conference on Pattern Recognition, Cancun, Mexico, 4–8 December 2016.
23. Li, M.; Mourikis, A.I. 3-D motion estimation and online temporal calibration for camera-IMU systems. In Proceedings of the IEEE International Conference on Robotics and Automation, Karlsruhe, Germany, 6–10 May 2013.
24. Xu, H.; Wang, X. Camera calibration based on perspective geometry and its application in LDWS. *Phys. Procedia* **2012**, *33*, 1626–1633. [CrossRef]
25. Prakash, C.D.; Akhbari, F.; Karam; L. J. Robust obstacle detection for advanced driver assistance systems using distortions of inverse perspective mapping of a monocular camera. *Robot. Auton. Syst.* **2019**, *114*, 172–186. [CrossRef]
26. Wang, H.; Cai, Y.; Lin, G.; Zhang, W. A novel method for camera external parameters online calibration using dotted road line. *Adv. Robot.* **2014**, *28*, 1033–1042. [CrossRef]
27. De Paula, M.B.; Jung, C.R.; da Silveira, L.G., Jr. Automatic on-the-fly extrinsic camera calibration of onboard vehicular cameras. *Expert Syst. Appl.* **2014**, *41*, 1997–2007. [CrossRef]
28. Lee, J.H.; Lee, D.-W. A hough-space-based automatic online calibration method for a side-rear-view monitoring system. *Sensors* **2020**, *20*, 3407. [CrossRef]
29. Choi, K.; Jung, H.G.; Suhr, J.K. Automatic calibration of an around view monitor system exploiting lane markings. *Sensors* **2018**, *18*, 2956. [CrossRef] [PubMed]
30. Lee, Y.H.; Kim, W.-Y. An automatic calibration method for AVM cameras. *IEEE Access* **2020**, *8*, 192073–192086. [CrossRef]
31. Bradley, D.; Roth, G. Adapting thresholding using the integral image. *J. Graph. Tools* **2007**, *12*, 13–21. [CrossRef]
32. Kong, T.Y.; Rosenfeld, A. *Topological Algorithms for Digital Image Processing*; Elsevier Science: Amsterdam, The Netherlands, 1996.
33. Haralick, R.M.; Shapiro, L.G. *Computer and Robot Vision*; Addison-Wesley: Boston, MA, USA, 1992.
34. Lloyd, S. Least squares quantization in PCM. *IEEE Trans. Inf. Theory* **1982**, *28*, 129–137. [CrossRef]
35. PIXPRO SP360. KODAK Company. Available online: https://kodakpixpro.com/cameras/360-vr/sp360-4k/ (accessed on 3 February 2021).
36. SONATA. HYUNDAI MOTOR Company. Available online: https://www.hyundai.com/kr/en/sedan/sonata/ (accessed on 3 February 2021).
37. Farneback, G. Two-Frame Motion Estimation Based on Polynomial Expansion. In Proceedings of the Scandinavian Conference on Image Analysis, Halmstad, Sweden, 29 June–2 July 2003.

Article

MRI-Based Brain Tumor Classification Using Ensemble of Deep Features and Machine Learning Classifiers

Jaeyong Kang [1], Zahid Ullah [1] and Jeonghwan Gwak [1,2,3,4,*]

1   Department of Software, Korea National University of Transportation, Chungju 27469, Korea; kjysmu@ut.ac.kr (J.K.); zahid@ut.ac.kr (Z.U.)
2   Department of Biomedical Engineering, Korea National University of Transportation, Chungju 27469, Korea
3   Department of AI Robotics Engineering, Korea National University of Transportation, Chungju 27469, Korea
4   Department of IT Convergence (Brain Korea PLUS 21), Korea National University of Transportation, Chungju 27469, Korea
*   Correspondence: jgwak@ut.ac.kr; Tel.: +82-43-841-5852

Abstract: Brain tumor classification plays an important role in clinical diagnosis and effective treatment. In this work, we propose a method for brain tumor classification using an ensemble of deep features and machine learning classifiers. In our proposed framework, we adopt the concept of transfer learning and uses several pre-trained deep convolutional neural networks to extract deep features from brain magnetic resonance (MR) images. The extracted deep features are then evaluated by several machine learning classifiers. The top three deep features which perform well on several machine learning classifiers are selected and concatenated as an ensemble of deep features which is then fed into several machine learning classifiers to predict the final output. To evaluate the different kinds of pre-trained models as a deep feature extractor, machine learning classifiers, and the effectiveness of an ensemble of deep feature for brain tumor classification, we use three different brain magnetic resonance imaging (MRI) datasets that are openly accessible from the web. Experimental results demonstrate that an ensemble of deep features can help improving performance significantly, and in most cases, support vector machine (SVM) with radial basis function (RBF) kernel outperforms other machine learning classifiers, especially for large datasets.

Keywords: deep learning; ensemble learning; brain tumor classification; machine learning; transfer learning

check for
updates

Citation: Kang, J.; Ullah, Z.; Gwak, J. MRI-Based Brain Tumor Classification Using Ensemble of Deep Features and Machine Learning Classifiers. *Sensors* **2021**, *21*, 2222. https://doi.org/10.3390/s21062222

Academic Editors: Changho Lee and Changhan Yoon

Received: 10 February 2021
Accepted: 17 March 2021
Published: 22 March 2021

Publisher's Note: MDPI stays neutral with regard to jurisdictional claims in published maps and institutional affiliations.

1. Introduction

In the human body, the brain is an enormous and complex organ that controls the whole nervous system, and it contains around 100-billion nerve cells [1]. This essential organ is originated in the center of the nervous system. Therefore, any kind of abnormality that exists in the brain may put human health in danger. Among such abnormalities, brain tumors are the most severe ones. Brain tumors are uncontrolled and unnatural growth of cells in the brain that can be classified into two groups such as primary tumors and secondary tumors. The primary tumors present in the brain tissue, while the secondary tumors expand from other parts of the human body to the brain tissue through the bloodstream [2]. Among the primary tumors, glioma and meningioma are two lethal types of brain tumors, and they may lead a patient to death if not diagnosed at an early stage [3]. In fact, the most common brain tumor in humans is glioma [4].

According to the World Health Organization (WHO), brain tumors can be classified into four grades [1]. The grade 1 and grade 2 tumors describe lower-level tumors (e.g., meningioma), while grade 3 and grade 4 tumors consist of more severe ones (e.g., glioma). In clinical practice, the incidence rates of meningioma, pituitary, and glioma tumors are approximately 15%, 15%, and 45%, respectively.

89

There are different ways to treat brain tumors depends on the tumor location, size, and type. Presently, the most common treatment for brain tumors is surgery as it has no side effects on the brain [5]. Different types of medical imaging technologies such as computed tomography (CT), positron emission tomography (PET), and magnetic resonance imaging (MRI) are available that are used to observe the internal parts of the human body conditions. Among all these imaging modalities, MRI is considered most preferable as it is the only non-invasive and non-ionizing modality that offers valuable information in 2D and 3D formats about brain tumor type, size, shape, and position [6]. However, manually reviewing these images is time-consuming, hectic, and even prone to error due to the influx of patients [7]. To address this problem, the development of an automatic computer-aided diagnosis (CAD) system is required to alleviate the workload of the classification and diagnosis of brain MRI and act as a tool for helping radiologists and doctors.

Several efforts have been made to develop a highly accurate and robust solution for the automatic classification of brain tumors. However, due to high inter and intra shape, texture, and contrast variations, it remains a challenging problem. The traditional machine learning (ML) techniques rely on handcrafted features, which restrains the robustness of the solution. Whereas the deep learning-based techniques automatically extract meaningful features which offer significantly better performance. However, deep learning-based techniques require a large amount of annotated data for training, and acquiring such data is a challenging task. To overcome these issues, in this study, we proposed a hybrid solution that exploits (1) various pre-trained deep convolutional neural networks (CNNs) as feature extractors to extract powerful and discriminative deep features from brain magnetic resonance (MR) images, and (2) various ML classifiers to identify the normal and abnormal brain MR images. Also, to investigate the benefits of combining features from different pre-trained CNN models, we designed the novel feature ensemble method for the MRI-based brain tumor classification task. We proposed the novel feature evaluation and selection mechanism where the deep features from 13 different pre-trained CNNs are evaluated using 9 different ML classifiers and selected based on our proposed feature selection criteria. In our proposed framework, we concatenated the selected top three deep features from three different CNNs to form a synthetic feature. The concatenation process integrates the information from different CNNs to create a more discriminative feature representation than using the feature extracted from a single CNN model since different CNN architectures can capture diverse information in brain MR images. An ensemble of deep features is then fed into several ML classifiers to predict the final output, whereas most of the previous works have employed traditional feature extraction techniques [8]. In our experiment, we provided an extensive evaluation using 13 different pre-trained deep convolutional neural networks and 9 different ML classifiers on three different datasets: (1) BT-small-2c, the small dataset with 2 classes (normal/tumor), (2) BT-large-2c, the large dataset with 2 classes (normal/tumor), and (3) the large dataset with 4 classes (normal, glioma tumor, meningioma tumor, and pituitary tumor) for brain tumor classification. Our experiment results demonstrate that the ensemble of deep features can help improving performance significantly. In summary, our contributions are listed as follows:

- We designed and implemented a fully automatic hybrid scheme for brain tumor classification, which uses both (1) the pre-trained CNN models to extract the deep features from brain MR images and (2) ML classifiers to classify brain tumor type effectively.
- We proposed a novel method which consists of three steps: (1) extract deep features using pre-trained CNN models for meaningful information extraction and better generalization, (2) select the top three performing features using fined-tuned several ML models for our task, and (2) combine them to build the ensemble model to achieve state-of-the-art performance for brain tumor classification in brain MR images.
- We conducted extensive experiments on 13 different pre-trained CNN models and 9 different ML classifiers to compare the effectiveness of each pre-trained CNN model and each ML classifier on three different brain MRI datasets: (1) BT-small-2c, the small dataset with 2 classes (normal/tumor), (2) BT-large-2c, the large dataset with

2 classes (normal/tumor), and (3) the large dataset with 4 classes (normal, glioma tumor, meningioma tumor, and pituitary tumor) for brain tumor classification.

The layout of this study is organized as follows: The related work is given in Section 2. The proposed method is presented in Section 3. The experimental settings and results are shown in Section 4. The conclusion section is described in Section 5.

2. Related Work

Numerous techniques have been proposed for automatic brain MRI classification based on traditional ML and deep learning methods as shown in Table 1.

The traditional ML methods are comprised of several steps: pre-processing, feature extraction, feature reduction, and classification. In traditional ML methods, feature extraction is a core step as the classification accuracy relies on extracted features. There are two main types of feature extraction. The first type of feature extraction is low-level (global) features, for instance, texture features and intensity, first-order statistics (e.g., mean, standard deviation, and skewness), and second-order statistics such as gray-level co-occurrence matrix (GLCM), wavelet transform (WT), Gabor feature, and shape. For instance, Selvaraj et al. [9] employed first-order and second-order statistics using least square support vector machine (SVM) and develop a binary classifier to classify the normal and abnormal brain MR images. John et al. [10] used GLCM and discrete wavelet transformation-based methods for tumor identification and classification. The low-level features represent the image efficiently; however, the low-level features and their representation capacity are limited since most brain tumors have similar appearances such as texture, boundary, shape, and size. Ullah et al. [8] extracted the approximation and detail coefficient of level-3 decomposition using DWT, reduced the coefficient by employing color moments (CM), and finally employed a feed-forward artificial neural network to identify the normal and abnormal brain MR images.

The second type of feature extraction is the high-level (local) features, such as fisher vector (FV), scale-invariant feature transformation (SIFT), and bag-of-words (BoW). Different researchers have employed BoW for medical image retrieval and classification. Such as the classification of breast tissue density in mammograms [11], X-ray images retrieval and classification on pathology and organ levels [12], and content-based retrieval of brain tumor [13]. Cheng et al. [14] employed FV to retrieve the brain tumor. The statistical features extracted from SIFT, FV, and BoW are high-level features formulated on a local scale that does not consider spatial information. Hence, it is noticeable that in the traditional ML method, there are two main problems in the feature extraction stage. First, it only focuses on either high-level or low-level features. Second, the traditional ML method depends on handcrafted features, which need strong prior information such as the location or position of the tumor in an image, and there are high chances of human errors. Therefore, it is essential to develop a method to combine both high-level and low-level features without using handcrafted features.

Most of the existing works in medical MR imaging refers to automatic segmentation of tumor region. Recently, Numerous researchers have proposed different techniques to detect and segment the tumor region in MR images [15–17]. Once the tumor in MRI is segmented, these tumors need to be classified into different grades. In previous research studies, binary classifiers have been employed to identify the benign and malignant classes [8,18,19]. For instance, Ullah et al. [8] proposed a hybrid scheme for the classification of brain MR images into normal and abnormal using histogram equalization, Discrete wavelet transform, and feed-forward artificial neural network, respectively. Kharrat et al. [18] categorize the brain tumor into normal and abnormal using a genetic algorithm and support vector machine. Besides, Papageorgiou et al. [19] categorized the high-grade and low-grade gliomas based on fuzzy cognitive maps and attained 93.22% and 90.26% accuracy for high-grade and low-grade brain tumors, respectively.

Table 1. Work Related to Brain Tumor Classification.

Author	Type of Solution	Classification Method	Objective	Dataset	Feature Extraction Method	Accuracy
Rajan and Sundar, 2019		Support vector machine (SVM)	Tumor detection and segmentation	41 magnetic resonance (MR) images	Adaptive Gray-Level Co-Occurrence Matrix (AGLCM)	98%
Kharrat et al., 2010		Hybrid method-Genetic algorithm with SVM	Classification of brain tumor into normal, malignant, and benign tumor	83 MR images	Wavelet-based features	98.14%
Shree and Kumar, 2018		Probabilistic neural network (PNN)	Classification of brain MRI into normal and abnormal	650 MR images	Gray level co-occurrence matrix	95%
Arunachalam and Royappan, 2017	Classical Machine Learning-based Solutions	Feed-forward back propagation neural network	Classification of brain MRI into normal and abnormal	230 MR images	Gabor, GLCM, and discrete wavelet transform (DWT)	99.8%
Ullah et al., 2020		Feed-forward neural network	Classification of brain MRI intonormal and abnormal	71 MR images	DWT	95.8%
B. Ural, 2018		PNN	Brain tumor detection	25 MR images	k-mean with fuzzy c-mean (KMFCM)	90%
Preethi and Ashwarya, 2019		Deep neural network (DNN)	Classification of tumor and non-tumor image	20 MR images	GLCM + Wavelet GLCM	99.3%
Francisco et al., 2021		Multi-pathway convolutional neural network (CNN)	Brain tumor classification	3064 MR images	CNN	97.3%
Deepak and Ameer, 2019		Deep transfer learning	Classification of glioma, meningioma, and pituitary tumors	3064 MR images	GoogleNet	98%
Ahmet and Mohammad, 2020		CNN models	Brain tumor detection and classification	253 MR images	CNN	97.2%
Das et al., 2019	Advanced Deep Learning-based Solutions	CNN	Brain tumor classification	3064 MR images	CNN	94.39%
Saed et al., 2017		CNN	Classification of brain MRI into normal and abnormal	587 MR images	CNN	91.16%
Saxena et al., 2019		CNN networks with transfer learning	Binary classification of brain tumor into normal and abnormal	253 MR images	CNN	95%
Paul et al., 2017		Fully connected and CNN	Brain tumor classification	3064 MR images	CNN	91.43%
Hemanth et al., 2019		CNN	MR brain image classification into normal and abnormal	220 MR images	CNN	94.5%

Shree and Kumar [20] divided the brain MRI into two classes: normal and abnormal. They used GLCM for feature extraction, while a probabilistic neural network (PNN) classifier has been employed to classify the brain MR image into normal and abnormal and obtained 95% accuracy. Arunachalam and Savarimuthu [21] proposed a model to categorize the normal and abnormal brain tumor in brain MR images. Their proposed model comprised enhancement, transformation, feature extraction, and classification. First, they have enhanced the brain MR image using shift-invariant shearlet transform (SIST). Then, they extracted the features using Gabor, grey level co-occurrence matrix (GLCM), and discrete wavelet transform (DWT). Finally, these extracted features were then fed into feed-forward backpropagation neural network and obtained a high accuracy rate. Rajan and Sundar [22] proposed a hybrid energy-efficient method for automatic tumor detection and segmentation. Their proposed method is comprised of seven long phases and reported 98% accuracy. The main drawback of their proposed model is high computation time due to the use of numerous techniques.

Since the last decade, deep learning methods have been widely used for brain MRI classification [23,24]. The deep learning method does not need handcrafted (manually) extracted features as it embedded the feature extraction and classification stage in self-learning. The deep learning method requires a dataset where sometimes a pre-processing operation needs to be done, and then salient features are determined in a self-learning manner [25]. In MR imaging classification, a key challenge is to reduce the semantic gap between the high-level visual information perceived by the human evaluator and the low-level visual information captured by the MR imaging machine. To reduce the semantic gap, the convolutional neural networks (CNNs), one of the famous deep learning techniques for image data, can be used as a feature extractor to capture the relevant features for the classification task. Feature maps in the initial layers and higher layers of CNNs models extract low-level features and high-level content (domain) specific features, respectively. Feature maps in the earlier layer construct simple structural information, for instance, shape, textures, and edges, whereas higher layers combine these low-level features to construct (encode) efficient representation, which integrates global and local information.

Recently, different researchers have used CNNs for brain MRI classification and validated their proposed methodology on brain tumor classification datasets [26–28]. Deepak and Ameer [29] used a pre-trained GoogLeNet to extract features from brain MR images with deep CNN to classify three types of brain tumor and obtained 98% accuracy. Ahmet and Muhammad [30] used different CNN models such as GoogLeNet, Inception V3, DenseNet-201, AlexNet, and ResNet-50 to classify the brain MR images and obtained reasonable accuracies. They modified pre-trained ResNet-50 CNN by removing its last 5 layers and added new 8 layers, and obtained 97.2% accuracy with this model, which is the highest accuracy among all pre-trained models. Khwaldeh et al. [31] proposed a CNN model to classify the normality and abnormality of brain MR images as well as high-grade and low-grade glioma tumors. They have modified the AlexNet CNN model and used it as their network architecture, and they obtained 91% accuracy. Despite the valuable works being done in this area, developing a robust and practical method still requires more effort to classify brain MR images. Saxena et al. [32] used Inception V3, ResNet-50, and VGG-16 models with transfer learning methods to classify brain tumor data. The ResNet-50 model obtained the highest accuracy rate with 95%. In studies [33,34] CNN architectures have been introduced to classify brain tumors. In these architectures, the convolution neural network extracts the features from brain MRI using convolution and pooling operations. The main purpose of these proposed models is to find the best deep learning model that accurately classifies the brain MR images. Francisco et al. [35] presented a multi-pathway CNN architecture for automatic brain tumor segmentation such as glioma, meningioma, and pituitary tumor. They have evaluated their proposed model using a publicly available T1-weighted contrast-enhanced MRI dataset and obtained 97.3% accuracy. However, their training procedure is quite expensive. Raja et al., [36] proposed a hybrid deep autoencoder (DAE) for brain tumor classification using the Bayesian fuzzy clustering (BFC) approach. Initially, they have used a non-local mean filter to remove the noise from the image. Then the BFC approach is employed for brain tumor segmentation. Furthermore, some robust features were extracted using scattering transform (ST), information-theoretic measures, and wavelet packet Tsallis entropy (WPTE). Eventually, a hybrid scheme of DAE is employed for brain tumor classification and achieved high accuracy. The main drawback of this approach is, it requires high computation time due to the complex proposed model.

In summary, as observed from the above research studies, the acquired accuracies using deep learning techniques for brain MRI classification are significantly high as compared to traditional ML techniques. However, the deep learning models require a massive amount of data for training in order to perform better than traditional ML techniques.

It is clearly seen from recently published studies that deep learning techniques have become one of the mainstream of expert and intelligent systems and medical image analysis. Furthermore, the techniques mentioned earlier have certain limitations which should be considered while working with brain tumor classification and segmentation. The major drawback of the previously proposed systems is that they only consider binary classification (normal and abnormal) MR image dataset and ignore the multi-class dataset [37]. In the pre-screening stage of a patient, binary class classification is required for physicians and radiologists, where the physicians take further action based on binary class classification. Preethi and Aishwarya [38] proposed a model to classify the brain tumor based on multiple stages. They combined the wavelet-based gray-level co-occurrence matrix and GLCM to produce the feature matrix. The extracted features were further reduced using the oppositional flower pollination algorithm (OFPA). Finally, the deep neural network is employed to classify the MR brain image based on the selected features and obtained 92% accuracy. Ural [39] initially enhanced the brain MRI using different image processing techniques. Also, different segmentation process has been mixed for boosting the performance of the solution. Further, the PNN method is employed to detect and localize the tumor area in the brain. The computational time of their proposed method is quite low and also the acquired accuracy rate is quite reasonable.

3. Proposed Methods

In this section, the overall architecture of our proposed method is first described. After that, we describe the details of four key components in the following subsections.

The architecture of our proposed method for brain tumor classification is illustrated in Figure 1. First, input MR images are pre-processed (e.g., brain cropping, resize, and augmentation) before feeding into the model (Section 3.1). Second, the pre-processed images are used as the input of pre-trained CNN models as feature extractors (Section 3.2). The extracted features from pre-trained CNN models are evaluated by several ML classifiers. (Section 3.3). The top three deep features are selected based on evaluation results from the classifiers (Section 3.4). The top three deep features are concatenated in our ensemble module, and the concatenated deep features are further used as an input to ML classifiers to predict final output (Section 3.5).

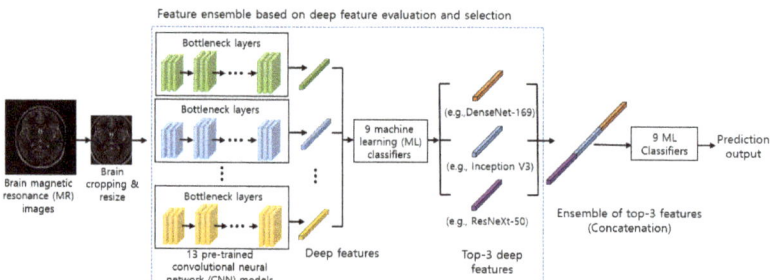

Figure 1. Architecture of our proposed model using feature ensemble based on deep feature evaluation and selection.

3.1. Image Pre-Processing

Almost every image in our brain MRI datasets contains undesired spaces and areas, leading to poor classification performance. Hence, it is necessary to crop the images to remove unwanted areas and use only useful information from the image. We use the cropping method in [40] which uses extreme point calculation. The step to crop the MR images using extreme point calculation is shown in Figure 2. First, we load the original MR images for pre-processing. After that, we apply thresholding to the MR images to convert them into binary images. Also, we perform the dilation and erosions operations to remove the noise of images. After that, we selected the largest contour of the threshold images and calculated the four extreme points (extreme top, extreme bottom, extreme right, and extreme left) of the images. Lastly, we crop the image using the information of contour and extreme points. The cropped tumor images are resized by bicubic interpolation. The specific reason to choose the bicubic interpolation is that it can create a smoother curve than other interpolation methods such as bilinear interpolation and is a better choice for MR images since there is a large amount of noise along the edges.

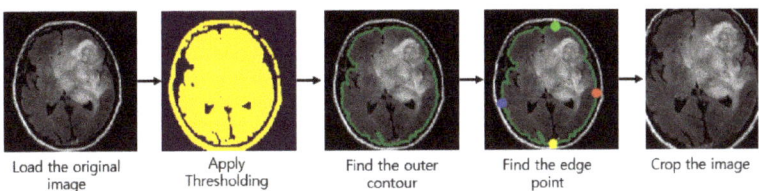

| Load the original image | Apply Thresholding | Find the outer contour | Find the edge point | Crop the image |

Figure 2. Step to crop the magnetic resonance (MR) images.

Also, we used image augmentation since the size of our MRI dataset is not very large. Image augmentation is the technique that creates an artificial dataset by modifying the original dataset. It is known as the process of creating multiple copies of the original image with different scales, orientation, location, brightness, and so on. It is reported that the classification accuracy of the model can be improved by augmenting the existing data rather than collecting new data.

In our image augmentation step, we used 2 augmentation strategies (rotation and horizontal flipping) to generate new training sets. The rotation operation used for data augmentation is done by randomly rotating the input by 90 degrees zero or more times. Also, we applied horizontal flipping to each of the rotated images.

Since the MR images in our dataset are of different width, height, and sizes, it is recommended to resize them to equal width and height to get optimum results. In this work, we resize the MR images to the size of either 224 × 224 (or 299 × 299) pixels since input image dimensions of pre-trained CNN models are 224 × 224 pixels except for the Inception V3, which requires the input images with size 299 × 299.

3.2. Deep Feature Extraction Using Pre-Trained CNN Models

3.2.1. Convolutional Neural Network

CNN is a class of deep neural networks that uses the convolutional layers for filtering inputs for useful information. The convolutional layers of CNN apply the convolutional filters to the input for computing the output of neurons that are connected to local regions in the input. It helps in extracting the spatial and temporal features in an image. A weight-sharing method is used in the convolutional layers of CNN to reduce the total number of parameters [41,42].

CNN is generally comprised of three building blocks: (1) a convolutional layer to learn the spatial and temporal features, (2) a subsampling (max-pooling) layer to reduce or downsample the dimensionality of an input image, and (3) a fully connected (FC) layer for classifying the input image into various classes. The architecture of CNN is shown in Figure 3.

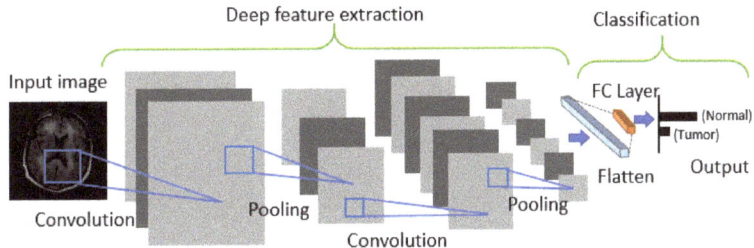

Figure 3. Architecture of Convolutional Neural Networks.

3.2.2. Transfer Learning

Generally, CNN has better performance in a larger dataset than a smaller one. Transfer learning can be used when it is not feasible to create a large training dataset. The concept of transfer learning can be depicted in Figure 4, where the model pre-trained on large benchmark datasets (e.g., ImageNet [43]) can be used as a feature extractor for the different task with a relatively smaller dataset such as an MRI dataset. In recent years, transfer learning technique has been successfully applied in various domains, such as medical image classification and segmentation, and X-ray baggage security screening [44–47]. This reduces the long training time that is normally required for training deep learning models from scratch and also removes the requirement of having a large dataset for the training model [48,49].

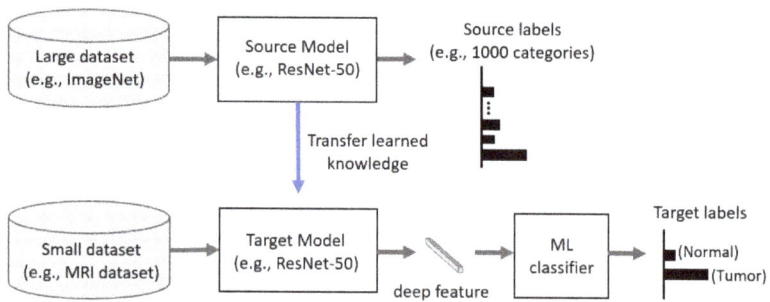

Figure 4. Concept of transfer learning.

3.2.3. Deep Feature Extraction

In this study, we use a CNN-based model as a deep learning-based feature extractor since it can capture the important features without any human supervision. Also, we use a transfer learning-based approach to build our feature extractor since our MRI dataset is not very large and training and optimizing deep CNN such as DenseNet-121 from scratch is often not feasible. Hence, we use the fixed weights of each CNN model pre-trained on a large ImageNet dataset to extract the deep features of brain MR images.

The pre-trained CNN models used in our study are ResNet [50], DenseNet [51], VGG [52], AlexNet [53], Inception V3 [54], ResNeXt [55], ShuffleNet V2 [56], MobileNet V2 [57], and MnasNet [58]. The extracted deep features are then fed into the ML classifiers, including neural networks with a FC layer as a traditional deep learning approach using CNN as shown in Figure 3 to predict the output.

3.3. Machine Learning Classifiers for Brain Tumor Classification

The extracted deep features from pre-trained CNN models are used as an input of several ML classifiers, including neural networks with an FC layer, Gaussian Naïve Bayes (Gaussian NB), Adaptive Boosting (AdaBoost), K-Nearest Neighbors (k-NN), Random forest (RF), SVM with three different kernels: linear, sigmoid, and radial basis function (RBF), Extreme Learning Machine (ELM). We implemented these ML classifiers using the scikit-learn ML library [59]. These ML classifiers and their hyper-parameter settings used in our experiments for brain tumor classification are discussed in the following subsections.

3.3.1. Fully Connected Layer

In neural networks with an FC layer, which is the traditional deep learning approach, the loss function is defined to calculate the loss, which is a prediction error of the neural network. The loss is used to calculate the gradients to update the weights of the neural network as a training step. In our training step of the FC classifier, we use the cross-entropy loss function, which is the most commonly used loss function for CNN and other neural networks. It calculates the loss between the soft target estimated by the softmax function and the ground-truth label to learn our model parameters as follows:

$$L(y, z) = \sum_{i=0}^{M} -y_i \log\left(\frac{z_i}{\sum_j exp(z_i)}\right) \tag{1}$$

where M is the total number of class, for instance, M is set to 2 when the classifier is trained on the two MRI datasets, BT-small-2c and BT-large-2c, which contain two classes (normal and tumor) of MR images, and M is set to 4 when the classifier is trained on the MRI dataset, BT-large-4c, which contains four classes (normal, glioma tumor, meningioma tumor, and pituitary tumor) of MR images (See Section 4.1 for the details of these datasets), y is a one-hot encoded vector representing the ground-truth label of the training set as 1 and all other elements as 0, and z_i is the logit which is the output of the last layer for the i-th class of the model.

In this work, we update the weight of the layers via Adaptive Moment Estimation (Adam), the optimizer that calculates the adaptive learning rates of every parameter. The learning rate is set to 0.001. We run each of the methods for 100 epochs. We collect the highest average accuracy for our test dataset for each run.

3.3.2. Gaussian Naïve Bayes

Naïve Bayes classifier is the ML classifier with the assumption of conditional independence between the attributes given the class. In this work, we use Gaussian NB classifier as one of our ML classifiers for brain tumor classification. In Gaussian NB classifier, the conditional probability $P(y \mid X)$ is calculated as a product of the individual conditional probabilities using the naïve independence assumption as follows:

$$P(y|X) = \frac{P(y)P(X|y)}{P(X)} = \frac{P(y)\prod_{i=1}^{n} P(x_i|y)}{P(X)} \tag{2}$$

where X is given data instance (extracted deep feature from brain MR image) which is represented by its feature vector $(x_1, ..., x_n)$, y is a class target (type of brain tumor) with two classes (normal and tumor) for two MRI datasets, BT-small-2c and BT-large-2c, or four classes (normal, glioma tumor, meningioma tumor, and pituitary tumor) for BT-large-4c dataset. Since $P(X)$ is constant, the given data instance can be classified as follows:

$$\hat{y} = \arg\max_y P(y) \prod_{i=1}^{n} P(x_i|y) \tag{3}$$

where $(x_i|y)$ is calculated assuming that the likelihood of features to be Gaussian as follows:

$$P(x_i|y) = \frac{1}{\sqrt{2\pi\sigma_y^2}} exp(\frac{(x_i - \mu_y)^2}{2\sigma_y^2}) \tag{4}$$

where the parameters μ_y and σ_y are estimated using maximum likelihood.

In this work, the smoothing variable representing the portion of the largest variance of all features that are added to variances for calculation stability is set to 10^{-9}, the default value of the scikit-learn ML library.

3.3.3. AdaBoost

AdaBoost, proposed by Freund and Schapire [60], is an ensemble learning algorithm that combines multiple classifiers to improve performance. AdaBoost classifier builds a well-performing strong classifier by combining multiple weak classifiers using the iterative ensemble method. The underlying idea of Adaboost is to set the weights of classifiers and train the data sample in each boosting iteration to accurately predict a class target (a type of brain tumor) of a given data instance (extracted deep feature from brain MR image) with two classes (normal and tumor) for two MRI datasets, BT-small-2c and BT-large-2c, or four classes (normal, glioma tumor, meningioma tumor, and pituitary tumor) for BT-large-4c dataset. Any ML classifier that accepts the weights on the training set can be used as a base classifier.

In this work, we adopt the decision tree classifier as our base classifier since it is a commonly used base classifier for AdaBoost. Also, the number of the estimator is set to 150.

3.3.4. K-Nearest Neighbors

k-NN is one of the simplest classification techniques. It performs predictions directly from the training set that is stored in the memory. For instance, to classify a new data instance (a deep feature from brain MR image), k-NN chooses the set of k objects from the training instances that are closest to the new data instance by calculating the distance and assigns the label with two classes (normal or tumor) or four classes (normal, glioma, meningioma, and pituitary tumor) and does the selection based on the majority vote of its k neighbors to the new data instance.

Manhattan distance and Euclidean distance are the most commonly used to measure the closeness of the new data instance with the training data instances. In this work, we used the Euclidean distance measure for the k-NN algorithm. Euclidean distance d between data point x and data point y are calculated as follows:

$$d(x, y) = \sqrt{\left(\sum_{i=1}^{N}(x_i - y_i)^2\right)} \qquad (5)$$

The brief summary of k-NN algorithm is illustrated below:

- First select a suitable distance metric.
- Store all the training data set P in pairs in the training phase as follows:

$$P = (y_i, c_i), i = 1, ..., n \qquad (6)$$

where in the training dataset, y_i is a training pattern, n is the amount of training patterns and c_i is its corresponding class.

- In the testing phase, compute the distances between the new features vector and the stored (training data) features, and classify the new class example by a majority vote of its k neighbors.

The correct classification given in the test phase is used to evaluate the accuracy of the algorithm. If the result is not satisfactory, the k value can be adjusted until a reasonable level of accuracy is obtained. It is noticeable here that we set the number of neighbors from 1 to 4 and selected the one with the highest accuracy.

3.3.5. Random Forest

RF, proposed by Breiman [61], is an ensemble learning algorithm that builds multiple decision trees using the bagging method to classify new data instance (a deep feature of brain MR image) to a class target (a type of brain tumor) with two classes (normal and tumor) for two MRI datasets, BT-small-2c and BT-large-2c, or four classes (normal, glioma tumor, meningioma tumor, and pituitary tumor) for BT-large-4c dataset. RF selects random n attributes or features to find the optimal split point using the Gini index as a cost function while creating the decision trees. This random selection of the attributes or features can reduce the correlation among the trees and have lower ensemble error rates. The new observation is fed into all classification trees of the RF for predicting a class target (a type of brain tumor) of the new incoming data instance. RF counts the numbers of predictions for each class and selects the class with the largest number of votes as the class label for the new data instance.

In this work, the number of features to consider when looking for the best split is set to the square root of the total number of features. Also, we set the number of trees from 1 to 150 and selected the one with the highest accuracy.

3.3.6. Support Vector Machine

SVM, proposed by Vapnik [62], is one of the most powerful classification algorithms. SVM uses the kernel function, $K(x_n, x_i)$, to transform the original data space into an another space with a higher dimension. The hyperplane function for separating the data can be defined as follows:

$$f(x_i) = \sum_{n=1}^{N} \alpha_n y_n K(x_n, x_i) + b \tag{7}$$

where x_n is support vector data (deep features from brain MR image), α_n is Lagrange multiplier, and y_n represent a target class of these three datasets employed in this paper, such that the two datasets are binary (normal and abnormal) class datasets, while the third dataset has four different classes (normal, glioma, meningioma, and pituitary tumor) with $n = 1, 2, 3, ..., N$.

In this work, we used the most commonly used kernel functions at the SVM algorithm: (1) linear kernel, (2) sigmoid kernel, and (3) RBF kernel. Table 2 shows the details of three kernels. Also, SVM has two key hyper-parameters, C and Gamma. C is the hyper-parameter for the soft margin cost function that controls the influence of each support vector. Gamma is the hyper-parameter that decides how much curvature we want in a decision boundary. We set the gamma and C values to [0.00001, 0.0001, 0.001, 0.01] and [0.1, 1, 10, 100, 1000, 10000], respectively, and selected the combination of gamma and C values with the highest accuracy.

Table 2. Kernel types and their required parameters.

Kernel	Equation	Parameters
Linear	$K(x_n, x_i) = (x_n, x_i)$	-
Sigmoid	$K(x_n, x_i) = tanh(\gamma(x_n, x_i) + C)$	γ, C
RBF	$K(x_n, x_i) = exp(-\gamma\|x_n - x_i\|^2 + C)$	γ, C

3.3.7. Extreme Learning Machine (ELM)

Extreme Learning Machine (ELM) is a simple learning algorithm for single-hidden layer feed-forward neural networks (SLFNs). ELM was initially proposed by Huang et al. [63] to overcome the limitations of traditional SLFNs learning algorithms, such as poor generalization effectiveness, irrelevant parameter tuning, and slow learning speed. ELM has shown a considerable ability for regression and classification tasks with good generalization performance.

In ELM, the output of a SLFN with $\tilde{N}$ hidden nodes given N distinct training samples, can be represented as follows:

$$o_j = \sum_{i=1}^{\tilde{N}} \beta_i f_i(x_j) = \sum_{i=1}^{\tilde{N}} \beta_i f(x_j; a_i, b_i), j = 1, ..., N \tag{8}$$

where o_j is the output vector of the SLFN, which represents the probability of the input sample x_i (deep features from brain MR image) belonging to a class target (type of brain tumor) with two classes (normal and tumor) for two MRI datasets, BT-small-2c and BT-large-2c, or four classes (normal, glioma tumor, meningioma tumor, and pituitary tumor) for BT-large-4c dataset, a_i and b_i are learning parameters generated randomly of the j-th hidden node, respectively, β_i is the link connecting the j-th hidden node and the output nodes, and $f(x_j; a_i, b_i)$ is the activation function of ELM.

The ELM learning algorithm can be explained in 3 steps. First, the parameters (weights and biases) of all neurons are randomly initialized. Second, the hidden layer output matrix of the neural network H is calculated. Third, the output weight, β is calculated as follows:

$$\beta = H'T \tag{9}$$

where H' is the Moore-Penrose generalized inverse of matrix H (the hidden layer output matrix), which can be obtained by minimum-norm least-squares solution, and T is the target matrix corresponding to H.

In this work, the number of the hidden layer is set to [5000, 6000, 7000, 8000, 9000, 10,000], and select the one with the highest accuracy.

3.3.8. Discussion

Several efforts have been made to develop a highly accurate and robust solution for MRI-based brain tumor classification using various ML classifiers: neural network classifier [8,21,64], Naïve Bayes classifier [65], AdaBoost classifier [66], k-NN classifier [64], RF classifier [64,67], SVM classifier [18,22], and ELM classifier [68]. However, there have been no studies done on evaluating the effectiveness of ML classifiers for the MRI-based brain tumor classification task. Hence, in our study, we use 9 well-known different ML classifiers to examine which ML classifier performs well for the MRI-based brain tumor classification task.

Since the performance of ML classifiers are highly dependent on input feature map, designing a method to produce a discriminative and informative feature from brain MR images plays a key role to successfully build the model for MRI-based brain tumor classification. In recent years, several studies proposed deep-learning-based feature extraction methods for MRI-based brain tumor classification using pre-trained deep CNN models: ResNet-50 [69,70], ResNet-101 [71], DenseNet-121 [70,72], VGG-16 [69,70], VGG-19 [70,73], AlexNet [74], Inception V1 (GoogLeNet) [29], Inception V3 [69,75], and MobileNet V2 [76]. However, no study has been carried out to evaluate the effectiveness of several pre-trained deep CNN models as a feature extractor for MRI-based brain tumor classification task. Hence, we use 13 different pre-trained deep CNN models to examine which pre-trained CNN models are useful as a feature extractor for MRI-based brain tumor classification task.

3.4. Deep Feature Evaluation and Selection

We evaluate each deep feature extracted from 13 different pre-trained CNNs using 9 different ML classifiers (FC, Gaussian NB, AdaBoost, k-NN, RF, SVM-linear, SVM-sigmoid, SVM-RBF, and ELM) described in Section 3.3 and choose the top three deep features based on the average accuracy of 9 different ML classifiers for each of our 3 different MRI datasets. In case the accuracy of two or more deep features is the same, we choose the one with the lowest standard deviation. Also, if there are more than 2 deep features extracted from two homogeneous pre-trained models (e.g., DenseNet-121 and DenseNet-169) among the top three features, we exclude the one with lower accuracy and choose the next best deep feature. The reason for doing this is that the deep features extracted from two homogeneous models share similar feature spaces. Hence, the ensemble of these features has redundant feature space and a lack of diversity. The top three deep features are fed into our ensemble module described in the following sub-section.

3.5. Ensemble of Deep Features

Ensemble learning aims at improving the performance and prevents the risk of using a single feature extracted from one model with a poor performance by combining multiple features from several different models into one predictive feature. Ensemble learning can be divided into feature ensemble and classifier ensemble depending on integration level. Feature ensemble involves integrating feature sets that are further fed to the classifier for final output, while classifier ensemble involves integrating output sets from classifiers where voting methods determine the final output. Since the feature set contains richer information about the MR images than the output set of each classifier, integration at this level is expected to provide better classification results. Hence, in this work, we use feature ensemble as our ensemble learning.

In our ensemble module, we concatenate the top three deep features from three different pre-trained CNNs as one sequence. For instance, in Figure 1, the top three deep features are DenseNet-169, Inception V3, and ResNeXt-50, and these features are concatenated into one sequence as our feature-level ensemble step. The concatenated deep feature is further fed to ML classifiers for predicting the final output. Also, we concatenate all the possible combinations of two features from the top three features, which is further fed to ML classifiers to compare with the model using the ensemble of the top three features in our experiments.

4. Experiments and Results

4.1. Dataset

We perform a set of experiments on three different brain MRI datasets which are publicly available for the tasks of brain tumor classification. The first dataset of brain MR images was downloaded from the Kaggle website [77], and for our simplicity, we named this dataset BT-small-2c. The BT-small-2c dataset comprises 253 images, out of which 155 images contain tumors while the remaining 98 images are without tumors. The second dataset was also downloaded from the Kaggle website, namely Brain Tumor Detection 2020 [78], and we call it BT-large-2c. This database comprises 3000 images, out of which 1500 images contain tumors while the remaining 1500 images are without tumors. The third dataset consists of 3064 T1-weighted images containing three different types of brain tumors such as gliomas, meningiomas, and pituitary tumors. The dataset was acquired from the Kaggle website [37], and we named this dataset as BT-large-4c. The BT-small-2c and BT-large-2c datasets contain brain MR images with two classes (normal and tumor). The BT-large-4c dataset contains brain MR images with four classes (normal, glioma tumor, meningioma tumor, and pituitary tumor). Each dataset is subdivided into a training set (80% of the total dataset) and a test set (20% of the total dataset). Table 3 shows details of the dataset used in our experiments. The examples of brain MR images in BT-small-2c, BT-large-2c, and BT-large-4c datasets are shown in Figure 5.

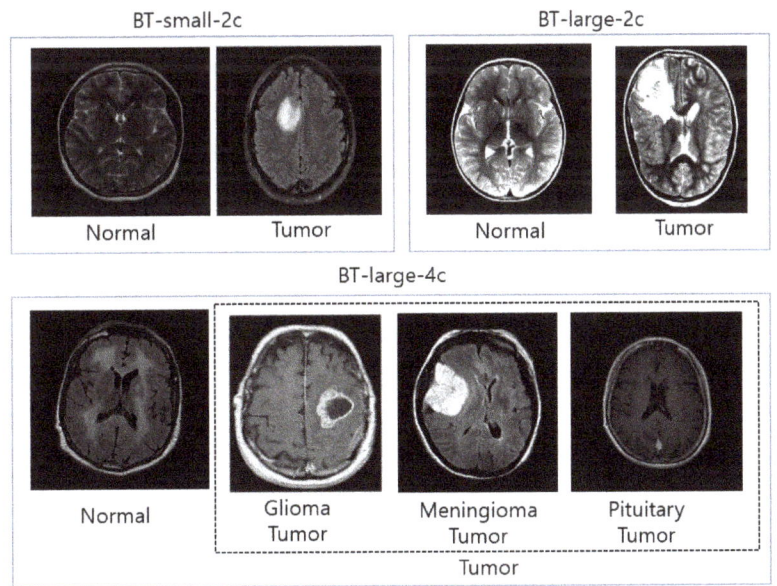

Figure 5. The examples of brain MR images in BT-small-2c, BT-large-2c, and BT-large-4c datasets.

Table 3. Details of the dataset.

Types	Number of Class	Training Set	Test Set
BT-small-2c	2	202	51
BT-large-2c	2	2400	600
BT-large-4c	4	2611	653

4.2. Experimental Setting

In our experiment, we use 13 different pre-trained deep convolutional neural networks as a feature extractor: ResNet-50, ResNet-101, DenseNet-121, DenseNet-169, VGG-16, VGG-19, AlexNet, Inception V3, ResNext-50, ResNext-101, ShuffleNet, MobileNet, MnasNet. We freeze the weight of bottleneck layers of deep CNN models pre-trained on the ImageNet [79] dataset. Also, we use 9 different ML classifiers: FC layer, Gaussian NB, AdaBoost, k-NN, RF, SVM with three different kernels (linear, sigmoid, and RBF), ELM. Before the training step, we pre-processed the input images as described in Section 3.1. Also, we converted the images to the size 224 × 224 (or 299 × 299) pixels as the pre-trained networks used in our experiments require the input images with size 224 × 224 except for the Inception V3, which requires the input images with size 299 × 299. All experiments were performed on a PC with an NVIDIA GeForce GTX 1070 Ti GPU.

4.3. Results

The empirical results were obtained for three different datasets (BT-small-2c, BT-large-2c, and BT-large-4c) for the tasks of the brain tumor classification. The first experiment is designed to compare the several different pre-trained CNN networks with several different ML classifiers. The second experiment is designed to show the effectiveness of the ensemble of top 2 or 3 deep features selected by the results from the first experiment with several different ML classifiers. The results of the first experiments on BT-small-2c, BT-large-2c, and BT-large-4c datasets are shown in Tables 4–6, respectively. As shown in Table 4, DenseNet-169 feature, Inception V3 feature, and ResNeXt-50 feature are selected as the top three deep features on BT-small-2c dataset. As shown in Table 5, DenseNet-121 feature, ResNeXt-101 feature, and MnasNet feature are selected as the top three deep features on BT-small-4c dataset. Also in Table 6, DenseNet-169 feature, ShuffleNet V2 feature, and MnasNet feature are selected as the top three deep features on BT-large-4c dataset.

Table 4. Accuracies of pre-trained CNN models with ML classifiers on BT-small-2c dataset (⋆ : top-3 features based on average accuracy).

Deep Feature from the Pre-Trained CNN Model	ML Classifier—Accuracy									
	FC	Gaussian NB	AdaBoost	k-NN	RF	SVM (Linear)	SVM (Sigmoid)	SVM (RBF)	ELM	Average
ResNet-50 feature	0.9216	0.8431	0.8431	0.8627	0.8824	0.8235	0.8824	0.9020	0.9020	0.8736
ResNet-101 feature	0.9216	0.8824	0.8431	0.8235	0.9020	0.8235	0.8824	0.9020	0.8824	0.8736
DenseNet-121 feature	0.9216	0.7647	0.8235	0.9216	0.8824	0.8431	0.8824	0.8627	0.9020	0.8671
DenseNet-169 feature ⋆	0.9608	0.8039	0.8627	0.9020	0.9412	0.9608	0.9608	0.9804	0.9412	**0.9237**
VGG-16 feature	0.8431	0.7451	0.7451	0.7059	0.8431	0.8627	0.8627	0.8039	0.8039	0.8017
VGG-19 feature	0.8235	0.6863	0.7843	0.6863	0.8235	0.8235	0.8235	0.8235	0.9020	0.7974
AlexNet feature	0.9216	0.7255	0.8431	0.7843	0.9020	0.8235	0.8627	0.9020	0.9020	0.8519
Inception V3 feature ⋆	0.9216	0.8824	0.9020	0.8235	0.9412	0.9020	0.9020	0.9020	0.9020	0.8976
ResNeXt-50 feature ⋆	0.9412	0.9020	0.9020	0.9020	0.9216	0.9216	0.9216	0.9216	0.9216	0.9172
ResNeXt-101 feature	0.9216	0.8039	0.8235	0.8235	0.9020	0.8627	0.9020	0.9216	0.9216	0.8758
ShuffleNet V2 feature	0.8431	0.7647	0.9216	0.8627	0.9020	0.9412	0.9412	0.9412	0.9412	0.8954
MobileNet V2 feature	0.8824	0.8431	0.7843	0.8431	0.8824	0.8627	0.8824	0.8824	0.8627	0.8584
MnasNet feature	0.9216	0.7843	0.8235	0.8235	0.9216	0.8431	0.8627	0.8627	0.9020	0.8606
Average	**0.9035**	0.8024	0.8386	0.8281	0.8959	0.8688	0.8899	0.8929	0.8989	

The bold text represents the highest average accuracy of all ML classifier or all deep features.

Also, the results of the second experiments on BT-small-2c, BT-large-2c, and BT-large-4c datasets are shown in Tables 7–9, respectively. Also, the computational complexity of ensemble models is compared based on the inference time on a test set of the BT-large-4c dataset as shown in Table 10. From these results, five observations were made.

Table 5. Accuracies of pre-trained CNN models with ML classifiers on BT-large-2c dataset (⋆ : top-3 features based on average accuracy).

Deep Feature from the Pre-Trained CNN Model	ML Classifier—Accuracy									
	FC	Gaussian NB	AdaBoost	k-NN	RF	SVM (Linear)	SVM (Sigmoid)	SVM (RBF)	ELM	Average
ResNet-50 feature	0.9767	0.8117	0.9600	0.9767	0.9400	0.9750	0.9750	0.9817	0.9667	0.9515
ResNet-101 feature	0.9767	0.8250	0.9433	0.9733	0.9567	0.9750	0.9733	0.9800	0.9717	0.9528
DenseNet-121 feature ⋆	0.9750	0.8383	0.9600	0.9817	0.9683	0.9683	0.9683	0.9833	0.9817	0.9583
DenseNet-169 feature	0.9750	0.8400	0.9650	0.9783	0.9633	0.9667	0.9650	0.9800	0.9800	0.9570
VGG-16 feature	0.9550	0.7383	0.8833	0.9617	0.9283	0.9517	0.9500	0.9650	0.9550	0.9209
VGG-19 feature	0.9550	0.7067	0.8850	0.9600	0.9300	0.9550	0.9550	0.9633	0.9450	0.9172
AlexNet feature	0.9633	0.7067	0.9200	0.9550	0.9500	0.9400	0.9500	0.9750	0.9633	0.9248
Inception V3 feature	0.9817	0.8317	0.9567	0.9800	0.9567	0.9750	0.9733	0.9883	0.9800	0.9581
ResNeXt-50 feature	0.9717	0.8600	0.9550	0.9817	0.9550	0.9700	0.9683	0.9833	0.9750	0.9578
ResNeXt-101 feature ⋆	0.9783	0.8583	0.9633	0.9833	0.9617	0.9717	0.9717	0.9817	0.9817	**0.9613**
ShuffleNet V2 feature	0.9433	0.8533	0.9533	0.9700	0.9517	0.9617	0.9617	0.9783	0.9700	0.9493
MobileNet V2 feature	0.9667	0.8400	0.9367	0.9700	0.9450	0.9617	0.9617	0.9783	0.9633	0.9470
MnasNet feature ⋆	0.9817	0.8550	0.9467	0.9750	0.9567	0.9700	0.9733	0.9817	0.9833	0.9581
Average	0.9692	0.8127	0.9406	0.9728	0.9510	0.9647	0.9651	**0.9785**	0.9705	

The bold text represents the highest average accuracy of all ML classifier or all deep features.

Table 6. Accuracies of pre-trained CNN models with ML classifiers on BT-large-4c dataset (⋆ : top-3 features based on average accuracy).

Deep Feature from the Pre-Trained CNN Model	ML Classifier—Accuracy									
	FC	Gaussian NB	AdaBoost	k-NN	RF	SVM (Linear)	SVM (Sigmoid)	SVM (RBF)	ELM	Average
ResNet-50 feature	0.8760	0.6937	0.6570	0.8576	0.8530	0.8744	0.8760	0.8989	0.8591	0.8273
ResNet-101 feature	0.8867	0.7228	0.6799	0.8438	0.8499	0.8897	0.8897	0.9081	0.8683	0.8377
DenseNet-121 feature	0.8913	0.7106	0.7198	0.8943	0.8744	0.8698	0.8729	0.9158	0.8760	0.8472
DenseNet-169 feature ⋆	0.8959	0.7228	0.7335	0.8821	0.8652	0.8652	0.8729	0.9204	0.8806	0.8487
VGG-16 feature	0.8760	0.6677	0.7106	0.8331	0.8300	0.8606	0.8606	0.8744	0.8423	0.8173
VGG-19 feature	0.8683	0.5942	0.6309	0.8346	0.8377	0.8606	0.8606	0.8790	0.8453	0.8013
AlexNet feature	0.8637	0.6340	0.6554	0.8714	0.8453	0.8652	0.8683	0.9066	0.8361	0.8162
Inception V3 feature	0.8652	0.6708	0.6830	0.8300	0.8132	0.8591	0.8591	0.8867	0.8438	0.8123
ResNeXt-50 feature	0.8744	0.7152	0.6891	0.8775	0.8346	0.8560	0.8576	0.8959	0.8560	0.8285
ResNeXt-101 feature	0.8851	0.6692	0.7198	0.8714	0.8346	0.8744	0.8744	0.8989	0.8744	0.8336
ShuffleNet V2 feature	0.8637	0.7152	0.7381	0.8637	0.8576	0.8989	0.8989	0.9112	0.8606	0.8453
MobileNet V2 feature	0.8928	0.6983	0.7136	0.8897	0.8423	0.8851	0.8851	0.9158	0.8729	0.8440
MnasNet feature ⋆	0.8851	0.6922	0.7458	0.8928	0.8515	0.8959	0.8959	0.9127	0.8775	**0.8499**
Average	0.8788	0.6851	0.6982	0.8648	0.8453	0.8735	0.8748	**0.9019**	0.8610	

The bold text represents the highest average accuracy of all ML classifier or all deep features.

- *Observation 1.* SVM with RBF kernel outperforms other ML classifiers on two large datasets (BT-large-2c and BT-large-4c).
- *Analysis.* Tables 5 and 6 show that the SVM with RBF kernel outperforms other ML classifiers on two large datasets (BT-large-2c and BT-large-4c). This is because SVM with RBF kernel can find a more effective and complex set of decision boundaries than other ML classifiers. However, as you can see in Table 4, SVM with RBF kernel does not outperform other ML classifiers on the small dataset. This is because SVM tends to underperform when the number of features for each data point is larger than the number of training data samples.
- *Observation 2.* Gaussian NB performs worst among other ML classifiers on three datasets.
- *Analysis.* Tables 4–6 show that Gaussian NB performs worst among other ML classifiers on three datasets. This is because Gaussian NB assumes the features are

independent. However, it is almost impossible that the extracted features from the pre-trained models are completely independent.

- *Observation 3.* The deep feature from DenseNet architectures performs well than the deep features from other pre-trained CNN networks, while the deep features from VGG perform worse than the deep features from other pre-trained CNN networks on three different datasets.

- *Analysis.* Tables 4–6 show that the deep feature from DenseNet architectures performs well than the deep features from other pre-trained CNN networks on three different datasets. This is because the features extracted from DenseNet have all complexity levels. Hence, it tends to give more smooth decision boundaries, which can predict well when training data is insufficient. On the other hand, the deep feature from VGG performs worse than the deep features from other pre-trained CNN networks on three different datasets. This is because VGG is a more basic architecture that uses no residual blocks than other pre-trained CNN networks.

- *Observation 4.* Using the ensemble of deep features from two or three pre-trained CNN models is effective for all ML classifiers on a large dataset. However, the ensemble of deep features is effective for only ML classifiers on a small dataset.

- *Analysis.* Tables 8 and 9 show that the model with the ensemble of deep features from two or three pre-trained CNN models achieves higher accuracy than the model with a deep feature from an individual pre-trained CNN model. This is because the ensemble model takes advantages of well-performing top-2 or 3 deep features by concatenating them, and also the concatenation of these deep features has a set of features that are capable of representing the data present in the images in a different way which benefits to improve the performance of ML classifiers. However, Table 7 shows that the ensemble of deep features is effective for only a few ML classifiers on a small dataset. This is because the number of the training sample is not enough in the small dataset to learn the complex set of the ensemble of deep features.

- *Observation 5.* k-NN classifier takes the longest time for inference on a test set while FC, Gaussian NB, and RF take a shorter inference time.

- *Analysis.* Table 10 shows that the k-NN classifier takes the longest time for inference on a test set among other ML classifiers while FC, Gaussian NB, and RF classifiers take a very short time for inference on a test set. This is because the k-NN classifier has to look at all the data points to make a single prediction, whereas other ML classifiers are not dependent on the number of training data points on the predict phase. On the other hand, the Gaussian NB classifier uses the Bayes equation to compute the posterior probabilities for inference. This involves trivial arithmetic operations such as multiplication and addition. Also, normalization is done by simple division operations. In the fully connected layer (FC), the entire matrix calculation for inference can be done by fast GPU. RF classifier leverages the power of multiple decision trees, which are simple and fast for making decisions. Therefore, these three classifiers achieved significantly less computation time for inference than other ML classifiers.

Table 7. Accuracies of ensemble of pre-trained CNN models with ML classifiers on BT-small-2c dataset.

Deep Feature from the Pre-Trained CNN Model	ML Classifier—Accuracy								
	FC	Gaussian NB	AdaBoost	k-NN	RF	SVM (Linear)	SVM (Sigmoid)	SVM (RBF)	ELM
DenseNet-169 feature	**0.9608**	0.8039	0.8627	0.9020	**0.9412**	**0.9608**	**0.9608**	**0.9804**	**0.9412**
Inception V3 feature	0.9216	0.8824	**0.9020**	0.8235	**0.9412**	0.9020	0.9020	0.9020	0.9020
ResNeXt-50	0.9412	**0.9020**	0.9020	0.9020	0.9216	0.9216	0.9216	0.9216	0.9216
(DenseNet-169 + Inception V3) feature	0.9412	0.8627	0.9020	0.8824	0.9020	0.9412	0.9412	0.9608	**0.9412**
(DenseNet-169 + ResNeXt-50) feature	0.9412	**0.9020**	**0.9216**	0.8627	0.9216	0.9216	0.9412	0.9412	**0.9412**
(Inception V3 + ResNeXt-50) feature	0.9412	**0.9020**	0.8824	**0.9412**	**0.9412**	0.9216	0.9412	0.9412	**0.9412**
(DenseNet-169 + Inception V3 + ResNeXt-50) feature	0.9412	**0.9020**	**0.9216**	0.9020	0.9216	0.9020	0.9412	0.9412	0.9216

The bold text represents the highest accuracy for each ML classifier.

Table 8. Accuracies of ensemble of pre-trained CNN models with ML classifiers on BT-large-2c dataset.

Deep Feature from the Pre-Trained CNN Model	ML Classifier—Accuracy								
	FC	Gaussian NB	AdaBoost	k-NN	RF	SVM (Linear)	SVM (Sigmoid)	SVM (RBF)	ELM
DenseNet-121 feature	0.9750	0.8383	0.9600	0.9817	0.9683	0.9683	0.9683	0.9833	0.9817
ResNeXt-101 feature	0.9783	0.8583	0.9633	0.9833	0.9617	0.9717	0.9717	0.9817	0.9817
MnasNet feature	0.9817	0.8550	0.9467	0.9750	0.9567	0.9700	0.9733	0.9817	0.9833
(DenseNet-121 + ResNeXt-101) feature	0.9800	0.8733	0.9700	0.9817	0.9667	0.9783	0.9783	0.9833	0.9850
(DenseNet-121 + MnasNet) feature	0.9817	0.8767	0.9633	**0.9850**	0.9683	0.9700	0.9717	0.9783	0.9767
(ResNeXt-101 + MnasNet) feature	**0.9883**	0.8700	0.9633	**0.9850**	0.9667	**0.9850**	**0.9850**	**0.9850**	0.9850
(DenseNet-121 + ResNeXt-101 + MnasNet) feature	**0.9883**	**0.8800**	0.9750	0.9817	**0.9717**	0.9783	0.9800	**0.9850**	**0.9867**

The bold text represents the highest accuracy for each ML classifier.

Table 9. Accuracies of ensemble of pre-trained CNN models with ML classifiers on BT-large-4c dataset.

Deep Feature from the Pre-Trained CNN Model	ML Classifier—Accuracy								
	FC	Gaussian NB	AdaBoost	k-NN	RF	SVM (Linear)	SVM (Sigmoid)	SVM (RBF)	ELM
DenseNet-169 feature	0.8959	0.7228	0.7335	0.8821	0.8652	0.8652	0.8729	0.9204	0.8806
Shufflenet feature	0.8637	0.7152	0.7381	0.8637	0.8576	0.8989	0.8989	0.9112	0.8606
MnasNet feature	0.8851	0.6922	0.7458	0.8928	0.8515	0.8959	0.8959	0.9127	0.8775
(DenseNet-169 + Shufflenet) feature	0.8959	**0.7504**	0.7427	0.8821	0.8668	0.8668	0.8714	0.9204	0.8744
(DenseNet-169 + MnasNet) feature	0.9142	0.7259	0.7274	**0.9096**	0.8668	**0.9020**	**0.9096**	**0.9372**	0.8790
(Shufflenet + MnasNet) feature	0.8913	0.7305	0.7397	0.8943	**0.8790**	0.8974	0.8974	0.9127	0.8637
(DenseNet-169 + Shufflenet + MnasNet) feature	**0.9158**	0.7397	**0.7534**	**0.9096**	0.8760	**0.9020**	**0.9096**	**0.9372**	**0.8851**

The bold text represents the highest accuracy for each ML classifier.

Table 10. Computational complexity of ensemble of pre-trained CNN models with ML classifiers on BT-large-4c dataset.

Deep Feature from the Pre-Trained CNN Model	ML Classifier—Accuracy								
	FC	Gaussian NB	AdaBoost	k-NN	RF	SVM (Linear)	SVM (Sigmoid)	SVM (RBF)	ELM
(DenseNet-169 + Shufflenet) feature	0.0222	0.0214	0.2709	5.0436	0.0148	1.7390	1.9813	2.2653	0.1831
(DenseNet-169 + MnasNet) feature	0.0225	0.0232	0.3070	5.5191	0.0187	1.9650	2.1004	2.5780	0.2184
(Shufflenet + MnasNet) feature	0.0224	0.0186	0.2403	4.3021	0.0170	1.4580	1.4725	2.2544	0.1784
(DenseNet-169 + Shufflenet + MnasNet) feature	0.0229	0.0312	0.4133	7.4238	0.0247	2.6586	2.8507	3.4730	0.2772

5. Conclusions

In summary, we presented a brain tumor classification method using the ensemble of deep features from pre-trained deep convolutional neural networks with ML classifiers. In our proposed framework, we use several pre-trained deep convolutional neural networks to extract deep features from brain MR images. The extracted deep features are then evaluated by several ML classifiers. The top three deep features which perform well on several ML classifiers are selected and concatenated as an ensemble of deep feature which is then fed into several ML classifiers to predict the final output. In our experiment, we provided an extensive evaluation using 13 different pre-trained deep convolutional neural networks and nine different ML classifiers on three different datasets (BT-small-2c, BT-large-2c, and BT-large-4c) for brain tumor classification. Our experiment results indicate that from our architecture, (1) DenseNet-169 deep feature alone is a good choice in case the size of the MRI dataset is very small and the number of classes is 2 (normal, tumor), (2) the ensemble of DenseNet-169, Inception V3, and ResNeXt-50 deep features is a good choice in case the size of MRI dataset is large and the number of classes is 2 (normal, tumor) and (3) the ensemble of DenseNet-169, ShuffleNet V2, and MnasNet deep features is a good choice in case the size of MRI dataset is large and there are four classes (normal, glioma tumor, meningioma tumor, and pituitary tumor). Also, in most cases, SVM with RBF kernel outperforms other ML classifiers for the MRI-based brain tumor classification task. In summary, our proposed novel feature ensemble method helps to overcome the limitations of a single CNN model and produces superior and robust performance, especially for large datasets. These results indicated that our proposed method using an ensemble of deep features and ML classifiers is suitable for the classification of brain tumors. Although the performance of our proposed method is promising, further research needs to be done to reduce the size of the model to deploy on a real-time medical diagnosis system using knowledge distillation approaches.

Author Contributions: Conceptualization, J.G.; methodology, J.K. and J.G.; software, J.K. and J.G.; validation, J.K., Z.U. and J.G.; formal analysis, J.K., Z.U. and J.G.; investigation, J.K., Z.U. and J.G.; resources, J.G.; data curation, J.K.; writing—original draft preparation, J.K. and Z.U.; writing—review and editing, J.G.; visualization, J.K.; supervision, J.G.; project administration, J.G.; funding acquisition, J.G. All authors have read and agreed to the published version of the manuscript.

Funding: This research was supported by the Basic Science Research Program through the National Research Foundation of Korea (NRF) funded by the Ministry of Education (Grant No. NRF-2020R1I1A3074141) and the Brain Research Program through the NRF funded by the Ministry of Science, ICT and Future Planning (Grant No. NRF-2019M3C7A1020406).

Institutional Review Board Statement: Not applicable.

Informed Consent Statement: Not applicable.

Data Availability Statement: Data are available in publicly accessible repositories which are described in Section 4.1.

Acknowledgments: The authors would like to thank the editors and all the reviewers for their valuable comments on this manuscript.

Conflicts of Interest: The authors declare no conflict of interest.

References

1. Louis, D.N.; Perry, A.; Reifenberger, G.; Deimling, A.V.; Figarella-Branger, D.; Cavenee, W.K.; Ohgaki, H.; Wiestler, O.D.; Kleihues, P.; Ellison, D.W. The 2016 World Health Organization classification of tumors of the central nervous system: A summary. *Acta Neuropathol.* **2016**, *131*, 803–820. [CrossRef]
2. Tandel, G.S.; Biswas, M.; Kakde, O.G.; Tiwari, A.; Suri, H.S.; Turk, M.; Laird, J.R.; Asare, C.K.; Ankrah, A.A.; Khanna, N.N.; et al. A review on a deep learning perspective in brain cancer classification. *Cancers* **2019**, *11*, 111. [CrossRef]
3. Anaraki, A.K.; Ayati, M.; Kazemi, F. Magnetic resonance imaging-based brain tumor grades classification and grading via convolutional neural networks and genetic algorithms. *Biocybern. Biomed. Eng.* **2019**, *39*, 63–74. [CrossRef]
4. Liu, J.; Pan, Y.; Li, M.; Chen, Z.; Tang, L.; Lu, C.; Wang, J. Applications of deep learning to MRI images: A survey. *Big Data Min. Anal.* **2018**, *1*, 1–18.
5. Mehrotra, R.; Ansari, M.A.; Agrawal, R.; Anand, R.S. A Transfer Learning approach for AI-based classification of brain tumors. *Mach. Learn. Appl.* **2020**, *2*, 10–19. [CrossRef]
6. Pereira, S.; Pinto, A.; Alves, V.; Silva, C.A. Brain tumor segmentation using convolutional neural networks in MRI images. *IEEE Trans. Med. Imaging* **2018**, *35*, 1240–1251. [CrossRef] [PubMed]

7. Popuri, K.; Cobzas, D.; Murtha, A.; Jägersand, M. 3D variational brain tumor segmentation using Dirichlet priors on a clustered feature set. *Int. J. Comput. Assist. Radiol. Surg.* **2012**, *7*, 493–506. [CrossRef] [PubMed]
8. Ullah, Z.; Farooq, M.U.; Lee, S.H.; An, D. A Hybrid Image Enhancement Based Brain MRI Images Classification Technique. *Med. Hypotheses* **2020**, *143*, 109922. [CrossRef]
9. Selvaraj, H.; Selvi, S.T.; Selvathi, D.; Gewali, L. Brain MRI slices classification using least squares support vector machine. *Int. J. Intell. Comput. Med. Sci. Image Process.* **2007**, *1*, 21–33. [CrossRef]
10. John, P. Brain tumor classification using wavelet and texture based neural network. *Int. J. Sci. Eng. Res.* **2012**, *3*, 1–7.
11. Bosch, A.; Munoz, X.; Oliver, A.; Marti, J. Modeling and classifying breast tissue density in mammograms. In Proceedings of the 2006 IEEE Computer Society Conference on Computer Vision and Pattern Recognition (CVPR'06), New York, NY, USA, 17–22 June 2006; Volume 2, pp. 1552–1558.
12. Avni, U.; Greenspan, H.; Konen, E.; Sharon, M.; Goldberger, J. X-ray categorization and retrieval on the organ and pathology level, using patch-based visual words. *IEEE Trans. Med. Imaging* **2010**, *30*, 733–746. [CrossRef]
13. Yang, W.; Lu, Z.; Yu, M.; Huang, M.; Feng, Q.; Chen, W. Content-based retrieval of focal liver lesions using bag-of-visual-words representations of single-and multiphase contrast-enhanced CT images. *J. Digit. Imaging* **2012**, *25*, 6. [CrossRef]
14. Cheng, J.; Yang, W.; Huang, M.; Huang, W.; Jiang, J.; Zhou, Y.; Yang, R.; Zhao, J.; Feng, Y.; Feng, Q.; et al. Retrieval of brain tumors by adaptive spatial pooling and fisher vector representation. *PLoS ONE* **2016**, *11*, e157112. [CrossRef] [PubMed]
15. Mohammad, H.; Axel, D.; Warde, F. Brain tumor segmentation with deep neural networks. *Med. Image Anal.* **2017**, *35*, 18–31.
16. Prastawa, M.; Bullitt, E.; Moon, N.; Van, L.; Gerig, G. Automatic brain tumor segmentation by subject specific modification of atlas priors1. *Acad. Radiol.* **2003**, *10*, 1341–1348. [CrossRef]
17. Ateeq, T.; Majeed, M.; Nadeem, A.; Syed, M.; Maqsood, M.; Rehman, Z.; Lee, J.W.; Muhammad, K.; Shuihua, B.; Sung, W.; et al. Ensemble-classifiers-assisted detection of cerebral microbleeds in brain MRI. *Comput. Electr. Eng.* **2018**, *69*, 768–781. [CrossRef]
18. Kharrat, A.; Gasmi, K.; Messaoud, M.; Ben, N.B.; Abid, M. A hybrid approach for automatic classification of brain MRI using genetic algorithm and support vector machine. *Leonardo J. Sci.* **2010**, *17*, 71–82.
19. Papageorgiou, E.; Spyridonos, P.; Glotsos, D.; Stylios, C.; Ravazoula, P.; Nikiforidis, G.; Groumpos, P. Brain tumor characterization using the soft computing technique of fuzzy cognitive maps. *Appl. Soft Comput.* **2008**, *8*, 820–828. [CrossRef]
20. Shree, N.V.; Kumar, T.N.R. Identification and classification of brain tumor MRI images with feature extraction using DWT and probabilistic neural network. *Brain Inform.* **2018**, *5*, 23–30. [CrossRef] [PubMed]
21. Arunachalam, M.; Royappan, S.S. An efficient and automatic glioblastoma brain tumor detection using shift-invariant shearlet transform and neural networks. *Int. J. Imaging Syst. Technol.* **2017**, *27*, 216–226. [CrossRef]
22. Rajan, P.G.; Sundar, C. Brain tumor detection and segmentation by intensity adjustment. *J. Med. Syst.* **2019**, *43*, 1–13. [CrossRef] [PubMed]
23. Kleesiek, J.; Urban, G.; Hubert, A.; Schwarz, D.; Maier-Hein, K.; Bendszus, M.; Biller, A. Deep MRI brain extraction: A 3D convolutional neural network for skull stripping. *NeuroImage* **2016**, *129*, 460–469. [CrossRef]
24. Paul, J.S.; Plassard, A.J.; Landman, B.A.; Fabbri, D. Deep learning for brain tumor classification. *Med. Imaging Biomed. Appl. Mol. Struct. Funct. Imaging* **2017**, *10137*, 1013710.
25. Abiwinanda, N.; Hanif, M.; Hesaputra, S.T.; Handayani, A.; Mengko, T.R. Brain tumor classification using convolutional neural network. In Proceedings of the World Congress on Medical Physics and Biomedical Engineering 2018, Prague, Czech Republic, 3–8 June 2019; pp. 183–189.
26. Seetha, J.; Raja, S.S. Brain tumor classification using convolutional neural networks. *Biomed. Pharmacol. J.* **2018**, *11*, 3. [CrossRef]
27. Hemanth, D.J.; Anitha, J.; Naaji, A.; Geman, O.; Popescu, D.E. A modified deep convolutional neural network for abnormal brain image classification. *IEEE Access* **2018**, *7*, 4275–4283. [CrossRef]
28. Balasooriya, N.M.; Nawarathna, R.D. A sophisticated convolutional neural network model for brain tumor classification. In Proceedings of the IEEE International Conference on Industrial and Information Systems (ICIIS), Roorkee, India, 15–16 December 2017; pp. 1–5.
29. Deepak, S.; Ameer, P.M. Brain tumor classification using deep CNN features via transfer learning. *Comput. Biol. Med.* **2019**, *111*, 103345. [CrossRef]
30. Çinar, A.; Yıldırım, M. Detection of tumors on brain MRI images using the hybrid convolutional neural network architecture. *Med. Hypotheses* **2020**, *139*, 109684. [CrossRef]
31. Khawaldeh, S.; Pervaiz, U.; Rafiq, A.; Alkhawaldeh, R.S. Noninvasive grading of glioma tumor using magnetic resonance imaging with convolutional neural networks. *Appl. Sci.* **2018**, *8*, 27. [CrossRef]
32. Saxena, P.; Maheshwari, A.; Maheshwari, S. Predictive modeling of brain tumor: A Deep learning approach. *arXiv* **2019**, arXiv:1911.02265.
33. Xuesong, Y.; Yong, F. Feature extraction using convolutional neural networks for multi-atlas based image segmentation. *Med. Imaging Image Process.* **2018**, *10574*, 1057439.
34. Wicht, B. Deep Learning Feature Extraction for Image Processing. Ph.D. Thesis, éditeur non Identifié, The University of Fribourg, Fribourg, Switzerland, 2017.
35. Francisco, J.P.; Mario, Z.M.; Miriam, R.A. A Deep Learning Approach for Brain Tumor Classification and Segmentation Using a Multiscale Convolutional Neural Network. *Healthcare* **2021**, *9*, 153.

36. Raja, P.M.S.; Antony, V.R. Brain tumor classification using a hybrid deep autoencoder with Bayesian fuzzy clustering-based segmentation approach. *Biocybern. Biomed. Eng.* **2020**, *40*, 440–453. [CrossRef]
37. Bhuvaji, S.; Kadam, A.; Bhumkar, P.; Dedge, S.; Kanchan, S. Brain Tumor Classification (MRI) Dataset. Available online: https://www.kaggle.com/sartajbhuvaji/brain-tumor-classification-mri (accessed on 1 August 2020).
38. Preethi, S.; Aishwarya, P. Combining Wavelet Texture Features and Deep Neural Network for Tumor Detection and Segmentation Over MRI. *J. Intell. Syst.* **2019**, *28*, 571–588. [CrossRef]
39. Ural, B. A computer-based brain tumor detection approach with advanced image processing and probabilistic neural network methods. *J. Med. Biol. Eng.* **2018**, *38*, 867–879. [CrossRef]
40. Zhang, X.; Zhou, X.; Lin, M.; Sun, J. Finding Extreme Points in Contours with OpenCV. In PyImageSearch. Available online: https://www.pyimagesearch.com/2016/04/11/finding-extreme-points-in-contours-with-opencv (accessed on 10 August 2020).
41. Goyal, M.; Goyal, R.; Lall, B. Learning Activation Functions: A New Paradigm of Understanding Neural Networks. *arXiv* **2019**, arXiv:1906.09529.
42. Albawi, S.; Mohammed, T.A.; Al-Zawi, S. Understanding of a convolutional neural network. In Proceedings of the 2017 International Conference on Engineering and Technology (ICET), Antalya, Turkey, 21–23 August 2017; pp. 1–6.
43. Krizhevsky, A.; Sutskever, I.; Hinton, G.E. Imagenet classification with deep convolutional neural networks. *Adv. Neural Inf. Process. Syst.* **2012**, *25*, 1097–1105. [CrossRef]
44. Akçay, S.; Kundegorski, M.E.; Devereux, M.; Breckon, T.P. Transfer learning using convolutional neural networks for object classification within x-ray baggage security imagery. In Proceedings of the 2016 IEEE International Conference on Image Processing (ICIP), Phoenix, AZ, USA, 25–28 September 2016; pp. 1057–1061.
45. Baltruschat, I.M.; Nickisch, H.; Grass, M.; Knopp, T.; Saalbach, A. Comparison of deep learning approaches for multi-label chest X-ray classification. *Sci. Rep.* **2019**, *9*, 6381. [CrossRef] [PubMed]
46. Christodoulidis, S.; Anthimopoulos, M.; Ebner, L.; Christe, A.; Mougiakakou, S. Multisource transfer learning with convolutional neural networks for lung pattern analysis. *IEEE J. Biomed. Health Inform.* **2016**, *21*, 76–84. [CrossRef]
47. Kang, J.; Gwak, J. Ensemble of instance segmentation models for polyp segmentation in colonoscopy images. *IEEE Access* **2019**, *7*, 26440–26447. [CrossRef]
48. Tajbakhsh, N.; Shin, J.Y.; Gurudu, S.R.; Hurst, R.T.; Kendall, C.B.; Gotway, M.B.; Liang, J. Convolutional neural networks for medical image analysis: Full training or fine tuning? *IEEE Trans. Med. Imaging* **2016**, *35*, 1299–1312. [CrossRef]
49. Pan, S.J.; Yang, Q. A survey on transfer learning. *IEEE Trans. Knowl. Data Eng.* **2009**, *22*, 1345–1359. [CrossRef]
50. He, K.; Zhang, X.; Ren, S.; Sun, J. Deep residual learning for image recognition. In Proceedings of the IEEE Conference on Computer Vision and Pattern Recognition, Las Vegas, NV, USA, 27–30 June 2016; pp. 770–778.
51. Huang, G.; Liu, Z.; Maaten, L.V.D.; Weinberger, K.Q. Densely connected convolutional networks. In Proceedings of the IEEE Conference on Computer Vision and Pattern Recognition, Honolulu, HI, USA, 21–26 July 2017; pp. 4700–4708.
52. Simonyan, K.; Zisserman, A. Very Deep Convolutional Networks for Large-Scale Image Recognition. *arXiv* **2014**, arXiv:1409.1556.
53. Krizhevsky, A. One Weird Trick for Parallelizing Convolutional Neural Networks. *arXiv* **2014**, arXiv:1404.5997.
54. Szegedy, C.; Vanhoucke, V.; Ioffe, S.; Shlens, J.; Wojna, Z. Rethinking the inception architecture for computer vision. In Proceedings of the IEEE Conference on Computer Vision and Pattern Recognition, Las Vegas, NV, USA, 27–30 June 2016; pp. 2818–2826.
55. Xie, S.; Girshick, R.; Dollár, P.; Tu, Z.; He, K. Aggregated residual transformations for deep neural networks. In Proceedings of the IEEE Conference on Computer Vision and Pattern Recognition, Honolulu, HI, USA, 21–26 July 2017; pp. 1492–1500.
56. Ma, N.; Zhang, X.; Zheng, H.T.; Sun, J. Shufflenet v2: Practical guidelines for efficient cnn architecture design. In Proceedings of the European Conference on Computer Vision (ECCV), Munich, Germany, 8–14 September 2018; pp. 116–131.
57. Sandler, M.; Howard, A.; Zhu, M.; Zhmoginov, A.; Chen, L.C. Mobilenetv2: Inverted residuals and linear bottlenecks. In Proceedings of the IEEE Conference on Computer Vision and Pattern Recognition, Salt Lake City, UT, USA, 18–22 June 2018; pp. 4510–4520.
58. Tan, M.; Chen, B.; Pang, R.; Vasudevan, V.; Sandler, M.; Howard, A.; Le, Q.V. Mnasnet: Platform-aware neural architecture search for mobile. In Proceedings of the IEEE Conference on Computer Vision and Pattern Recognition, Long Beach, CA, USA, 15–20 June 2019; pp. 2820–2828.
59. Pedregosa, F.; Varoquaux, G.; Gramfort, A.; Michel, V.; Thirion, B.; Grisel, O.; Blondel, M.; Prettenhofer, P.; Weiss, R.; Dubourg, V.; et al. Scikit-learn: Machine learning in Python. *J. Mach. Learn. Res.* **2011**, *12*, 2825–2830.
60. Freund, Y.; Schapire, R.E. A decision-theoretic generalization of on-line learning and an application to boosting. *J. Comput. Syst. Sci.* **1997**, *55*, 119–139. [CrossRef]
61. Breiman, L. Random forests. *Mach. Learn.* **2001**, *45*, 5–32. [CrossRef]
62. Cortes, C.; Vapnik, V. Support-vector networks. *Mach. Learn.* **1995**, *20*, 273–297. [CrossRef]
63. Huang, G.B.; Zhu, Q.Y.; Siew, C.K. Extreme learning machine: A new learning scheme of feedforward neural networks. In Proceedings of the 2004 IEEE International Joint Conference on Neural Networks, Budapest, Hungary, 25–29 July 2004; Volume 2, pp. 985–990.
64. Kaplan, K.; Kaya, Y.; Kuncan, M.; Ertunç, H.M. Brain tumor classification using modified local binary patterns (LBP) feature extraction methods. *Med. Hypotheses* **2020**, *139*, 109696. [CrossRef] [PubMed]
65. Kaur, G.; Oberoi, A. Novel Approach for Brain Tumor Detection based on Naïve Bayes Classification. In *Data Management, Analytics and Innovation*; Springer: Singapore, 2020; pp. 451–462.

66. Minz, A.; Mahobiya, C. MR image classification using adaboost for brain tumor type. In Proceedings of the 2017 IEEE 7th International Advance Computing Conference (IACC), Hyderabad, India, 5–7 January 2017; pp. 701–705.
67. Anitha, R.; Siva, S.; Raja, D. Development of computer-aided approach for brain tumor detection using random forest classifier. *Int. J. Imaging Syst. Technol.* **2018**, *28*, 48–53. [CrossRef]
68. Gumaei, A.; Hassan, M.M.; Hassan, M.R.; Alelaiwi, A.; Fortino, G. A hybrid feature extraction method with regularized extreme learning machine for brain tumor classification. *IEEE Access* **2019**, *7*, 36266–36273. [CrossRef]
69. Khan, H.A.; Jue, W.; Mushtaq, M.; Mushtaq, M.U. Brain tumor classification in MRI image using convolutional neural network. *Math. Biosci. Eng.* **2020**, *17*, 6203–6216. [CrossRef] [PubMed]
70. Polat, Ö; Güngen, C. Classification of brain tumors from MR images using deep transfer learning. *J. Supercomput.* **2021**. [CrossRef]
71. Ghosal, P.; Nandanwar, L.; Kanchan, S.; Bhadra, A.; Chakraborty, J.; Nandi, D. Brain tumor classification using ResNet-101 based squeeze and excitation deep neural network. In Proceedings of the 2019 Second International Conference on Advanced Computational and Communication Paradigms (ICACCP), Sikkim, India, 25–28 February 2019; pp. 1–6.
72. Zhou, Y.; Li, Z.; Zhu, H.; Chen, C.; Gao, M.; Xu, K.; Xu, J. Holistic brain tumor screening and classification based on densenet and recurrent neural network. In Proceedings of the International MICCAI Brainlesion Workshop, Granada, Spain, 16 September 2018; pp. 208–217.
73. Saba, T.; Mohamed, A.S.; El-Affendi, M.; Amin, J.; Sharif, M. Brain tumor detection using fusion of hand crafted and deep learning features. *Cogn. Syst. Res.* **2020**, *59*, 221–230. [CrossRef]
74. Ezhilarasi, R.; Varalakshmi, P. Tumor detection in the brain using faster R-CNN. In Proceedings of the 2018 2nd International Conference on I-SMAC (IoT in Social, Mobile, Analytics and Cloud)(I-SMAC), Palladam, India, 30–31 August 2018; pp. 388–392.
75. Soumik, M.F.I.; Hossain, M.A. Brain Tumor Classification With Inception Network Based Deep Learning Model Using Transfer Learning. In Proceedings of the 2020 IEEE Region 10 Symposium (TENSYMP), Dhaka, Bangladesh, 5–7 June 2020; pp. 1018–1021.
76. Lu, S.Y.; Wang, S.H.; Zhang, Y.D. A classification method for brain MRI via MobileNet and feedforward network with random weights. *Pattern Recognit. Lett.* **2020**, *140*, 252–260. [CrossRef]
77. Chakrabarty, N. Brain MRI Images for Brain Tumor Detection Dataset. Available online: https://www.kaggle.com/navoneel/brain-mri-images-for-brain-tumor-detection (accessed on 1 August 2020).
78. Hamada, A. Br35H Brain Tumor Detection 2020 Dataset. Available online: https://www.kaggle.com/ahmedhamada0/brain-tumor-detection (accessed on 1 August 2020).
79. Krizhevsky, A.; Sutskever, I.; Hinton, G.E. Imagenet classification with deep convolutional neural networks. *Commun. ACM* **2017**, *60*, 84–90. [CrossRef]

Article

IoT-Based Research Equipment Sharing System for Remotely Controlled Two-Photon Laser Scanning Microscopy

Eunwoo Park [1,2], Jaehyun Lim [2], Byung Cheol Park [3] and Daekeun Kim [2,*]

[1] Advanced Photonics Research Institute, Gwangju Institute of Science and Technology, Gwangju 61005, Korea; statice13@gist.ac.kr
[2] Department of Mechanical Engineering, Dankook University, Yongin 16890, Korea; LOMM_jhlim@dankook.ac.kr
[3] Department of Dermatology, College of Medicine, Dankook University, Cheonan 31116, Korea; 4exodus@dankook.ac.kr
* Correspondence: dkim@dankook.ac.kr; Tel.: +82-31-8005-3568

Abstract: In this study, two-photon laser scanning microscopy (TPLSM) based on the internet of things (IoT) is proposed as a remote research equipment sharing system, which enables the remote sharing economy. IoT modules, where data are transmitted to and received from the remote users in the web service via IoT, instead of a data acquisition (DAQ) system embedded in the conventional TPLSM, are installed in the IoT-based TPLSM (IoT-TPLSM). The performance for each IoT module is evaluated independently, and it is confirmed that it works well even in a personal computer-free environment. In addition, a message queuing telemetry transport (MQTT) protocol is applied to the DAQ interface in the web service, and a graphic user interface for enabling the remote users to operate IoT-TPLSM remotely is also designed and implemented. For the image acquisition demonstration, the stained cellular images and the autofluorescent tissue images are obtained in IoT-TPLSM. Lastly, it is confirmed that the comparable performance is provided with the conventional TPLSM by evaluating the imaging conditions and qualities of the three-dimensional image stacks processed in IoT-TPLSM.

Keywords: IoT; remote control; remote operation; remote sharing economy; research equipment sharing; two-photon laser scanning microscopy; MQTT

Citation: Park, E.; Lim, J.; Park, B.C.; Kim, D. IoT-Based Research Equipment Sharing System for Remotely Controlled Two-Photon Laser Scanning Microscopy. *Sensors* **2021**, *21*, 1533. https://doi.org/10.3390/s21041533

Academic Editor: Francesco Lamonaca

Received: 31 December 2020
Accepted: 19 February 2021
Published: 23 February 2021

1. Introduction

With the advent of the Fourth Industrial Revolution (4IR) worldwide, the integration of diverse networks between industries has formed a hyper-connected society. All industries converge with each other and develop together, as described in "Industry 4.0" [1,2]. For example, in manufacturing, a smart factory where a series of processes is linked to each other, unlike conventional automation only applied to an individual unit process, is aimed to optimize the operation by upgrading the production process and ensuring flexibility [3]. Accordingly, the national policies on the 4IR are presented, and its utilization methods in various fields are actively discussed. One of the representative 4IR applications is the sharing economy or the on-demand economy that connects social demand and supply, such as car-sharing or home-sharing. It has spread out through a digital platform to establish a new economic structure and maximize resource usage [4].

The concept of a sharing economy can be applied to research equipment, which is essential to do research but too expensive to own. Before the sharing economy arose, several trial systems to share research equipment had been developed. At the government level, a system for joint use of equipment, such as "e-Tube" or "ZEUS", has been established in South Korea to utilize national research facilities and idle equipment [5]. Similarly, the UK operates an equipment sharing policy called "Equipment.data", which promotes sharing of equipment between universities [6,7]. Both systems provide the database for basic information, current status, and services on the list of sharable equipment, but the user should ask the usage schedule to the equipment operator individually and visit onsite with a sample to use it [8]. At the company level, a German company provides an inventory software solution that tracks samples, specimens, consumables, and chemicals in the laboratory [9]. It was only for management purposes used to schedule the equipment or check its condition, and it is impossible to check the equipment conditions from outside since users must be in the same network to share content. At the university level, research equipment sharing services have been initiated in the Netherlands and the United States [10,11] but are operated closed, according to each university's internal policy. Eventually, users should be onsite with a sample since these systems or services are mostly limited to provide information for sharable equipment only. In the joint equipment support facilities, users can utilize equipment, but they have the hassle of visiting the facility in person to utilize the equipment. Moreover, the use of equipment not provided by these facilities must be additionally approved for both the visit and the use by individually contacting the institution that owns it. One possible suggestion to resolve these problems is to send a sample to be processed or measured to the facility or the institution. However, the equipment operators have specialized knowledge for the equipment itself but are not familiar with users' samples, resulting in compelling equipment users to still visit the facility or the institution with samples.

Currently, the concept of a remote sharing economy is proposed, which is similar to remote surgery. The only difference is that users located at a distance share equipment during the teleoperation, whereas surgeons operate on a patient located at a distance during the remote surgery. The user sends a sample to the institution that owns the necessary equipment, and the institution operator loads the sample into the equipment according to schedule. Then, the remote user can obtain the results by operating the equipment as desired. Equipment sharing based on remote operation gives several advantages over the conventional one. First, all the users can use sharable equipment provided by the facilities and the institutions worldwide once the sample is delivered and ready to use. Second, they do not have extra travel to conduct the experiment, resulting in saving time. Third, there are not any security issues from access by outsiders when they visit facilities or institutions.

There have been several studies on remote equipment operation over the Internet [12,13]. These studies used their equipment-specific protocols to operate equipment remotely and have a limitation to expand to other equipment. Recently, the Internet of Things (IoT) [14], one of the representative 4IR technologies, has been introduced and is extensively used for exchanging data between devices, mostly by monitoring signals from various sensors [15–17]; it also has been employed in controlling systems [18]. Nevertheless, IoT-based remote operation has not been applied to share equipment up to now, and demand on standard protocol still exists for sharing pieces of equipment.

In this paper, an IoT-based remote operation system for sharing equipment is proposed. As a piece of sharable equipment, two-photon laser scanning microscopy (TPLSM) is selected, which have IoT capability via a message queuing telemetry transport (MQTT), the standard protocol for IoT messaging between the equipment and its users. It is designed and built for a computer-independent system by combining independent IoT-based modules for the actuators and sensors. A web service for IoT-based TPLSM (IoT-TPLSM) is also implemented to manage and operate it remotely, including a database and MQTT broker. Its performance was evaluated from the image quality, and its potential for sharing equipment via standard protocols was confirmed by remotely acquiring 3D images for biological samples. The proposed remote sharing system gives remote users a high degree of freedom of operation in a stable network via IoT, providing a realistic solution for the sharing of equipment.

2. Materials and Methods

2.1. Hardware Design

The overall hardware configuration for IoT-TPLSM, a 3D fluorescence microscope with IoT capability, is shown in Figure 1. A customized TPLSM consists of four major components, which are the variable laser attenuator, galvanometer, Z positioner, and photon detector: the variable laser attenuator adjusts the laser power manually. The galvanometer and z positioner move the focal spot laterally and axially, respectively. Photon detector identifies optical pulse. Its users operate the TPLSM with their personal computer (PC) with the data acquisition (DAQ) system embedded in the PC and obtained 3D image stacks in the graphic user interface (GUI) implemented on the PC. On the other hand, IoT-TPLSM functions with microcontroller unit (MCU) development boards (WeMos D1 mini pro, WeMos Electronics) for WiFi MCU (ESP8266, Espressif, Shanghai, China) in a PC-free environment, instead of a DAQ system and GUI of a PC; that is, the existing DAQ system is replaced with these MCUs, and a GUI is substituted with web services via IoT. These MCUs are combined with an analog-to-digital converter (ADC), a digital-to-analog converter (DAC), or a counter (CNTR) circuit, as shown in Figures A1 and A2. These IoT-based modules include a laser power controller, a 3D scanner, and a photon counter, as explained in Figure 1a.

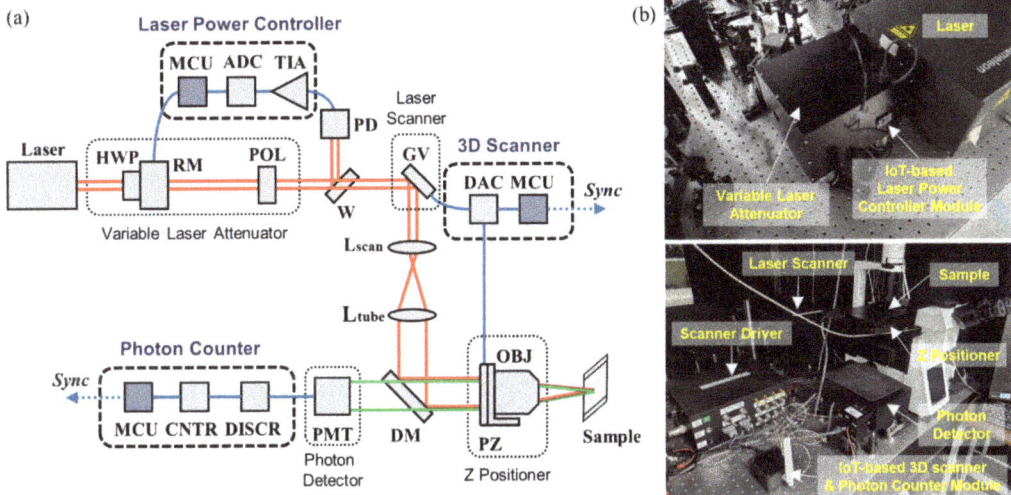

Figure 1. (a) Hardware configuration of the Internet of Things two-photon laser scanning microscopy (IoT-TPLSM). (b) Photograph of the customized modules. HWP: half-wave plate; RM: rotary motor; W: optical window; PD: photodiode; TIA: trans-impedance amplifier; GV: galvanometer mirror; L: optical lens; DM: dichroic mirror; PZ: piezo stage; OBJ: objective lens; PMT: photomultiplier tube; DISCR: discriminator; CNTR: pulse counter; MCU: microcontroller unit.

2.1.1. Laser Power Controller Module

A tunable Ti:Sapphire femtosecond pulsed laser (Chameleon Vision II, Coherent, Santa Clara, CA, USA) was used as the light source for TPLSM. It has a repetition rate of 80 MHz, and its average optical power was up to about 3.0 W at a wavelength of 800 nm without a power attenuator. Since photodamage may occur when excessive light is irradiated on the sample [19], a laser optical attenuator should be additionally required to adjust the proper light intensity on the sample. The customized variable laser attenuator unit was designed based on the light polarization. The laser output was horizontally polarized, and a half-wave plate (10RP02-46, Newport, Irvine, CA, USA) was manually rotated to change the light polarization direction, resulting in controlling the light intensity through a polarizer (GL10-B, Thorlabs, Newton, NJ, USA).

The laser power controller module was designed to regulate laser power automatically to the reference input power value given via IoT in a variable laser attenuator unit. It measured a certain percentage (about 0.19% at a wavelength of 800 nm) of the light reflected by a laser window (WG11010-B, Thorlabs) at a Si-photodiode (FDS10X10, Thorlabs) and actuated a rotary motor (T-RSW60C, Zaber Technologies, Vancouver, BC, Canada) where a half-wave plate was mounted. A closed-loop controller for laser power was implemented with MCU. The customized trans-impedance amplifier (TIA) circuit in photoconductive mode converted the photocurrent I_D into the output voltage V_{TIA} according to Equation (1):

$$V_{TIA} = V_{CC} \cdot \left(\frac{R_3}{R_2 + R_3} \right) - R_F \cdot I_D \tag{1}$$

where V_{TIA} is a measured voltage indicating the light intensity, V_{CC} is a supply voltage of a circuit, I_D is a photocurrent generated by the photodiode, and R is resistance at each point in Figure A1a. The resistance values were selected so that the output voltage represented the light intensity used in the TPLSM. The voltage V_{TIA} converted from the light intensity was transferred to the MCU through a 16-bit ADC (ADS1115, Texas Instruments), as seen in Figure A1b. The feedback loop in the laser power controller module was constructed so that the MCU made the measured light intensity reach the target one given by the remote users in the web service via IoT by controlling a rotary motor.

2.1.2. 3D Scanner Module

In the conventional TPLSM, the lateral raster scan and the axial scan are cross-repeated to obtain a 3D volumetric image with the DAQ system in a PC. For the lateral raster scan, the galvanometer XY mirrors (6210H, Cambridge Technology, Bedford, MA, USA) in the laser scanner unit steer the focal spot into the lateral position. The maximum scanner driving voltage (V_{Scan}) to cover the field of view (FOV) was determined according to Equation (2):

$$V_{Scan} = \left\{ \arctan \left(\frac{FOV}{2000} \cdot \frac{FL_{Tube}}{FL_{Obj} + FL_{Scan}} \right) \right\} \times \left\{ \frac{180°}{\pi} \cdot C_{Mirror} \cdot C_{Obj} \right\}. \tag{2}$$

The term enclosed in the first brace is for calculating the optical angle from an optical lens configuration. *FOV* is the FOV on the image plane in μm, and *FL* in mm is a focal length of the lens that corresponds to the subscript. The focal lengths of the scan lens, the tube lens, and the objective lens (UPLFLN 40XO, Olympus Life Science, Waltham, MA, USA) were 50 mm, 300 mm, and 5 mm, respectively. The term enclosed in the second brace is for converting the mechanical angle in the scanning mirrors to the driving voltage. *C* is a constant corresponding to the subscript, and the values for the galvanometer mirror and the objective lens are 0.25 and 2.2287, respectively. For the axial scan, the Z positioner unit moves with the objective lens, resulting in shifting the focal spot axially. As a Z positioner, a piezo objective positioner (MIPOS-250, Piezosystem Jena, Jena, Germany) is driven with a piezo controller (NV 40/1 CLE, Piezosystem Jena), and its driving voltage to the axial position is set as shown in Equation (3):

$$V_{Z(l)} = (Z_0 + \Delta Z \cdot l) \times C_Z \tag{3}$$

where $V_{Z(l)}$ in mV is a driving voltage of Z positioner, Z_0 in μm is an initial axial position, ΔZ in μm is an axial step size, l is a layer number to be scanned, and C_Z in mV/μm is a constant that converts the axial position value into the driving voltage. The constant is set to 0.05, taking into account the movement of 1 μm per 50 mV. In the proposed IoT-TPLSM, the 3D scanner module controlled both the laser scanner unit and Z positioner unit, which consisted of an MCU and DAC instead of PC-based DAQ system. The 3D imaging information, such as the image size in pixel numbers, the imaging area in μm, number of axial layers, and axial step size in μm, was transmitted to the MCU from the remote users via IoT, and the MCU generated the waveforms of sawtooth with different periods for three-axis scanning, based on this information. In the 3D scanner module, driving voltages calculated from Equations (2) and (3) were converted via a 16-bit DAC (AD5764R, Analog Devices, Wilmington, MA, USA), as seen in Figure A2a, and these were delivered to the laser scanner and the Z positioner, respectively, to operate IoT-TPLSM remotely.

2.1.3. Photon Counter Module

In traditional TPLSM, as shown in Figure 2, the fluorescence signal emitted from the sample is detected using a photomultiplier tube (PMT, H10682-01, Hamamatsu Photonics, Hamamatsu, Japan). As a single photon is transformed to an electron and it is multiplied in the PMT, a series of current pulses is generated. In the amplifier, it is converted and amplified to a series of voltage pulses. Only pulses above the preset threshold voltage level (V_{TH}) are altered to digital pulses in the comparator. The intensity of the fluorescence signal can be quantified by counting them through the DAQ system and it is stored in PC. However, In the IoT-TPLSM, the photon counter module quantitated the number of photons detected with a customized photon discriminator and a pulse counter (LS7366R, LSI computer systems, Melville, NY, USA), and it was transferred to remote users via IoT to construct 3D image stacks in the web service.

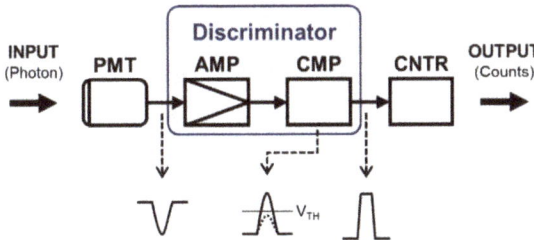

Figure 2. Configuration of the photon counter module. PMT: photomultiplier tube; AMP: amplifier; CMP: comparator; CNTR: counter.

Besides, a single image was created based on the number of photons detected during pixel residence time at each pixel by raster scan in the laser scanner unit, and the 3D scanner module should be synchronized with the photon counter module to do it. Therefore, handshaking was established by transmitting a 2-bit flag wired into two digital input/output pins in each MCU.

2.2. Software Design

The software configuration for IoT-TPLSM is shown in Figure 3. It is divided into two main parts: IoT module programming and web service programming. The MCU in each IoT module has a serial communication program that enables data exchange between the MCU and DAQ system and MQTT client program that enables the communication between the IoT module and web service with MQTT protocol. A web service functions users to send commands to the IoT actuator modules and receive data from the IoT sensor modules via web browsers in the terminal or PC. It includes an MQTT broker, web server, and database. The MQTT broker connects the web server and MQTT clients, the web server operates the web pages, and the database stores the information for the web pages and the raw data from the DAQ system.

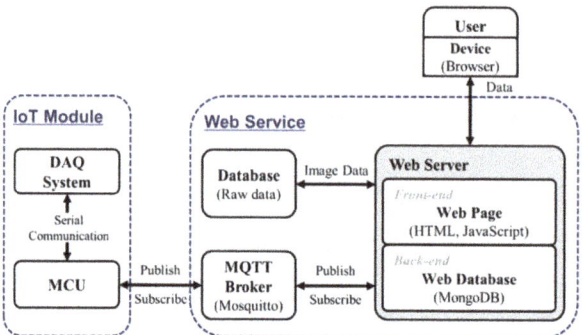

Figure 3. Software configuration of the IoT-TPLSM.

2.2.1. Serial Communication

In this study, an MCU was programmed in the Arduino integrated development environment, and data were transmitted and received by the main device of each module through serial communication. In the laser power controller module, ADC transmitted the voltage readout for the laser power to the MCU through the inter-integrated circuit (I2C) communication, and the MCU delivered the position command as a feedback control signal to the rotary motor via the universal asynchronous receiver/transmitter (UART) RS232 communication. In the 3D scanner module, position command was given to DAC using the serial peripheral interface (SPI) communication. In the photon counter module, the number of photons was passed to the MCU through SPI communication.

2.2.2. MQTT Broker

The MQTT back-end had a data transfer through the Mosquitto broker, specifying the host address and the port to be accessed by the server and MCUs. Datasets were transmitted (publish) and received (subscribe) in the JavaScript object notation (JSON) format on separate channels (topics) for the function of each module and server. The JSON format was used in various programming languages and platforms, and it was easy to exchange data between different systems through parsing [20]. Table 1 shows examples of the dataset formats used in this study.

Table 1. Examples of a message set in JSON format.

Topic	{device ID}/mirror/cmd	{device ID}/mirror/data
Clients	Publish client: Server Subscribe client: Device	Publish client: Device Subscribe client: Server

JSON format dataset	{
	"command": "axialscan",
	"options": {
	"xPixel": number,
	"yPixel": number,
	"xFov": number,
	"yFov": number,
	"imagingSpeed": number,
	"ch1": boolean,
	"ch2": boolean,
	"ch3": boolean,
	"axialStepSize": number,
	"axialZero": number,
	"numOfZLayer": number
	}
	}
		{
		"message": "axialdata",
		"data": {
		"channel": string,
		"totalPages": number,
		"page": number,
		"xPixel": number,
		"yPixel": number,
		"line": number,
		"imagingData": Uint16Array
		}
		}

The MQTT protocol provides three levels for quality of service (QoS) to ensure communication stability: from Level 0, which does not guarantee the QoS, to Level 2, which guarantees the highest QoS [21]. However, Level 2 has a disadvantage in speed performance since it tracks the handshaking process of the messages. In this study, QoS Level 1 was selected, and stability was guaranteed by including page and line information in the dataset.

2.2.3. Web Service

Figure 4 shows a detailed configuration of the web service. In this study, Amazon web service (AWS EC2, Amazon Web Services, Inc., Seattle, WA, USA) was selected as a web service, which is operated on the Ubuntu OS. The web server performs server-side scripting with the code written using JavaScript and was built through the Node.js-based Express web framework. Express is extensible, so there is no unnecessary interference in writing code and can be easily extended to third-party libraries, and an application programming interface (API) can be created or called quickly and easily through hypertext transfer protocol (HTTP) utility methods and middleware. WebSocket is a hypertext markup language 5 (HTML5) protocol that forms a dynamic two-way connection channel between a user's browser and a server. It is possible to send a message to the server through the WebSocket API and receive a response without a request. However, HTML5 may not be supported by older browsers. In consideration of compatibility issues between the browsers or with previous versions, the cross-platform WebSocket API, Socket.IO, was used to transmit data messages from the web server. At the front-end of the web server, HTML and JavaScript pages were constructed through the Angular framework. The Angular framework, which is Google's open-source JavaScript framework for single-page application (SPA) development, has most of the functions required for front-end development of not only web applications but also mobile environments and desktop applications.

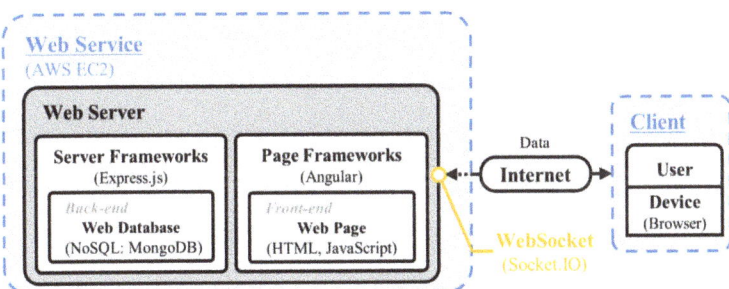

Figure 4. Configuration of the web service.

In the back-end, a general information database was created using MongoDB, which is a Not Only SQL (NoSQL) database. MongoDB can process most queries quickly with its powerful indexing function, and its processing time is faster than that of MySQL in terms of read and write [22]. Since all data is stored in JSON format, it is very easy to use with MQTT, which transmits and receives data. In addition, The MQTT connection was restricted by receiving user information and reservation information. This was to implement a minimal security system at the web server level because MQTT does not have a separate security system. Finally, result images were saved as data files and could be viewed through the server.

3. Results

3.1. Functional Validation at the Module Level

Figure 5 shows the results of the independent online operation for each module constituting the hardware. Here, modules were operated using the extension program

"MQTTBox" to directly transfer the JSON-formatted datasets. The laser power controller module was controlled through a web browser and the results were validated using a calibrated laser power meter (PowerMax PM10, Coherent). The laser optical power in Figure 5a was measured from 20 mW to 1200 mW in 20 mW steps. The coefficient of determination in the linear regression, i.e., the R^2, was 0.9990, showing a high correlation. Although the measurement is performed according to Equation (1), the value is saturated when the photocurrent exceeds a certain level. Therefore, the appropriate range should be set by adjusting the resistance value. In this study, we focused on 3D optical microscopy and set the range to low power to expand into biopsy or in-vivo studies. To ensure stability, the input outside the range was processed to output the target value as the default value of 100 mW.

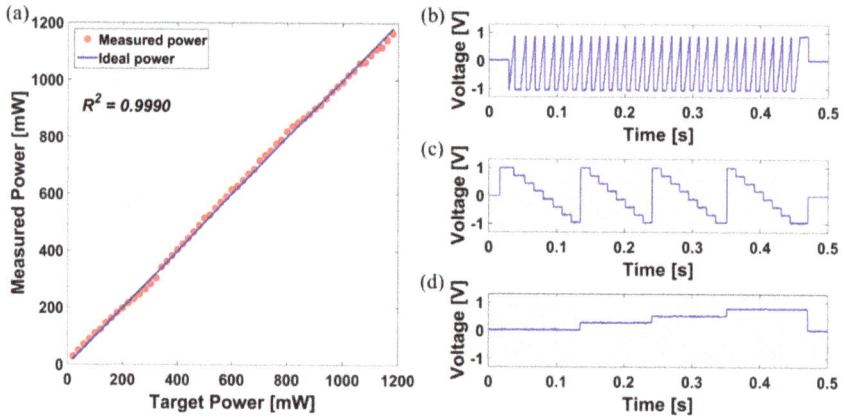

Figure 5. Verification graphs of (**a**) the laser power controller and the 3D scanner for the (**b**) X−axis, (**c**) Y−axis, and (**d**) Z−axis.

Synchronized 3D scanning was performed using units of the laser scanner and the Z positioner. It was run briefly only for functional verification of remote operation. The FOV of 50 μm × 50 μm was set to 8 × 8 pixels, imaging speed per pixel was set to 10 μs, and depth was set to 4 layers in 5 μm increments. The synchronized drive following Equation (2) was confirmed as shown in Figure 5b–d.

3.2. Performance Comparison for IoT-TPLSM at the System Level
3.2.1. Web Service for IoT-TPLSM

In order to operate IoT-TPLSM remotely, the users need to access a web service for microscopes via a web browser, and the procedure for imaging biological samples is demonstrated as follows. The users are supposed to login first through the login and signup page shown in Figure 6a. Then, users choose which microscope they use and reserve which dates they will image a sample with the selected microscope on the web page that appears in Figure 6b. On the date when the microscope is reserved, they image a sample by controlling the microscope remotely on the page presented in Figure 6c, which provides functions such as setting parameters, monitoring images, and saving the 3D image stack. After imaging, all the information for the 3D image stack log is displayed, as shown in Figure 6d. The 3D image stacks stored in the cloud service can be retrieved later. The detailed descriptions about user interface panels for the web service are in Appendix B. Such a whole remote imaging procedure was confirmed by acquiring the following images step-by-step.

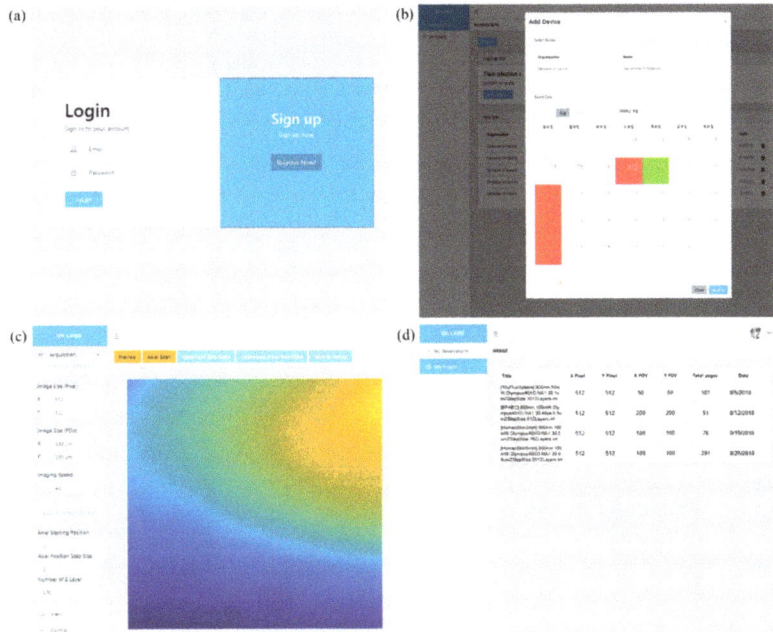

Figure 6. Page screenshots of the web service user interface: (**a**) login and sign up page; (**b**) equipment selection page; (**c**) equipment operation page; and (**d**) data log page.

3.2.2. Precision Comparison with Fluorescent Microsphere Imaging

As a standard for 3D fluorescence imaging, yellow-green fluorescent microspheres (F8836, Molecular Probes, Eugene, OR, USA) with a nominal diameter of 10 μm were imaged to evaluate the image pixel precision. The 3D image stack was obtained up to a depth of 100 μm with 1 μm steps by setting the laser power to 50 mW at a wavelength of 800 nm and FOV of 50 μm × 50 μm, which corresponds to 512 × 512 pixels. Some images extracted with a 5 μm step are presented in Figure 7a–f.

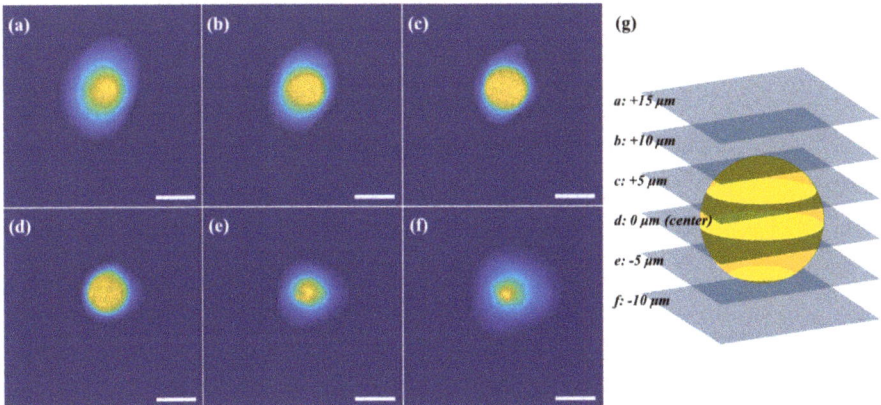

Figure 7. The images for a fluorescent microsphere at different depths from the center: (**a**) +15 μm; (**b**) +10 μm; (**c**) +5 μm; (**d**) 0 μm; (**e**) −5 μm; and (**f**) −10 μm. (scale bar: 10 μm). (**g**) Illustration of a 3D image of a fluorescent microsphere.

Images extracted in the layer for the microsphere center in the axial direction are shown in Figure 8i for quantitative comparison between images acquired online and offline. The normalized intensity profiles along the x and y direction passing through the center of a microsphere are plotted in Figure 8ii,iii, respectively. The microsphere's diameter was expressed in terms of full width at half maximum (FWHM) of the intensity profile after applying the piecewise cubic Hermite interpolation. In the IoT-TPLSM, its diameters were measured as 11.1861 μm and 11.5812 μm on the x and y direction, respectively. In the conventional TPLSM as a control, they were measured as 11.4126 μm and 11.7462 μm on the x and y direction, respectively. The error on the x-direction was 1.98%, and that on the y direction was 1.40%, which confirmed that similar precision was maintained between the online (IoT-TPLSM) and offline (TPLSM) imaging results.

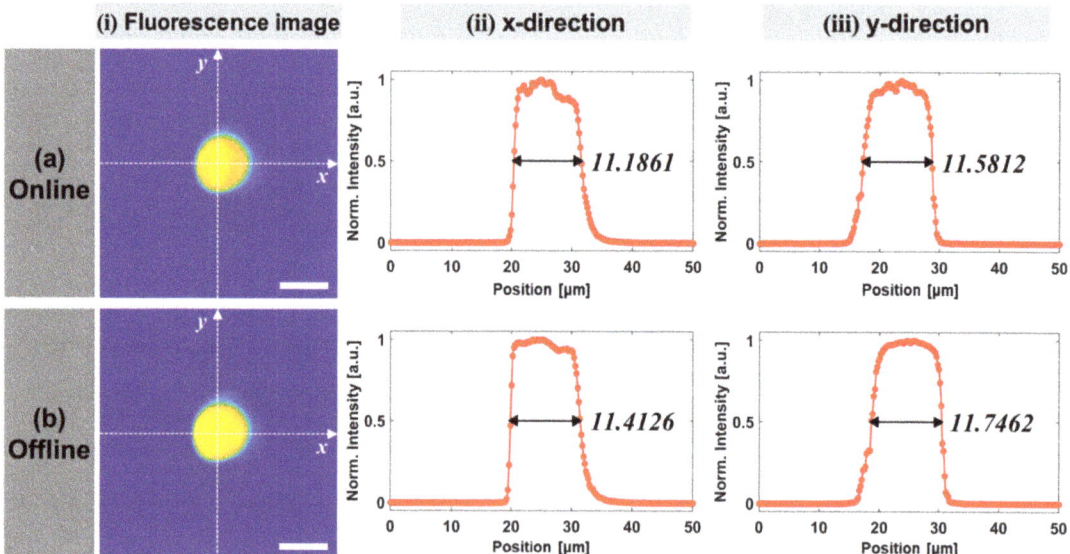

Figure 8. Quantitative comparison of the performance of (**a**) online and (**b**) offline systems. (**i**) Fluorescent microsphere images, and the corresponding diameter of the (**ii**) x-direction and (**iii**) y-direction intensity profiles (scale bar: 10 μm).

3.3. Demonstration of IoT-TPLSM

3.3.1. 3D Fluorescence Imaging at the Cellular Level

As an application of a biological sample, a 3D image stack for the stained bovine pulmonary artery endothelial (BPAE) cells (F36924, Molecular Probes), was obtained up to a depth of 25 μm, with 0.5 μm steps with the laser power at 100 mW at a wavelength of 800 nm and FOV of 200 μm × 200 μm, which corresponds to 512 × 512 pixels. Some images extracted with 2 μm steps are displayed in Figure 9a–f. Although BPAE cells were stained using three fluorescent dyes, only F-actin stained with Alexa Fluor 488 phalloidin and nuclei stained with DAPI were clearly identified. Mitochondria stained with MitoTracker Red CMXRos was not detected because the laser wavelength is out of range on its excitation wavelengths. Operating the IoT-TPLSM remotely, the 3D fluorescent image at the cellular level was obtained with high similarity to the offline system.

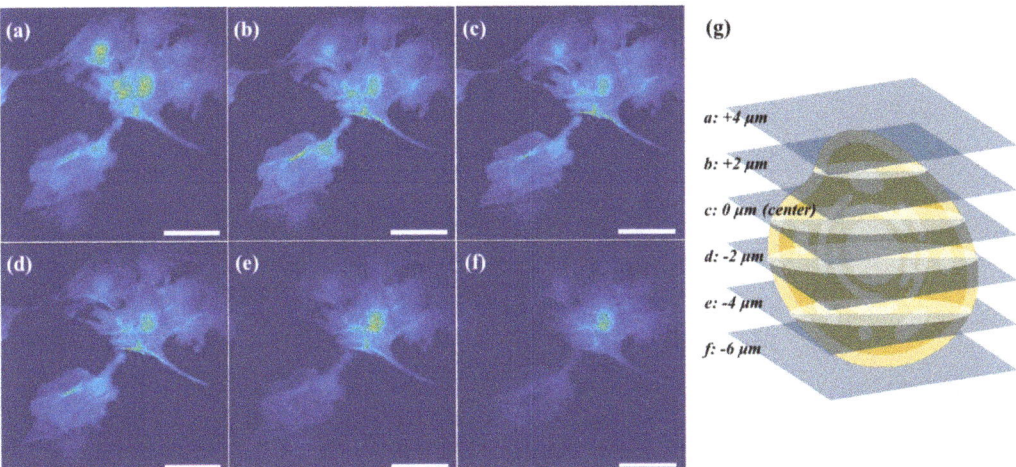

Figure 9. The images for the stained bovine pulmonary artery endothelial (BPAE) cells at different depths from the center: (**a**) +4 μm; (**b**) +2 μm; (**c**) 0 μm; (**d**) −2 μm; (**e**) −4 μm; and (**f**) −6 μm (scale bar: 10 μm). (**g**) Illustration of the 3D image of BPAE cells.

3.3.2. 3D Autofluorescence Imaging at the Tissue Level

Ex-vivo human skin tissue provided from Dankook University Hospital was also imaged as a 3D autofluorescence image stack, which is expressed in the unstained tissue. The 3D images were acquired up to a depth of 100 μm with a 2 μm step by setting the laser power of 100 mW at a wavelength of 800 nm and FOV of 100 μm × 100 μm corresponds to 512 × 512 pixels. Some images extracted at different depths are represented in Figure 10a–i. Starting from the stratum corneum without nuclei on the surface of the skin, the keratinocyte in the epidermis, the dermal-epidermal junction, and the collagenous fiber tissue in the dermis were definitely recognized. It was observed that the stratum corneum without nuclei existed at 17 μm; nuclei in the epidermal cells began to appear at 22 μm. It was noticed that the keratinocyte nuclei were distributed as the granular layer at 45 μm, cell membranes were maintained at 54 μm, polygonal keratinocytes as the stratum spinosum at 60 μm, cubic basal cells at 66 μm, and the dermal–epidermal junction where the dermal fibrous tissue and some cells were mixed at 76 μm. It was also found that the collagenous fiber tissue and the amorphous collagen tissue in the dermis were located at 82 μm and 96 μm, respectively. It was shown that the 3D image of the label-free ex-vivo human skin tissue was obtained successfully by operating IoT-TPLSM remotely.

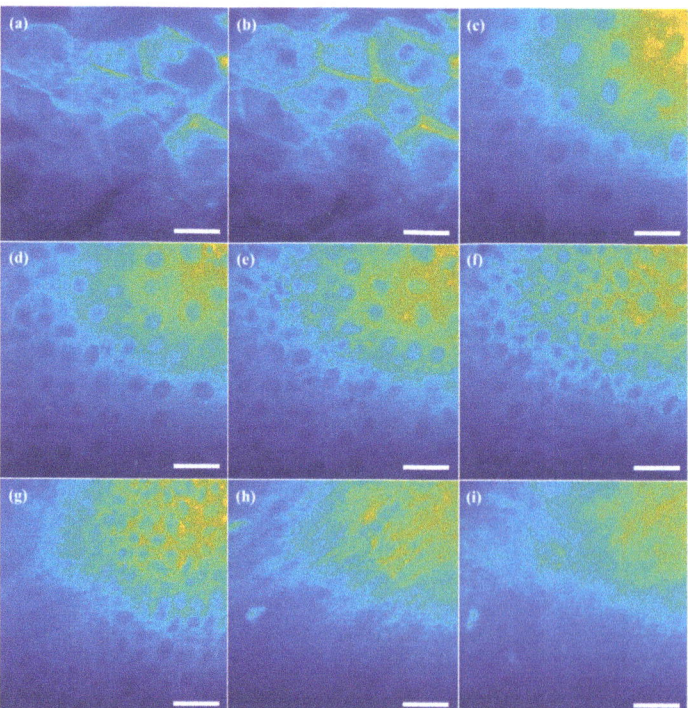

Figure 10. The images of the human skin tissue at different depths from the surface: (**a**) stratum corneum without nuclei at 17 μm, (**b**) epidermal cells with nuclei 17 μm, (**c**) granular layer at 45 μm, (**d**) cell membranes at 54 μm, (**e**) stratum spinosum at 60 μm, (**f**) cubic basal cells at 66 μm, (**g**) dermal fibrous tissue at 76 μm, (**h**) collagenous fiber tissue at 82 μm, and (**i**) amorphous collagen tissue at 96 μm (scale bar: 20 μm).

4. Discussion and Conclusions

In this study, an IoT-based remote control system for shared research equipment was proposed and implemented. The offline system for equipment was expanded to the IoT convergence platform and cloud, resulting in transforming the online system. Moreover, a single synchronized system with independently configured MCUs and the web service interface for a customized DAQ were completed. Using the remote full-duplex, it was confirmed that the remote operation for various research equipment can be additionally and alternatively utilized in diverse research fields. It is also expected that the IoT-based research equipment sharing system allows researchers at a remote site to set up an experiment as well as check and save the result at their own will.

By taking IoT-TPLSM as an example application, the stained cellular images and the autofluorescent tissue images were obtained. As a result, it was confirmed that performances for the online system, such as the image acquisition time, the image quality, and GUI for image acquisition, were almost the same as those for the offline one. The image distortion shown under 2% can be easily corrected with the calibration for driving voltages. Besides, as the proposed remote sharing system used the web service and MCU, access for the IoT module was fully granted to the remote users to operate shared equipment freely. Simultaneously, since the shared equipment working with the MQTT protocol via IoT was independent of the computer itself, unexpected OS problems were eliminated, and the operating stability was more secured.

The remote operation of research equipment was executed through a wireless network. However, various attempts to overcome its vulnerability are needed, since wireless networks are relatively insecure compared to wired networks. As the MQTT protocol supports QoS, the optimal QoS for real-time communication can be set [23]. In addition, while using the MQTT, a standby database can be placed between the gateway and the server [24], and the current protocol can be upgraded or attached parallel to other wireless communication protocols to address network failures [25–27]. By applying such stabilization to the system in this study, it is believed it would ensure the rapid and stable remote operation of shared equipment, even using a wireless network.

The IoT-based remote sharing system is expected to provide a realistic solution for equipment utilization and thus can be used as a basic technology in many industries. In the manufacturing field, it can be applied to a smart factory or for hybrid manufacturing implemented with remote robot systems [28]. In the biomedical field, the remote robot system could enable automatic sample replacement and remote experiments in a single queue, and the remote operation system can be extended to telemedicine with deep learning to aid in disease diagnosis in the clinic [29–31]. The proposed remote sharing system is also expected to serve as a window for network formation and integration between researchers in various fields through the remote sharing of various research equipment and to open a new chapter in research and development areas.

Author Contributions: Conceptualization, D.K.; methodology, E.P. and D.K.; software, E.P. and J.L.; validation, E.P., J.L., B.C.P., and D.K.; formal analysis, E.P.; investigation, E.P.; resources, J.L., B.C.P., and D.K.; data curation, E.P.; writing—original draft preparation, E.P. and J.L.; writing—review and editing, E.P., J.L., B.C.P., and D.K.; visualization, E.P. and J.L.; supervision, D.K.; project administration, D.K.; funding acquisition, D.K. All authors have read and agreed to the published version of the manuscript.

Funding: The present research was conducted by the research fund of Dankook University in 2019.

Institutional Review Board Statement: The human tissue used in this study is obtained from a genomic analysis study during the staged development of skin cancer, which is previously approved by the Dankook University Hospital Ethics Committee (IRB No. DKUH 2019-08-005).

Informed Consent Statement: Not applicable.

Data Availability Statement: The hardware data presented in this study are available in the article. The software data presented in this study are available on request from the corresponding author. Source codes are not publicly available due to the security policy of the funding organization.

Conflicts of Interest: The authors declare no conflict of interest.

Appendix A. Electronic Circuits for Modules

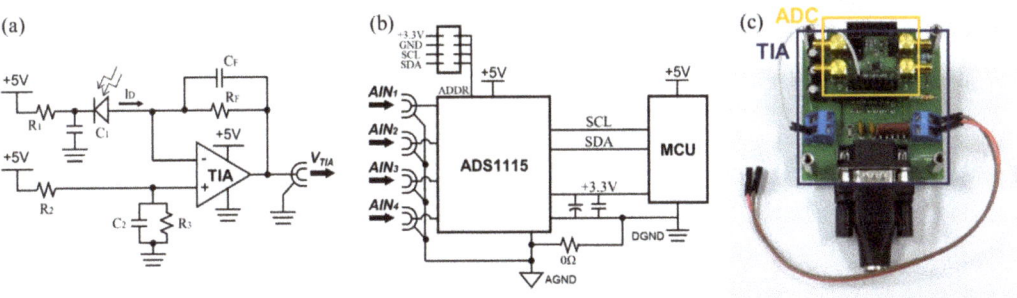

Figure A1. Electronic schematics for (**a**) the trans-impedance amplifier (TIA) circuit and (**b**) the analog-to-digital converter (ADC) circuit. (**c**) A photograph of the combined circuit boards for the laser power controller module.

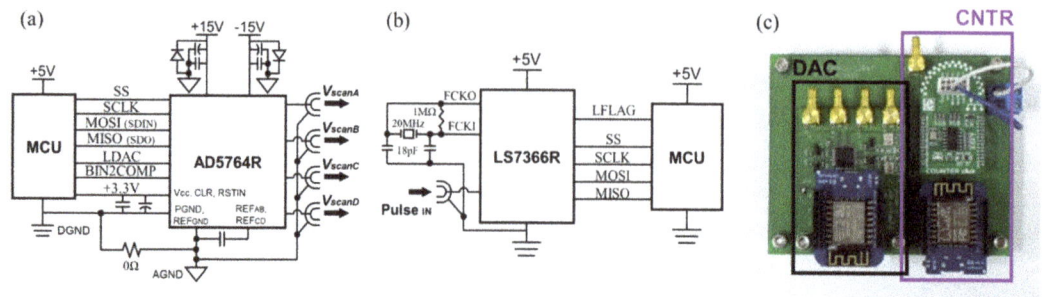

Figure A2. Electronic schematics for (**a**) the digital-to-analog converter (DAC) circuit and (**b**) the pulse counter (CNTR) circuit. (**c**) A photograph of the combined circuit boards for the 3D scanner module and the photon counter module.

Appendix B. IoT-TPLSM Web Service User Interfaces

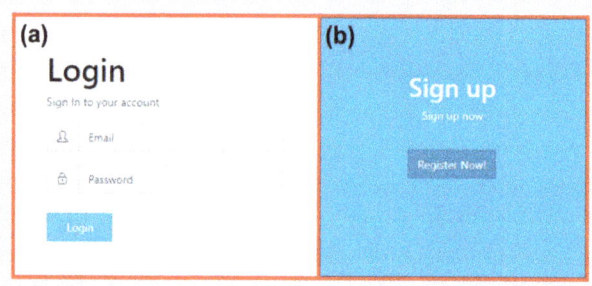

Figure A3. The user interface of the login and sign up page: (**a**) login panel and (**b**) sign-up panel.

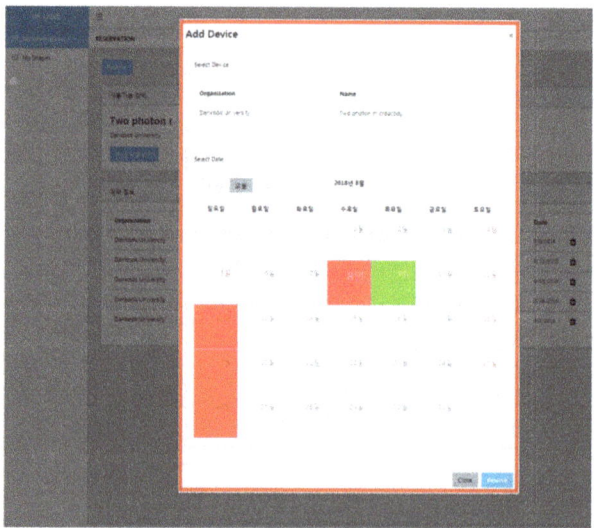

Figure A4. The user interface of the equipment selection page. The red box is a calendar for reservations.

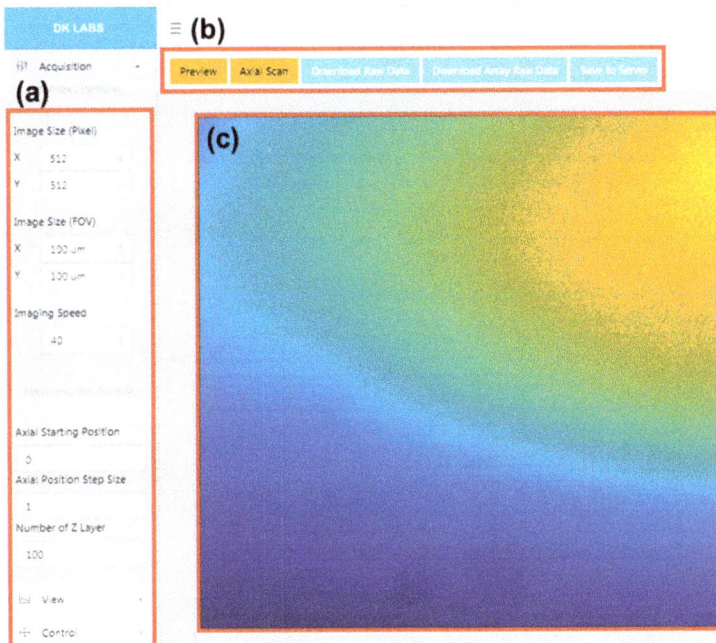

Figure A5. The user interface of the equipment operation page: (**a**) parameter-setting panel; (**b**) operation button panel; and (**c**) result image panel.

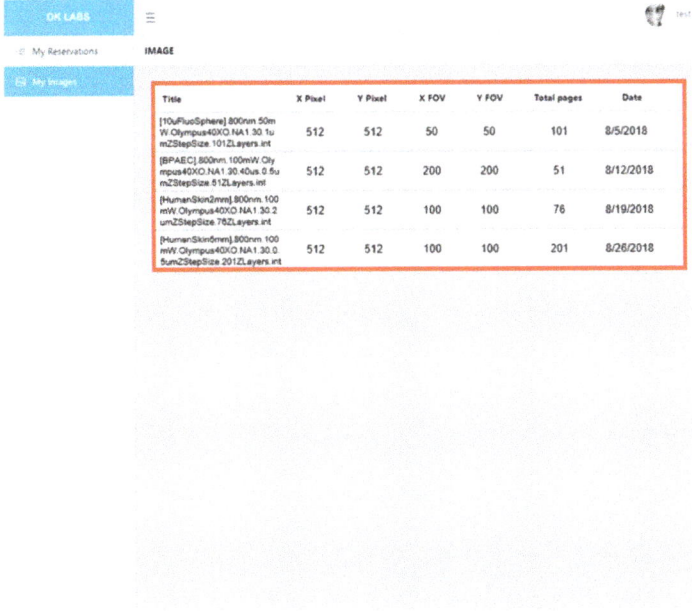

Figure A6. The user interface of the data log page. The red box is a result data log panel of shared equipment.

References

1. Schwab, K. *The Fourth Industrial Revolution*; Penguin Books Ltd.: London, UK, 2016; ISBN 978-024-130-075-6.
2. Lee, S.H. *Policy Trends of the 4th Industrial Revolution in Major Advanced Countries*; Institute for Information & Communications Technology Promotion: Seoul, Korea, 2016.
3. Radziwon, A.; Bilberg, A.; Bogers, M.; Madsen, E.S. The smart factory: Exploring adaptive and flexible manufacturing solutions. *Procedia Eng.* **2014**, *69*, 1184–1190. [CrossRef]
4. Chung, K.P.; Cho, K.B.; Kim, S.W. The study of availability and factor analysis on car-sharing for sharing economy. *Korean Comp. Gov. Rev.* **2015**, *19*, 105–124. [CrossRef]
5. Kwon, K.H.; Kim, H.J.; Shin, S.H.; Kim, B.H.; Seo, I.S.; No, S.C. *2015 National Research Facility Equipment Operation Management Survey Report*; Ministry of Science, ICT and Future Planning: Daejeon, Korea, 2015.
6. University of Cambridge Research Operations Office. Available online: https://www.research-operations.admin.cam.ac.uk/policies/equipment-sharing (accessed on 5 August 2020).
7. Equipment.data. Available online: http://equipment.data.ac.uk/ (accessed on 5 August 2020).
8. UNIVERSITIES, U.K. *Efficiency, Effectiveness and Value for Money*; Universities UK: London, UK, 2015.
9. Eppendorf AG. Available online: https://online-shop.eppendorf.us/US-en/Freezers-44537/Sample-Management-376575/eLABInventory-software-PF-372725.html (accessed on 28 July 2020).
10. Wageningen University & Research. Available online: https://www.wur.nl/en/Value-Creation-Cooperation/Facilities/Wageningen-Shared-Research-Facilities.htm (accessed on 28 July 2020).
11. The Ohio State University College of Veterinary Medicine. Available online: https://vet.osu.edu/research/shared-resources-and-equipment (accessed on 28 July 2020).
12. Sánchez, J.; Dormido, S.; Pastor, R.; Morilla, F. A Java/Matlab-based environment for remote control system laboratories: Illustrated with an inverted pendulum. *IEEE Trans. Educ.* **2004**, *47*, 321–329. [CrossRef]
13. Ann, Y.H.; Kang, J.S.; Jung, H.J.; Kim, H.S.; Jung, H.S.; Han, H.; Jeong, J.M.; Gu, J.E.; Lee, S.D.; Lee, J.S.; et al. Remote access and data acquisition system for high voltage electron microscopy. *Appl. Microsc.* **2006**, *36*, 7–16.
14. Greengard, S. *The Internet of Things*, 1st ed.; MIT Press: Cambridge, MA, USA, 2015; ISBN 978-15-3450-774-6.
15. Jung, H.; Park, C.W. Design and Implementation of MQTT Based Real-time HVAC Control Systems. *J. Korea Inst. Inf. Commun. Eng.* **2015**, *19*, 1163–1172. [CrossRef]
16. Susila, I.P.; Istofa; Kusuma, G.; Sukandar; Isnaini, I. Development of IoT based meteorological and environmental gamma radiation monitoring system. In Proceedings of the 4th International Conference on Engineering, Technology, and Industrial Application (ICETIA2017), Surakarta, Indonesia, 13–14 December 2017; p. 060004.
17. Antony, J.; Mathuria, D.S.; Datta, T.S.; Maity, T. Development of intelligent instruments with embedded HTTP servers for control and data acquisition in a cryogenic setup—The hardware, firmware, and software implementation. *Rev. Sci. Instrum.* **2015**, *86*, 125003. [CrossRef] [PubMed]
18. Seo, J.O.; Kim, C.W. Design and Implementation of Realtime Things Control System Using MQTT and WebSocket in IoT Environment. *J. Korea Inst. Electron. Commun. Sci.* **2018**, *13*, 517–524. [CrossRef]
19. Owen, T. *Fundamentals of Modern UV-Visible Spectroscopy Principles and Applications of UV-Visible Spectroscopy*; Agilent Technologeis: Santa Clara, CA, USA, 1996.
20. Nurseitov, N.; Paulson, M.; Reynolds, R.; Izurieta, C. Comparison of JSON and XML data interchange formats: A case study. *Caine* **2009**, *9*, 157–162.
21. Behnel, S.; Fiege, L.; Muhl, G. On quality-of-service and publish-subscribe. In Proceedings of the 26th IEEE International Conference on Distributed Computing Systems Workshops (ICDCSW'06), Lisboa, Portugal, 4–7 July 2006; p. 20.
22. Nyati, S.S.; Pawar, S.; Ingle, R. Performance evaluation of unstructured NoSQL data over distributed framework. In Proceedings of the IEEE International Conference on Advances in Computing, Communications and Informatics (ICACCI), Mysore, India, 22–25 August 2013; pp. 1623–1627.
23. Chang, B.; Zhao, G.; Imran, M.A.; Chen, Z.; Li, L. Dynamic wireless QoS analysis for real-time control in URLLC. In Proceedings of the 2018 IEEE Globecom Workshops (GC Wkshps), Abu Dhabi, United Arab Emirates, 9–13 December 2018; pp. 1–5.
24. Zhang, Z.; Jin, Y. Design of Temperature Remote Monitoring System Based on STM32. In Proceedings of the 2020 IEEE International Conference on Artificial Intelligence and Computer Applications (ICAICA), Dalian, China, 27–29 June 2020; pp. 757–759.
25. Lee, K.-H.; Park, K.Y. Overall Design of Satellite Networks for Internet Services with QoS Support. *Electronics* **2019**, *8*, 683. [CrossRef]
26. Marabissi, D.; Mucchi, L.; Caputo, S.; Nizzi, F.; Pecorella, T.; Fantacci, R.; Nawaz, T.; Seminara, M.; Catani, J. Experimental Measurements of a Joint 5G-VLC Communication for Future Vehicular Networks. *J. Sens. Actuator Netw.* **2020**, *9*, 32. [CrossRef]
27. Güldenring, J.; Gorczak, P.; Patchou, M.; Arendt, C.; Tiemann, J.; Wietfeld, C. SKATES: Interoperable Multi-Connectivity Communication Module for Reliable Search and Rescue Robot Operation. In Proceedings of the 2020 16th International Conference on Wireless and Mobile Computing, Networking and Communications (WiMob), Thessaloniki, Greece, 12–14 October 2020; pp. 7–13.
28. Fu, S.; Bhavsar, P.C. Robotic arm control based on internet of things. In Proceedings of the 2019 IEEE Long Island Systems, Applications and Technology Conference (LISAT), Farmingdale, NY, USA, 3 May 2019; pp. 1–6.

29. Iqbal, F.M.; Lam, K.; Joshi, M.; Khan, S.; Ashrafian, H.; Darzi, A. Clinical outcomes of digital sensor alerting systems in remote monitoring: A systematic review and meta-analysis. *NPJ Digit. Med.* **2021**, *4*, 1–12. [CrossRef] [PubMed]

30. Kindle, R.D.; Badawi, O.; Celi, L.A.; Sturland, S. Intensive care unit telemedicine in the era of big data, artificial intelligence, and computer clinical decision support systems. *Crit. Care Clin.* **2019**, *35*, 483–495. [CrossRef] [PubMed]

31. Wijesinghe, I.; Gamage, C.; Perera, I.; Chitraranjan, C. A smart telemedicine system with deep learning to manage diabetic retinopathy and foot ulcers. In Proceedings of the 2019 Moratuwa Engineering Research Conference (MERCon), Moratuwa, Sri Lanka, 3–5 July 2019; pp. 686–691.

sensors

MDPI

Communication

Integrated Quad-Scanner Strategy-Based Optical Coherence Tomography for the Whole-Directional Volumetric Imaging of a Sample

Sm Abu Saleah [1,†], Daewoon Seong [1,†], Sangyeob Han [1,2], Ruchire Eranga Wijesinghe [3], Naresh Kumar Ravichandran [4], Mansik Jeon [1,*] and Jeehyun Kim [1,*]

1 School of Electronic and Electrical Engineering, College of IT Engineering, Kyungpook National University, 80, Daehak-ro, Buk-gu, Daegu 41566, Korea; abu.saleah@knu.ac.kr (S.A.S.); smc7095@knu.ac.kr (D.S.); syhan850224@knu.ac.kr (S.H.)
2 Institute of Biomedical Engineering, School of Medicine, Kyungpook National University, 80, Daehak-ro, Buk-gu, Daegu 41566, Korea
3 Department of Materials and Mechanical Technology, Faculty of Technology, University of Sri Jayewardenepura, Pitipana, Homagama 10200, Sri Lanka; erangawijesinghe@sjp.ac.lk
4 Center for Scientific Instrumentation, Korea Basic Science Institute, 169-148, Gwahak-ro Yuseong-gu, Daejeon 34133, Korea; nareshr9169@kbsi.re.kr
* Correspondence: msjeon@knu.ac.kr (M.J.); jeehk@knu.ac.kr (J.K.)
† These authors contributed equally to this work.

Citation: Saleah, S.A.; Seong, D.; Han, S.; Wijesinghe, R.E.; Ravichandran, N.K.; Jeon, M.; Kim, J. Integrated Quad-Scanner Strategy-Based Optical Coherence Tomography for the Whole-Directional Volumetric Imaging of a Sample. *Sensors* **2021**, *21*, 1305. https://doi.org/10.3390/s21041305

Academic Editor: Manuchehr Soleimani

Received: 15 January 2021
Accepted: 8 February 2021
Published: 11 February 2021

Publisher's Note: MDPI stays neutral with regard to jurisdictional claims in published maps and institutional affiliations.

Abstract: Whole-directional scanning methodology is required to observe distinctive features of an entire physical structure with a three dimensional (3D) visualization. However, the implementation of whole-directional scanning is challenging for conventional optical coherence tomography (OCT), which scans a limited portion of the sample by utilizing unidirectional and bidirectional scanning methods. Therefore, in this paper an integrated quad-scanner (QS) strategy-based OCT method was implemented to obtain the whole-directional volumetry of a sample by employing four scanning arms installed around the sample. The simultaneous and sequential image acquisition capabilities are the conceptual key points of the proposed QS-OCT method, and were implemented using four precisely aligned scanning arms and applied in a complementary way according to the experimental criteria. To assess the feasibility of obtaining whole-directional morphological structures, a roll of Scotch tape, an ex vivo mouse heart, and kidney specimens were imaged and independently obtained tissue images at different directions were delicately merged to compose the 3D volume data set. The results revealed the potential merits of QS-OCT-based whole-directional imaging, which can be a favorable inspection method for various discoveries that require the dynamic coordinates of the whole physical structure.

Keywords: optical coherence tomography; quad-scanner scanning strategy; whole-directional scanning; full-directional imaging

1. Introduction

Whole-directional (i.e., full-directional) scanning has been widely applied and utilized for various medical imaging techniques to identify distinctive features of samples [1,2]. Magnetic resonance imaging (MRI) and computed tomography (CT) are representative existing imaging techniques that have been actively applied for the whole-body imaging of samples [3,4]. Moreover, positron emission tomography (PET) and PET/CT techniques have also been demonstrated to obtain the whole-body imaging of samples [5]. These imaging systems are widely used for the diagnosis of cancer because whole-body imaging enables the inspection of the entire sample through a single scanning attempt. In accordance with the development of high-resolution whole-body imaging at the human level, full-body scanning methods have been developed to match the sample characteristics, such as

129

acquiring tractography for whole mouse heart [6], brain [7], and the quantitative analysis of embryos [8]. The aforementioned imaging modalities are suitable for full-body imaging at the human level because of their deep penetration depth. However, these imaging techniques are limited by their low resolution; therefore, an optical imaging modality such as optical coherence tomography (OCT) could be a suitable solution for acquiring high-resolution morphological images of a 3D sample at the tissue level.

OCT is a non-invasive optical imaging technique used to obtain high-resolution, cross-sectional images of inner microstructures in materials and biological tissues [9]. OCT has been widely applied in various fields, including medical diagnosis [10,11], dentistry [12,13], and industrial applications [14,15]. The single-scanner-based unidirectional scanning of conventional OCT has been applied in numerous applications [16,17]. Unidirectional scanning OCT has been widely utilized for high-resolution imaging and has been utilized in various system designs, such as optical Doppler tomography [18], OCT angiography [19], and polarization-sensitive OCT [20]. However, conducting whole-directional volumetric screening is challenging with these systems because it involves the single-side imaging of a 3D sample. Therefore, the bidirectional scanning method was implemented to compensate for the limitation of single-side scanning methods by imaging dual sides of a sample, which is necessary for measuring the thickness and overlapping morphological structures of thin and high-refractive-index samples [21,22]. However, the imaging results of bidirectional scanning are affected by the sample shape and thickness, limiting its applicability for full-directional imaging.

To assess the full-directional morphological structure of a target sample, the rotational imaging (RI) strategy was demonstrated by rotating the sample stage multiple times. RI-OCT was initially implemented for in situ embryonic imaging to obtain structural information from different angles [23,24]. However, the RI-OCT system has some drawbacks, as shifting the sample direction by rotating the sample stage and adjusting the imaging focus after every rotation requires a long image acquisition time. To reduce the acquisition time, a parallel imaging scheme was introduced by scanning multiple locations of a sample concurrently. Parallel images were obtained simultaneously from multiple locations of a sample by adding optical path length delays that are longer than the light penetration depth of tissue in a parallel channel [25]. Space-division multiplexing (SDM) OCT is a representative parallel-imaging technique that achieves an improved imaging speed [26]. The SDM-OCT system was demonstrated to achieve the efficient imaging of a single-directional morphological structure using 8-beam multiplexing [27] and has been indicated for a clinical feasibility analysis of its ophthalmic applications [25].

In this study, we demonstrated a QS methodology-based OCT imaging technique to obtain whole-directional tomographic images of a sample without the rotation of the sample stage. To satisfy the different imaging criteria, such as whole-directional simultaneous imaging and unidirectional sequential imaging, we propose applying two different types of QS-OCT concepts: simultaneous and sequential modes. In simultaneous QS-OCT, the whole-directional imaging of a 3D sample was scanned from four different directions concurrently using the SDM technique of parallel imaging, which enhanced the imaging speed of the system. To verify the possibility of simultaneous QS-OCT implementation, a rolled Scotch tape sample was imaged from the whole-direction of the sample concurrently. In addition, sequential QS-OCT was demonstrated to obtain whole-directional volumetric images of a 3D sample using four scanners operated in a successive order and to address the drawbacks of the power loss of the simultaneous QS-OCT approach. To verify the capability of the whole-directional volumetric imaging of the sequential QS-OCT system, ex vivo mouse heart and kidney specimens were imaged and merged. Therefore, the QS-OCT concept in simultaneous and sequential modes is a whole-directional imaging modality that supports different imaging criteria and can obtain whole-directional morphological images without any rotation of the sample.

2. Materials and Methods

2.1. Optical Configuration and Duty Cycle Illustration of Simultaenous QS-OCT

The schematic of the optical configuration for simultaneous QS-OCT is shown in Figure 1a. The system was equipped with a broadband light source (EXS210090-01, Exalos, Zurich, Swiss) with a central wavelength of 840 nm, a full width at half maximum bandwidth of 48 nm, and an average output power of 15 mW. Four one-axis galvanometerscanners (GVS001, Thorlabs, Newton, NJ, USA) were mounted at four sides of a sample stage to cover the full-directional scanning of the sample. In each sample arm, a 2-inch object lens (AC508-100-B, Thorlabs, Newton, NJ, USA) was used for large-area scanning. Four reference arms, identically designed with a collimator (F260APC-B, Thorlabs, Newton, NJ, USA), lens (AC254-030-B, Thorlabs, Newton, NJ, USA), and mirror (PF10-03-P01, Thorlabs, Newton, NJ, USA), were used for the four sample arms in this system. The power of each interferometer was equally divided and maintained at the saturation level of the detector. The ratio of all fiber couplers (TW850R5A2, Thorlabs, Newton, NJ, USA) utilized in this system was 50:50. Polarization controllers (FPC023, Thorlabs, Newton, NJ, USA) were utilized in each reference arm and sample arm to regulate the polarized state of the transmitted light. To apply the SDM technique for obtaining whole-directional images concurrently, the optical path length of each interferometer was accurately controlled. Each interference signal obtained by four different interferometers was transferred to a customized spectrometer, whose configuration was described in detail in [28]. A frame grabber (PCIe-1433, National Instruments, Austin, TX, USA) and data acquisition board (DAQ, PCIe-6323, National Instruments, Austin, TX, USA) were employed to precisely control the hardware compositions. A linear motor stage (M-403, PI, Karlsruhe, Germany) was used to move the sample stage up and down to acquire the 3D volumetric imaging of the sample.

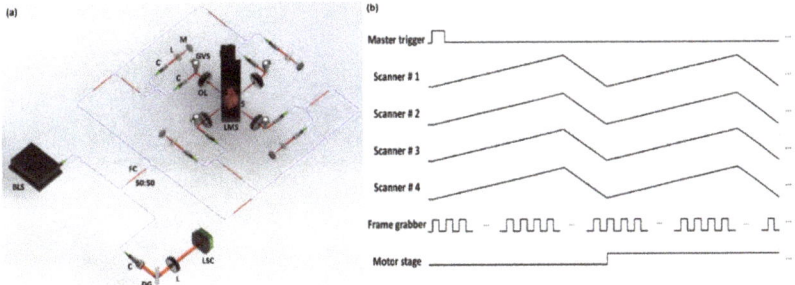

Figure 1. The optical configuration and duty cycle illustration of the simultaneous quad-scanner (QS)-OCT system. (**a**) Simultaneous QS-OCT with space-division multiplexing for simultaneous whole-directional imaging. (**b**) The duty cycles of simultaneous QS-OCT system operation. BLS, broadband light source; C, collimator; DG, diffraction grating; FC, fiber coupler; GVS, galvanometer scanner; L, lens; LMS, linear motor stage; LSC, line-scan camera; OL, objective lens; S, sample.

The synchronized operation of different hardware instruments (four scanners, frame grabber, and linear motor stage) is essential to precisely obtain whole-directional morphological images of a sample at the same time. Therefore, the duty cycle illustration explains the sequence of operation, which was employed in simultaneous QS-OCT system setups, as shown in Figure 1b. In accordance with the rising edge of the main trigger from DAQ, four scanners and a frame grabber were synchronized concurrently for scanning and grabbing. Although four sample arms were implemented independently, each scanner was controlled with identical operating timing by sawtooth waves. Moreover, the frame grabber was precisely controlled according to the grabbing timing, which was determined by the utilized A-scan rate of the scanner. In the proposed QS-OCT (both simultaneous and sequential), the operations of the scanners started with 20 kHz A-line rates following the rising edge of the main trigger signal. In addition, the frame grabber and motor stage were started simultaneously, with intervals for grabbing and moving identically set as 50 μs to be matched with the A-scan rate. After obtaining single B-scan images, the motor stage was moved immediately as much as the preset step size for measuring the whole volume of the sample. The aforementioned scanning process was continued for the total range of the sample and the whole-directional volumetric raw data were obtained using the proposed simultaneous QS-OCT system.

2.2. Optical Configuration and Duty Cycles of Sequential QS-OCT

Figure 2a shows the schematic of the optical configuration of the sequential QS-OCT for whole-directional volumetric imaging that resolves the power limitations of the simultaneous method. In the case of a simultaneous strategy, SDM was implemented for parallel imaging to obtain whole-directional images simultaneously. Although the simultaneous system offered a fast scanning speed, the power of the detected interference signal was comparably low, leading to the degradation of image intensity at high frequencies (depth direction in the imaging window), because four different scanners were each equipped with one interferometer and the source power was divided into four interferometers. In contrast, as a feature of the sequential QS-OCT system, each scanner was operated successively in a systematically fixed order to improve the transferred power of each sample arm. The optical components used in the sequential system (including source, fiber coupler, sample arm, reference arm, and spectrometer) were the same as those used in the simultaneous system. Unlike the simultaneous method, one common reference arm was used for four sample arms in the sequential QS-OCT system to reduce the number of fiber couplers used in the simultaneous QS-OCT system. A motorized linear stage, utilized in the simultaneous method as well, was implemented for scanning the whole-directional volumetric imaging where the switching between individual scanners was performed manually.

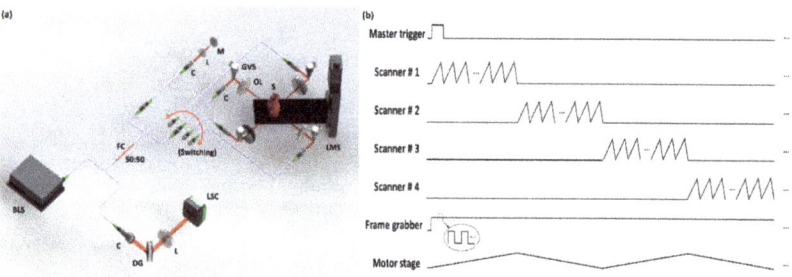

Figure 2. The optical configuration and duty cycle diagram of sequential quad-scanner (QS)-OCT system. (**a**) Sequential QS-OCT for successive operation of each scanner. (**b**) The duty cycle illustration of sequential QS-OCT system operation. BLS, broadband light source; C, collimator; DG, diffraction grating; FC, fiber coupler; GVS, galvanometer scanner; L, lens; LMS, linear motor stage; LSC, line-scan camera; OL, objective lens; S, sample.

Likewise, a synchronization process for operating the simultaneous system and the hardware compositions of sequential QS-OCT method were controlled according to precisely synchronized timing, as shown in Figure 2b. The overall operating timing is identical to that of the simultaneous method; however, the scanners operated independently according to the switching sequence to enhance the power of the system. After acquiring a single B-scan image, the motor stage was moved to the next point according to the timing of a single B-scan acquisition, but the direction of movement was conversely changed for efficient scanning (i.e., upward (#1 and #3) and downward (#2 and #4)). As a result of whole-direction imaging by following the precisely controlled operating order, 3D volumetric raw data were obtained and processed by customized merging software.

2.3. The Description and the Flow Chart of the Image Processing Algorithm

To obtain a whole-directional volumetric data set, we developed a LabVIEW (National Instruments, Austin, TX, USA) -based customized merging software and the operational flowchart of this software (described in Figure 3). The first step of the software was used to classify the images acquired from four different scanners using simultaneous and sequential methods, respectively. In the case of the simultaneous method, the input image separation process was required, since the obtained B-scan image was composed with different signals of 4 independent scanners by employing SDM for the enhancement of imaging speed. Separated images matched with each scanner were obtained by replacing the value to zero, excepting the signals of each region-of-interest. In contrast, the separation process for the input image is not required in sequential QS-OCT, which scans four different directions successively. Four different images obtained from each direction acquired by both simultaneous and sequential QS-OCT—as shown in Figure 3a—were transferred to the post-processing part, including pixel rescaling, image placement, and fine merging. Pixel rescaling, the second step of the customized algorithm, was applied for adjusting the pixel resolution difference between the lateral (14.7 μm) and axial (4.69 μm) direction. The difference in pixel resolution between the lateral and axial direction affected the accuracy of the merging process, which requires a comparison of every pixel's position. To match the pixel resolution, linear interpolation was applied to the lateral direction according to the ratio difference between the lateral and axial direction. Next, interpolated images, which are shown in Figure 3b, were placed at each direction to conduct the image merging process. Prior to the imaging of QS-OCT, a 3D-printed cylindrical sample, which had features around the surface to indicate the overlapped position, was imaged to obtain a reference value of image merging. Therefore, as shown in Figure 3c, interpolated images of each scanner were placed in accordance with the pre-acquired reference value of pixel movement. In addition, to finely merge the whole-directional images, we compared intensities and selected higher values for the overlapped regions, based on the fact that the intensity of OCT signal, measured at the focal point, is higher than at other positions. To compare the quality of merged images between step 3 (image placement) and step 4 (fine merging), two representative regions (red and yellow squares) were selected and magnified. The result of applying the fine merging method, as shown in Figure 3d, demonstrates a distinctive outer and inner structure and smoothly connected edge lines, proving the validity of selecting a higher intensity for overlapped regions. By utilizing the demonstrated customized merging software, we obtained whole-directional volumetric images of the sample shown in the results section.

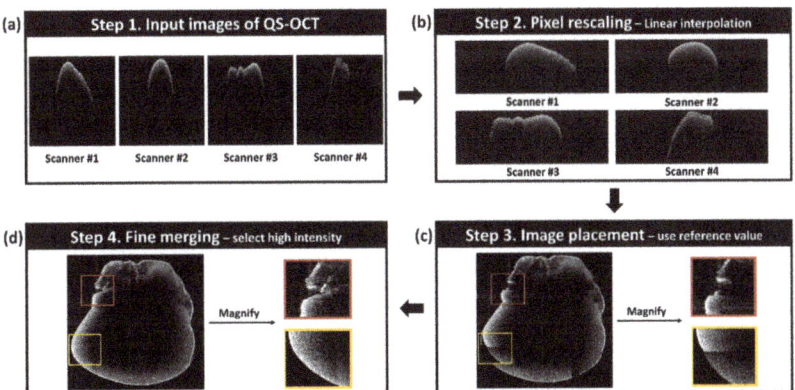

Figure 3. The description and flow chart of the image processing algorithm for obtaining the whole-directional volumetric data of the sample. (**a–d**) demonstrate each step of customized algorithm for merging cross-sectional OCT images obtained from quad-scanner methods.

2.4. Collection of Mouse Heart and Kidney Specimens

The mouse heart and kidney specimens utilized in this study were extracted from a 6-week-old BALB/c mouse and were harvested immediately after sacrificing the mouse. The extracted heart and kidney were preserved in 10% neutral-buffered formalin at room temperature for five days. All the animal experimental procedures were proceeded in conformity with a laboratory animal protocol approved by the Institutional Animal and Human Care and Use Committee of Kyungpook National University (No. KNU-2020-0025).

3. Results

3.1. Quantitative Analysis of Performance and Alignment of Scanners

To quantitatively analyze the performance and alignment of each scanner, which are crucial factors for obtaining whole-directional volumetric imaging, a rolled Scotch tape sample was scanned using four scanners from four directions, the acquired cross-sectional OCT images of which are shown in Figure 4a–d. While analyzing the scanner performance, the power of each scanner was equally controlled to objectively compare the image quality. All seven layers of the Scotch tape roll were distinguished in each cross-sectional image (Figure 4a–d) that was obtained using sequential QS-OCT. Moreover, to quantitatively assess the performance of each scanner, A-scan profiling was conducted, as shown in Figure 4e, and was acquired from the middle position of the Scotch tape cross-sectional images, indicated by the red dashed line in Figure 4a–d. The black, green, blue, and red profiles in Figure 4e represent the A-scans of each scanner from #1 to #4, respectively. The acquired A-scan result of the rolled Scotch tape revealed distinguishable internal layers along with information about the peak intensity, with an approximately similar peak height. According to the A-scan profiling results, the quality of each image acquired from different directions using QS-OCT was identical and reliable for conducting whole-directional imaging and merging to create a 3D volumetric image.

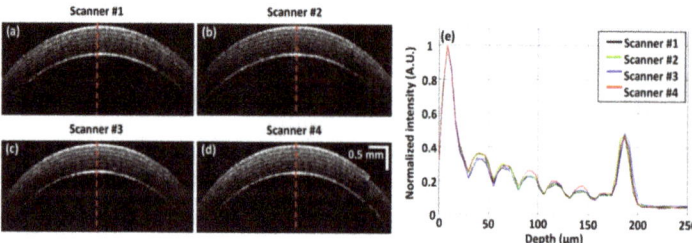

Figure 4. Quantitative performance assessment of each sample arm in quad-scanner (QS)-OCT; (**a–d**) are B-scan images of a Scotch tape roll obtained by sequentially switching the scanner; (**e**) demonstrates the A-scan profiling results of images (**a–d**) centered on the red-dashed line to quantitatively analyze performance.

The performance difference between simultaneous and sequential QS-OCT and the measured intensity fall-off graph are shown in Figure 5. Figure 5a,b show the cross-sectional OCT images of rolled scotch tape which were obtained from the simultaneous and sequential QS-OCT system, respectively. The red dotted boxes in Figure 5a,b indicate the ROIs, from where the depth intensity profiles were measured. Figure 5c shows the depth intensity profiles of both systems. A total of 150 A-lines were taken from the ROI, and then these A-lines were averaged to form the depth intensity profiles of the simultaneous and sequential QS-OCT system, where it is visualized that the depth intensity profile of sequential QS-OCT is higher than the depth intensity profile of simultaneous QS-OCT. Though internal layers of the Scotch tape were distinguished in both simultaneous and sequential QS-OCT, the overall intensities of each layer were found to be higher in the sequential method. In addition, as a case of sensitivity roll-off of the proposed system, the backscattered intensity was measured at every 100 pixels (from 100th to 900th) using a mirror as a target sample. As shown in Figure 5d, 28 dB was dropped at a depth of 4.1 mm compared to the top surface, which was caused by the characteristic of SD-OCT. Moreover, the lateral resolution was measured as 15.6 µm utilizing a resolution target (USAF 1951, Edmund Optics, Barrington, NJ, USA), and the axial resolution was calculated as 6.47 µm in air.

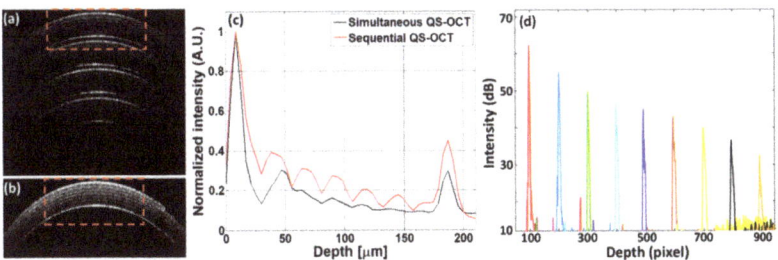

Figure 5. The performance difference between the simultaneous and sequential quad-scanner (QS) OCT system and the measured intensity fall-off of the proposed system. (**a**) is the cross-sectional OCT image of the simultaneous QS-OCT system. (**b**) is the cross-sectional OCT image of the sequential QS-OCT system. (**c**) is the depth intensity profiles of the simultaneous and sequential QS-OCT systems. (**d**) is the measured intensity fall-off graph of the proposed system obtained every 100 pixels.

3.2. Imaging Process and Measured Data Using Simultaneous QS-OCT

To explain the functional process of simultaneous QS-OCT, cross-section and merged OCT images are shown in Figure 6, obtained from the rolled Scotch tape sample used in Figure 4. The cross-sectional image of the Scotch tape roll, as shown in Figure 6a, was obtained simultaneously from whole-directions using the SDM technique with simultaneous QS-OCT. The Scotch tape cross-sectional images acquired from four scanners are indicated by red arrows. Because the difference between the path length of the sample and reference arm gradually increased towards the bottom of the imaging window, the measured intensity reduced according to the depth direction. Figure 6b exhibits the merged cross-sectional OCT images of the rolled Scotch tape that were shown in Figure 6a. The cross-sectional OCT images of the rolled Scotch tape sample shown in Figure 6a,b indicate the parallel imaging capability of simultaneous QS-OCT with the SDM imaging technique, as well as the drawback of power loss in the depth direction of the imaging window, leading to the degradation of image intensity. Although the intensity of the rolled Scotch tape image decreased in the depth direction due to the imaging depth limitation of SD-OCT, the possibility of using the proposed simultaneous QS-OCT method to concurrently obtain whole-directional images was demonstrated and is one of the proposed proofs-of-concept for whole-directional imaging.

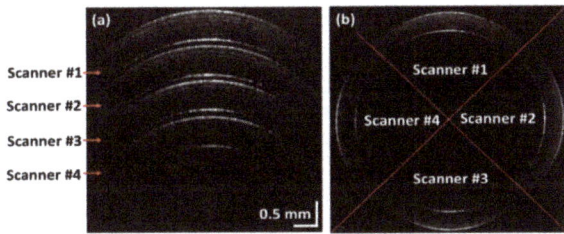

Figure 6. Simultaneously obtained images of a Scotch tape roll with the space-division multiplexing technique applied using a simultaneous quad-scanner (QS)-OCT. (**a**) is a representative B-scan image of the rolled Scotch tape imaged from four different directions with simultaneous QS-OCT, (**b**) is a merged image of the four QS-OCT images shown in (**a**).

3.3. The Examination of Merged 3D Data of Ex Vivo Mouse Kideny and Heart Obtained by Sequential

Sequential QS-OCT was developed to overcome the aforementioned limitation of power loss that occurs in the simultaneous QS-OCT system. Sequential QS-OCT utilizes a common reference arm for four different scanners to compensate for the power loss. To verify the performance of sequential QS-OCT in assessing the whole-directional imaging of biological samples, we obtained whole-directional morphological images of a mouse heart and kidney as ex vivo specimens. Following the set value of the A-scan rate (20 kHz), the volumetric imaging of each scanner consumed 36.25 s with 0.0125 μm step sizes of motor movement. Figure 7 shows the cross-sectional OCT images and their merged enface images of mouse heart and kidney specimens. The cross-sectional OCT images shown in Figures 7b–e and 7g–j were obtained using four scanners through whole-directional imaging. These four cross-sectional OCT images were merged using an image processing algorithm, described in Figure 3, to form the enface images shown in Figure 7a,f. All the characteristic features that are seen in the cross-sectional OCT images of the mouse heart and kidney specimens are retained in their respective enface images.

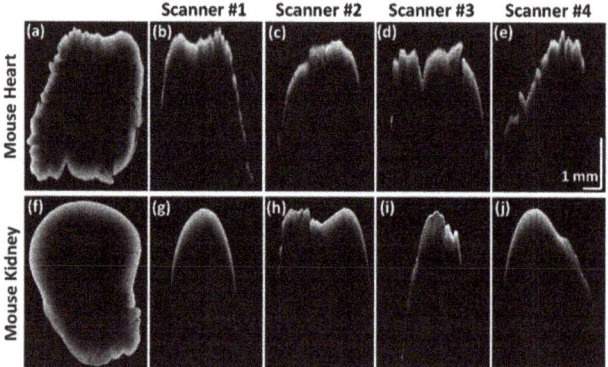

Figure 7. Enface and cross-sectional OCT images of mouse heart and kidney specimens. (**a**) and (**f**) are the enface images of mouse heart and kidney, respectively. (**b–e**), and (**g–j**) are the cross-sectional OCT images of mouse heart and kidney, respectively.

Figure 8 shows the 3D volumetric and enface images of mouse heart and kidney specimens acquired in sequential QS-OCT. Full-directional 3D rendering morphological images of the mouse heart and kidney are shown in Figure 8a,e, respectively. Every side of the sample was scanned vertically using a linear motor stage for volumetric imaging (1000 × 2048 × 725 pixels). Independently obtained volumetric images from four different sample arms were merged to obtain enface images with customized rendering software. Figure 8b–d,f–h are the representative enface images that were obtained from three different layers of 3D volumetric imaging of mouse heart and kidney specimens, shown in Figure 8a,e, respectively. The enface images shown in Figure 8b–d,f–h demonstrate the proper merging process for obtaining whole-directional volumetric data of the sample. The obtained 3D volumetric and enface images of the mouse heart and kidney specimens verify the whole-directional imaging capability of QS-OCT.

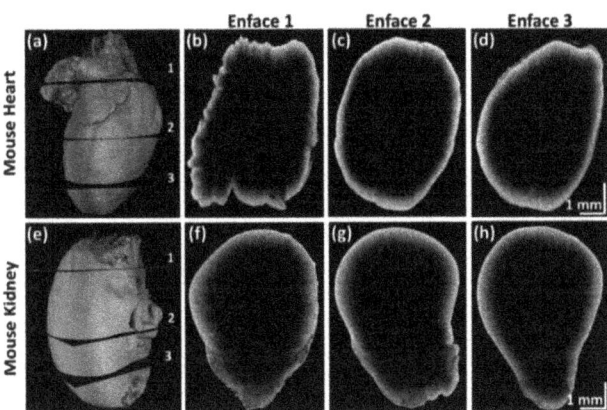

Figure 8. 3D volumetric and representative enface images of mouse heart and kidney using quad-scanner (QS)-OCT. (**a**,**e**) are full-directional 3D rendering morphological images of a mouse heart and kidney; (**b**–**d**) and (**f**–**h**) are the selected enface images obtained at the three different layers shown in (**a**,**e**), respectively.

4. Discussion

In this study, the whole-directional volumetric imaging strategy for full-directional morphological assessment of a 3D sample without additional sample rotation was successfully demonstrated in QS-OCT, where the assessment was challenged in conventional unidirectional and bidirectional scanning-based OCT. Conventional unidirectional scanning OCT has mostly been used for acquiring single-side cross-sectional images of a sample [16,17]. In addition, bidirectional imaging, which is based on the simultaneous top and bottom surface scanning of the sample, is limited by the sample thickness and outer shape [21]. In contrast, the proposed QS-OCT methods (simultaneous and sequential modes) scan from the whole-directions of the sample, which makes dynamic assessment possible for the entire position of the sample. Rotational imaging is a whole-directional imaging method in which additional sample handling such as the rotation of the sample stage and the adjustment of focus after every rotation are essential to continue the imaging capture. In the case of QS-OCT, however, the path length of each scanner was accurately controlled throughout the entire imaging process.

In simultaneous QS-OCT, whole-directional images were acquired simultaneously to enhance the imaging speed by applying the SDM technique. However, the division of source power into four reference arms and four sample arms was the major drawback of the simultaneous QS-OCT system, leading to degradation of the image intensity at high frequencies. Averaged irradiation power to sample of each scanner, utilizing simultaneous and sequential method, were measured as 6.79% and 36.4% compared with the source, which verified the significant decrement of power in simultaneous method. The drawback of source power division into multiple reference and sample arms can be compensated using a high-power light source. The sequential QS-OCT imaging concept has been demonstrated to overcome the power loss caused by the simultaneous method. In sequential QS-OCT, a common reference arm was used for the four sample arms to maintain the high power needed to conduct whole-directional imaging. The mutual compensation characteristics of simultaneous and sequential methods of the proposed QS-OCT concept can be utilized based on application requirements for scanning speed and division of source power. In the case of simultaneous QS-OCT, this method could be a suitable solution for the dynamic whole-directional volumetric assessment of a sample using a fast scanning speed. In contrast, sequential QS-OCT could be a suitable technique for obtaining the whole-directional volumetric imaging of a sample with an enhanced image quality.

As an aspect of SDM applied to simultaneous QS-OCT, the degradation of the image intensity in depth direction, as shown in Figure 6, was caused by the short coherence length of SD-OCT. In addition, the SD-OCT based quad-scanner with SDM has the restraint of applications, which is proper to utilize for thin samples because of the limited number of camera pixels. The limitation of a short coherence length while using SD-OCT can be compensated by using swept-source OCT, which has a comparatively longer coherence length to capture multiple images in the imaging depth [29]. However, the proposed simultaneous QS-OCT method was applied to obtain the volumetric imaging of a sample using the SDM technique with four scanners, whereas the SDM technique was conventionally used using a single scanner in other applications [26].

5. Conclusions

The proposed methodology of whole-directional scanning serves as a proof of concept for the whole-directional volumetric imaging of a 3D sample using QS-OCT in simultaneous or sequential order. Simultaneous QS-OCT was developed with the SDM technique by mounting four precisely aligned sample arms around the sample stage to achieve whole-directional volumetric imaging with a fast imaging speed. A sequential QS-OCT concept has also been demonstrated to improve image quality while using fewer fiber couplers than the simultaneous QS-OCT system by compromising the imaging speed. The performance of the scanners was quantitatively analyzed by imaging the Scotch tape with an A-scan profiling result. The applicability of the proposed simultaneous and sequential QS-OCT concepts was demonstrated with merged and volumetric images of the Scotch tape roll and biological tissues (heart and kidney of a mouse), respectively. In conclusion, the possibility of QS-OCT-based whole-directional imaging was achieved, and the proposed methods could be useful in various fields that require a dynamic assessment of the entire position of a sample, such as material testing and optical inspection, to detect product defects.

Author Contributions: Conceptualization, M.J. and J.K.; methodology, S.A.S., D.S., and N.K.R.; software, D.S. and S.H.; validation, R.E.W.; formal analysis, S.A.S. and D.S.; investigation, S.H. and N.K.R.; resources, M.J. and J.K.; data curation, S.A.S. and D.S.; writing—original draft preparation, S.A.S. and D.S.; writing—review and editing, R.E.W.; visualization, S.A.S., D.S., and R.E.W.; supervision, M.J. and J.K.; project administration, M.J. and J.K.; funding acquisition, M.J. and J.K. All authors have read and agreed to the published version of the manuscript.

Funding: This work was funded in part by the Basic Science Research Program under the National Research Foundation of Korea (NRF) through the MSIP under grant 2017M3A9E2065282. Also, this work was supported by the Korean government (the Ministry of Science and ICT, the Ministry of Trade, Industry and Energy, the Ministry of Health & Welfare, Republic of Korea, the Ministry of Food and Drug Safety) through the Korea Medical Device Development Fund under grant 202011C13.

Institutional Review Board Statement: The study was conducted according to the guidelines of the Declaration of Helsinki, and approved by the Institutional Review Board of the Institutional Animal and Human Care and Use Committee of Kyungpook National University (No. KNU-2020-0025).

Informed Consent Statement: Not applicable.

Data Availability Statement: Not applicable.

Conflicts of Interest: The authors declare no conflict of interest.

References

1. Vaughan, J.T.; Snyder, C.J.; DelaBarre, L.J.; Bolan, P.J.; Tian, J.; Bolinger, L.; Adriany, G.; Andersen, P.; Strupp, J.; Ugurbil, K. Whole-body imaging at 7T: Preliminary results. *Magn. Reson. Med.* **2008**, *61*, 244–248. [CrossRef]
2. Yang, M.; Baranov, E.; Li, X.-M.; Wang, J.W.; Jiang, P.; Li, L.; Moossa, A.; Penman, S.; Hoffman, R.M. Whole-body and in-travital optical imaging of angiogenesis in orthotopically implanted tumors. *Proc. Natl. Acad. Sci. USA* **2001**, *98*, 2616–2621. [CrossRef]
3. Gu, J.; Chan, T.; Zhang, J.; Leung, A.Y.H.; Kwong, Y.L.; Khong, P.-L. Whole-Body Diffusion-Weighted Imaging: The Added Value to Whole-Body MRI at Initial Diagnosis of Lymphoma. *Am. J. Roentgenol.* **2011**, *197*, W384–W391. [CrossRef]

4. Seifert, S.; Barbu, A.; Zhou, S.K.; Liu, D.; Feulner, J.; Huber, M.; Suehling, M.; Cavallaro, A.; Comaniciu, D. Hierarchical parsing and semantic navigation of full body CT data. *SPIE Med. Imaging* **2009**, *7259*, 725902. [CrossRef]
5. Antoch, G.; Vogt, F.M.; Freudenberg, L.S.; Nazaradeh, F.; Goehde, S.C.; Barkhausen, J.; Dahmen, G.; Bockisch, A.; Debatin, J.F.; Ruehm, S.G. Whole-Body Dual-Modality PET/CT and Whole-Body MRI for Tumor Staging in Oncology. *JAMA* **2003**, *290*, 3199–3206. [CrossRef] [PubMed]
6. Wang, Y.; Yao, G. Optical tractography of the mouse heart using polarization-sensitive optical coherence tomography. *Biomed. Opt. Exp.* **2013**, *4*, 2540–2545. [CrossRef] [PubMed]
7. Susaki, E.A.; Tainaka, K.; Perrin, D.; Kishino, F.; Tawara, T.; Watanabe, T.M.; Yokoyama, C.; Onoe, H.; Eguchi, M.; Yamaguchi, S.; et al. Whole-Brain Imaging with Single-Cell Resolution Using Chemical Cocktails and Computational Analysis. *Cell* **2014**, *157*, 726–739. [CrossRef] [PubMed]
8. Yokomizo, T.; Yamada-Inagawa, T.; Yzaguirre, A.D.; Chen, M.J.; Speck, N.A.; Dzierzak, E. Whole-mount three-dimensional imaging of internally localized immunostained cells within mouse embryos. *Nat. Protoc.* **2012**, *7*, 421–431. [CrossRef]
9. Fujimoto, J.G.; Pitris, C.; Boppart, S.A.; Brezinski, M.E. Optical coherence tomography: An emerging technology for bio-medical imaging and optical biopsy. *Neoplasia* **2000**, *2*, 9–25. [CrossRef] [PubMed]
10. Adhi, M.; Duker, J.S. Optical coherence tomography—Current and future applications. *Curr. Opin. Ophthalmol.* **2013**, *24*, 213–221. [CrossRef]
11. Seong, D.; Kwon, J.; Jeon, D.; Wijesinghe, R.E.; Lee, J.; Ravichandran, N.K.; Han, S.; Lee, J.; Kim, P.; Jeon, M. In Situ Characterization of Micro-Vibration in Natural Latex Membrane Resembling Tympanic Membrane Functionally Using Optical Doppler Tomography. *Sensors* **2020**, *20*, 64. [CrossRef] [PubMed]
12. Lee, J.; Abu Saleah, S.; Jeon, B.; Wijesinghe, R.E.; Lee, D.-E.; Jeon, M.; Kim, J. Assessment of the Inner Surface Roughness of 3D Printed Dental Crowns via Optical Coherence Tomography Using a Roughness Quantification Algorithm. *IEEE Access* **2020**, *8*, 133854–133864. [CrossRef]
13. Iacov-Crăiţoiu, M.M.; Popescu, S.M.; Amărăscu, M.; Scrieciu, M.; Osiac, E.; Mercuţ, V. Optical Coherence Tomography (OCT) Applications in Conventional Fixed Prosthodontics. *Int. J. Med. Dent.* **2019**, *23*, 112–120.
14. Kim, K.; Kim, P.; Lee, J.; Kim, S.; Park, S.; Choi, S.H.; Hwang, J.; Lee, J.H.; Lee, H.; Wijesinghe, R.E. Non-destructive identification of weld-boundary and porosity formation during laser transmission welding by using optical coherence tomography. *IEEE Access* **2018**, *6*, 76768–76775. [CrossRef]
15. Saccon, F.A.M.; Muller, M.; Fabris, J.L. Optical fiber characterization by optical coherence tomography. In Proceedings of the 2009 SBMO/IEEE MTT-S International Microwave and Optoelectronics Conference (IMOC), Belem, Brazil, 3–6 November 2009; pp. 625–628.
16. Shimada, Y.; Sadr, A.; Burrow, M.F.; Tagami, J.; Ozawa, N.; Sumi, Y. Validation of swept-source optical coherence tomography (SS-OCT) for the diagnosis of occlusal caries. *J. Dent.* **2010**, *38*, 655–665. [CrossRef] [PubMed]
17. Ravichandran, N.K.; Lee, S.-Y.; Jung, H.-Y.; Jeon, M.; Kim, J. Optical inspection and monitoring of moisture content in Pleurotus eryngii during storage life by refrigeration. *Int. J. Appl. Eng. Res.* **2017**, *12*, 5011–5015.
18. Chen, Z.; Milner, T.E.; Srinivas, S.; Wang, X.; Malekafzali, A.; Van Gemert, M.J.C.; Nelson, J.S. Noninvasive imaging of in vivo blood flow velocity using optical Doppler tomography. *Opt. Lett.* **1997**, *22*, 1119–1121. [CrossRef]
19. Moult, E.; Choi, W.; Waheed, N.K.; Adhi, M.; Lee, B.; Lu, C.D.; Jayaraman, V.; Potsaid, B.; Rosenfeld, P.J.; Duker, J.S. Ultrahighspeed swept-source OCT angiography in exudative AMD. *Ophthalmic Surg. Lasers Imaging Retin.* **2014**, *45*, 496–505. [CrossRef]
20. De Boer, J.F.; Milner, T.E.; Van Gemert, M.J.C.; Nelson, J.S. Two-Dimensional birefringence imaging in biological tissue using phase and polarization sensitive optical coherence tomography. *Adv. Opt. Imaging Photon Migr.* **1998**, *22*, 934–936. [CrossRef]
21. Ravichandran, N.K.; Wijesinghe, R.E.; Shirazi, M.F.; Park, K.; Jeon, M.; Jung, W.; Kim, J. Depth enhancement in spectral domain optical coherence tomography using bidirectional imaging modality with a single spectrometer. *J. Biomed. Opt.* **2016**, *21*, 76005. [CrossRef]
22. Wu, Q.; Wang, X.; Liu, L.; Mo, J. Dual-side view optical coherence tomography for thickness measurement on opaque materials. *Opt. Lett.* **2020**, *45*, 832–835. [CrossRef] [PubMed]
23. Sudheendran, N.; Wu, C.; Larina, I.V.; Dickinson, M.E.; Larin, K.V. Rotational imaging OCT for full-body embryonic imaging. In Proceedings of the 2014 Optical Coherence Tomography and Coherence Domain Optical Methods in Biomedicine XVIII, San Francisco, CA, USA, 3–5 February 2014; p. 89342.
24. Sudheendran, N.; Wu, C.; Dickinson, M.E.; Larina, I.V.; Larin, K.V. Enhancing imaging depth by multi-angle imaging of embryonic structures. In *Optical Methods in Developmental Biology II, 2014*; International Society for Optics and Photonics: Bellingham, WA, USA, 2014; p. 895306.
25. Jerwick, J.; Huang, Y.; Dong, Z.; Slaudades, A.; Brucker, A.J.; Zhou, C. Wide-field ophthalmic space-division multiplexing optical coherence tomography. *Photon. Res.* **2020**, *8*, 539–547. [CrossRef]
26. Huang, Y.; Jerwick, J.; Liu, G.; Zhou, C. Full-range space-division multiplexing optical coherence tomography angiography. *Biomed. Opt. Exp.* **2020**, *11*, 4817–4834. [CrossRef] [PubMed]
27. Huang, Y.; Badar, M.; Nitkowski, A.; Weinroth, A.; Tansu, N.; Zhou, C. Wide-field high-speed space-division multiplexing optical coherence tomography using an integrated photonic device. *Biomed. Opt. Express* **2017**, *8*, 3856–3867. [CrossRef]

28. Seong, D.; Han, S.; Jeon, D.; Kim, Y.; Wijesinghe, R.E.; Ravichandran, N.K.; Lee, J.; Lee, J.; Kim, P.; Lee, D.-E.; et al. Dynamic Compensation of Path Length Difference in Optical Coherence Tomography by an Automatic Temperature Control System of Optical Fiber. *IEEE Access* **2020**, *8*, 77501–77510. [CrossRef]
29. Zhou, C.; Alex, A.; Rasakanthan, J.; Ma, Y. Space-division multiplexing optical coherence tomography. *Opt. Express* **2013**, *21*, 19219–19227. [CrossRef] [PubMed]

sensors

MDPI

Article

Transrectal Ultrasound and Photoacoustic Imaging Probe for Diagnosis of Prostate Cancer

Jihun Jang [1], Jinwoo Kim [2], Hak Jong Lee [3] and Jin Ho Chang [2,*]

[1] Department of Electronic Engineering, Sogang University, Seoul 04107, Korea; jhjang@sogang.ac.kr
[2] Department of Information and Communication Engineering, Daegu Gyeongbuk Institute of Science and Technology, Daegu 42988, Korea; kimjw0319@dgist.ac.kr
[3] Department of Radiology, Seoul National University of Bundang Hospital, Seongnam-si 13620, Korea; hakjlee@snu.ac.kr
* Correspondence: jhchang@dgist.ac.kr; Tel.: +82-53-785-6330

Abstract: A combined transrectal ultrasound and photoacoustic (TRUS–PA) imaging probe was developed for the clear visualization of morphological changes and microvasculature distribution in the prostate, as this is required for accurate diagnosis and biopsy. The probe consisted of a miniaturized 128-element 7 MHz convex array transducer with 134.5° field-of-view (FOV), a bifurcated optical fiber bundle, and two optical lenses. The design goal was to make the size of the TRUS–PA probe similar to that of general TRUS probes (i.e., about 20 mm), for the convenience of the patients. New flexible printed circuit board (FPCB), acoustic structure, and optical lens were developed to meet the requirement of the probe size, as well as to realize a high-performance TRUS–PA probe. In visual assessment, the PA signals obtained with the optical lens were 2.98 times higher than those without the lens. Moreover, the in vivo experiment with the xenograft BALB/c (Albino, Immunodeficient Inbred Strain) mouse model showed that TRUS–PA probe was able to acquire the entire PA image of the mouse tight behind the porcine intestine about 25 mm depth. From the ex vivo and in vivo experimental results, it can be concluded that the developed TRUS–PA probe is capable of improving PA image quality, even though the TRUS–PA probe has a cross-section size and an FOV comparable to those of general TRUS probes.

Keywords: transrectal probe; optical lens; ultrasound imaging; photoacoustic imaging; prostate cancer

check for
updates

Citation: Jang, J.; Kim, J.; Lee, H.J.; Chang, J.H. Transrectal Ultrasound and Photoacoustic Imaging Probe for Diagnosis of Prostate Cancer. *Sensors* **2021**, *21*, 1217. https://doi.org/10.3390/s21041217

Academic Editor: Changho Lee

Received: 14 December 2020
Accepted: 5 February 2021
Published: 9 February 2021

Publisher's Note: MDPI stays neutral with regard to jurisdictional claims in published maps and institutional affiliations.

1. Introduction

Transrectal ultrasound (TRUS) has been used for the screening and diagnosis of prostate cancer, which is one of the most common cancers occurring in adult men [1]. For imaging, a TRUS probe is inserted into the rectum. Therefore, it is desirable that the size of TRUS probes should be as small as possible, to relieve of the patient's pain during imaging. In order to image the entire prostate, the field-of-view (FOV) of conventional TRUS probes should be as large as possible. These two restrictions limit the spatial and contrast resolutions of TRUS images, because aperture size is one factor determining the spatial resolution and signal-to-noise ratio of ultrasound (US) images, and physical conformation for wide FOV possibly degrades the sensitivity of US probes. For these reasons, TRUS imaging does not provide enough resolution and sensitivity to clearly identify and locate prostate cancers (especially early stage prostate cancers) and to accurately distinguish prostate cancers from benign prostatic hyperplasia [2]. In addition, the accuracy of TRUS-image-guided biopsy is only 20 to 30% [3], because optimal biopsy sites are not clearly shown on TRUS images, thus requiring repeated biopsies at the expense of the cost of diagnosis and the risk of complications.

Contrast-enhanced ultrasound (CEUS), in combination with TRUS imaging, has been used successfully to improve diagnostic accuracy of prostate cancer [4–6]. Since CEUS facilitates clear visualization of micro- and neo-vascularization, the success rate of TRUS

141

image-guided biopsy is increased. This is because hyper-vascularity is typically observed in the periphery of prostate cancer [5,7]. However, CEUS is less sensitive to small blood vessels and slow blood flow, even if US contrast agents are used [6]. Note that it is considered that those are the indicators associated with early stage cancers [8].

On the other hand, photoacoustic (PA) imaging is highly sensitive to blood vessels [9] and biopsy needles [10]. Therefore, combined TRUS and PA (TRUS–PA) imaging can be a solution to the problems of CEUS and TRUS imaging if high performance TRUS–PA probes are available. The first feasible study was conducted to demonstrate that TRUS–PA imaging can be used for accurate diagnosis of prostate cancer due to the clear visualization of microvasculature distribution in the prostate [11,12]. For the pilot study, a TRUS–PA probe with a wide FOV of 160° was developed; it consisted of a 128-element, 6.5 MHz TRUS array transducer and two convex-shaped optical modules for irradiated light to cover the wide FOV [11]. Note that no detailed technical information about the specifications of the TRUS transducer, the design of the optical modules, the integration of the TRUS, and optical modules could be found. As another approach, it has recently been reported that a 64-element, 5 MHz linear capacitive micromachined ultrasonic transducer (CMUT) could be used for TRUS–PA imaging [13]. The CMUT array was a side-looking transducer with a FOV of 40°, and three optical fiber bundles were placed on three sides of the CMUT array, to create dark field light illumination. However, the CMUT-based TRUS–PA probe should be further improved, because general TRUS probes are forward-looking transducers and have a wide FOV larger than 130°, to ensure diagnostic efficiency. In addition, both types of the TRUS–PA probes have a maximum cross-sectional size of 25 mm or more, and that is larger than general TRUS probes.

Since a TRUS transducer should be integrated with an optical module for TRUS–PA imaging of the prostate, it is challenging that the TRUS–PA probe is similar in size to general TRUS probes but has a large FOV. In this paper, we report a recently developed TRUS–PA probe that meets both requirements of size and FOV; the objective of the development was that the TRUS–PA had a size and FOV similar to conventional TRUS probes, with which high-quality PA images could be obtained. To achieve the development goal, particularly, the optical lens was designed to have a concave–convex shape in the lateral-axial plane for divergence and a planar–oblique shape in the elevation-axial plane for refraction. The TRUS–PA probe developed here consisted of a miniaturized 128-element 7 MHz convex array transducer with a FOV of 134.5°, a bifurcated optical fiber bundle, and two optical lenses; the maximum cross-sectional size of the TRUS–PA probe was about 20.5 mm, which is similar to that of the commercial TRUS transducers. From ex vivo and in vivo experiments, it was ascertained that the developed optical lens facilitates efficient delivery of light to the imaging plane (i.e., lateral-axial plane). In this study, additionally, light penetration through the porcine intestine was measured as a function of wavelength, to determine an optimal wavelength for PA imaging of the prostate. This was necessary because radiated light should penetrate the wall of the rectum, to reach the prostate, and it is known that light absorption highly occurs in the rectal wall.

2. Transrectal Ultrasound and Photoacoustic Imaging Probe

The developed TRUS–PA probe consists of an optical module, a TRUS array transducer, and a housing. The goal in developing the TRUS–PA probe was to make its diameter similar to that of general TRUS probes (i.e., about 20 mm), for the convenience of the patients. Note that the patients generally complain of great pain when a TRUS probe is inserted into the rectum for prostate imaging; the smaller the probe size, the better. Moreover, the probe was designed to have an imaging plane covering the prostate gland volume that typically measures 30 mm (anteroposterior) × 30 mm (width) × 50 mm (longitudinal) [7,14]. The goal could be achieved by developing a miniaturized convex ultrasound array and an optical lens, as shown in Figure 1; the cross-sectional size of the front part of the developed TRUS–PA probe was 14 mm × 15 mm. To the best of our knowledge, this size is the smallest of the reported TRUS–PA probes. Each component of the developed TRUS–PA is described here.

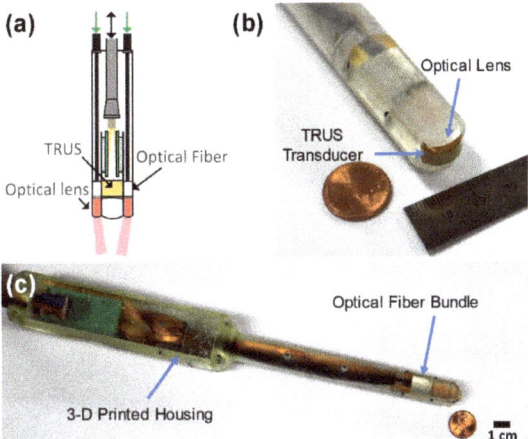

Figure 1. (**a**) Schematic of the developed transrectal ultrasound–photoacoustic (TRUS–PA) probe consisting of a TRUS array transducer, two optical fiber bundles, and two optical lenses. (**b,c**) Photographs of the developed TRUS–PA probe.

2.1. Optical Module

For high-quality PA images, light should be delivered to an imaging plane efficiently (see Figure 2). A simple and general way is to place two optical fiber bundles on each side of an US transducer and to tilt the fiber bundles at a certain angle, so that the beams overlap at the desired depth in an imaging plane [15–17]. As another way, optical reflectors attached to one side of an US transducer can be used to deliver light to the imaging plane [18,19]. However, these methods inevitably result in increasing the size of a US–PA probe, thus being not suitable for a TRUS–PA probe. For the sake of small size, optical fiber bundles can be simply attached parallel to each side of an US transducer. If the outlets of optical fiber bundles have a large numerical aperture, emitted light can spread at a large angle, so that the light can cover the region of interest (ROI) in the desired imaging plane. Although the divergent beam may be a feasible solution for a TRUS–PA probe with small external size and large FOV, the light fluence delivered in ROI is too small to be suitable for high-quality PA imaging. This is because emitted light can suitably spread in the lateral-axial plane, but much of the light cannot reach an imaging plane, due to no focusing on the elevation-axial plane. Note that PA signal intensity is linearly proportional to light fluence. Additionally, undesired PA signals are possibly generated from the off-axis of an imaging plane and received by a US transducer, thus degrading PA image quality.

For large FOV and light focus on ROI, while minimizing the size of the TRUS–PA probe, we designed an optical lens, as shown in Figure 2; the desired optical lens should produce a divergent beam in the lateral-axial plane that is equal to the imaging plane (Figure 2b) and a refracted beam in the elevation-axial plane (Figure 2c). To obtain the properties, the optical lens should have a concave–convex shape in the lateral-axial plane for divergence and a planar–oblique shape in the elevation-axial plane for refraction. The optical lens was designed, using the ray-tracing technique [20,21], to determine key parameters for fabrication of the lens: radius of curvature of the concave and convex boundaries for the concave–convex lens, and inclination angle of the oblique boundary for the planar–oblique lens. For the sake of simplifying the design, we assumed that a ray was a collimated beam (i.e., light diffraction was not considered) and ignored the law of reflection. Since the output aperture of the optical fiber bundle used for this study was configured as a 13 mm × 2 mm rectangle, the width of the collimated beam was set to be 13 mm in the lateral-axial plane and 2 mm in the elevation-axial plane. Therefore, the lens thickness in the elevation direction was selected to be 2 mm. Note that the custom-made optical fiber bundle had a numerical aperture (NA) of 0.22, so that the emitted light could be approximately considered as a collimated beam. The focal length of the lens was determined to be 25 mm, considering the longitudinal size of the prostate. Note that most prostate cancers occur at a depth of less than 30 mm from the rectal wall [22]. As a lens material, we selected Epotek-301 (Epoxy Technologies, Billerica, MA, USA), because the optical transparency of the material is 0.95 in the 382–1640 nm range [23]; its refractive index is 1.519. The equations derived for the optical lens design based on the ray-tracing method and numerical-simulation results can be found in Appendix A.

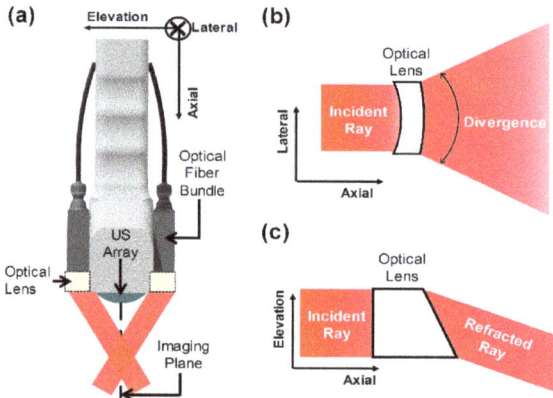

Figure 2. Conceptual illustration of (**a**) the developed TRUS–PA probe with two optical lenses placed on each side of the ultrasound (US) array transducer and the desired optical lens with (**b**) a concave–convex shape in the lateral-axial plane and (**c**) a planar–oblique shape in the elevation-axial plane.

As a result, the concave–convex radii were selected to be 8 and 11.5 mm (see Figure 3a). In this case, an FOV of 105° in the lateral-axial plane was expected. Moreover, the inclination angle of the oblique boundary for the planar–oblique lens was determined to be 80°. With these parameters, a positive mold for the optical lens was designed by using a 3D CAD (Computer Aided Design) program and created by using a 3D printer (Form 2, Formlabs, MA, USA), as shown in Figure 3a,b. Glass plates surrounded the positive mold, to construct dams, and Room-Temperature-Vulcanizing (RTV) silicone rubber (RTV664, Momentive Performance Material Inc., Waterford, NY, USA) was poured into the positive mold and cured at room temperature, for 24 h. The negative RTV mold was prepared after removing the positive mold (Figure 3c). Epotek-301 resin and hardener were mixed at a ratio of 4:1, and the epoxy mixture was degassed for 10 min. The epoxy mixture was poured into the negative RTV mold and cured at room temperature, overnight, in a dry box. Finally, the completed optical lens was separated from the RTV mold, as shown in Figure 3d.

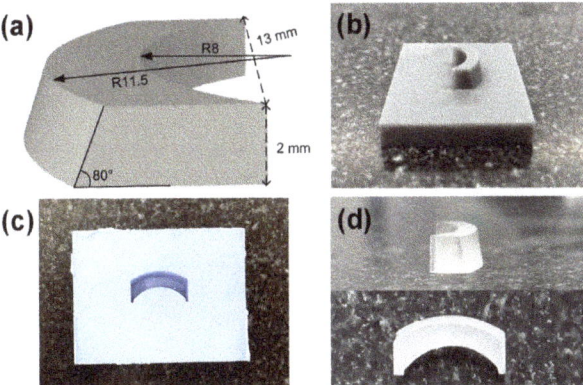

Figure 3. (**a**) Optical lens designed by a 3D CAD (Computer Aided Design) program; (**b**) positive optical lens mold constructed by using a 3D printer; (**c**) negative optical lens mold constructed by using Room-Temperature Vulcanizing (RTV), to fabricate the optical lens with the lens material, i.e., Epotek-301; and (**d**) completed optical lens of which inner and outer radii were 8 and 11.5 mm, respectively, and oblique angle was 80°.

2.2. Transrectal Ultrasound Array

We designed and fabricated a miniaturized 7 MHz TRUS array transducer, of which the footprint was 11.4 mm (lateral) × 5 mm (elevation): 128 elements, 30 mm elevational focal length, and 134.5° FOV. Geometric focus in the elevation direction, instead of lens focus, was employed to avoid ultrasound attenuation in an acoustic lens material. Therefore, the array transducer had a saddle-shaped aperture (Figure 4a). The first and second acoustic matching layers were 2–3.5 μm silver-loaded epoxy and mixture of Insulcast 502 and Insulcure 9. The backing block was constructed by using Epotek-301. To obtain high transmission and reception efficiency, additionally, a PZT-5H-based 1–3 piezocomposite was designed and fabricated (Figure 4b). For a small-sized TRUS probe, FPCB (flexible printed circuit board) should be completely bent perpendicular to the convex surface. For this, a new structure of FPCB, with several strain relief slits between the signal trace groups, was developed (Figure 4c). The center frequency and −6 dB fractional bandwidth of the fabricated TRUS array were measured at 6.75 MHz and 66%, respectively. The detailed fabrication process and imaging performance of the TRUS array developed for the TRUS–PA probe can be found in Reference [24].

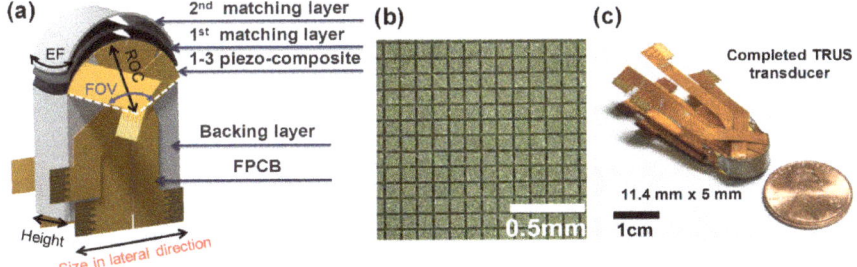

Figure 4. (**a**) Description of acoustic structure of the developed TRUS probe, (**b**) photograph of the finished 1–3 piezocomposite, and (**c**) photograph of the finished TRUS probe. Reprinted with modification from Reference [24].

2.3. Housing

The TRUS–PA probe housing was designed using a 3D CAD software to integrate the miniaturized TRUS array, optical lens, and bifurcated optical fiber bundles (Figure 5). The cross-section size of the front part of the housing, that is inserted into the patient's rectum for imaging, was determined to be 14 mm × 15 mm, considering the usefulness in the diagnosis and the alleviation of the patient's pain during imaging (Figure 5b). The housing had two grooves for mounting the fabricated optical lenses. Moreover, the outlets of the optical fiber bundles were fixed on the aligners in the housing (Figure 5c). A prototype of the TRUS–PA probe housing was constructed using the 3D printer, and the material of the housing was biocompatible photopolymer resin. Figure 1 shows the photographs of the completed TRUS–PA probe. The remarkable fact is that the maximum cross-sectional size of the developed TRUS–PA probe was about 20.5 mm, which was comparable to the commercial TRUS transducers although the probe contained both acoustic and optical modules.

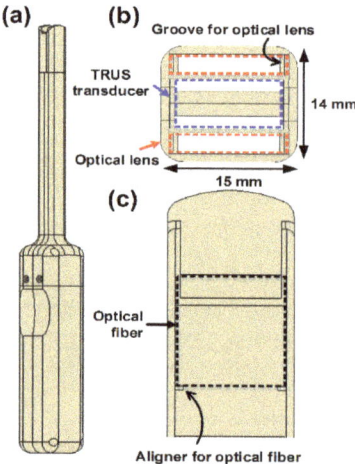

Figure 5. (**a**) Schematic illustration of the custom-designed housing of the TRUS–PA probe: (**a**) side view and (**b**) front view of the housing, and (**c**) cross-section view of the housing tip.

3. Performance Evaluation and Discussion

3.1. Light-Intensity Distribution

The performance of the developed optical lens was evaluated by measuring the light-intensity distribution as a function of depth. A continuous wave (CW) laser system (Nova Pro., RGB Photonics GmbH, Kelheim, Germany) was used to deliver a CW laser with a wavelength of 520 nm to a custom-made bifurcated optical fiber bundle with an NA of 0.66 (see Figure 6). Since irradiated light is scattered in biological media, we selected the fiber bundle with a relatively large NA; otherwise, the light hardly reached an imaging plane without the developed optical lens when the outlets of the fiber bundle were parallel to light propagation direction. For evaluating the performance of the optical lenses, we placed the optical lenses as close as possible to the bundle outlets, because it was assumed that a collimated beam entered the optical lens. Light intensity was measured after an optical screen was placed at a desired distance from the outlets of the fiber bundle, i.e., 10 to 60 in 10 mm increments. The light-intensity distribution on the screen was detected and recorded, using a charge-coupled device (CCD) camera (CoolSNAP MYO, Photometrics, Tucson, AZ, USA) equipped with an optical lens (Micro-Nikkor 105 f/2.8, Inc., Rochester, NY, USA).

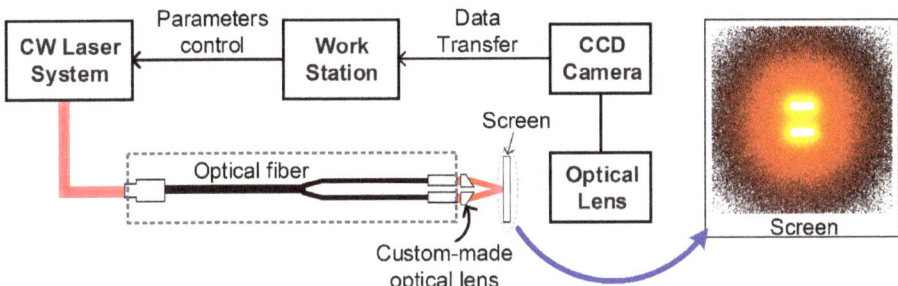

Figure 6. Illustration of the experimental setup for light-intensity distribution measurement as a function of depth. CW, continuous wave; CCD, charge-coupled device.

As shown in Figure 7a,c, the light-intensity distribution without the optical lens was naturally diffused, because an uncollimated beam (i.e., an NA of 0.66) was irradiated. The diffusion in biological media may be beneficial to a small-sized TRUS–PA probe in which optical fiber bundles are simply attached parallel to each side of an US transducer. In this particular experiment, the light beams irradiated from two optical fiber bundle outlets were separated from one another at depths of 10 and 20 mm, and these began to overlap after a depth of 30 mm (see Figure 7a). Since the irradiated light beams did not overlap completely in the imaging plane, the light intensity was weak in the imaging plane (i.e., lateral-axial plane), and the FOV of PA images was predicted to be narrow, as shown in the top panel of Figure 8. In this depth, additionally, the light intensity was strong in the off-axis of an imaging plane, thus resulting in reducing spatial and contrast resolutions of PA images; the adverse effect occurs for a similar reason that the spatial and contrast resolutions of US images are reduced due to large slice thickness (i.e., elevation resolution) [25]. In contrast, the light beams passing through the optical lenses overlapped from a depth of 10 mm (Figure 7b), and the light intensity at depths of 10 and 20 mm was about 5.3 and 4.6 times higher than that of the light delivered without the optical lens (the top panel of Figure 8). Additionally, the light-intensity distribution was wider in the imaging plane when the optical lens was used. This is because the lens had the ability to spread the irradiated light in the lateral-axial plane and refract it in the elevation-axial plane. The full-width at half maximum (FWHM) of the irradiated light through the lens was 20.2, 25.5, and 28.9 mm at depths of 10, 20, and 30 mm, whereas that of the light without the lens was 17.4, 12.3, and 15.8 mm. The maximum intensity of the light through the lens was similar to that without the lens at a depth of 30 mm as shown in Figure 8. After this depth, the maximum intensity of the light through the lens decreased slightly with depth, because the focal length of the planar–oblique lens in the elevation-axial plane was 25 mm and the light continued to spread in the lateral-axial plane; however, the FWHM also continued to broaden, i.e., 31.1, 32.1, and 32.5 mm at depths of 40, 50, and 60 mm, whereas the FWHM of the light irradiated without the lens was 21.8, 25.0, and 27.8 mm at depths of 40, 50, and 60 mm (the bottom panel of Figure 8). Note that moving averaging filtering with a length of 30 was performed for smoothing the pixel data indicated by the black lines.

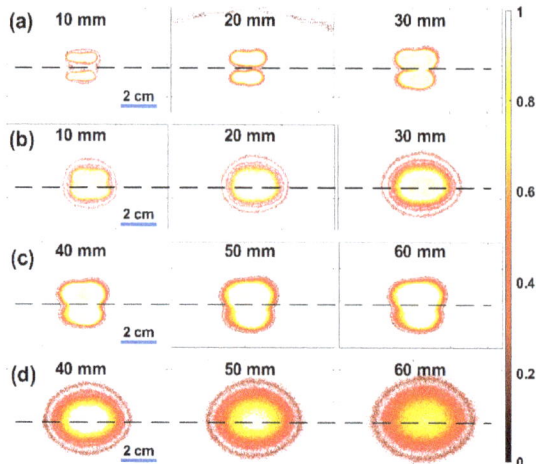

Figure 7. Measured light-intensity distributions at depths of 10 to 60 mm in 10 mm increments: (**a,c**) without the optical lenses and (**b,d**) with the optical lenses. The black dashed lines on each image indicate the imaging plane that is the center position of the US array transducer.

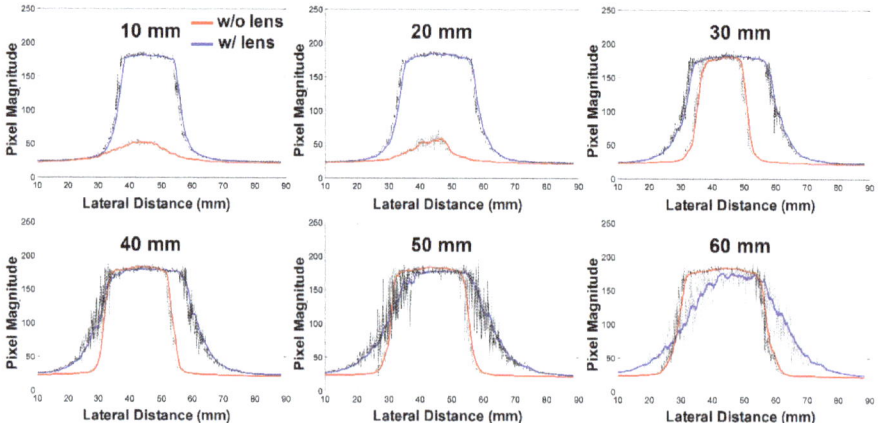

Figure 8. Light-intensity profiles measured along horizontal axis (i.e., imaging depth), indicated by the black dashed lines in Figure 7. The red and blue solid lines represent the moving averaged light intensities, without and with the optical lenses, respectively. The black lines indicate the measurement data.

The experimental results implied that the developed optical lens was predicted to be beneficial for PA image quality improvement and wide FOV. However, the performance may be different in biological media in which irradiated light spreads rapidly due to optical scattering, depending on the type of biological media [26]. In the results of Monte Carlo simulation (see Appendix C Figure A3), it was observed that the direction of the light scattering is dominated by the energy distribution of the initially irradiated light. Therefore, the developed optical lens was also expected to play an important role in increasing FOV and improving PA image quality in biological media. This was confirmed through the following experiments conducted to evaluate imaging performance.

3.2. Imaging Performance

The effect of the developed optical lens on FOV and PA signal intensity was ascertained through PA imaging of tungsten wires that were placed radially; each wire with a diameter of 100 μm was positioned at -75° to 75° at 15° angular intervals, and 5 to 55 mm at 10 mm radius intervals. The wire phantom was immersed into a container filled with 3% milk solution that served as optical scatterers. For imaging, laser pulses with a length of 7 ns and a wavement of 720 nm were generated by a Nd:YAG laser excitation system (Surelite III-10, Continuum Inc., Santa Clara, CA, USA), followed by an optical parametric oscillator (Surelite OPO Plus, Continuum Inc.). The developed TRUS–PA probe was connected to a commercial US imaging system (Vantage Research Ultrasound System, Verasonics Inc., Kirkland, WA, USA), to acquire PA image data. PA images were reconstructed, using an adaptive beamforming algorithm on MATLAB (MathWorks Inc., Natick, MA, USA) [27], and these were logarithmically compressed with a dynamic range of 35 dB. Note that a laser induced the noise signals that appeared on the PA images (Figure 9) in the dynamic range. The noise can be considerably reduced when electromagnetic interference shielding methods are applied to the housing and connector of the TRUS–PA probe for the purpose of commercialization.

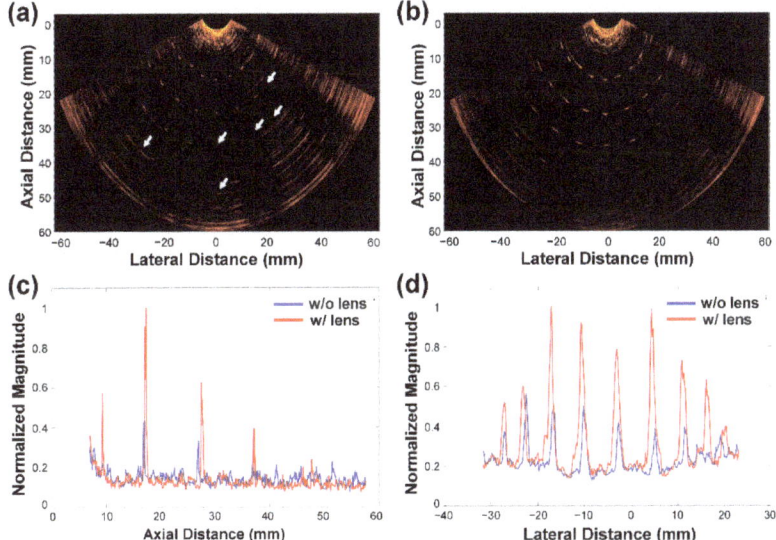

Figure 9. PA images of tungsten wire targets, (**a**) without and (**b**) with the developed optical lens. The white arrows in (**a**) indicate the invisible wire targets on the PA image in (**a**), but visible in (**b**). (**c**,**d**) Normalized envelope profiles of the wire target images along (**c**) the axial direction at the center position in the lateral direction and (**d**) the lateral direction at a depth of 28 mm.

In visual assessment, it was seen that the TRUS–PA probe with the developed optical lens provided a higher-quality PA image than without the optical lens (Figure 9a,b). Without the developed optical lens (Figure 9a), the wires located at 38 and 48 mm barely appeared on the image because PA signal intensity was similar to the noise. Note that the distance between the probe and the front wires was about 5 mm. In addition, there were some invisible wire images even at 25 mm, which were indicated by the white arrows. When the optical lens was used, in contrast, the wire images positioned up to 35 mm were clearly observed and some wires located at 45 mm also appeared; however, the wire images on the edge were not visible. This is possible because the outer scanlines of both US and PA images were generally formed by using fewer channel datasets than the middle scanlines.

For example, only 32 channel datasets are available for the outermost scanline, whereas the center scanline is formed by using 64 channel datasets. Based on the position of the wire images, it was found that the FOV of the developed TRUS–PA probe was about 120° at a depth of 25 mm, and it was 90° at 35 mm. Note that the optical lens was designed to have an FOV of 105° in the lateral-axial plane and a focal length of 25 mm for the planar–oblique lens in the elevation-axial plane. To assess the effect of the lens on PA signal intensity, the envelope signals generated from the center wires (Figure 9c) and the wires along the arc of the circle with a radius of 25 mm (Figure 9d) were obtained. The PA signals acquired with the optical lens were much higher than those without the lens (i.e., 2.98 times higher on average). From the experimental results, it could be concluded that the developed optical lens was effective in focusing irradiated laser onto the imaging plane, even in the scattering medium.

3.3. Combined US and PA Imaging of Targets Behind the Procine Intestine

The prostate is positioned behind the wall of the rectum. Therefore, we measured light penetration through the porcine intestine, to predict the effect of the rectal wall on the PA imaging of the prostate and to determine an optimal wavelength for PA imaging of the prostate. This experiment was necessary because some researchers have reported Monte Carlo simulation results that light intensity passing through the rectal wall is limited for transrectal PA imaging due to the high light absorption in the rectal wall. Based on the simulation results, they asserted that the transrectal approach for PA imaging of the prostate might not be suitable, and it would be difficult to achieve sufficient imaging depth, spatial resolution, and FOV for the prostate PA imaging [28,29].

For the attenuation measurement to explore the possibility of the TRUS–PA imaging of the prostate, the Nd:YAG laser excitation system, followed by the optical parametric oscillator, was used to generate 7 nm laser pulses, as shown in Figure 10a. The laser energy delivered by the bifurcated optical fiber bundle was measured by using an energy meter (MAESTRO, Gentec-EO Inc., Quebec, QC, Canada) and recorded. The laser energy measured without the porcine intestine served as a reference at a given laser wavelength. After placing the porcine intestine between the optical fiber bundles and the energy meter, the laser passing through the porcine intestine was measured. The thickness of the porcine intestine was about 3 mm, which is similar to the median human rectal wall thickness [30]. A ratio of laser energy penetration was calculated by dividing the measured laser energy by the reference. This process was repeated by changing the wavelength from 650 to 975 nm, at 25 nm intervals. Note that the experiments were performed four times, with different porcine intestines. As shown in Figure 10b, the highest mean ratio of the laser energy penetration was 26.3% at a wavelength of 780 nm, and the average of the mean ratios at all the wavelengths was 21.9%.

The feasibility of combined US and PA imaging through the porcine intestine was investigated. For this, five graphite rods with a diameter of 0.5 mm were embedded diagonally in chicken breast specimens covered by the porcine intestine, as shown in Figure 11a. For the PA imaging, a wavelength of 780 nm was selected. Despite the presence of the porcine intestine, the graphite targets were well distinguished from the speckle pattern in the US image of the chicken breast tissue, which were indicated by the white arrows in Figure 11b; the PA intensity decreased 2.4 times on average when the porcine intestine was covered, compared to that without the porcine intestine cover (see Appendix C Figure A4). Note that the measurement of the laser penetration shown in Figure 10 was conducted without the developed optical lens. Due to the beam focus on the imaging plane by the optical lens, the reduction ratio in the imaging test was smaller than the direct measurement. Unlike the previously reported simulation results, the experimental results showed the possibility of acquiring a combined US and PA image of the prostate through the human rectum intestine. The similar results were also obtained in vivo, as shown in Figure 11d.

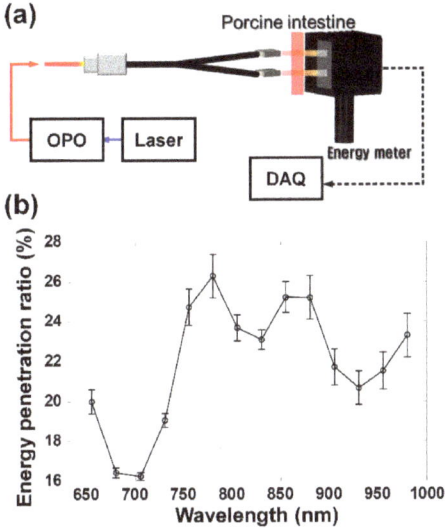

Figure 10. (**a**) Illustration of the experimental setup for measuring laser penetration through the porcine intestine and (**b**) ratio of laser energy penetration as a function of wavelength, which was obtained after measuring laser energy, both with and without the porcine intestine. The circles and the error bars indicate the mean and the variation. OPO and DAQ stand for optical parametric oscillator and data acquisition, respectively.

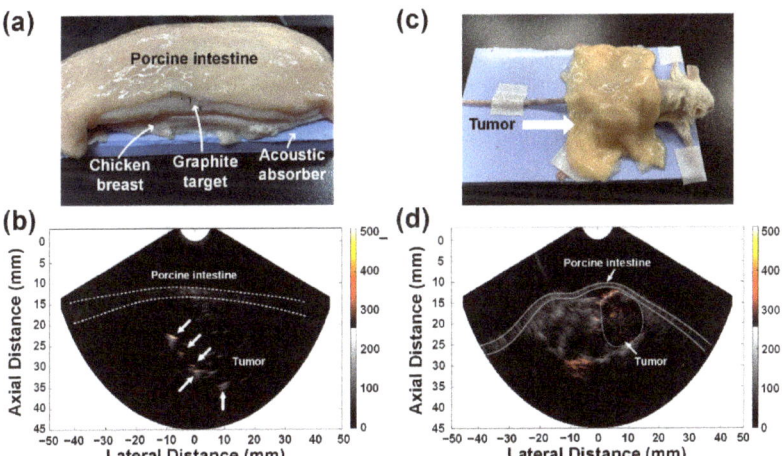

Figure 11. (**a**) Photograph of the imaging target used for the ex vivo experiments and (**b**) combined US and PA image of the five graphite rods in the chicken breast tissue covered by the porcine intestine. The US and PA images were logarithmically compressed with a dynamic range of 55 and 25 dB, respectively. (**c**) Photograph of the xenograft BALB/c (Albino, Immunodeficient Inbred Strain) mouse covered by the porcine intestine for the in vivo experiments and (**d**) combined US and PA images of the tumor site on the mouse. The US and PA images were logarithmically compressed with dynamic ranges of 45 and 25 dB, respectively.

For the in vivo experiment, the xenograft BALB/c (Albino, Immunodeficient Inbred Strain) mouse model, in which PC-3 prostate cancer cells were implanted around the thigh, was prepared. The animal experiment was conducted in accordance with the guidelines and regulations approved by the Institutional Animal Care and Use Committee of Seoul National University Bundang Hospital, South Korea. The mouse model was fixed on the acoustic absorber, and the porcine intestine was placed on the back of the mouse, enough to cover the tumor, as shown in Figure 11c. The laser wavelength was set to 780 nm for the PA imaging. Figure 11d shows the combined US and PA image of the PC-3 tumor mouse model. The white dashed line in this image represents the tumor boundary, and the two solid lines indicate the porcine intestine boundary. Note that suspicious tumors appear hypoechoic in US images [31]. The PA signals were observed around and inside the tumor, which may be evidence of the neovascularization for tumor cell growth [12,32]. Additionally, it was seen that the developed TRUS–PA probe was able to acquire the entire PA image of the mouse thigh behind the porcine intestine (i.e., about 25 mm depth from the porcine intestine), even though no contrast agent was used.

4. Conclusions

The primary challenge in accurate diagnosis of prostate cancer is to locate micro- and neo-vascularization accurately, as well as to delineate the cancer boundary clearly. Combined US and PA imaging is the most feasible way to achieve the goal because of high-sensitivity PA imaging of blood vessels in conjunction with US anatomic imaging; this emerging method is analogous to combined CEUS and US B-mode imaging that is less sensitive to small blood vessels and slow blood flow even if US contrast agents are used. Additionally, it is well-known that PA imaging is able to provide clear visualization of a biopsy needle. As a result, the diagnosis of prostate cancer can be another candidate for clinical application of combined US and PA imaging. This can be realized by a combined US and PA imaging system equipped with a high-performance hybrid imaging probe. Based on the ex vivo and in vivo experimental results, we believe that the FPCB, acoustic structure, and optical lens developed in this study can contribute to the realization of a high-performance TRUS–PA probe for accurate diagnosis of prostate cancer, because these features enable the developed TRUS–PA probe to improve PA image quality, as well as to have a cross-section size and a field of view comparable to those of general TRUS probes.

Author Contributions: Conceptualization, J.H.C.; methodology, J.J. and J.K.; validation, J.J., J.K., H.J.L. and J.H.C.; formal analysis, J.J., J.K., H.J.L. and J.H.C.; writing—original draft preparation, J.J.; writing—review and editing, J.J., J.K., H.J.L. and J.H.C.; visualization, J.J.; supervision, J.H.C.; funding acquisition, J.H.C. and H.J.L. All authors have read and agreed to the published version of the manuscript.

Funding: This work was supported by the DGIST Start-Up Fund Program of the Ministry of Science and ICT (2020030086) and the Research and Development Program of MOTIE/KEIT, South Korea, through the Development of Transrectal Probes for Combined Ultrasound and Photoacoustic Imaging, under Grant 10060071.

Institutional Review Board Statement: The study was conducted according to the guidelines of the Declaration of Helsinki, and approved by the Institutional Animal Care and Use Committee (IACUC) of Seoul National University Bundang Hospital, South Korea (BA-1908-278-070-01).

Informed Consent Statement: Not applicable.

Data Availability Statement: Not applicable.

Conflicts of Interest: The authors declare no conflict of interest.

Appendix A. Optical Lens Design

The optical lenses were designed and modeled by using the ray tracing technique [20,21]. Rays, idealized models of light, can be obtained by selecting a line that actually indicates the direction of energy flow perpendicular to light wavefront. Rays were used to model light propagation through optical systems, such as optical lenses. Ray tracing is achieved by dividing a light irradiation field into discrete rays that can be used to estimate the path of light through an optical system. Ray tracing is described by three equations, i.e., refraction, reflection, and transfer equations.

Based on the ray tracing technique, a desired optical lens could be simply designed. For the sake of simplification, we assumed that a ray was a collimated beam (i.e., light diffraction was not considered) and ignored the law of refection. The desired optical lens should produce a divergent beam in the lateral-axial plane (Appendix A Figure A1a) and a refracted beam in the elevation-axial plane (Appendix A Figure A1b); the optical lens should have a concave–convex shape in the lateral-axial plane for divergence and a planar–oblique shape in the elevation-axial plane for refraction.

For a divergent ray in the lateral-axial plane, we considered collimated rays that pass through a concave boundary. In this case, Snell's law leads to the following:

$$sin(\theta_2) = \frac{n_1}{n_2} sin(\theta_1),$$
(A1)

where θ_1 and θ_2 are the angles of incident and refracted rays; n_1 and n_2 are the refractive indices of air and the lens material (see Figure A1c). Therefore, θ_1 and θ_2 can be calculated by using the following equation:

$$\theta_1 = sin^{-1}\left(\frac{x_1}{R_1}\right),$$
(A2)

$$\theta_2 = sin^{-1}\left(\frac{n_1}{n_2}\frac{x_1}{R_1}\right),$$
(A3)

where x_1 is the radius of the incident beam, and R_1 is the radius of curvature of the concave boundary. The geometry yields are as follows:

$$\phi_1 = \theta_1 - \theta_2 = sin^{-1}\left(\frac{x_1}{R_1}\right) - sin^{-1}\left(\frac{n_1}{n_2}\frac{x_1}{R_1}\right)$$
(A4)

and

$$sin(\phi_2) = \frac{x_2}{R_2},$$
(A5)

where R_2 is the radius of curvature of the convex boundary, and x_2 is determined by the following:

$$x_2 = x_1 + tan(\phi_1)[R_2 cos(\phi_2) - R_1 cos(\theta_2)],$$
(A6)

By using Equations (A4) and (A5), at the convex boundary, θ_3 can be expressed as follows:

$$\theta_3 = \phi_2 - \phi_1 = sin^{-1}\left(\frac{x_2}{R_2}\right) - sin^{-1}\left(\frac{x_1}{R_1}\right) + sin^{-1}\left(\frac{n_1}{n_2}\frac{x_1}{R_1}\right),$$
(A7)

When passing through a convex boundary, the ray also experiences Snell's law, which is given by the following:

$$sin(\theta_4) = \frac{n_2}{n_1} sin(\theta_3) = \frac{n_2}{n_1} sin\left[sin^{-1}\left(\frac{x_2}{R_2}\right) - sin^{-1}\left(\frac{x_1}{R_1}\right) + sin^{-1}\left(\frac{n_1}{n_2}\frac{x_1}{R_1}\right)\right],$$
(A8)

Finally, the divergent angle, ϕ_3, can be expressed as follows:

$$\phi_3 = \phi_2 - \theta_4.$$
(A9)

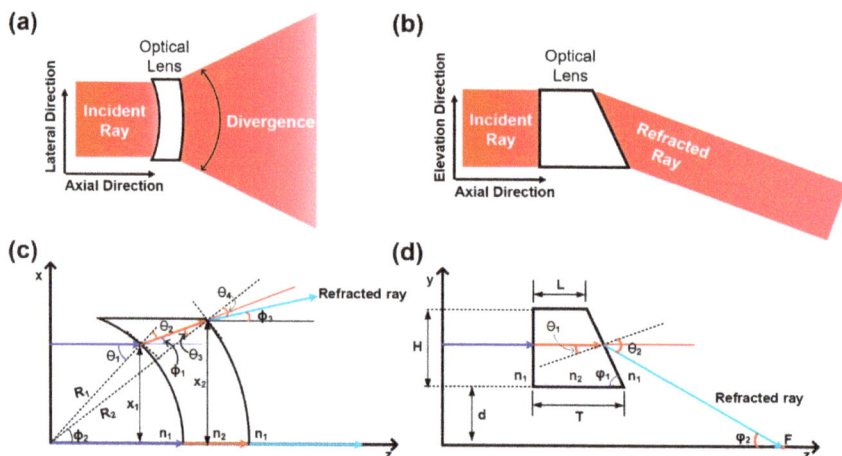

Figure A1. Conceptual illustration of the desired optical lens in (**a**) the lateral-axial plane and (**b**) elevation-axial plane. The models for divergence and refraction in the lens are shown in (**c**,**d**), respectively.

For the transrectal ultrasound–photoacoustic (TRUS–PA) imaging, the size of incident light (i.e., x_1) is determined by the height of an optical fiber bundle. After selecting the desired divergent angle, ϕ_3, the design parameters R_1 and R_2 can be determined by using Equations (A5), (A8), and (A9). Note that if the radius of curvature of the concave boundary, R_1, is much longer than the incident beam radius, x_1, Equation (A9) can be approximately expressed as follows:

$$\phi_3 \approx \left(1 - \frac{n_2}{n_1}\right)\frac{x_2}{R_2} + \frac{n_2}{n_1}\left(1 - \frac{n_1}{n_2}\right)\frac{x_1}{R_1}, \tag{A10}$$

because it is valid that $sin(\theta) \approx \theta$ if $\theta < 14°$, at which an error rate is 1%.

For converging light illumination in the elevation-axial plane, collimated rays meet the planar boundary at which normal incidence occurs (see Appendix A Figure A1d). The rays are only refracted at the oblique boundary. The angle of refraction is as follows:

$$\theta_2 = sin^{-1}\left(\frac{n_1}{n_2}sin(\theta_1)\right), \tag{A11}$$

where n_1 and n_2 are the refractive indices of air and the lens material, and θ_1 can be obtained by the following:

$$\theta_1 = 90 - \varphi_1. \tag{A12}$$

Note that φ_1 is the inclination angle of the oblique boundary. Moreover, the angle of the refracted ray to the z-axis is derived as follows:

$$\varphi_2 = \varphi_1 + \theta_2 - 90. \tag{A13}$$

Finally, the focal length from the surface of an ultrasound transducer can be derived as follows:

$$F = L + \frac{0.5H}{tan(\varphi_1)} + \frac{d + 0.5H}{tan(\varphi_2)}, \tag{A14}$$

where L is the length of the short base of the planar–oblique lens, H is the height of the lens, and d is the gap between the focal depth and the lens in the elevation direction. Note that d is approximately equal to half the height of an ultrasound transducer.

Appendix B. Numerical Simulation

The equations derived for the optical lens design cannot be expressed as a closed form. The target parameters for the concave/convex-shaped lens are the radius curvatures (i.e., R_1 and R_2), whereas the parameter for the planar/oblique-shaped lens is the inclination angle of the oblique boundary, φ_1. When other parameters were given, the target parameters were found by numerical simulation.

As an optical lens material, we chose Epotek-301 (Epoxy Technologies, Billerica, MA, USA) because the optical transparency of the material is 0.99. Its refractive index (i.e., n_2) is 1.519. We assumed that other media was air, of which the refractive index is 1.0003 (i.e., n_1). Since the length of the optical fiber bundle used for this study was 13 mm, the radius of the collimated beam (i.e., x_1) was set to be 6.5 mm. For efficient photoacoustic (PA) imaging of the prostate, the field of view (FOV) in the lateral-axial plane should be as wide as possible; our target FOV was wider than 100°, so the desired divergent angle, ϕ_3, in Equation (A9) should be larger than 50°. With the given parameters, we conducted iterative numerical simulation to find the radius curvatures (i.e., R_1 and R_2) of the concave/convex-shaped lens. Finally, we selected a R_1 of 8 mm and a R_2 of 11.5 mm. In this case, FOV was expected to be 105° (see Appendix B Figure A2a).

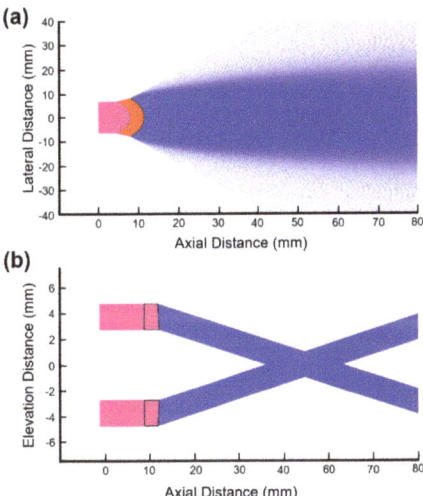

Figure A2. Results of numerical simulation based on the equations derived by using the ray tracing method: (**a**) concave/convex-shaped lens in the lateral-axial plane and (**b**) planar/oblique-shaped lens in the elevation-axial plane. Note that the magenta lines are collimated incoming rays, and the red lines are the refracted rays.

For the planar/oblique-shaped lens, the height of the lens (i.e., H) was set to be 2 mm, the height of the optical fiber bundle was 2 mm, and the focal length (i.e., F) was chosen to be 25 mm. Since the height of the TRUS transducer in the elevation direction was 5.5 mm, the gap between the focal depth and the lens in the elevation direction (i.e., d) was set to be 2.75 mm. Moreover, the length of the short base of the planar–oblique lens L was 3.5 mm, which was the difference between R_1 and R_2. From the iterative numerical simulation, finally, the inclination angle of the oblique boundary (i.e., φ_1) was determined to be 80° (see Appendix B Figure A2b).

Appendix C. Monte Carlo Simulation

The simulation was conducted by using a Monte Carlo light-scattering program (available from the Oregeon Medical Laser Center, https://omlc.org/software (accessed on 10 December 2020)), to ascertain the effect of optical scattering on the performance of the developed optical lens (see Appendix C Figure A3). The results imply that the direction of light scattering is determined by initial optical intensity distribution. In other words, it is not significant that the optical energy is widened by the natural diffusion, because the amount of the optical energy or energy distribution is mainly determined by the initial direction of irradiated light.

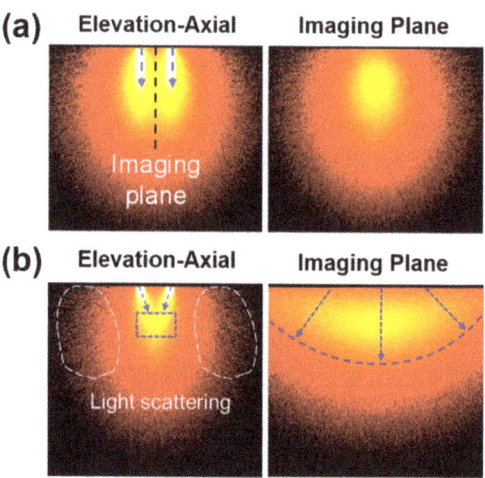

Figure A3. Results of the Monte Carlo simulation for light propagation in optical scattering media; light is delivered (**a**) in parallel and (**b**) at an oblique angle.

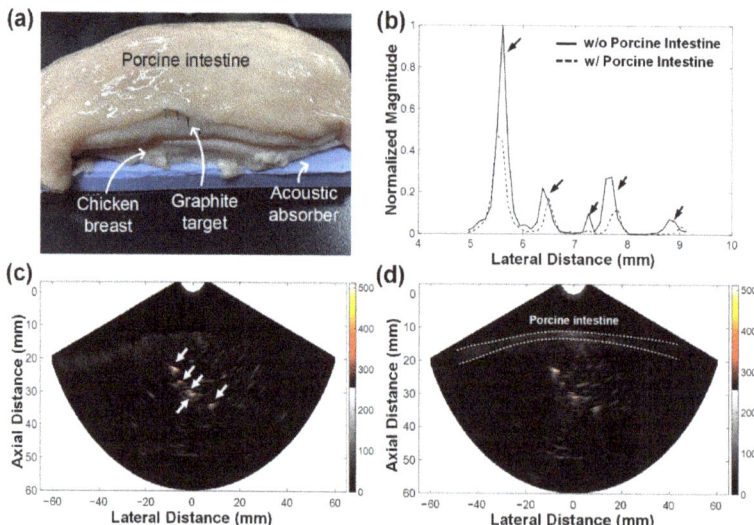

Figure A4. (**a**) Photograph of the imaging target used for the ex vivo experiments and (**b**) lateral beam profiles of the five graphite rods in the chicken breast tissue, without (solid line) or with (dashed line) the porcine intestine cover. (**c,d**) Combined ultrasound and photoacoustic images of the image target, (**c**) without and (**d**) with the porcine intestine cover. The US and PA images were logarithmically compressed with a dynamic range of 55 and 25 dB, respectively.

References

1. Center, M.M.; Jemal, A.; Lortet-Tieulent, J.; Ward, E.; Ferlay, J.; Brawley, O.; Bray, F. International variation in prostate cancer incidence and mortality rates. *Eur. Urol.* **2012**, *61*, 1079–1092. [CrossRef] [PubMed]
2. Sedelaar, J.P.; Van Roermund, J.G.; Van Leenders, G.L.; Hulsbergen-van de Kaa, C.A.; Wijkstra, H.; De la Rosette, J.J. Three-dimensional grayscale ultrasound: Evaluation of prostate cancer compared with benign prostatic hyperplasia. *Urology* **2001**, *57*, 914–920. [CrossRef]
3. Lopez-Corona, E.; Ohori, M.; Wheeler, T.M.; Reuter, V.E.; Scardino, P.T.; Kattan, M.W.; Eastham, J.A. Prostate Cancer Diagnosed After Repeat Biopsies Have a Favorable Pathological Outcome but Similar Recurrence Rate. *J. Urol.* **2006**, *175*, 923–928. [CrossRef]
4. Sano, F.; Terao, H.; Kawahara, T.; Miyoshi, Y.; Sasaki, T.; Noguchi, K.; Kubota, Y.; Uemura, H. Contrast-enhanced ultrasonography of the prostate: Various imaging findings that indicate prostate cancer. *BJU Int.* **2011**, *107*, 1404–1410. [CrossRef]
5. Shigeno, K.; Igawa, M.; Shiina, H.; Wada, H.; Yoneda, T. The role of colour Doppler ultrasonography in detecting prostate cancer. *BJU Int.* **2000**, *86*, 229–233. [CrossRef]
6. Sauvain, J.L.; Palascak, P.; Bourscheid, D.; Chabi, C.; Atassi, A.; Bremon, J.M.; Palascak, R. Value of power doppler and 3D vascular sonography as a method for diagnosis and staging of prostate cancer. *Eur. Urol.* **2003**, *44*, 21–30. [CrossRef]
7. Mitterberger, M.; Horninger, W.; Aigner, F.; Pinggera, G.M.; Steppan, I.; Rehder, P.; Frauscher, F. Ultrasound of the prostate. *Cancer Imaging* **2010**, *10*, 30–48. [CrossRef] [PubMed]
8. Jiang, J.; Chen, Y.Q.; Zhu, Y.K.; Yao, X.H.; Qi, J. Factors influencing the degree of enhancement of prostate cancer on contrast-enhanced transrectal ultrasonography: Correlation with biopsy and radical prostatectomy specimens. *Br. J. Radiol.* **2012**, *85*, e979–e986. [CrossRef]
9. Liu, W.; Yao, J. Photoacoustic microscopy: Principles and biomedical applications. *Biomed. Eng. Lett.* **2018**, *8*, 203–213. [CrossRef] [PubMed]
10. Wu, K.W.; Wang, Y.A.; Li, P.C. Laser Generated Leaky Acoustic Waves for Needle Visualization. *IEEE Trans. Ultrason. Ferroelectr. Freq. Control* **2018**, *65*, 546–556. [CrossRef] [PubMed]
11. Horiguchi, A.; Tsujita, K.; Irisawa, K.; Kasamatsu, T.; Hirota, K.; Kawaguchi, M.; Shinchi, M.; Ito, K.; Asano, T.; Shinmoto, H.; et al. A pilot study of photoacoustic imaging system for improved real-time visualization of neurovascular bundle during radical prostatectomy. *Prostate* **2016**, *76*, 307–315. [CrossRef]
12. Horiguchi, A.; Shinchi, M.; Nakamura, A.; Wada, T.; Ito, K.; Asano, T.; Shinmoto, H.; Tsuda, H.; Ishihara, M. Pilot Study of Prostate Cancer Angiogenesis Imaging Using a Photoacoustic Imaging System. *Urology* **2017**, *108*, 212–219. [CrossRef]
13. Kothapalli, S.R.; Sonn, G.A.; Choe, J.W.; Nikoozadeh, A.; Bhuyan, A.; Park, K.K.; Cristman, P.; Fan, R.; Moini, A.; Lee, B.C.; et al. Simultaneous transrectal ultrasound and photoacoustic human prostate imaging. *Sci. Transl. Med.* **2019**, *11*, eaav2169. [CrossRef] [PubMed]
14. Zhang, S.J.; Qian, H.N.; Zhao, Y.; Sun, K.; Wang, H.Q.; Liang, G.Q.; Li, F.H.; Li, Z. Relationship between age and prostate size. *Asian J. Androl.* **2013**, *15*, 116–120. [CrossRef] [PubMed]
15. Kang, J.; Chang, J.H.; Wilson, B.C.; Veilleux, I.; Bai, Y.; DaCosta, R.; Kim, K.; Ha, S.; Lee, J.G.; Kim, J.S.; et al. A prototype hand-held tri-modal instrument for in vivo ultrasound, photoacoustic, and fluorescence imaging. *Rev. Sci. Instrum.* **2015**, *86*, 034901. [CrossRef]
16. Kang, J.; Chang, J.H.; Kim, S.M.; Lee, H.J.; Kim, H.; Wilson, B.C.; Song, T.K. Real-time sentinel lymph node biopsy guidance using combined ultrasound, photoacoustic, fluorescence imaging: In vivo proof-of-principle and validation with nodal obstruction. *Sci. Rep.* **2017**, *7*, 45008. [CrossRef] [PubMed]
17. Zhou, Y.; Li, G.; Zhu, L.; Li, C.; Cornelius, L.A.; Wang, L.V. Handheld photoacoustic probe to detect both melanoma depth and volume at high speed in vivo. *J. Biophoton.* **2015**, *8*, 961–967. [CrossRef]
18. Li, M.; Liu, C.; Gong, X.; Zheng, R.; Bai, Y.; Xing, M.; Du, X.; Liu, X.; Zeng, J.; Lin, R.; et al. Linear array-based real-time photoacoustic imaging system with a compact coaxial excitation handheld probe for noninvasive sentinel lymph node mapping. *Biomed. Opt. Express* **2018**, *9*, 1408–1422. [CrossRef] [PubMed]
19. Upputuri, P.K.; Pramanik, M. Recent advances toward preclinical and clinical translation of photoacoustic tomography: A review. *J. Biomed. Opt.* **2016**, *22*, 041006. [CrossRef]
20. Moore, D.T. Ray tracing in gradient-index media. *J. Opt. Soc. Am.* **1975**, *65*, 451–455. [CrossRef]
21. Hanrahan, P.; Cook, R.L.; Arvo, J.; Kirk, D.; Heckbert, P.S. *An Introduction to Ray Tracing*; Glassner, A.S., Ed.; Academic Press: New York, NY, USA, 1989; ISBN 978-0122861604.
22. Shaikhibrahim, Z.; Lindstrot, A.; Ellinger, J.; Rogenhofer, S.; Buettner, R.; Perner, S.; Wernert, N. The peripheral zone of the prostate is more prone to tumor development than the transitional zone: Is the ETS family the key? *Mol. Med. Rep.* **2012**, *5*, 313–316. [CrossRef]
23. Park, S.; Kang, S.; Chang, J.H. Optically Transparent Focused Transducers for Combined Photoacoustic and Ultrasound Microscopy. *J. Med. Biol. Eng.* **2020**, *40*, 707–718. [CrossRef]
24. Jang, J.; Chang, J.H. Design and Fabrication of a Miniaturized Convex Array for Combined Ultrasound and Photoacoustic Imaging of the Prostate. *IEEE Trans. Ultrason. Ferroelectr. Freq. Control* **2018**, *65*, 2086–2096. [CrossRef] [PubMed]
25. Lee, J.; Jang, J.; Chang, J.H. Oblong-Shaped-Focused Transducers for Intravascular Ultrasound Imaging. *IEEE Trans. Biomed. Eng.* **2017**, *64*, 671–680. [CrossRef]

26. Kim, H.; Chang, J.H. Increased light penetration due to ultrasound-induced air bubbles in optical scattering media. *Sci. Rep.* **2017**, *7*, 16105. [CrossRef] [PubMed]
27. Yoon, C.; Kang, J.; Han, S.; Yoo, Y.; Song, T.K.; Chang, J.H. Enhancement of photoacoustic image quality by sound speed correction: Ex Vivo evaluation. *Opt. Express* **2012**, *20*, 3082–3090. [CrossRef]
28. El-Gohary, S.H.; Metwally, M.K.; Eom, S.; Jeon, S.H.; Byun, K.M.; Kim, T.S. Design study on photoacoustic probe to detect prostate cancer using 3D Monte Carlo simulation and finite element method. *Biomed. Eng. Lett.* **2014**, *4*, 250–257. [CrossRef]
29. Lediju Bell, M.A.; Guo, X.; Song, D.Y.; Boctor, E.M. Transurethral light delivery for prostate photoacoustic imaging. *J. Biomed. Opt.* **2015**, *20*, 036002. [CrossRef]
30. Rasmussen, S.N.; Riis, P. Rectal Wall Thickness Measured by Ultrasound in Chronic Inflammatory Diseases of the Colon. *Scand. J. Gastroenterol.* **1985**, *20*, 109–114. [CrossRef]
31. Lee, F.; Torp-Pedersen, S.; Littrup, P.J.; McLeary, R.D.; McHugh, T.A.; Smid, A.P.; Stella, P.J.; Borlaza, G.S. Hypoechoic lesions of the prostate: Clinical relevance of tumor size, digital rectal examination, and prostate-specific antigen. *Radiology* **1989**, *170*, 29–32. [CrossRef]
32. Ishihara, M.; Horiguchi, A.; Shinmoto, H.; Tsuda, H.; Irisawa, K.; Wada, T.; Asano, T. Comparison of transrectal photoacoustic, Doppler, and magnetic resonance imaging for prostate cancer detection. In Proceedings of the Photons Plus Ultrasound: Imaging and Sensing, San Francisco, CA, USA, 13–18 February 2016; Volume 9708, p. 970852.

Article

Pilot Study: Quantitative Photoacoustic Evaluation of Peripheral Vascular Dynamics Induced by Carfilzomib In Vivo

Thi Thao Mai [1], Manh-Cuong Vo [2], Tan-Huy Chu [2], Jin Young Kim [3], Chulhong Kim [3], Je-Jung Lee [2,4], Sung-Hoon Jung [4,*] and Changho Lee [1,5,*]

[1] Department of Artificial Intelligence Convergence, Chonnam National University, 77 Yongbong-ro, Buk-gu, Gwangju 61186, Korea; 196286@jnu.ac.kr

[2] Research Center for Cancer Immunotherapy, Chonnam National University Hwasun Hospital, 264, Seoyang-ro, Hwasun-eup, Hwasun-gun, Jeollanam-do 58128, Korea; cuong44cnsh@yahoo.com (M.-C.V.); huychutan2010@gmail.com (T.-H.C.); drjejung@chonnam.ac.kr (J.-J.L.)

[3] Department of Creative IT Engineering and Electrical Engineering, Pohang University of Science and Technology (POSTECH), 77 Cheongam-ro, Nam-gu, Pohang, Gyeongbuk-do 37673, Korea; ronsan@postech.ac.kr (J.Y.K.); chulhong@postech.edu (C.K.)

[4] Department of Hematology-Oncology, Chonnam National University Hwasun Hospital, Hwasun, Jeollanam-do 58128, Korea

[5] Department of Nuclear Medicine, Chonnam National University Medical School & Hwasun Hospital, Hwasun, Jeollanam-do 58128, Korea

* Correspondence: shglory@hanmail.net (S.-H.J.); ch31037@jnu.ac.kr (C.L.); Tel.: +82-2-6986-1820 (S.-H.J.); +82-61-379-2885 (C.L.)

Citation: Mai, T.T.; Vo, M.-C.; Chu, T.-H.; Kim, J.Y.; Kim, C.; Lee, J.-J.; Jung, S.-H.; Lee, C. Pilot Study: Quantitative Photoacoustic Evaluation of Peripheral Vascular Dynamics Induced by Carfilzomib In Vivo. *Sensors* 2021, 21, 836. https://doi.org/10.3390/s21030836

Academic Editor: Ruben Specogna

Received: 26 December 2020
Accepted: 23 January 2021
Published: 27 January 2021

Publisher's Note: MDPI stays neutral with regard to jurisdictional claims in published maps and institutional affiliations.

Abstract: Carfilzomib is mainly used to treat multiple myeloma. Several side effects have been reported in patients treated with carfilzomib, especially those associated with cardiovascular events, such as hypertension, congestive heart failure, and coronary artery disease. However, the side effects, especially the manifestation of cardiovascular events through capillaries, have not been fully investigated. Here, we performed a pilot experiment to monitor peripheral vascular dynamics in a mouse ear under the effects of carfilzomib using a quantitative photoacoustic vascular evaluation method. Before and after injecting the carfilzomib, bortezomib, and PBS solutions, we acquired high-resolution three-dimensional PAM data of the peripheral vasculature of the mouse ear during each experiment for 10 h. Then, the PAM maximum amplitude projection (MAP) images and five quantitative vascular parameters, i.e., photoacoustic (PA) signal, diameter, density, length fraction, and fractal dimension, were estimated. Quantitative results showed that carfilzomib induces a strong effect on the peripheral vascular system through a significant increase in all vascular parameters up to 50%, especially during the first 30 min after injection. Meanwhile, bortezomib and PBS do not have much impact on the peripheral vascular system. This pilot study verified PAM as a comprehensive method to investigate peripheral vasculature, along with the effects of carfilzomib. Therefore, we expect that PAM may be useful to predict cardiovascular events caused by carfilzomib.

Keywords: carfilzomib; peripheral vasculature; photoacoustic microscopy; quantitative analysis

1. Introduction

Carfilzomib is a second-generation proteasome inhibitor, mainly used to treat multiple myeloma (MM) [1]. Combined with dexamethasone or lenalidomide and dexamethasone, carfilzomib has proven to significantly improve the survival outcomes of relapsed refractory MM patients in randomized phase 3 clinical trials [2]. Although carfilzomib is generally well tolerated, it is associated with adverse cardiovascular events, including hypertension, congestive heart failure, and coronary artery disease. In a small prospective study, patients treated with carfilzomib exhibited more cardiovascular events than those treated with bortezomib (51% vs. 17%, $P = 0.002$), and patients who experienced cardiovascular events exhibited significantly inferior progression-free survival ($P = 0.01$) and overall survival

Sensors 2021, 21, 836. https://doi.org/10.3390/s21030836　　　　　　　　　　　　　https://www.mdpi.com/journal/sensors

($P < 0.001$) [3]. While the mechanisms of cardiovascular adverse events have not been completely elucidated, one study suggests that endothelial dysfunction in the coronary vasculature caused by proteasome inhibition impacted the development of cardiovascular adverse events [4]. Generally, endothelial dysfunction is not limited to coronary vasculature and may affect peripheral vasculature, which could predict cardiovascular events [5]. Therefore, we hypothesize that proteasome inhibitors change the peripheral vasculature, and the identification of these changes may be a useful and convenient method to predict cardiovascular events using carfilzomib.

Various imaging modalities already contribute to the investigation and monitoring of the peripheral vasculature and its changes. Computed tomography (CT) and magnetic resonance imaging (MRI) can visualize the peripheral vasculature by injecting contrast agents [6,7]. Unfortunately, these are limited due to their relatively low resolution, radiation exposure, exogenous contrast, large system size, high post-processing time, and high cost. Ultrasound imaging (USI), with the Doppler effect, is also utilized to obtain vasculature information. However, Doppler USI is limited to the visualization of microvasculatures. Recently, USI has been used to visualize the microvasculatures of a mouse brain using microbubble-contrast agents and fast GPU processing of big data [8]. Optical imaging techniques, such as confocal microscopy (CM), multiphoton microscopy (MPM), and optical coherence tomography (OCT), have been utilized to visualize microvessels, based on a focused beam with a spatial resolution less than 10 μm [9–14]. Nevertheless, due to light scattering, CM and MPM are limited by a penetration depth in the range of hundreds of micrometers. Furthermore, these methods require additional contrast agents. Although OCT enables a better imaging depth (approximately 1 mm), it still requires complex signal processing.

One of the most notable imaging methods spotlighted in recent times is photoacoustic microscopy (PAM). Based on the photoacoustic (PA) effect that generates ultrasonic waves from light absorption, it inherits the hybrid imaging property of optical imaging and USI [15–17]. Owing to less scattering of ultrasound in the tissue, PAM can achieve a penetration depth of up to several centimeters and completely break the optical transport mean free path (i.e., ~1 mm) of pure optical imaging techniques. Depending on the crucial imaging setup, PAM can also provide multi-scale resolution from nano- to micro-scales by maintaining its optical absorption contrast. These benefits overcome the limited resolution of USI [18–20]. Moreover, PAM is a compact, inexpensive, non-ionized, label-free, functional, and real-time imaging modality [21–25]. It particularly aids the visualization of the distribution of intrinsic molecules in the body, such as hemoglobin, lipid, melanin, and proteins [26–31].

Additionally, PAM is widely utilized to visualize vasculatures and monitor their changes. For peripheral vascular visualization, B. Rao et al. reported an optical resolution-PAM (OR-PAM) system to visualize mouse ear microvasculatures [32]. A dual-modality imaging system combining PAM with USI was reported by Y. Tang [33] to achieve anatomical and functional information of a mouse's hind paw. Moreover, aiming towards human implementation, a PAM-based system for human peripheral arteries was developed [34]. For drug monitoring, Y. Liu et al. used the PAM system to assess norepinephrine via cerebral vessels [35]. Additionally, R. Bi et al. investigated orthotopic glioma blood vessels under the effect of combretastatin A4 phosphate [25]. L. Nie et al. proved PAM's effectiveness in monitoring nanocarrier-enhanced chemotherapy response in the early stage of brain tumor treatment [36].

However, most of these peripheral vascular visualization studies are interpreted qualitatively. To improve the accuracy in assessing the abnormal vascular, vessel quantitative parameters need to be extracted such as diameter, density, length fraction, and fractal dimension. One of the most common parameters of a blood vessel, the diameter, represents its size. A change in temperature causes a change in the vascular diameter [37]. The phases of perfusion of NaCl are also reflected through variation in brain vessel diameter [38]. Furthermore, a change in retinal vessel diameter has been shown to be a sign of stroke [39–41].

The density and length fraction of vessels implies the value of the whole vessel area and the total vessel length, respectively. These parameters play an important role in evaluating blood vessel abnormalities of the retina such as hyperoxia or glaucoma [42–45]. In addition, the vascular density was monitored for malignant and non-malignant skin [46,47]. Fractal dimension describes how completely a vascular network fills a space [48]. On the other hand, it indicates the vessel tortuosity and branching [49]. Hence, fractal dimension is used to describe the complexity of biological structures, including the coronary [50], parafoveal capillary network [51] and tumor vascular network [48]. Therefore, these parameters reveal the characteristic features of a blood vessel, which are especially useful in monitoring abnormalities that occur in the blood vessels.

In this study, we focus on photoacoustically monitoring the dynamics of peripheral vasculatures in mouse ears under the effect of carfilzomib for 10 h using high-resolution label-free OR-PAM. The same follow-up imaging process was repeated with the injection of bortezomib and PBS. After acquiring three-dimensional OR-PAM data for all the cases, the OR-PAM MAP images were reconstructed and analyzed. For a comprehensive assessment, the peripheral vasculature's structural morphology was also monitored, and the quantitative vasculature evaluation process, including diameter, density, length fraction, fractal dimension, and PA signal, was implemented.

2. Materials and Methods

2.1. Carfilzomib Solution

The proteasome inhibitors carfilzomib (Kyprolis; Onyx Pharmaceuticals, San Francisco, CA, USA) and bortezomib (Velcade; Millennium Pharmaceuticals, Cambridge, MA, USA) were dissolved in sterile, 0.9% (v/v) normal saline immediately before use. Carfilzomib and bortezomib, at a dose of 10 and 0.5 mg/kg in 100 μL PBS, respectively, were administrated to the mice by intravenous injection. The doses of carfilzomib and bortezomib were determined based on previous studies [52]. In clinical practice, the bortezomib dose is constant. However, the dose of carfilzomib depends on the combination of drugs or the administration schedule. We selected a higher dose of carfilzomib in this experiment, as cardiovascular events are associated with higher carfilzomib doses [53].

2.2. Animal Preparing

All experimental animal procedures followed laboratory animal protocols approved by the Institutional Animal care and use committee of Chonnam National University Hwasun Hospital (CNU IACUC-H-2018-68). Healthy six-to-eight-week-old female BALB/c (H-2d) mice, weighing ~20 g, were purchased from Orient Bio (Iksan, Korea) and maintained under specific pathogen-free conditions. Each mouse was anesthetized with an intraperitoneal injection of Ketamine (80 mg/kg)/Xylazine (12 mg/kg). After removing the downy hairs on its ear, the mouse was placed on a homemade animal holder. An isoflurane system (Luna Vaporiser, NorVap international LTD, Barrowford, UK) for gaseous anesthetization and a temperature maintain bed were used in long-term in vivo observation (10 h) to maintain stable body conditions of the mouse. The energy of the illuminated laser pulse on the mouse skin was approximately 5 mJ/cm^2 below the American National Standards Institute (ANSI) safety limit (20 mJ/cm^2).

2.3. Optical-Resolution Photoacoustic Microscopy

Figure 1 describes the schematic of the OR-PAM system. Trigger signals from the data acquisition (DAQ) board (PCIe-6321, NI instruments, Austin, TX, USA) were sent to a diode laser (SPOT-10-200-532, Elforlight, Daventry, UK) to operate the primary laser beam at a wavelength of 532 nm. The fired laser beam, with a duration of 6 ns, was spatially filtered by an iris (SM1D12, Thorlabs, Newton, NJ, USA). The reshaped laser beam was reflected between a pair of mirrors to transfer it to a collimator (F280APC-A, Thorlabs, NJ, USA; f = 18.07 mm, NA = 0.15). After coupling into a single-mode optical fiber (P1-405BPM-FC-1, Thorlabs, NJ, USA), the laser beam was efficiently delivered to the second collimator

(F260APC-A, Thorlabs, NJ, USA, f = 15.01 mm, NA = 0.17). An objective lens (AC254-060-A, Thorlabs, NJ, USA), with a focal length of 60 mm, was utilized to focus the collimated laser beam. A high-precision zoom housing (SM1ZM, Thorlabs, NJ, USA), which could adjust the optical focal plane, with a maximum of 4.1 mm along the z-axis, was used to mount the objective lens. The focused beam was passed through a correction lens and made to penetrate a hand-made beam combiner. Inside the combiner, the beam was reflected by a layer created by a normal and an aluminum-coated prism. Before the focused beam illuminated the sample, the focused beam was redirected by reflecting it on the mirror of the MEMS scanner (OptichoMS-001, Opticho Inc., Ltd., Pohang, Korea) and passed through a tank. The confocal and co-axial alignment of the incident laser beam and ultrasound occurred between the combiner and the MEMS scanner's mirror. The omnidirectional photoacoustic wave was generated immediately after the illumination of the sample. With the focused support of a concave lens located on the right side of the combiner, the PA wave easily passed through the combiner and was detected by a high-frequency ultrasonic transducer (V214-BC-RM, 50 MHz, Olympus, Tokyo, Japan). The signal acquisition part aimed to improve and convert the PA waves into image information. Two RF-amplifiers (ZX60-3018G-S+, Mini-Circuit, Brooklyn, NY, USA) were used to amplify the released PA wave from the sample. Then, a high-speed digitizer (ATS9371, AlazarTech, Pointe-Claire, QC, Canada) digitalized the signal to convert it into primary image information. A linear stepper motor stage (L-509-10SD00, Physik Instrumente, Karlsruhe, Germany) on the y-axis was associated with the MEMS scanner on the x-axis to obtain the 3D image data. This integrated scan system was used to drive the beam to scan the object's surface, with a speed of 25 Hz B-scan. The data acquisition time for a mouse ear with an area of 10 × 12 mm is 180 s. The measured lateral and axial resolutions were 12 and 45 μm, respectively [21]. A LabVIEW program (National Instruments, Austin, TX, USA) was used to operate the OR-PAM system. The released image information was reconstructed and analyzed with MATLAB (R2016a, Mathworks, Natick, MA, USA).

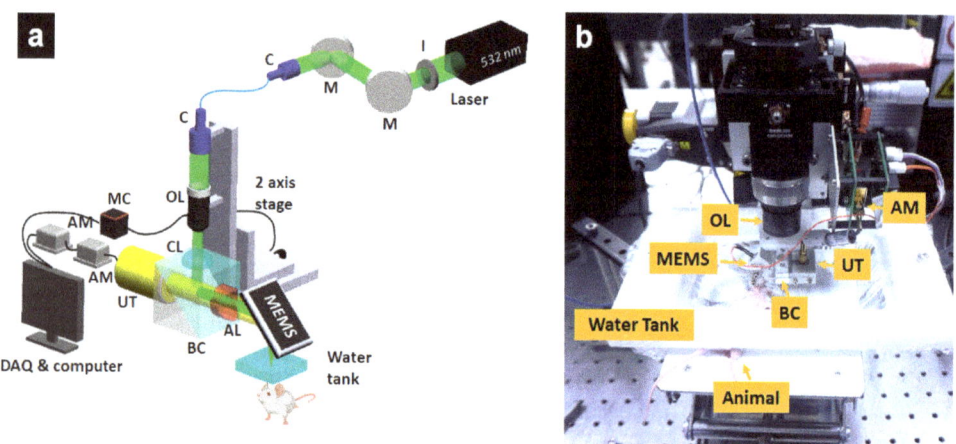

Figure 1. Optical-resolution photoacoustic microscopy (OR-PAM) system to monitor the peripheral vasculature of the mouse ear. (**a**) Schematic of OR-PAM, (**b**) Photograph of the OR-PAM probe part. I, iris; M, mirror; C, collimator; OL, objective lens, CL, correction lens, TR, transducer; BC, beam combiner; AL, acoustic lens; UT, ultrasound transducer; AM, amplifier; MC, motion controller; MEMS, MEMS scanner.

2.4. Quantification Evaluation Process of OR-PAM Image

Figure 2 illustrates the quantitative evaluation process of the OR-PAM images and presents the two main steps involved: (1) image segmentation and (2) quantitative parameter extraction. In the segmentation stage, the multi-scale Hessian filter, intended specifically for vascular objects, was used. The Hessian filter could separate the boundaries between blood vessels and the background based on the second-order gradient of the image [54]. By utilizing multiple scales, multi-size vessels were segmented. The binary image was obtained using the adaptive threshold method [55]. Finally, morphological operations were used to obtain a skeleton map. A skeleton map showing only the centerline of the object was also obtained. These two maps were used as input sources for the next step, i.e., extraction. The four required parameters were extracted by implementing some methods on the binary and skeleton maps. Table 1 summarizes the formulas of the parameters.

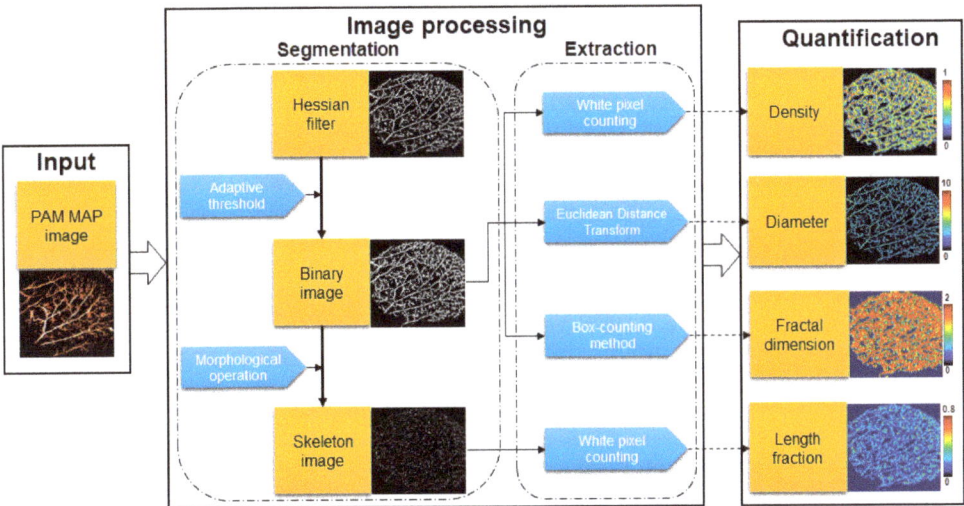

Figure 2. Flow chart of the quantitative OR-PAM image evaluation process.

Table 1. The formulas of quantitative parameters.

Parameters	Formula	
Diameter	$\text{Diameter} = \frac{\sum_{i=1}^{m}\sum_{j=1}^{n} E(i,j)}{S(i,j)}$ (1)	$E(i,j)$: Euclidean distance transform $S(i,j)$: White pixel of skeleton map
Density	$\text{Density} = \frac{\sum_{i=1}^{m}\sum_{j=1}^{n} B(i,j)}{T(i,j)}$ (2)	$B(i,j)$: White pixel of the binary map $T(i,j)$: Total pixels appeared on the binary map
Length fraction	$\text{Length fraction} = \frac{\sum_{i=1}^{m}\sum_{j=1}^{n} S(i,j)}{T(i,j)}$ (3)	$S(i,j)$: White pixel of skeleton map $T(i,j)$: Total pixels of skeleton map
Fractal dimension	$\text{Fractal dimension} = \frac{\log(N_r)}{\log(\frac{1}{r})}$ (4)	r : Size of the unit box N_r : Number of boxes
PA signal	$\text{PA signal} = \frac{\sum_{i=1}^{m}\sum_{j=1}^{n} I(i,j)}{T(i,j)}$ (5)	$I(i,j)$: Intensity at point (i,j) of the MAP image $T(i,j)$: Total pixels of the MAP image

The shortest distance calculated by Euclidean transform for a certain vertical section of a vessel was assumed as the diameter of that section. The diameter of a single vessel was measured as the average distance transform along the skeleton of that vessel. Thus, only the vessel area will contribute to the average diameter vessel of an image. Equation (1) indicates the average diameter vessel for an image, where the Euclidean distance transform was implemented on the corresponding binary map. In Equations (2) and (3), the percentage area covered by the vessel was used to define the density and length fraction parameters. Length fraction only considers the existence of the vessel by counting the number of white pixels on the skeleton map to figure out the ratio of white pixels and total pixels.

Thus, dilation or constriction of the vessel does not impact length fraction. In contrast, density was influenced by both the diameter and length fraction of the vessel. To compute density, we counted the number of white pixels on the binary image and divided it by the number of the total pixels of the image. The fractal dimension was calculated by the box-counting method [56]. The binary map was divided into square boxes of the same size, called the unit boxes. The number of unit boxes needed to cover a vessel was counted. The logarithm of the unit box size with the corresponding number of boxes produces a curve; the fractal dimension is determined as the absolute value of that curve (Equation (4)). The PA signal was also monitored to comprehensively evaluate the change in blood vessels comprehensively. PA signal describes the average incident photon energy that was absorbed by blood vessels. For all formula, m and n are sizes of the image (MAP image, binary map and skeleton map share same size); *i* and *j* are the calculated coordinates of the pixel. Total pixels $T(i, j) = m \times n$.

3. Results

3.1. In Vivo OR-PAM Observation for the Peripheral Vasculatures after Carfilzomib Solution Injection

The mouse ear's peripheral vasculature (Figure 3k) was monitored for 10 h before and after injecting the carfilzomib solution. Figure 3a–j shows the acquired OR-PAM MAP images with the image color scale fixed on all images to investigate signal change monitoring. Figure 3a is a control image, showing that the main vasculatures and microvessels are generally well distributed. The OR-PAM MAP image presented in Figure 3b was obtained 10 min after carfilzomib solution injection. The bright color of the vessels indicates a significant increase in the PA signal intensity in these vessels. Several bleeding spots were formed, especially along the large blood vessels, and several new microvasculatures, that had appeared suddenly, had formed. The effect of the carfilzomib seemed to have peaked at 30 min, based on the image presented in Figure 3c. Although the PA signal in the major vasculatures had only marginally changed compared to that at the previous imaging timepoint, the areas where the microvasculatures were present continued to expand. Figure 3d–j presents the OR-PAM MAP images, obtained 1, 2, 3, 4, 6, 8, and 10 h after the injection, respectively. During this time period, the PA signal decreased gradually. Specifically, at the 4 h mark, the microvessels had begun to disappear slowly and completely disappeared at the 10 h mark. The PA signal gradually recovered to a steady state, similar to that shown in the control image.

For more detailed analysis, we randomly monitored nine small regions representing large vasculatures (ROI 1 and ROI 8) and microvasculatures (ROI 2, ROI 3, ROI 4, ROI 5, ROI 6, ROI 7, and ROI 9), as shown in Figures S1–S3 (Supplementary Information 1). For the large vasculatures, ROI 8, on the margins of the mouse ear, showed the least amount of change during the follow-up imaging. In contrast, ROI 1, which included bleeding points, exhibited an increase in the signal approximately 30 min after the injection and a gradual decrease thereafter. However, this increase was only approximately 50% to 100% compared to the initial observation value. Overall, the variation in the signal due to carfilzomib's effect was observed in all areas of the small blood vessels. Within the first 30 min, it was observed that the closer the ROI was to the center, the larger the extent of changes. For example, ROI 2, ROI 7, ROI 6, and ROI 4 showed changes in the PA signal ranging between 150% and 400%, and for the other parameters, only 50–120% changes were present. Meanwhile, the ROIs in the far center (ROI 3 and ROI 5) exhibited an increase in the PA signal by approximately 50–100%, and only 10–50% changes were exhibited for the other parameters. Located at the edge of the ear, where there were very few blood vessels, ROI 9 exhibited changes that were distinct from those of the rest of the ROIs. However, all ROIs began to stabilize at the 4-h mark.

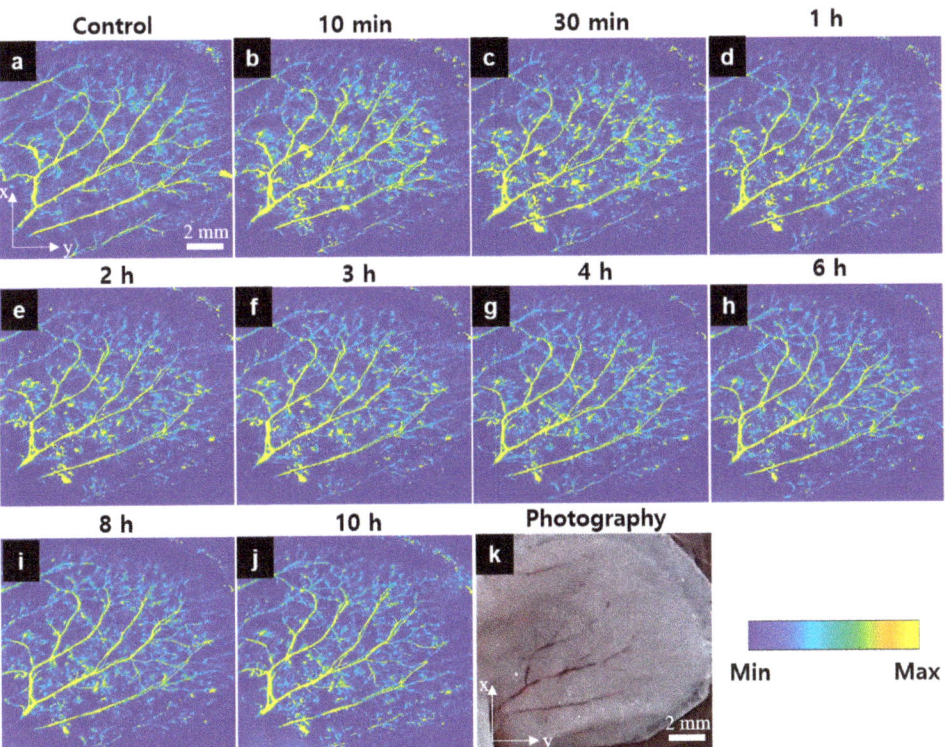

Figure 3. In vivo OR-PAM MAP images of the peripheral vasculatures of a mouse ear after carfilzomib solution injection (Movie 1). (**a**) OR-PAM MAP image before carfilzomib injection, (**b**–**j**) OR-PAM MAP images after carfilzomib solution injection during 10 h of observation, (**k**) Photograph of the corresponding mouse ear.

3.2. In Vivo OR-PAM Observation for the Peripheral Vasculatures after Bortezomib Solution Injection

The experiment was repeated by substituting carfilzomib with bortezomib. The OR-PAM MAP image of the control presented in Figure 4a, obtained before injecting the bortezomib solution, clearly shows all the vasculatures not shown in Figure 4k. Unlike carfilzomib, bortezomib did not significantly increase the vessel's signal after the first 10 min of the injection (Figure 4b). The mouse ear was monitored for approximately 10 h with the OR-PAM MAP images. Figure 4c–j depicts the mouse ear 30 min and 1, 2, 3, 4, 6, 8, and 10 h after the bortezomib injection. The same color was observed in the OR-PAM MAP images with the big vasculatures, indicating no increase in the PA signal. Moreover, no bleeding area was detected, and no small vessel appeared. This indicates that steady state was maintained for 6 h after injection. In Figures S4–S6, the MAP images and quantification results for nine small ROIs are shown, in which ROI 1, ROI 2, ROI 3, ROI 4, and ROI 9 indicate small vessel areas, and ROI 5, ROI 6, ROI 7, and ROI 8 describe large vessel areas. The stability, monitored using MAP images, maintained during 6 h post-injection, was observed at all the ROIs. Furthermore, variation in the quantitative values of all the parameters was only approximately 20% compared to those at the control time. Peak value at the 8 h mark was only noticeable at ROI 3, ROI 8, and ROI 9, which were near the mouse ear's edges. However, the increase was only approximately 50–60%, which decreased after 10 h of monitoring.

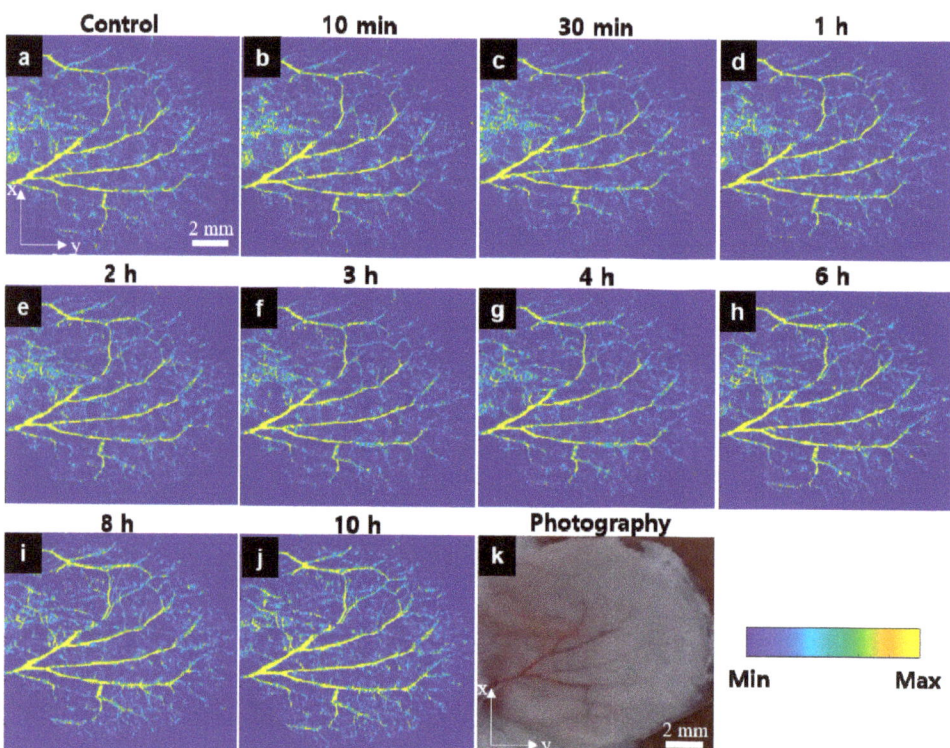

Figure 4. In vivo OR-PAM MAP images for the peripheral vasculatures of the mouse ear after bortezomib solution injection (Movie 2). (**a**) OR-PAM MAP image before bortezomib injection, (**b–j**) OR-PAM MAP images after bortezomib solution injection within 10 h, and (**k**) Photograph of the corresponding mouse ear.

3.3. In Vivo OR-PAM Observation for the Peripheral Vasculatures of after PBS Injection

An experiment with PBS injection is required to standardize and accurately and objectively evaluate the effectiveness of carfilzomib and bortezomib. Figure 5 displays the OR-PAM MAP images of the monitoring process. Figure 5a shows the condition before the injection, Figure 5b–j shows the observations from the 10 min to the 10 h mark. Figure 5k shows the photograph of the mouse's ear. Under stable mouse conditions, the PBS did not seem to affect the stability of the large blood vessels and small capillaries before and after the injection. As depicted in Figures S7–S9, for OR-PAM MAP images and graphs of quantification results, there is no general trend in the change in signals at all the ROIs. The highest variations, smaller than 100%, belong to ROI 3 and ROI 5 for the length fraction and PA signal, respectively. For the remaining ROIs, the values were always less than 50% for all parameters.

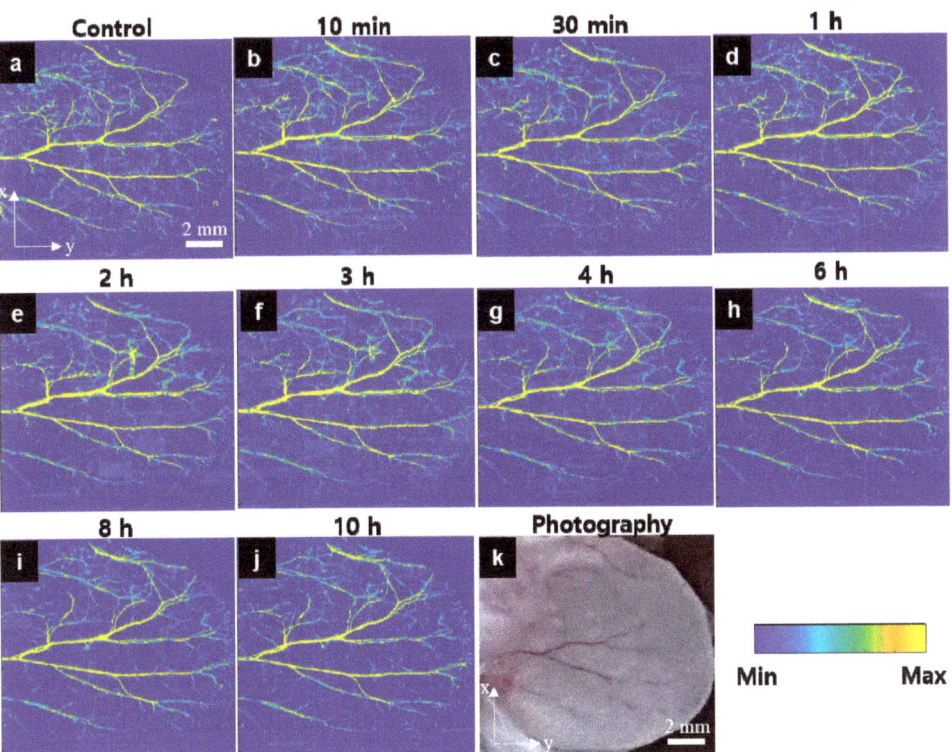

Figure 5. In vivo OR-PAM MAP images for the peripheral vasculatures of the mouse ear after PBS injection (Movie 3). (**a**) OR-PAM MAP image before PBS, (**b–j**) OR-PAM MAP images after PBS solution injection within 10 h, and (**k**) Photograph of the corresponding mouse ear.

3.4. Quantitative Evaluation of OR-PAM Data

A quantitative assessment is a necessary supplement, as only comparing images is insufficient to accurately assess the effects of carfilzomib and bortezomib on the peripheral vasculature of the mouse ear. The quantitative process presented in Section 2.4, for the carfilzomib, bortezomib, and PBS injection cases, was applied. After obtaining five parameters for each case, the values of all three cases were compared for each parameter: (1) PA signal, (2) diameter, (3) density, (4) length fraction, and (5) fractal dimension. The values were presented as a percentage difference from the control value to unit normalize all the parameters. Figure 6 shows the results of this process. For all parameters, the variation in the signal, caused by carfilzomib, always gives the maximum value, 45% (PA signal at the 10-min timepoint) or 48% (density at 3-h timepoint). Meanwhile, the values corresponding to bortezomib and PBS were almost lower than 20%. The PA signal comparison shown in Figure 6a highlights that the value corresponding to carfilzomib tends to increase rapidly 10 min after the injection and then gradually decrease for 10 h. The mean value with carfilzomib is found to be 18.2%, almost eight times higher than that observed with bortezomib (2.3%) and 5.5 times higher than that observed with PBS (3.3%). This proves that image-based judgment is plausible. The graphs comparing diameter (Figure 6b), density (Figure 6c), length fraction (Figure 6d), and fractal dimension (Figure 6e) also show a similar trend. With the exception of the 3-h mark and 6-h mark, the values observed in the bortezomib case did not change significantly. In the PBS case, the values were stable at all timepoints. Specifically, at the 30-min timepoint, the higher values observed in the carfilzomib case compared to those of the bortezomib and PBS cases were 51.7% and 54.2%

for the density value, 30.7% and 36.7% for the length fraction value, and 19.7% and 16.8% for the diameter, respectively. The effect of carfilzomib is evident in Figure 6e, depicting the fractal dimension, which is approximately 10 times higher than that observed in the bortezomib and PBS case.

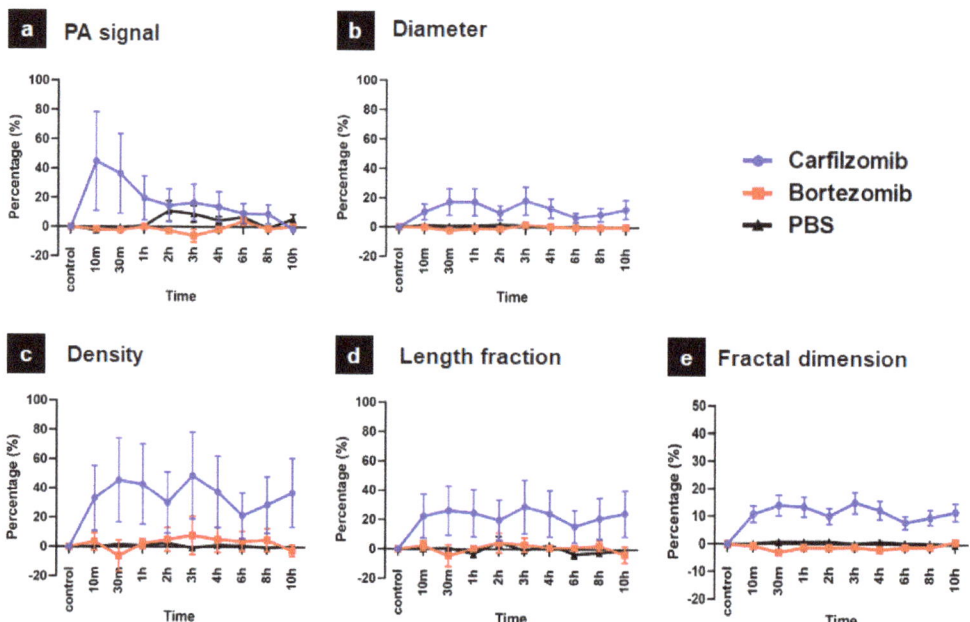

Figure 6. Quantitative evaluation results among carfilzomib, bortezomib, and PBS. (**a**) PA signal, (**b**) Diameter, (**c**) Density, (**d**) Fraction length, (**e**) Fractal dimension.

4. Discussion

We successfully monitored the effects of carfilzomib on the peripheral vasculature for over 10 h using a hybrid process of OR-PAM imaging and quantitative evaluation. The results show the mouse ear's peripheral vasculature dynamic in morphological and quantitative evaluations. Among the morphological aspects, the vascular changes were observed 10 min after injecting the carfilzomib solution, with the appearance of bleeding spots along with vessels. The emergence and strong spread of microvasculatures between 10 and 30 min indicate the occurrence of the peak effect of carfilzomib during the first 30 min after the injection. It was observed that the expansion of the microvasculature area slowly decreased, and the mouse ear vascular network recovered to the steady state after 10 h. The quantitative evaluation was used to evaluate five blood vessel parameters, including the PA signal, diameter, density, length fraction, and fractal dimension. A vascular abnormality could be identified based on the changes in all the vessel properties, which were closely related. The diameter and length fraction present the size and the perfusion of the vessel, respectively. The density value, influenced by the diameter and length fraction, gives an overview of the general situation of an area. The fractal dimension indicates the complexity of the vascular network. Microvasculatures are typically more complex than large vasculatures [57]. The results of this study show that the diameter, density, length fraction, and fractal dimension follow a similar trend. These parameters significantly increased and peaked 30 min after the injection. This seems to indicate an increase in blood pressure resulting from the effect of carfilzomib. When the blood vessels dilated, the perfusion increased and expanded the area of the blood vessels simultaneously.

Additionally, an increase in the length fraction with the fractal dimension implies the appearance of new capillaries. Beginning from the 1 h mark, values of all the parameters started to decrease slightly and then maintained a stable value until the 10 h mark. All these changes were relatively consistent with the morphological visualization captured in the OR-PAM MAP images. Unlike other parameters, the PA signal is not a vasculature morphological property as it is reflected by the intensity of the photons absorbed by the blood. The value of the PA signal exhibited a significant increase in the first 10 min, after which it rapidly decreased. At a wavelength of 532 nm, the strong light absorption of the blood made our system highly sensitive to changes in the blood. Hence, the maximum value change and the post-peak reduction rate of the PA signal were always the highest.

In addition to carfilzomib, we conducted the same experiment using bortezomib and PBS. The average values calculated for all the parameters using carfilzomib were eight times higher than those observed with bortezomib and 5.5 times higher than those with PBS, emphasizing that carfilzomib has a stronger effect on the peripheral vasculature than bortezomib and PBS. However, especially at the 30-min mark, the higher values in the values observed with carfilzomib with respect to those with bortezomib and PBS were 51.7% and 54.2% for the density value, 30.7% and 36.7% for the length fraction value, and 19.7% and 16.8% for the diameter, respectively. As mentioned in Section 2.4, the change in density depends on the changes in diameter and length fraction. Thus, the contribution to the density value in this experiment was mainly from the length fraction, indicating that the length fraction changes the most under the effect of carfilzomib. It also reflects that carfilzomib has a stronger impact on the capillaries than large blood vasculatures, because diameter depends on large blood vessels, while length is determined by small blood vessels.

In myeloma cells, the anti-myeloma effect induced by the proteasome inhibitor results from the accumulation of regulatory proteins within the endoplasmic reticulum, which further induces the apoptosis cascade [58]. Likewise, proteasome inhibition in myocardial cells leads to an abnormal accumulation of ubiquitinated proteins and can result in cardiac damage and heart failure [59]. Additionally, endothelial dysfunction caused by proteasome inhibition in the vasculature is associated with cardiovascular adverse events, such as hypertension and myocardial ischemia. Proteasome inhibition affects signaling in vascular smooth muscle endothelium and leads to increased vascular tone and coronary resistance [4]. Therefore, all proteasome inhibitors can theoretically cause cardiovascular adverse events. However, these effects are different when using bortezomib and carfilzomib, owing to differences in their pharmacodynamics and pharmacokinetic characteristics. The effect of carfilzomib is irreversible, and it is a more selective inhibitor for the β5 domain, with chymotrypsin-like activity of the 20s subunit of the proteasome, compared to bortezomib. Efentakis P et al. [60] demonstrated that bortezomib did not affect cardiac function. However, carfilzomib caused deterioration of the left ventricular function through increased PP2A activity and inhibition of the AMPKα pathway in the in vivo mouse model. In clinical studies, carfilzomib cases had a higher incidence of cardiovascular adverse events than bortezomib cases [3,61,62]. However, the effect of carfilzomib on the peripheral vasculature has not been reported.

Despite the success in monitoring the effects of carfilzomib on the peripheral vasculature with OR-PAM, our approach has certain limitations. First, the limit of the low penetration depth of its system (~1 mm) is the main hindrance to OR-PAM to become a clinical imaging method even it has super high resolution. Hence, for human applications, a proper handheld probe and the near-infrared laser should be implemented [63–65]. Second, although it is able to detect the concentration of hemoglobin, the use of single wavelength laser (i.e., 532 nm) is limited to provide further functional information including oxygenated (HbO_2) and deoxygenated (HbR) hemoglobin. Thus, our system cannot directly provide the functional information of vessel such as oxygen saturation (SO_2) and cerebral metabolic rate of oxygen ($CMRO_2$) in the peripheral vasculature. By combining different wavelength lasers, it becomes able to offer metabolism information directly [66].

5. Conclusions

In conclusion, we implemented PAM for the mouse ear peripheral vasculature visualization within 10 h after carfilzomib, bortezomib, and PBS injection. Not only morphological, but also vascular parameters such as PA signal, diameter, density, length fraction, and fractal dimension, were successfully evaluated. Carfilzomib induces a strong effect on the peripheral vascular system during the first 30 min after injection, which can be which can be qualitatively visualized by the appearance of bleeding spots and capillary in the MAP images. Moreover, for the quantitative results, all vascular parameters significantly increase up to 50%. In contrast, bortezomib and PBS do not have much impact on the peripheral vascular system. As a pilot study, we monitored the effects of carfilzomib on the peripheral vascular system with the PAM technique and its quantitative analysis. Therefore, we expect PAM to be able to use an important tool to predict cardiovascular events triggered by carfilzomib.

Supplementary Materials: The following are available online at https://www.mdpi.com/1424-8 220/21/3/836/s1, Figure S1: Selected small 9 ROIs after carfilzomib solution injection., Figure S2: OR-PAM MAP images of 9 ROIs during 10 h after carfilzomib solution injection. Figure S3: Quantitative evaluation of 9 ROIs within 10 h after carfilzomib injection. Figure S4: Selected small 9 ROIs after bortezomib solution injection. Figure S5: OR-PAM MAP images of 9 ROIs during 10 h after bortezomib solution injection. Figure S6: Quantitative evaluation of 9 ROIs within 10 h after bortezomib solution injection. Figure S7: Selected small 9 ROIs after PBS injection Figure S8: OR-PAM MAP images of 9 ROIs during 10 h after PBS injection. Figure S9: Quantitative evaluation of 9 ROIs within 10 h after PBS injection.

Author Contributions: Conceptualization, S.-H.J. and C.L.; methodology, T.T.M., J-J.L., S.-H.J. and C.L.; software, J.Y.K, C.K. and T.T.M.; validation, T.T.M., M.-C.V., T.-H.C., and C.L.; formal analysis, J.K., T.T.M. and C.L.; writing—original draft preparation, T.T.M., S.-H.J., and C.L.; writing—review and editing, T.T.M., S.-H.J., and C.L.; supervision, S.-H.J. and C.L. All authors have read and agree to the published version of the manuscript.

Funding: NRF grant funded by the Korea government (MSIT) (NRF-2019R1F1A1062948), the Pioneer Research Center Program through the NRF of Korea funded by the Ministry of Science, ICT & Future Planning (No. 2015M3C1A3056407) Basic Science Research Program through NRF funded by the Ministry of Education (2020R1A6A1A03047902), and Chonnam National University (No.2020-0358).

Conflicts of Interest: C. Kim and J. Y. Kim have financial interests in Opticho, which, however, did not support this work.

References

1. Choueiri, T.K.; Escudier, B.; Powles, T.; Tannir, N.M.; Mainwaring, P.N.; Rini, B.I.; Hammers, H.J.; Donskov, F.; Roth, B.J.; Peltola, K. Cabozantinib versus everolimus in advanced renal cell carcinoma (METEOR): Final results from a randomised, open-label, phase 3 trial. *Lancet Oncol.* **2016**, *17*, 917–927. [CrossRef]
2. Stewart, A.K.; Rajkumar, S.V.; Dimopoulos, M.A.; Masszi, T.; Špička, I.; Oriol, A.; Hájek, R.; Rosiñol, L.; Siegel, D.S.; Mihaylov, G.G. Carfilzomib, lenalidomide, and dexamethasone for relapsed multiple myeloma. *N. Engl. J. Med.* **2015**, *372*, 142–152. [CrossRef] [PubMed]
3. Cornell, R.F.; Ky, B.; Weiss, B.M.; Dahm, C.N.; Gupta, D.K.; Du, L.; Carver, J.R.; Cohen, A.D.; Engelhardt, B.G.; Garfall, A.L. Prospective study of cardiac events during proteasome inhibitor therapy for relapsed multiple myeloma. *J. Clin. Oncol.* **2019**, *37*, 1946–1955. [CrossRef] [PubMed]
4. Chen-Scarabelli, C.; Corsetti, G.; Pasini, E.; Dioguardi, F.S.; Sahni, G.; Narula, J.; Gavazzoni, M.; Patel, H.; Saravolatz, L.; Knight, R. Spasmogenic effects of the proteasome inhibitor carfilzomib on coronary resistance, vascular tone and reactivity. *EBioMedicine* **2017**, *21*, 206–212. [CrossRef] [PubMed]
5. Behroozian, A.; Beckman, J.A. Microvascular Disease Increases Amputation in Patients With Peripheral Artery Disease. *Arterioscler. Thromb. Vasc. Biol.* **2020**, *40*, 534–540. [CrossRef]
6. Haider, C.R.; Glockner, J.F.; Stanson, A.W.; Riederer, S.J. Peripheral vasculature: High-temporal-and high-spatial-resolution three-dimensional contrast-enhanced MR angiography. *Radiology* **2009**, *253*, 831–843. [CrossRef]
7. Mishra, A.; Bhaktarahalli, J.N.; Ehtuish, E.F. Imaging of peripheral arteries by 16-row multidetector computed tomography angiography: A feasible tool? *Eur. J. Radiol.* **2007**, *61*, 528–533. [CrossRef]
8. Errico, C.; Pierre, J.; Pezet, S.; Desailly, Y.; Lenkei, Z.; Couture, O.; Tanter, M. Ultrafast ultrasound localization microscopy for deep super-resolution vascular imaging. *Nature* **2015**, *527*, 499–502. [CrossRef]

9. Campisi, M.; Shin, Y.; Osaki, T.; Hajal, C.; Chiono, V.; Kamm, R.D. 3D self-organized microvascular model of the human blood-brain barrier with endothelial cells, pericytes and astrocytes. *Biomaterials* **2018**, *180*, 117–129. [CrossRef]
10. Dickie, R.; Bachoo, R.; Rupnick, M.; Dallabrida, S.; Deloid, G.; Lai, J.; DePinho, R.A.; Rogers, R. Three-dimensional visualization of microvessel architecture of whole-mount tissue by confocal microscopy. *Microvasc. Res.* **2006**, *72*, 20–26. [CrossRef]
11. Manconi, F.; Kable, E.; Cox, G.; Markham, R.; Fraser, I. Whole-mount sections displaying microvascular and glandular structures in human uterus using multiphoton excitation microscopy. *Micron* **2003**, *34*, 351–358. [CrossRef] [PubMed]
12. Mariampillai, A.; Standish, B.A.; Moriyama, E.H.; Khurana, M.; Munce, N.R.; Leung, M.K.; Jiang, J.; Cable, A.; Wilson, B.C.; Vitkin, I.A. Speckle variance detection of microvasculature using swept-source optical coherence tomography. *Opt. Lett.* **2008**, *33*, 1530–1532. [CrossRef] [PubMed]
13. Nesper, P.L.; Roberts, P.K.; Onishi, A.C.; Chai, H.; Liu, L.; Jampol, L.M.; Fawzi, A.A. Quantifying microvascular abnormalities with increasing severity of diabetic retinopathy using optical coherence tomography angiography. *Investig. Ophthalmol. Vis. Sci.* **2017**, *58*, BIO307–BIO315. [CrossRef]
14. Padera, T.P.; Stoll, B.R.; So, P.T.; Jain, R.K. Conventional and high-speed intravital multiphoton laser scanning microscopy of microvasculature, lymphatics, and leukocyte-endothelial interactions. *Mol. Imaging* **2002**, *1*, 9–15. [CrossRef] [PubMed]
15. Kim, J.Y.; Lee, C.; Park, K.; Han, S.; Kim, C. High-speed and high-SNR photoacoustic microscopy based on a galvanometer mirror in non-conducting liquid. *Sci. Rep.* **2016**, *6*, 34803. [CrossRef]
16. Wang, L.V.; Hu, S. Photoacoustic tomography: In vivo imaging from organelles to organs. *Science* **2012**, *335*, 1458–1462. [CrossRef]
17. Zhang, H.F.; Maslov, K.; Stoica, G.; Wang, L.V. Functional photoacoustic microscopy for high-resolution and noninvasive in vivo imaging. *Nat. Biotechnol.* **2006**, *24*, 848–851. [CrossRef]
18. Park, S.; Lee, C.; Kim, J.; Kim, C. Acoustic resolution photoacoustic microscopy. *Biomed. Eng. Lett.* **2014**, *4*, 213–222. [CrossRef]
19. Kim, J.; Kim, J.Y.; Jeon, S.; Baik, J.W.; Cho, S.H.; Kim, C. Super-resolution localization photoacoustic microscopy using intrinsic red blood cells as contrast absorbers. *Light Sci. Appl.* **2019**, *8*, 1–11. [CrossRef]
20. Baik, J.W.; Kim, J.Y.; Cho, S.; Choi, S.; Kim, J.; Kim, C. Super wide-field photoacoustic microscopy of animals and humans in vivo. *IEEE Trans. Med. Imaging* **2019**, *39*, 975–984. [CrossRef]
21. Kim, J.; Mai, T.T.; Kim, J.Y.; Min, J.-J.; Kim, C.; Lee, C. Feasibility Study of Precise Balloon Catheter Tracking and Visualization with Fast Photoacoustic Microscopy. *Sensors* **2020**, *20*, 5585. [CrossRef] [PubMed]
22. Lee, C.; Lee, D.; Zhou, Q.; Kim, J.; Kim, C. Real-time near-infrared virtual intraoperative surgical photoacoustic microscopy. *Photoacoustics* **2015**, *3*, 100–106. [CrossRef] [PubMed]
23. Park, E.-Y.; Lee, D.; Lee, C.; Kim, C. Non-Ionizing Label-Free Photoacoustic Imaging of Bones. *IEEE Access* **2020**, *8*, 160915–160920. [CrossRef]
24. Yao, J.; Maslov, K.I.; Zhang, Y.; Xia, Y.; Wang, L.V. Label-free oxygen-metabolic photoacoustic microscopy in vivo. *J. Biomed. Opt.* **2011**, *16*, 076003. [CrossRef] [PubMed]
25. Bi, R.; Balasundaram, G.; Jeon, S.; Tay, H.C.; Pu, Y.; Li, X.; Moothanchery, M.; Kim, C.; Olivo, M. Photoacoustic microscopy for evaluating combretastatin A4 phosphate induced vascular disruption in orthotopic glioma. *J. Biophotonics* **2018**, *11*, e201700327. [CrossRef]
26. He, Y.; Shi, J.; Pleitez, M.A.; Maslov, K.; Wagenaar, D.A.; Wang, L.V. Label-free imaging of lipid-rich biological tissues by mid-infrared photoacoustic microscopy. *J. Biomed. Opt.* **2020**, *25*, 106506. [CrossRef]
27. Park, E.; Lee, Y.-J.; Lee, C.; Eom, T.J. Effective photoacoustic absorption spectrum for collagen-based tissue imaging. *J. Biomed. Opt.* **2020**, *25*, 056002. [CrossRef]
28. Shi, J.; Wong, T.T.; He, Y.; Li, L.; Zhang, R.; Yung, C.S.; Hwang, J.; Maslov, K.; Wang, L.V. High-resolution, high-contrast mid-infrared imaging of fresh biological samples with ultraviolet-localized photoacoustic microscopy. *Nat. Photonics* **2019**, *13*, 609–615. [CrossRef]
29. Zhang, C.; Wang, L.V.; Cheng, Y.-J.; Chen, J.; Wickline, S.A. Label-free photoacoustic microscopy of myocardial sheet architecture. *J. Biomed. Opt.* **2012**, *17*, 060506. [CrossRef]
30. Kim, H.; Baik, J.W.; Jeon, S.; Kim, J.Y.; Kim, C. PAExM: Label-free hyper-resolution photoacoustic expansion microscopy. *Opt. Lett.* **2020**, *45*, 6755–6758. [CrossRef]
31. Park, B.; Bang, C.H.; Lee, C.; Han, J.H.; Choi, W.; Kim, J.; Park, G.S.; Rhie, J.W.; Lee, J.H.; Kim, C. 3D Wide-field Multispectral Photoacoustic Imaging of Human Melanomas In Vivo: A Pilot Study. *J. Eur. Acad. Dermatol. Venereol.* **2020**. [CrossRef] [PubMed]
32. Rao, B.; Li, L.; Maslov, K.; Wang, L. Hybrid-scanning optical-resolution photoacoustic microscopy for in vivo vasculature imaging. *Opt. Lett.* **2010**, *35*, 1521–1523. [CrossRef] [PubMed]
33. Tang, Y.; Liu, W.; Li, Y.; Zhou, Q.; Yao, J. Concurrent photoacoustic and ultrasound microscopy with a coaxial dual-element ultrasonic transducer. *Vis. Comput. Ind. Biomed. Art* **2018**, *1*, 3. [CrossRef] [PubMed]
34. Plumb, A.A.; Huynh, N.T.; Guggenheim, J.; Zhang, E.; Beard, P. Rapid volumetric photoacoustic tomographic imaging with a Fabry-Perot ultrasound sensor depicts peripheral arteries and microvascular vasomotor responses to thermal stimuli. *Eur. Radiol.* **2018**, *28*, 1037–1045. [CrossRef]
35. Liu, Y.; Yang, X.; Gong, H.; Jiang, B.; Wang, H.; Xu, G.; Deng, Y. Assessing the effects of norepinephrine on single cerebral microvessels using optical-resolution photoacoustic microscope. *J. Biomed. Opt.* **2013**, *18*, 076007. [CrossRef]

36. Nie, L.; Huang, P.; Li, W.; Yan, X.; Jin, A.; Wang, Z.; Tang, Y.; Wang, S.; Zhang, X.; Niu, G. Early-stage imaging of nanocarrier-enhanced chemotherapy response in living subjects by scalable photoacoustic microscopy. *ACS Nano* **2014**, *8*, 12141–12150. [CrossRef]

37. Norouzpour, A.; Hooshyar, Z.; Mehdizadeh, A. Autoregulation of blood flow: Vessel diameter changes in response to different temperatures. *J. Biomed. Phys. Eng.* **2013**, *3*, 63.

38. Postnov, D.D.; Tuchin, V.V.; Sosnovtseva, O. Estimation of vessel diameter and blood flow dynamics from laser speckle images. *Biomed. Opt. Express* **2016**, *7*, 2759–2768. [CrossRef]

39. Ikram, M.; De Jong, F.; Bos, M.; Vingerling, J.; Hofman, A.; Koudstaal, P.; De Jong, P.; Breteler, M. Retinal vessel diameters and risk of stroke: The Rotterdam Study. *Neurology* **2006**, *66*, 1339–1343. [CrossRef]

40. Owolabi, M.O.; Agunloye, A.M.; Ogunniyi, A. The relationship of flow velocities to vessel diameters differs between extracranial carotid and vertebral arteries of stroke patients. *J. Clin. Ultrasound* **2014**, *42*, 16–23. [CrossRef]

41. Gutierrez, J.; Cheung, K.; Bagci, A.; Rundek, T.; Alperin, N.; Sacco, R.L.; Wright, C.B.; Elkind, M.S. Brain arterial diameters as a risk factor for vascular events. *J. Am. Heart Assoc.* **2015**, *4*, e002289. [CrossRef] [PubMed]

42. Pechauer, A.D.; Jia, Y.; Liu, L.; Gao, S.S.; Jiang, C.; Huang, D. Optical Coherence Tomography Angiography ofPeripapillary Retinal Blood Flow Response to Hyperoxia. *Investig. Ophthalmol. Vis. Sci.* **2015**, *56*, 3287–3291. [CrossRef] [PubMed]

43. Lee, K.; Maeng, K.J.; Kim, J.Y.; Yang, H.; Choi, W.; Lee, S.Y.; Seong, G.J.; Kim, C.Y.; Bae, H.W. Diagnostic ability of vessel density measured by spectral-domain optical coherence tomography angiography for glaucoma in patients with high myopia. *Sci. Rep.* **2020**, *10*, 1–10. [CrossRef] [PubMed]

44. Li, Z.; Xu, Z.; Liu, Q.; Chen, X.; Li, L. Comparisons of retinal vessel density and glaucomatous parameters in optical coherence tomography angiography. *PLoS ONE* **2020**, *15*, e0234816. [CrossRef] [PubMed]

45. Seaman, M.E.; Peirce, S.M.; Kelly, K. Rapid analysis of vessel elements (RAVE): A tool for studying physiologic, pathologic and tumor angiogenesis. *PLoS ONE* **2011**, *6*, e20807. [CrossRef]

46. Toader, M.P.; Țăranu, T.; Toader, Ș.; Chirana, A.; Țăranu, T. Correlation between lymphatic vessel density and microvessel density in cutaneous malignant melanoma. *Rom. J. Morphol. Embryol.* **2014**, *55*, 141–145.

47. Rohrbach, D.J.; Salem, H.; Aksahin, M.; Sunar, U. Photodynamic therapy-induced microvascular changes in a nonmelanoma skin cancer model assessed by photoacoustic microscopy and diffuse correlation spectroscopy. In *Photonics*; MDPI: Basel, Switzerland, 2016; p. 48.

48. Vakoc, B.J.; Lanning, R.M.; Tyrrell, J.A.; Padera, T.P.; Bartlett, L.A.; Stylianopoulos, T.; Munn, L.L.; Tearney, G.J.; Fukumura, D.; Jain, R.K. Three-dimensional microscopy of the tumor microenvironment in vivo using optical frequency domain imaging. *Nat. Med.* **2009**, *15*, 1219–1223. [CrossRef]

49. Reif, R.; Qin, J.; An, L.; Zhi, Z.; Dziennis, S.; Wang, R. Quantifying optical microangiography images obtained from a spectral domain optical coherence tomography system. *Int. J. Biomed. Imaging* **2012**, *2012*, 509783. [CrossRef]

50. Zamir, M. Fractal dimensions and multifractility in vascular branching. *J. Theor. Biol.* **2001**, *212*, 183–190. [CrossRef]

51. Schmoll, T.; Singh, A.S.; Blatter, C.; Schriefl, S.; Ahlers, C.; Schmidt-Erfurth, U.; Leitgeb, R.A. Imaging of the parafoveal capillary network and its integrity analysis using fractal dimension. *Biomed. Opt. Express* **2011**, *2*, 1159–1168. [CrossRef]

52. Chauhan, D.; Singh, A.; Brahmandam, M.; Podar, K.; Hideshima, T.; Richardson, P.; Munshi, N.; Palladino, M.A.; Anderson, K.C. Combination of proteasome inhibitors bortezomib and NPI-0052 trigger in vivo synergistic cytotoxicity in multiple myeloma. *Blood* **2008**, *111*, 1654–1664. [CrossRef]

53. Waxman, A.J.; Clasen, S.; Hwang, W.-T.; Garfall, A.; Vogl, D.T.; Carver, J.; O'Quinn, R.; Cohen, A.D.; Stadtmauer, E.A.; Ky, B. Carfilzomib-associated cardiovascular adverse events: A systematic review and meta-analysis. *JAMA Oncol.* **2018**, *4*, e174519. [CrossRef] [PubMed]

54. Frangi, A.F.; Niessen, W.J.; Vincken, K.L.; Viergever, M.A. Multiscale vessel enhancement filtering. In Proceedings of the International Conference on Medical Image Computing and Computer-Assisted Intervention, Cambridge, MA, USA, 11–13 October 2020; pp. 130–137.

55. Singh, T.R.; Roy, S.; Singh, O.I.; Sinam, T.; Singh, K. A new local adaptive thresholding technique in binarization. *arXiv Prepr.* **2012**, arXiv:1201.5227.

56. Li, J.; Du, Q.; Sun, C. An improved box-counting method for image fractal dimension estimation. *Pattern Recognit.* **2009**, *42*, 2460–2469. [CrossRef]

57. Liu, T.; Sun, M.; Feng, N.; Wu, Z.; Shen, Y. Multiscale Hessian filter-based segmentation and quantification method for photoacoustic microangiography. *Chin. Opt. Lett.* **2015**, *13*, 091701.

58. Adams, J. The proteasome: A suitable antineoplastic target. *Nat. Rev. Cancer* **2004**, *4*, 349–360. [CrossRef] [PubMed]

59. Gavazzoni, M.; Vizzardi, E.; Gorga, E.; Bonadei, I.; Rossi, L.; Belotti, A.; Rossi, G.; Ribolla, R.; Metra, M.; Raddino, R. Mechanism of cardiovascular toxicity by proteasome inhibitors: New paradigm derived from clinical and pre-clinical evidence. *Eur. J. Pharmacol.* **2018**, *828*, 80–88. [CrossRef] [PubMed]

60. Efentakis, P.; Kremastiotis, G.; Varela, A.; Nikolaou, P.-E.; Papanagnou, E.-D.; Davos, C.H.; Tsoumani, M.; Agrogiannis, G.; Konstantinidou, A.; Kastritis, E. Molecular mechanisms of carfilzomib-induced cardiotoxicity in mice and the emerging cardioprotective role of metformin. *Blood* **2019**, *133*, 710–723. [CrossRef]

61. Hunt, B.J.; Parmar, K.; Horspool, K.; Shephard, N.; Nelson-Piercy, C.; Goodacre, S.; DiPEP Research Group. The Di PEP (Diagnosis of PE in Pregnancy) biomarker study: An observational cohort study augmented with additional cases to determine the diagnostic utility of biomarkers for suspected venous thromboembolism during pregnancy and puerperium. *Br. J. Haematol.* **2018**, *180*, 694–704. [CrossRef]

62. Li, W.; Garcia, D.; Cornell, R.F.; Gailani, D.; Laubach, J.; Maglio, M.E.; Richardson, P.G.; Moslehi, J. Cardiovascular and thrombotic complications of novel multiple myeloma therapies: A review. *JAMA Oncol.* **2017**, *3*, 980–988. [CrossRef]

63. Park, K.; Kim, J.Y.; Lee, C.; Jeon, S.; Lim, G.; Kim, C. Handheld photoacoustic microscopy probe. *Sci. Rep.* **2017**, *7*, 1–15. [CrossRef] [PubMed]

64. Jung, D.; Park, S.; Lee, C.; Kim, H. Recent progress on near-infrared photoacoustic imaging: Imaging modality and organic semiconducting agents. *Polymers* **2019**, *11*, 1693. [CrossRef] [PubMed]

65. Upputuri, P.K.; Pramanik, M. Photoacoustic imaging in the second near-infrared window: A review. *J. Biomed. Opt.* **2019**, *24*, 040901. [CrossRef] [PubMed]

66. Lee, C.; Jeon, M.; Jeon, M.Y.; Kim, J.; Kim, C. In vitro photoacoustic measurement of hemoglobin oxygen saturation using a single pulsed broadband supercontinuum laser source. *Appl. Opt.* **2014**, *53*, 3884–3889. [CrossRef] [PubMed]

Article

Mapping of Back Muscle Stiffness along Spine during Standing and Lying in Young Adults: A Pilot Study on Spinal Stiffness Quantification with Ultrasound Imaging

Christina Zong-Hao Ma †, Long-Jun Ren †, Connie Lok-Kan Cheng and Yong-Ping Zheng *

Department of Biomedical Engineering, The Hong Kong Polytechnic University, Hong Kong, China;
czh.ma@polyu.edu.hk (C.Z.-H.M.); longjun.ren@polyu.edu.hk (L.-J.R.);
lk-connie.cheng@polyu.edu.hk (C.L.-K.C.)
* Correspondence: yongping.zheng@polyu.edu.hk; Tel.: +852-2766-7664
† These authors contributed equally to this manuscript.

Received: 12 November 2020; Accepted: 16 December 2020; Published: 19 December 2020

Abstract: Muscle stiffness in the spinal region is essential for maintaining spinal function, and might be related to multiple spinal musculoskeletal disorders. However, information on the distribution of muscle stiffness along the spine in different postures in large subject samples has been lacking, which merits further investigation. This study introduced a new protocol of measuring bilateral back muscle stiffness along the thoracic and lumbar spine (at T3, T7, T11, L1 & L4 levels) with both ultrasound shear-wave elastography (SWE) and tissue ultrasound palpation system (TUPS) in the lying and standing postures of 64 healthy adults. Good inter-/intra-reliability existed in the SWE and TUPS back muscle stiffness measurements (ICC ≥ 0.731, $p < 0.05$). Back muscle stiffness at the L4 level was found to be the largest in the thoracic and lumbar regions ($p < 0.05$). The back muscle stiffness of males was significantly larger than that of females in both lying and standing postures ($p < 0.03$). SWE stiffness was found to be significantly larger in standing posture than lying among subjects ($p < 0.001$). It is reliable to apply SWE and TUPS to measure back muscle stiffness. The reported data on healthy young adults in this study may also serve as normative reference data for future studies on patients with scoliosis, low back pain, etc.

Keywords: back muscle stiffness; spine; elasticity; shear-wave elastography (SWE); tissue ultrasound palpation system (TUPS); reliability; Young's modulus

1. Introduction

Multiple musculoskeletal disorders can take place in the spine of human beings, including the spinal curvature deformity of adolescent idiopathic scoliosis [1] and chronic low back pain [2]. Both scoliosis [3,4] and low back pain [5] could lead to significant socioeconomic burdens and reduced quality of life in patients. Previous studies have reported that the imbalance of spinal muscles existed in scoliosis patients [6,7] and patients with low back pain [8,9].

The mechanical property of muscle stiffness is an essential factor for maintaining muscle function. It is related to muscle performance during exercise [10] and the joint constraint [11]. Abnormally increased muscle stiffness can be found in a number of musculoskeletal disorders, including spasticity [12], aging [13,14], spinal muscle atrophy [15], low back pain [16], adolescent idiopathic scoliosis [17,18], reduced joint range of motion [19], and reduced lumbar flexion during prolonged sitting [20]. Meanwhile, decreased muscle stiffness may lead to a higher risk of joint dislocation [21] and reduced muscle power [22]. Quantifying muscle stiffness can be helpful for understanding the mechanisms of these musculoskeletal symptoms and pathologies.

Muscle stiffness can be quantified by tissue elasticity (E) and measured by various technologies, including mechanical measurement, ultrasound indentation, and elastography. More recently, the non-invasive and real-time ultrasound shear-wave elastography (SWE) has become a popular and useful tool for assessing muscle stiffness [23,24]. For this technology, the ultrasound probe can induce a focused acoustic force and create a shear wave within the target tissue [25,26]. By capturing the propagation of the shear wave, the speed of the shear wave propagation (c) can be calculated, which can then be squared and multiped by three and the muscle mass density (ρ = 1000 kg/m^3) to calculate the muscle elasticity (assumed as E = $3\rho c^2$) [25]. With this rather new technology, a few previous studies have managed to measure the spinal muscle stiffness using the SWE; however, most of them were limited to the low back region, such as longissimus [27], multifidus [28], and erector spinae [28] of healthy subjects with the sample size being less than 24 participants.

The tissue ultrasound palpation system (TUPS) is another ultrasound-based instrument that can assess muscle stiffness. The probe of TUPS contained both force sensor and ultrasound transducer, which can record the real-time tissue deformation with the conventional B-mode ultrasound image to calculate Young's modulus or elasticity and acquire the tissue thickness [29]. The TUPS has been applied to evaluate the tissue stiffness of the foot [30] and scar thickness [31]. Two pilot studies have also used TUPS to measure the stiffness of back muscle at L4 level in 12 healthy subjects and 12 patients with low back pain [16], and at L1 and L4 levels in 10 patients with low back pain [29]. The latest updated TUPS system is a handheld and wireless version, which could eliminate the afference of the wires during measurement and make it more feasible to do the measurement in a clinical setting in the future [29].

To date and to the best of the authors' knowledge, none of the previous studies have used either the SWE or the TUPS to evaluate the distribution of back muscle stiffness, along the thoracic and lumbar spinal regions, in a large number of adults. While both SWE and TUPS have been used to evaluate back muscle stiffness, the sample size has been very small ($\leq$24 participants) and the evaluated region has been limited to the lower back of the spine. The generalization of these findings has been rather limited, and the normative data of muscle stiffness along the spine in different postures have been lacking. While both SWE and TUPS have been used to evaluate back muscle stiffness, the comparison of these two instruments on measuring back muscle stiffness was still scarce.

To address the above-mentioned issues, the current study aimed to (1) measure the back muscle stiffness along the spine at the levels of T3, T7, T11, L1, and L4 with both SWE and TUPS in both lying and standing postures in 64 healthy adults (*mapping the back muscle stiffness* with *large sample* size, the effect of *levels*); (2) compare the measured results of muscle stiffness between SWE and TUPS in terms of *reliability* and *relationship*; (3) identify the difference of distribution in muscle stiffness along the spine between the standing and lying postures (effect of *posture*), and (4) determine if the gender factor influenced the distribution of muscle stiffness along the spine in both standing and lying postures (effect of *gender*). The levels of T3, T7, T11, L1, and L4 were selected to reflect the muscle stiffness along the upper, middle, and lower thoracic and the upper and lower lumbar spinal regions [32].

2. Materials and Methods

2.1. Subjects

In total, sixty-four healthy young adults (32 males and 32 females) aged between 18 and 30 years were recruited. Subjects were excluded if they had low back pain within the last three months before the study; scoliosis; muscular disease of limbs or spine; and/or history of bone disease, fracture, surgery, or malformation at the spinal region. A Registered Physiotherapist verified the inclusion and exclusion criteria via subject interview and physical examination before the data collection.

Subjects were instructed not to have any vigorous exercise two days before the experiment, to avoid possible muscle fatigue and altered muscle stiffness that may affect the experimental results [33]. They should also avoid muscle relaxants and alcohol before the experiment, along with some other drugs. During the experiment, if subjects experienced any discomfort, the experiment would be stopped immediately with the condition been recorded. Ethical approval was granted by the authority of the local university (HSEARS20180122004). Written informed consent was signed and obtained from all subjects before the experiment.

2.2. Instruments for Measuring Back Muscle Stiffness

2.2.1. Wireless Hand-Held Tissue Ultrasound Palpation System (TUPS)

A newly-updated hand-held tissue ultrasound palpation system (TUPS), with a probe (7.5 MHz 128-elements ultrasound transducer with a 20 N in-series load cell) wirelessly connected to a laptop via Wi-Fi, was used to evaluate the back muscle stiffness (referred to as "TUPS stiffness" in this paper) and the thickness of soft tissues [29]. The probe sampled the ultrasound image and force data simultaneously and transmitted them to a laptop in real-time [29,34–36]. The frequency of the system was 12 Hz. For each measurement, five compression-release cycles within a duration of 10 s were performed to collect data. The thickness of soft tissue along the spine was also measured by the TUPS system.

2.2.2. Ultrasonic Scanner with Shear-Wave Elastography (SWE)

A commercially available multi-wave ultrasonic scanner (version 10.0; Super-Sonic Imagine, Aix-en-Provence, France) coupled with a convex probe (SuperCurved 6-1, Super-Sonic Imagine, Aix-en-Provence, France) in shear-wave elastography (SWE) and musculoskeletal (MSK) mode was used to evaluate the back muscle stiffness. For each measurement, a 10-s video with approximately 10 frames of ultrasound images was recorded and exported in "MP4" format, after the color map of stiffness was maintained as homogeneously as possible.

A developed Matlab script (Version 2016b, MathWorks, MA, USA) was used to process the stiffness data of the exported video. Firstly, the region of interest (ROI) was selected as the largest muscle area that avoided bone, fascia, or subcutaneous tissue. Secondly, the artifact pixel (showing no color in the color map or saturate at 300 kPa) within the ROI was excluded for data analysis. Thirdly, the Matlab image processing script converted each available pixel of the color map into a value of stiffness, based on the color scale. Finally, the mean value of the stiffness from the captured 10 frames was calculated to obtain the stiffness of each measurement (referred to as "SWE stiffness" in this paper) for further statistical analysis.

Pilot studies were conducted to evaluate the validity of the introduced data analyzing method with the developed Matlab image processing script. It revealed that the results generated by this Matlab script were similar to those from the Aixplorer scanner software (Q-BoxTM) as used in [27,28,37].

2.3. Experimental Procedure

Before the measurement, the spinal processes at the T3, T7, T11, L1, and L4 levels were located by palpation and ultrasound B-mode image and then marked with a water-insoluble eyeliner. The bilateral muscle belly that parallels with the spinal process was located for measuring the back muscle stiffness as suggested in [38].

During the experiment, all measurements were conducted in lying posture first, followed by the standing posture for both the instruments of TUPS and SWE. For the *lying* posture, subjects were prone with their faces in the hole of a massage bed and their upper limbs along the trunk. For the *standing* posture, subjects stood in front of a supporting frame to eliminate the possible influence generated by the compression-release of the TUPS' probe during the measurement on posture. Subjects were instructed not to resist the compression force generated by the TUPS voluntarily, but to reply on the supporting frame instead (Figure 1). Subjects were instructed to breathe naturally, put their weight equally on both feet, and maintain their heads in a neutral position during the measurement. A male assessor and two female assessors conducted the measurements for all male and female subjects, respectively. Assessors would instruct the subjects to adjust posture if subjects stood asymmetrically.

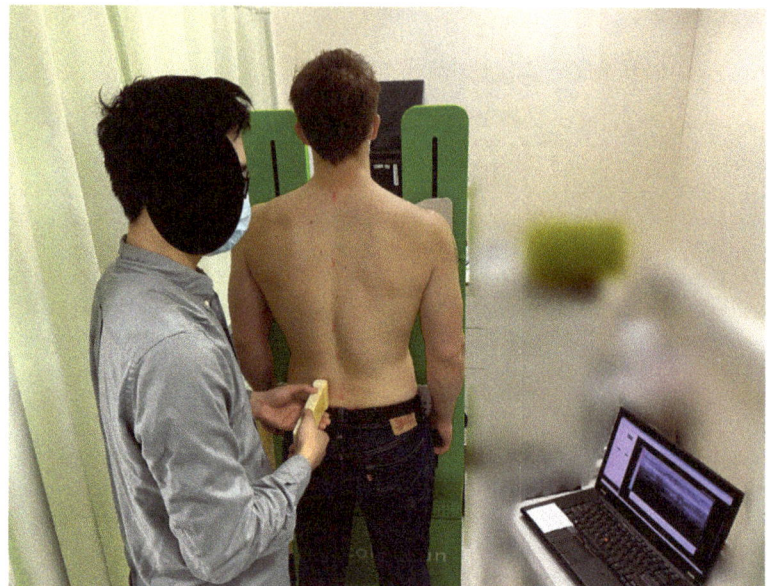

Figure 1. Illustration of the assessment in standing position with a supporting frame.

2.4. Reliability Test

Since three assessors (1 male and 2 females) conducted the measurements in this study, a reliability test was conducted to determine the reliability of the measurements prior to the start of the main experiment. The measurement and reliability tests were conducted on fourteen male subjects. During the test, the muscle stiffness of the left and right sides at the T7 and L1 levels were measured three times by each assessor in the lying position.

2.5. Data and Statistical Analysis

Statistical analysis was conducted using the SPSS (Version 24, SPSS Inc, Chicago, IL, USA). The muscle stiffness value of the left and right sides was averaged for statistical analysis. The percentage change of muscle stiffness from lying to standing posture was also calculated. Intra-rater and inter-rater reliability of the muscle stiffness measurements were examined with the Intraclass Correlation Coefficient (ICC (3,1)) with a 95% confidence interval (95% CI). Three-way mixed ANOVA with post-hoc pairwise comparison was conducted to determine the main effects of (1) *posture* (lying vs. standing), (2) *level* (T3, T7, T11, L1 vs. L4), and (3) *gender* (male vs. female), as well as the interaction effect on muscle stiffness. The Pearson correlation test was performed to examine the relationship between the two measurement techniques of SWE and TUPS. The significance level was set at 0.05. The effect size (η_p^2) for each parameter was also presented.

3. Results

A total of 64 subjects (32 males and 32 females, aged 23.5 ± 2.9 years, height 166.4 ± 8.5 cm, weight 60.1 ± 10.3 kg, and BMI 21.6 ± 2.5) participated in this study. Initially, seventy-five subjects were screened for this study. Among them, two subjects were excluded due to scoliosis, and four subjects were excluded due to uncomfortableness during the experiment. The data of five subjects were discarded due to the technological issue of hard disk failure where the data cannot be retrieved.

3.1. Intra-/Inter-Rater Reliability of Measurements

Good intra-rater reliability [*TUPS*: ICC = 0.822 (95% CI 0.632 to 0.962) at T7, and ICC = 0.905 (95% CI 0.704 to 0.969) at L1; *SWE*: ICC = 0.881 (95% CI 0.631 to 0.962) at T7, and ICC = 0.879 (95% CI 0.625 to 0.961) at L1] and good inter-rater reliability [*TUPS*: ICC = 0.742 (95% CI 0.368 to 0.910) at T7, and ICC = 0.836 (95% CI 0.602 to 0.943) at L1; *SWE*: ICC = 0.731 (95% CI 0.340 to 0.906) at T7, and ICC = 0.781 (95% CI 0.462 to 0.924) at L1] for both measurement instruments were identified in this study ($p < 0.05$). Higher intra-rater reliability of TUPS than SWE except at T7 level, and higher inter-rater reliability of TUPS than SWE were also found.

3.2. Muscle Stiffness Measured by TUPS

The measured muscle stiffness by TUPS is summarized in Table 1 and Figure 2. Significant main effects of gender ($p < 0.001$, $\eta_p^2 = 0.292$) and level ($p < 0.001$, $\eta_p^2 = 0.601$), and significant interaction effect among three factors ($p < 0.001$, $\eta_p^2 = 0.069$) were found.

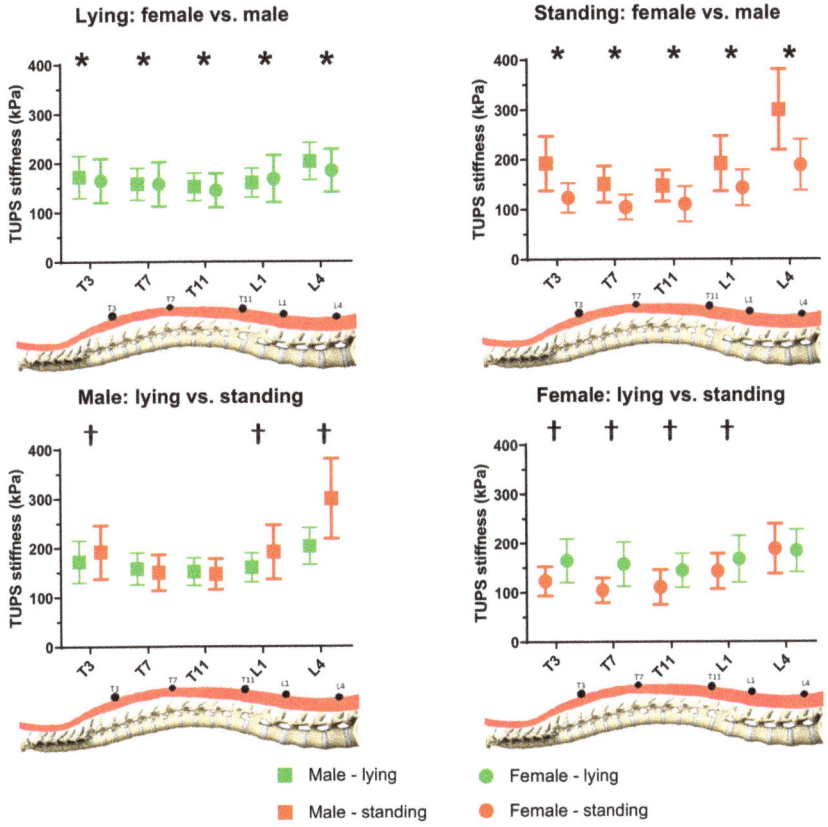

Figure 2. TUPS stiffness at different levels of two genders in lying and standing postures ($n = 64$).
* Significantly difference in gender; † Significant difference in posture.

Table 1. The tissue ultrasound palpation system (TUPS) stiffness in kPa (mean ± SD) (*n* = 64).

	Lying		Standing			
Level	Male (*n* = 32)	Female (*n* = 32)	Male (*n* = 32)	Female (*n* = 32)	*p*-Value	Partial Eta Squared, η_p^2
T3	172.0 ± 42.8 †	166.5 ± 46.4 †	192.3 ± 55.7 *†	122.7 ± 27.8 *†	Main effect:	
					Gender: <0.001	0.292
T7	157.8 ± 32.5	155.8 ± 45.2 †	150.8 ± 36.9 *	104.9 ± 26.7 *†	Posture: 0.707	0.002
					Level: <0.001	0.601
T11	153.3 ± 27.0	144.4 ± 36.1	148.3 ± 32.0 *	111.3 ± 37.6 *	Interaction effect:	
					Gender * level:<0.001	0.111
L1	162.1 ± 29.0 †	168.3 ± 48.4 †	194.7 ± 55.7 *†	137.5 ± 36.0 *†	Gender * posture: <0.001	0.481
					Posture * level: <0.001	0.355
L4	205.5 ± 36.2 †	182.7 ± 45.2 †	303.3 ± 80.6 *†	180.5 ± 46.7 *†	Gender * level * posture: <0.001	0.069

* Indicates significant gender difference at a certain level and posture. † Indicates significant posture difference at a certain level for each gender.

3.2.1. Effect of Gender

The results of post-hoc comparison revealed that the TUPS stiffness was significantly larger in male subjects than that of female subjects for all five levels ($p < 0.001$, $\eta_p^2 = 0.292$).

3.2.2. Effect of Level

The TUPS stiffness was found to be significantly different among the five different levels (T3: 163.4 ± 25.8 kPa, T7: 142.5 ± 20.9 kPa, T11: 138.1 ± 18.4 kPa, L1: 165.4 ± 25.6 kPa, and L4: 217.9 ± 33 kPa), except the two pairwise comparisons of T3 vs. L1 and T7 vs. T11. More specifically, the muscle stiffness significantly decreased from T3 to T7 level ($p < 0.05$), significantly increased from T11 to L1 level ($p < 0.05$), and significantly increased from L1 to L4 level ($p < 0.05$). The muscle stiffness at L4 was found to be significantly largest ($p < 0.05$); meanwhile, the muscle stiffness at T11 tended to be the smallest, but did not reach a significant level.

3.2.3. Effect of Posture

No significant difference in TUPS stiffness between the two postures was found (lying: 163.4 ± 25.8 kPa and standing: 163.4 ± 25.8 kPa, $p = 0.707$, $\eta_p^2 = 0.002$). Meanwhile, upon looking into the TUPS stiffness at each level, significantly larger TUPS stiffness at T3, L1, and L4 during standing posture than lying in male subjects existed, and a reversed trend of significantly larger TUPS stiffness at T3, T7, T11, and L1 during lying posture than standing in female subjects existed.

As shown in Figure 3, the percentage difference of changes in TUPS stiffness was significantly larger in male subjects than female subjects at all four levels ($p \leq 0.005$). More specifically, the percentage change at the L4 level appeared to be the largest (48.5% for males and 4.9% for females), and that of T7 was found to be the smallest (−2.9% for males and −30.4% for females).

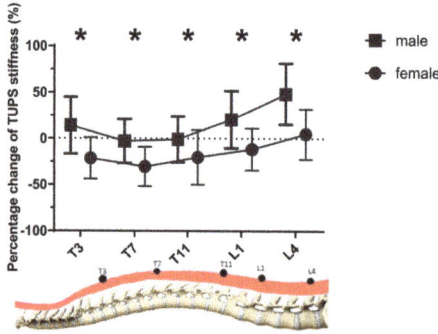

Figure 3. The percentage change of TUPS stiffness from lying to standing posture at different levels of two genders (*n* = 64). * Significant differences existed in gender.

3.3. Muscle Stiffness Measured by SWE

The measured muscle stiffness by SWE is summarized in Table 2 and Figure 4. Significant main effects of gender ($p = 0.030$, $\eta_p^2 = 0.074$) and posture ($p < 0.001$, $\eta_p^2 = 0.772$), and significant interaction effect only between gender and posture (gender*posture: $p = 0.016$, $\eta_p^2 = 0.090$) were found.

Table 2. The shear-wave elastography (SWE) stiffness in kPa (mean ± SD) ($n = 64$).

| Level | Lying | | Standing | | *p*-Value | Partial Eta Squared, η_p^2 |
	Male ($n = 32$)	Female ($n = 32$)	Male ($n = 32$)	Female ($n = 32$)		
T3	21.8 ± 8.3 *†	17.8 ± 6.8 *†	45.5 ± 20.0 *†	33.7 ± 17.4 *†	Main effect:	
					Gender: 0.030	0.074
T7	23.8 ± 5.3 †	23.1 ± 5.2 †	47.0 ± 16.2 *†	37.9 ± 13.6 *†	Posture: <0.001	0.772
					Level: 0.120	0.029
T11	22.5 ± 8.4 †	23.3 ± 7.1 †	43.7 ± 14.7 †	44.5 ± 15.1 †	Interaction effect:	
					Gender * level: 0.080	0.033
L1	21.1 ± 6.0 †	24.3 ± 7.9 †	45.2 ± 21.6 †	38.3 ± 18.8 †	Gender * posture: 0.016	0.090
					Posture * level: 0.171	0.025
L4	24.1 ± 9.3 †	23.3 ± 7.5 †	42.0 ± 18.9 †	35.2 ± 13.7 †	Gender * level * posture: 0.360	0.017

* Indicates a significant gender difference at certain levels and posture. † Indicates a significant posture difference at certain levels of each gender.

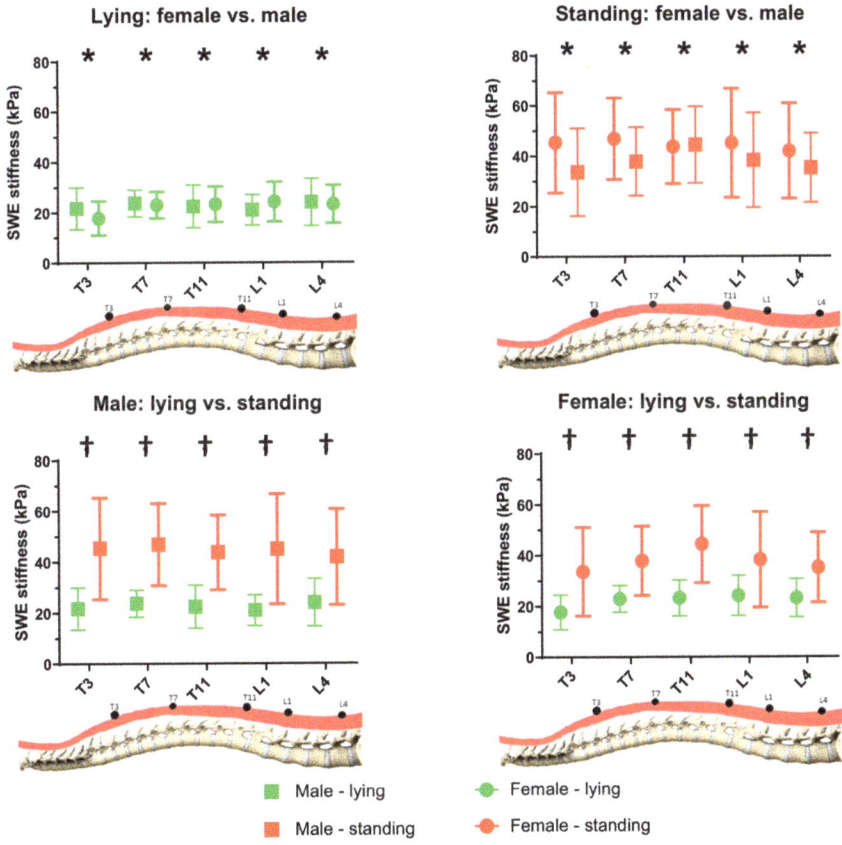

Figure 4. SWE stiffness at different levels of two genders in lying and standing postures ($n = 64$). * Significant differences existed in gender; † Significant differences existed in posture.

3.3.1. Effect of Gender

The results of post-hoc comparison revealed that the SWE stiffness was also significantly larger in male subjects than that of female subjects for all five levels ($p = 0.003$, $\eta_p^2 = 0.074$).

3.3.2. Effect of Posture

The SWE stiffness in lying posture was found to be significantly smaller than that of standing posture for both genders and for all five different levels (lying: 22.5 ± 1.6 kPa, and standing: 41.3 ± 2.2 kPa, $p < 0.001$, $\eta_p^2 = 0.772$).

As shown in Figure 5, the percentage difference of changes in SWE stiffness fluctuated, and no significant difference among different levels was found. While a significantly larger change in SWE stiffness in males than females at the L1 level was found ($p = 0.013$), no other significant difference between the two genders was found.

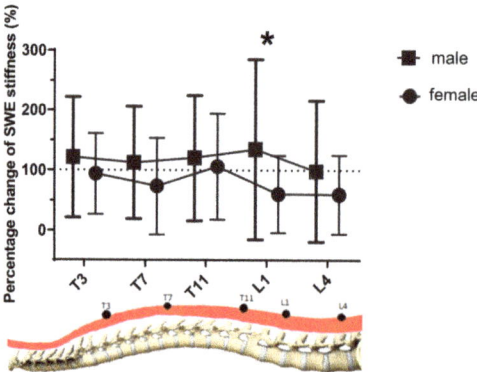

Figure 5. The percentage change of SWE stiffness from lying to standing posture at different levels of two genders ($n = 64$). * Significant differences existed in gender.

3.3.3. Effect of Level

No significant difference in SWE stiffness among the five different levels was found (T3: 29.7 ± 12.5 kPa; T7: 32.9 ± 11.6 kPa; T11: 33.5 ± 12.2 kPa; L1: 32.2 ± 11.4 kPa and L4: 31.1 ± 9.0 kPa).

3.4. Change of Soft Tissue Thickness from Lying to Standing Posture

As shown in Figure 6, a significantly moderate correlation between soft tissue thickness and posture was found at all five levels (r = 0.304, $p < 0.001$). Significant main effects of posture ($p = 0.001$), level ($p < 0.001$), and gender ($p < 0.001$), and no significant interaction effect among these factors were found. The soft tissue thickness was also found to be significantly larger in standing posture than lying posture at five levels ($p = 0.001$). Similar to the distribution of muscle stiffness at different levels, the soft tissue thickness was also found significantly decreased from T3 to T7 level, significantly increased from T11 to L1 level, and significantly increased from L1 to L4 level ($p < 0.001$). The soft tissue thickness at the L4 level was found to be significantly largest ($p < 0.001$), and the thickness at the T7 level was found to be the smallest ($p < 0.001$).

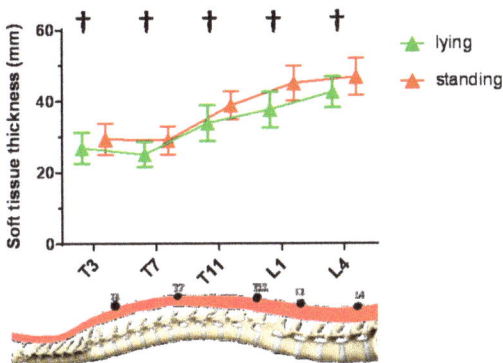

Figure 6. The soft tissue thickness at different levels in lying and standing postures among subjects (*n* = 64). † Significant differences existed in posture.

3.5. Relationship between the SWE and TUPS Measurement Techniques

Table 3 summarizes the results of Pearson's correlation coefficient (r) between the SWE and TUPS measurement techniques in lying and standing positions. Significantly moderate correlations between the SWE and TUPS measurement techniques were observed at T3 in the lying posture (r = −0.294, *p* = 0.018), L1 in the standing posture (r = 0.390, *p* = 0.001), and L4 in the standing posture (r = 0.358, *p* = 0.004). Low correlations between the two measurement techniques were observed at the remaining levels and postures, but the correlations were not significant.

Table 3. Relationship between the SWE and TUPS measurement techniques (*n* = 64).

	Pearson's Correlation Coefficient (r)	
Level	Lying	Standing
T3	−0.294 *	0.111
T7	0.196	0.232
T11	−0.141	0.166
L1	0.020	0.390 *
L4	−0.233	0.358 *

* Significantly correlation existed.

4. Discussions

To our knowledge, this is the very first study investigating the distribution of back muscle stiffness along the spine, with the stiffness measurements from both the SWE and TUPS, in healthy young adults. Several significant effects of posture, level, and gender on back muscle stiffness were identified in this study. The relationship and reliability of the SWE and TUPS measurement techniques were also investigated and established.

4.1. Intra-/Inter-Rater Reliability of Measurements

Good intra-rater and good inter-rater reliability for both measurement instruments of TUPS and SWE were observed in this study. This is in accordance with previous studies on the reliability of the TUPS [30,31,34] and SWE [39]. Additionally, the handheld and wireless setting of the TUPS also makes it more convenient to be applied in clinical settings [29] and not be restrained by the experiment [40] in the future.

4.2. Effect of Level (T3, T7, T11, L1 vs. L4)

This study has uncovered that regardless of the lying or standing posture, the back muscle at the L4 level has the largest TUPS stiffness value among the thoracic and lumbar regions that covered the commonly affected sites of scoliosis and low back pain. This helps explain why chronic low back pain has been commonly found in the L3-L4 region in previous studies [2,29]. The observed large stiffness values at L4 level in healthy adults in this study might help explain the cause/development of the low back pain pathology later on in middle-aged or older adults. It is likely that the greater muscle stiffness at the L4 level might be one potential cause of scoliosis or low back pain, however, future studies shall be conducted to explore and clarify this. Upon measuring the back muscle stiffness of both thoracic and lumbar regions in large samples (n = 64), the results of this study could also act as a normative database for future studies on scoliosis, low back pain, and other spinal musculoskeletal disorders.

The TUPS stiffness was found to decrease from T3 to T7, and then increase from T11 to L4 level. This is in line with previous studies suggesting that the muscle stiffness at L4 was larger than L1 in patients with low back pain [29], patients undergoing spinal surgery [41], and healthy young adults [27]. This could be explained by the curvature of the spine since the thoracic spine is in kyphosis position, which might stretch the back muscles; and the lumbar spine is in the lordosis region, which might compress the back muscles. Future studies, preferably using the real-time ultrasound images with large resolutions that can observe muscle fibers, are needed to further explore and look into the underlying mechanism of the observed different changes of back muscle stiffness at the thoracic and lumbar regions. It is also worthwhile to investigate if the observed change in muscle stiffness is a physiological phenomenon or the cause of pathology in future studies.

Unlike TUPS, no statistically significant difference in levels in SWE was observed in this study. This might be explained by the different underlying mechanisms of evaluating muscle stiffness by TUPS and SWE. While TUPS could not differentiate the multiple layers of soft tissues, SWE could specifically locate a certain muscle area by choosing the region of interest (ROI) on an ultrasound image. Most previous studies have compared the difference between lying and upright positions and/or between resting and contraction conditions of a single muscle group [42,43]. Unfortunately, few previous studies have used SWE to map the muscle stiffness along the spine, which makes it rather difficult to compare the current findings in this study with previous studies. The lack of previous findings might be because SWE is a rather new technology, and more studies shall be conducted in the future to enable the synthesis of information on spinal muscle stiffness as evaluated by SWE.

4.3. Effect of Gender (Female vs. Male)

Back muscle stiffness of male subjects was found to be larger than that of female subjects in both lying and standing postures as evaluated by TUPS and SWE. This accords with previous studies on muscle stiffness of knee extensor in healthy young athletes [44] and knee flexors in healthy young adults [45]. Male and female adults have different anatomical structures generally, such as the distribution and mass of body fat and muscles at the back of the trunk [46]. This might help explain the observed difference in back muscle stiffness between the two genders in this study. While most of the previous studies focused on limb muscle stiffness [44,45], this study provides more information and evidence about the effect of gender on back muscle stiffness along the thoracic and lumbar spine.

4.4. Effect of Posture (Lying vs. Standing)

SWE stiffness was found to be larger in standing posture than lying posture, while significantly larger TUPS stiffness at T3, L1, and L4 during standing in male subjects and that at T3, T7, T11, and L1 during lying in female subjects were found. Meanwhile, significant percentage changes in muscle stiffness from the lying to the standing posture was found for TUPS measurement, but not for the SWE measurement in this study, even the pattern/trend of changes appeared to be the same. The finding of significantly larger TUPS stiffness at the L4 level during standing in male subjects was in line with a previous study on male patients with low back pain [16]. The certain difference regarding the effect of posture between the SWE and TUPS measurements could be due to the different measuring mechanisms of these two instruments. The SWE assumed the muscle mass density ($\rho = 1000$ kg/m^3) to be a constant value when calculating the stiffness [25], and TUPS used the real-time tissue deformation with a conventional B-mode image to calculate the stiffness [29]. The soft tissue thickness has changed from lying to standing position, which might affect the muscle mass density and thus affect the SWE stiffness measurement. The SWE stiffness has also been suggested to be minimally associated with TUPS stiffness [47]. Additionally, the soft tissue thickness was found to be correlated with the different postures of lying and standing, and the changing pattern of soft tissue thickness also appeared to be similar to that of muscle stiffness at different levels as measured by TUPS in this study.

The different changes in TUPS stiffness from lying to standing postures between male and female subjects might be due to the different anatomical structures between the two genders. The TUPS used the tissue deformation during the compression-release cycles to acquire the stiffness [29]. It is reasonable to expect that the mass of body fat and breast may affect more of the back muscle elongation and contraction, especially in the upper trunk in female subjects than that of male subjects. Previous studies have also reported the increased anterior-posterior shear forces in females and decreased forces in males in response to stress [48], and the different trunk muscle geometry between the two genders [49]. All these findings may help explain the measured different changes from lying to standing posture between the SWE and TUPS measurements and between the two genders. Further studies, preferably in vitro studies with better control of the experimental setting for more robust results, are still needed to look into this issue and identify the exact cause.

4.5. Relationship between the SWE and TUPS Measurement Techniques

This study observed significantly moderate correlations between the SWE and TUPS measurement techniques at the T3 level in the lying position, L1 level in the standing position, and L4 level in the standing position only. This might be caused by the different underlying mechanisms to measure muscle stiffness by the two techniques. Additionally, while TUPS could not differentiate the multiple layers of soft tissues, SWE could specifically locate a certain muscle area by choosing the region of interest (ROI) on an ultrasound image. The muscle contraction and the overlying tissue might influence the SWE and TUPS stiffness measurements. The observed correlation may also contribute to the presented results and explain the difference in muscle stiffness as measured by the two techniques at the same location. For example, the TUPS stiffness values ranged between 104 kPa and 303 kPa, while SWE stiffness values ranged between 17 kPa and 37 kPa. This implies that when applying these two techniques to evaluate the muscle stiffness, the relative changes in measured values might be more meaningful for muscle assessment than that of absolute values.

4.6. Implications and Outlook

The results of this study will not only enlarge our understanding of back-muscle stiffness, but also provide us with insights about the possible cause and mechanism of scoliosis and low back pain. The findings of this study inspire future efforts to investigate if the observed change in muscle stiffness is a physiological phenomenon or the cause of pathologies of scoliosis and low back pain. The results may also serve as normative data about back muscle stiffness in healthy young adults, which could be useful in lots of healthcare areas in the future. Attention should be paid to the fact that the normative values might be different for different age groups, and only healthy young people aged between 18–30 years were evaluated in this study. The obtained values on muscle stiffness could be treated as a reference, but it might not be appropriate to directly generalize such values to patients with scoliosis or lower back pain. This study can be the first step for further studies investigating the distribution of back muscle stiffness along the spine in different populations, age groups, and patient groups. As the introduced experimental protocol has been evaluated to be reliable, this study also enables future studies to apply similar protocols on patients with scoliosis, low back pain, and other spinal musculoskeletal disorders.

4.7. Best Practice Recommendations for the SWE and TUPS Measurement Techniques

Good knowledge of the mechanism of the SWE and TUPS measurement techniques and careful consideration of the available experimental/clinical settings are needed to enable the best practice for the two techniques. It is recommended to apply the SWE to measure the muscle stiffness when differentiating and evaluating different layers of muscles and different regions of the same muscle, since the region of interest (ROI) of SWE can be easily customized to various shapes and sizes at various locations during the measurement. With the advantages of being portable, with a wireless connection, and occupying a small space, the TUPS would be more helpful to generally evaluate the muscle stiffness at certain anatomical locations, especially for screening purposes in labs/clinics with limited space or even in an outdoor environment [29,40]. The TUPS can be used to do the screenings first, and when abnormal muscle stiffness is identified, the SWE can then be applied to examine the exact cause by specifically locating various regions.

4.8. Limitations

While mapping the muscle stiffness at the thoracic and lumbar regions could already provide plenty of information, the data on cervical and sacral regions were not involved in this study. Future studies could consider expanding this mapping strategy to provide a more comprehensive picture of spinal muscle stiffness. During the experiment, subjects were instructed to stand symmetrically and put equal weight on both feet, supplemented by the observation of accessors. Future studies could consider putting pressure sensors and/or a pressure mat under the subject's feet to control the symmetric weight distribution more objectively. While subjects were instructed to breathe naturally during the experiment, it should be noted that the influence of breath dynamics on the measured results remained unclear. Future studies are needed to understand this issue. The muscle activity was unfortunately not measured as a covariable to explain muscle stiffness changes, which can also be considered in future investigations.

It shall be noted that the SWE and TUPS measurements might be sensitive to other factors besides the mechanical properties. A recent paper reported that indirect tissue stiffness measurements are often sensitive to the mechanical properties, geometrical dimensions, and tensional state of the tissue [50]. From a mechanical perspective, tissue elasticity is defined as a ratio of strain (strain = F/A) and elongation (dl/L), and can be measured directly using mechanical measurements. Meanwhile, ultrasound and elastography do not measure strain and elongation directly. These indirect measurements are indirect estimations of the stiffness. When standing, the muscles might contract and thereby influence the stiffness measurements. For the research community, it is essential to have a standard definition and understanding of muscle stiffness. Further studies are still needed to investigate, understand, and clarify this issue.

5. Conclusions

This study mapped the back muscle stiffness along the thoracic and lumbar spine, with both SWE and TUPS, in both lying and standing postures in healthy male and female adults. It identified that both SWE and TUPS are reliable in measuring back muscle stiffness. Significant effects of level, gender, and posture on back muscle stiffness were identified. This may provide some insights regarding the underlying mechanism of why the common sites of low back pain take place at a certain spinal region and facilitates future research on other spinal musculoskeletal disorders.

Sensors **2020**, *20*, 7317

Author Contributions: Conceptualization, C.Z.-H.M., L.-J.R., C.L.-K.C. and Y.-P.Z.; methodology, C.Z.-H.M., L.-J.R., C.L.-K.C. and Y.-P.Z.; software, C.Z.-H.M., L.-J.R. and C.L.-K.C.; validation, C.Z.-H.M., L.-J.R. and C.L.-K.C.; formal analysis, C.Z.-H.M., L.-J.R. and C.L.-K.C.; investigation, C.Z.-H.M., L.-J.R. and C.L.-K.C.; resources, Y.-P.Z.; writing—original draft preparation, C.Z.-H.M.; writing—review and editing, L.-J.R., C.L.-K.C. and Y.-P.Z.; visualization, C.Z.-H.M., L.-J.R. and Y.-P.Z.; supervision, Y.-P.Z.; funding acquisition, Y.-P.Z. All authors have read and agreed to the published version of the manuscript.

Funding: This research was funded by The Hong Kong Polytechnic University, grant number G-YBFV, G-YBRN, H-ZG4W.

Acknowledgments: The authors would like to thank Derek De Yang for developing the Matlab programming script for data analysis of this study. The authors would also like to thank all the subjects who volunteered to participate in this study.

Conflicts of Interest: The authors declare no conflict of interest.

References

1. Weinstein, S.L.; Dolan, L.A.; Cheng, J.C.; Danielsson, A.; Morcuende, J.A. Adolescent idiopathic scoliosis. *Lancet* **2008**, *371*, 1527–1537. [CrossRef]
2. Ganesan, S.; Acharya, A.S.; Chauhan, R.; Acharya, S. Prevalence and risk factors for low back pain in 1355 young adults: A cross-sectional study. *Asian Spine J.* **2017**, *11*, 610–617. [CrossRef] [PubMed]
3. Kim, Y.J.; Lenke, L.G.; Bridwell, K.H.; Cheh, G.; Whorton, J.; Sides, B. Prospective pulmonary function comparison following posterior segmental spinal instrumentation and fusion of adolescent idiopathic scoliosis: Is there a relationship between major thoracic curve correction and pulmonary function test improvement? *Spine* **2007**, *32*, 2685–2693. [CrossRef] [PubMed]
4. Martinez-Llorens, J.; Ramirez, M.; Colomina, M.J.; Bago, J.; Molina, A.; Caceres, E.; Gea, J. Muscle dysfunction and exercise limitation in adolescent idiopathic scoliosis. *Eur. Respir. J.* **2010**, *36*, 393–400. [CrossRef] [PubMed]
5. Wong, A.Y.; Parent, E.C.; Funabashi, M.; Kawchuk, G.N. Do changes in transversus abdominis and lumbar multifidus during conservative treatment explain changes in clinical outcomes related to nonspecific low back pain? A systematic review. *J. Pain* **2014**, *15*, 377.e1-35. [CrossRef] [PubMed]
6. Jiang, H.; Meng, Y.; Jin, X.; Zhang, C.; Zhao, J.; Wang, C.; Gao, R.; Zhou, X. Volumetric and fatty infiltration imbalance of deep paravertebral muscles in adolescent idiopathic scoliosis. *Med. Sci. Monit.* **2017**, *23*, 2089–2095. [CrossRef]
7. Liu, Y.; Pan, A.; Hai, Y.; Li, W.; Yin, L.; Guo, R. Asymmetric biomechanical characteristics of the paravertebral muscle in adolescent idiopathic scoliosis. *Clin. Biomech.* **2019**, *65*, 81–86. [CrossRef]
8. Hides, J.A.; Lambrecht, G.; Stanton, W.R.; Damann, V. Changes in multifidus and abdominal muscle size in response to microgravity: Possible implications for low back pain research. *Eur. Spine J.* **2015**, *25*, 175–182. [CrossRef]
9. Streisfeld, G.M.; Bartoszek, C.; Creran, E.; Inge, B.; McShane, M.D.; Johnston, T. Relationship between body positioning, muscle activity, and spinal kinematics in cyclists with and without low back pain: A systematic review. *Sports Health* **2017**, *9*, 75–79. [CrossRef]
10. Fouré, A.; Nordez, A.; McNair, P.; Cornu, C. Effects of plyometric training on both active and passive parts of the plantarflexors series elastic component stiffness of muscle–tendon complex. *Eur. J. Appl. Physiol.* **2011**, *111*, 539–548. [CrossRef]
11. Liew, B.X.; Netto, K.; Morris, S. Increase in leg stiffness reduces joint work during backpack carriage running at slow velocities. *J. Appl. Biomech.* **2017**, *33*, 347–353. [CrossRef] [PubMed]
12. Tisha, A.L.; Armstrong, A.A.; Johnson, A.W.; López-Ortiz, C. Skeletal muscle adaptations and passive muscle stiffness in cerebral palsy: A literature review and conceptual model. *J. Appl. Biomech.* **2019**, *35*, 68–79. [CrossRef] [PubMed]
13. Lacraz, G.; Rouleau, A.-J.; Couture, V.; Söllrald, T.; Drouin, G.; Veillette, N.; Grandbois, M.; Grenier, G. Increased stiffness in aged skeletal muscle impairs muscle progenitor cell proliferative activity. *PLoS ONE* **2015**, *10*, e0136217. [CrossRef] [PubMed]
14. Eby, S.F.; Cloud, B.A.; Brandenburg, J.E.; Giambini, H.; Song, P.; Chen, S.; Lebrasseur, N.K.; An, K.-N. Shear wave elastography of passive skeletal muscle stiffness: Influences of sex and age throughout adulthood. *Clin. Biomech.* **2015**, *30*, 22–27. [CrossRef] [PubMed]

15. Bailey, J.F.; Miller, S.L.; Khieu, K.; O'Neill, C.W.; Healey, R.M.; Coughlin, D.G.; Sayson, J.V.; Chang, D.G.; Hargens, A.R.; Lotz, J.C. From the international space station to the clinic: How prolonged unloading may disrupt lumbar spine stability. *Spine J.* **2018**, *18*, 7–14. [CrossRef]

16. Chan, S.-T.; Fung, P.-K.; Ng, N.-Y.; Ngan, T.-L.; Chong, M.-Y.; Tang, C.-N.; He, J.-F.; Zheng, Y.-P. Dynamic changes of elasticity, cross-sectional area, and fat infiltration of multifidus at different postures in men with chronic low back pain. *Spine J.* **2012**, *12*, 381–388. [CrossRef]

17. Mahaudens, P.; Detrembleur, C. Increase of passive stiffness in adolescent idiopathic scoliosis. *Comput. Methods Biomech. Biomed. Eng.* **2015**, *18*, 1992–1993. [CrossRef]

18. Linek, P.; Wolny, T.; Sikora, D.; Klepek, A. Intrarater reliability of shear wave elastography for the quantification of lateral abdominal muscle elasticity in idiopathic scoliosis patients. *J. Manip. Physiol. Ther.* **2020**, *43*, 303–310. [CrossRef]

19. Yamauchi, T.; Hasegawa, S.; Nakamura, M.; Nishishita, S.; Yanase, K.; Fujita, K.; Umehara, J.; Ji, X.; Ibuki, S.; Ichihashi, N. Effects of two stretching methods on shoulder range of motion and muscle stiffness in baseball players with posterior shoulder tightness: A randomized controlled trial. *J. Shoulder Elb. Surg.* **2016**, *25*, 1395–1403. [CrossRef]

20. Beach, T.A.; Parkinson, R.J.; Stothart, J.P.; Callaghan, J.P. Effects of prolonged sitting on the passive flexion stiffness of the in vivo lumbar spine. *Spine J.* **2005**, *5*, 145–154. [CrossRef]

21. Aslan, H.; Analan, P. Is there a correlation between Reimers' hip migration percentage and stiffness of hip muscles measured by shear wave elastography in children with cerebral palsy? *Ann. Phys. Rehabil. Med.* **2018**, *61*, e304. [CrossRef]

22. Ikezoe, T.; Asakawa, Y.; Fukumoto, Y.; Tsukagoshi, R.; Ichihashi, N. Associations of muscle stiffness and thickness with muscle strength and muscle power in elderly women. *Geriatr. Gerontol. Int.* **2012**, *12*, 86–92. [CrossRef] [PubMed]

23. Taniguchi, K.; Shinohara, M.; Nozaki, S.; Katayose, M. Acute decrease in the stiffness of resting muscle belly due to static stretching. *Scand. J. Med. Sci. Sports* **2015**, *25*, 32–40. [CrossRef] [PubMed]

24. Raiteri, B.J.; Hug, F.; Cresswell, A.G.; Lichtwark, G.A. Quantification of muscle co-contraction using supersonic shear wave imaging. *J. Biomech.* **2016**, *49*, 493–495. [CrossRef] [PubMed]

25. Nordez, A.; Hug, F. Muscle shear elastic modulus measured using supersonic shear imaging is highly related to muscle activity level. *J. Appl. Physiol.* **2010**, *108*, 1389–1394. [CrossRef] [PubMed]

26. Ling, Y.T.; Ma, C.Z.-H.; Shea, Q.T.K.; Zheng, Y.-P. Sonomechanomyography (SMMG): Mapping of skeletal muscle motion onset during contraction using ultrafast ultrasound imaging and multiple motion sensors. *Sensors* **2020**, *20*, 5513. [CrossRef]

27. Creze, M.; Nyangoh Timoh, K.; Gagey, O.; Rocher, L.; Bellin, M.F. Soubeyrand, M. Feasibility assessment of shear wave elastography to lumbar back muscles: A radioanatomic study. *Clin. Anat* **2017**, *30*, 774–780. [CrossRef]

28. Masaki, M.; Aoyama, T.; Murakami, T.; Yanase, K.; Ji, X.; Tateuchi, H.; Ichihashi, N. Association of low back pain with muscle stiffness and muscle mass of the lumbar back muscles, and sagittal spinal alignment in young and middle-aged medical workers. *Clin. Biomech.* **2017**, *49*, 128–133. [CrossRef]

29. Ren, L.J.; Wang, L.K.; Ma, C.Z.-H.; Yang, Y.X.; Zheng, Y.P. Effect of conventional physiotherapy on pain and muscle stiffness in patients with low back pain assessed by a wireless hand-held tissue ultrasound palpation system (TUPS). *Int. J. Phys. Med. Rehabil.* **2019**, *7*, 1–5.

30. Chao, C.Y.; Zheng, Y.-P.; Huang, Y.; Cheing, G.L.Y. Biomechanical properties of the forefoot plantar soft tissue as measured by an optical coherence tomography-based air-jet indentation system and tissue ultrasound palpation system. *Clin. Biomech.* **2010**, *25*, 594–600. [CrossRef]

31. Lau, J.C.; Li-Tsang, C.W.; Zheng, Y. Application of tissue ultrasound palpation system (TUPS) in objective scar evaluation. *Burns* **2005**, *31*, 445–452. [CrossRef] [PubMed]

32. Teraguchi, M.; Yoshimura, N.; Hashizume, H.; Muraki, S.; Yamada, H.; Minamide, A.; Oka, H.; Ishimoto, Y.; Nagata, K.; Kagotani, R.; et al. Prevalence and distribution of intervertebral disc degeneration over the entire spine in a population-based cohort: The Wakayama Spine Study. *Osteoarthr. Cartil.* **2014**, *22*, 104–110. [CrossRef] [PubMed]

33. Wang, D.; De Vito, G.; Ditroilo, M.; Delahunt, E. Effect of sex and fatigue on muscle stiffness and musculoarticular stiffness of the knee joint in a young active population. *J. Sports Sci.* **2017**, *35*, 1582–1591. [CrossRef] [PubMed]

34. Zheng, Y.-P.; Mak, A. An ultrasound indentation system for biomechanical properties assessment of soft tissues in-vivo. *IEEE Trans. Biomed. Eng.* **1996**, *43*, 912–918. [CrossRef]

35. Zhang, M.; Zheng, Y.P.; Mak, A.F.T. Estimating the effective Young's modulus of soft tissues from indentation tests–Nonlinear finite element analysis of effects of friction and large deformation. *Med. Eng. Phys.* **1997**, *19*, 512–517. [CrossRef]

36. Zheng, Y.; Mak, A.F.; Lue, B. Objective assessment of limb tissue elasticity: Development of a manual indentation procedure. *J. Rehabil. Res. Dev.* **1999**, *36*, 71–85.

37. Moreau, B.; Vergari, C.; Gad, H.; Sandoz, B.; Skalli, W.; Laporte, S. Non-invasive assessment of human multifidus muscle stiffness using ultrasound shear wave elastography: A feasibility study. *Proc. Inst. Mech. Eng. Part H J. Eng. Med.* **2016**, *230*, 809–814. [CrossRef]

38. Hu, X.; Lei, D.; Li, L.; Leng, Y.; Yu, Q.; Wei, X.; Lo, W.L.A. Quantifying paraspinal muscle tone and stiffness in young adults with chronic low back pain: A reliability study. *Sci. Rep.* **2018**, *8*, 14343. [CrossRef]

39. Hatta, T.; Giambini, H.; Uehara, K.; Okamoto, S.; Chen, S.; Sperling, J.W.; Itoi, E.; An, K.-N. Quantitative assessment of rotator cuff muscle elasticity: Reliability and feasibility of shear wave elastography. *J. Biomech.* **2015**, *48*, 3853–3858. [CrossRef]

40. Ma, C.Z.-H.; Ling, Y.T.; Shea, Q.T.K.; Wang, L.-K.; Wang, X.-Y.; Zheng, Y.-P. Towards wearable comprehensive capture and analysis of skeletal muscle activity during human locomotion. *Sensors* **2019**, *19*, 195. [CrossRef]

41. Ward, S.R.; Tomiya, A.; Regev, G.J.; Thacker, B.E.; Benzl, R.C.; Kim, C.W.; Lieber, R.L. Passive mechanical properties of the lumbar multifidus muscle support its role as a stabilizer. *J. Biomech.* **2009**, *42*, 1384–1389. [CrossRef] [PubMed]

42. Young, B.A.; Koppenhaver, S.L.; Timo-Dondoyano, R.M.; Baumann, K.; Scheirer, V.F.; Wolff, A.; Sutlive, T.G.; Elliott, J.M. Ultrasound shear wave elastography measurement of the deep posterior cervical muscles: Reliability and ability to differentiate between muscle contraction states. *J. Electromyogr. Kinesiol.* **2020**, *56*, 102488. [CrossRef] [PubMed]

43. Koppenhaver, S.L.; Scutella, D.; Sorrell, B.A.; Yahalom, J.; Fernández-de-las-Peñas, C.; Childs, J.D.; Shaffer, S.W.; Shinohara, M.J.C.B. Normative parameters and anthropometric variability of lumbar muscle stiffness using ultrasound shear-wave elastography. *Clin. Biomech.* **2019**, *62*, 113–120. [CrossRef] [PubMed]

44. Wang, D.; De Vito, G.; Ditroilo, M.; Fong, D.T.P.; Delahunt, E. A comparison of muscle stiffness and musculoarticular stiffness of the knee joint in young athletic males and females. *J. Electromyogr. Kinesiol.* **2015**, *25*, 495–500. [CrossRef] [PubMed]

45. Blackburn, J.T.; Riemann, B.L.; Padua, D.A.; Guskiewicz, K.M. Sex comparison of extensibility, passive, and active stiffness of the knee flexors. *Clin. Biomech.* **2004**, *19*, 36–43. [CrossRef]

46. Bawadi, H.; Hassan, S.; Zadeh, A.S.; Sarv, H.; Kerkadi, A.; Tur, J.A.; Shi, Z. Age and gender specific cut-off points for body fat parameters among adults in Qatar. *Nutr. J.* **2020**, *19*, 1–5. [CrossRef]

47. Eby, S.F.; Song, P.; Chen, S.; Chen, Q.; Greenleaf, J.F.; An, K.-N. Validation of shear wave elastography in skeletal muscle. *J. Biomech.* **2013**, *46*, 2381–2387. [CrossRef]

48. Marras, W.S.; Davis, K.G.; Heaney, C.A.; Maronitis, A.B.; Allread, W.G. The influence of psychosocial stress, gender, and personality on mechanical loading of the lumbar spine. *Spine* **2000**, *25*, 3045–3054. [CrossRef]

49. Marras, W.S.; Jorgensen, M.; Granata, K.; Wiand, B. Female and male trunk geometry: Size and prediction of the spine loading trunk muscles derived from MRI. *Clin. Biomech.* **2001**, *16*, 38–46. [CrossRef]

50. Sichting, F.; Kram, N.C. Phantom material testing indicates that the mechanical properties, geometrical dimensions, and tensional state of tendons affect oscillation-based measurements. *Physiol. Meas.* **2020**, *41*, 095010. [CrossRef]

Publisher's Note: MDPI stays neutral with regard to jurisdictional claims in published maps and institutional affiliations.

Article

In Vivo Evaluation of Plane Wave Imaging for Abdominal Ultrasonography

Sua Bae [1], Jintae Jang [1], Moon Hyung Choi [2,*] and Tai-Kyong Song [1,*]

[1] Department of Electronic Engineering, Sogang University, Seoul 04107, Korea;
suabae@sogang.ac.kr (S.B.); jterry123@sogang.ac.kr (J.J.)

[2] Department of Radiology, College of Medicine, The Catholic University of Korea, Seoul 03312, Korea

* Correspondence: choimh1205@gmail.com (M.H.C.); tksong@sogang.ac.kr (T.-K.S.)

Received: 15 September 2020; Accepted: 3 October 2020; Published: 5 October 2020

Abstract: Although plane wave imaging (PWI) has been extensively employed for ultrafast ultrasound imaging, its potential for sectorial B-mode imaging with a convex array transducer has not yet been widely recognized. Recently, we reported an optimized PWI approach for sector scanning that exploits the dynamic transmit focusing capability. In this paper, we first report the clinical applicability of the optimized PWI for abdominal ultrasonography by in vivo image and video evaluations and compare it with conventional focusing (CF) and diverging wave imaging (DWI), which is another dynamic transmit focusing technique generally used for sectorial imaging. In vivo images and videos of the liver, kidney, and gallbladder were obtained from 30 healthy volunteers using PWI, DWI, and CF. Three radiologists assessed the phantom images, 156 in vivo images, and 66 in vivo videos. PWI showed significantly enhanced ($p < 0.05$) spatial resolution, contrast, and noise and artifact reduction, and a 4-fold higher acquisition rate compared to CF and provided similar performances compared to DWI. Because the computations required for PWI are considerably lower than that for DWI, PWI may represent a promising technique for sectorial imaging in abdominal ultrasonography that provides better image quality and eliminates the need for focal depth adjustment.

Keywords: medical diagnostic imaging; ultrasonic imaging; abdominal ultrasound; plane wave imaging; diverging wave imaging; synthetic focusing

1. Introduction

Abdominal ultrasound (US) requires a large field-of-view with high image quality at all depths because abdominal organs examined by US imaging are of various sizes and located at various depths [1]. For example, the gallbladder and common bile duct, which are located at shallow depths (2–7 cm), need to be reconstructed with a sufficiently high spatial resolution to estimate the wall thickness, while the liver and kidney, which are located at both shallow and deep depths (5–20 cm), also require high spatial and contrast resolutions. When conventional focusing (CF) is used, however, the transmit beam has a fixed focal depth, which exhibits a higher spatial resolution and contrast of the image in the vicinity of the focal depth but lower image quality at other depths. Consequently, when scanning the entire abdomen, clinicians must constantly adjust the focal depth up and down to the region of interest with one hand while holding a transducer with the other hand. This constant manual adjustment of the focus prolongs the examination time.

A simple method of enhancing the image quality over various depths is to increase the number of transmit foci per scanline [2]. In the multifocus technique, multiple beams focused at different depths are successively transmitted to reconstruct a single scanline of an image. Consequently, the frame rate decreases inversely proportional to the number of foci. Although this technique is commonly used for linear array imaging with a short depth of view (<8 cm), it is difficult to apply for abdominal US imaging when the view depth is greater than 15 cm because of the long acquisition time. If three foci per scanline are used for a B-mode image with 256 scanlines and a depth of 20 cm, then the frame rate will be lowered to 5 Hz, which is extremely slow for real-time US scanning.

Another method is synthetic transmit focusing (STF), in which multiple low-resolution images obtained by transmitting an unfocused beam or a widely diverging beam before and after a tight focus are compounded coherently to achieve dynamic transmit focusing at every imaging point. Synthetic aperture (SA) imaging is the most investigated and widespread STF technique. In SA imaging, a virtual source (VS) is usually generated in front of the array transducer, and spherical wavefronts before and after the VSs are used for STF [3–5]. Because it was adapted for medical imaging in the 1980s–90s [3,6,7], numerous studies have been published and verified that SA imaging provides a high-resolution image over all depths compared to conventional focusing [8–11]. Consequently, it has been implemented as an innovative advanced beamformer on commercial high-end US systems, such as *n*SIGHT by Philips or cSound by GE [12,13]. However, this technique requires reconstruction of dozens of hundreds of more scanlines per frame compared to the CF method, which substantially increases the computational costs. In addition, it suffers from motion artifacts because high-quality SA imaging often requires approximately 100 emissions due to grating lobe problems [8,14], and tissue or hand motion could cause incoherence among the low-resolution images that are subsequently compounded and lower the performance of dynamic transmit focusing. Moreover, motion artifacts can be more problematic in abdominal ultrasonography due to the large field-of-view and long acquisition time, which hinders the fast scanning and capturing of various abdominal organs.

When VSs are set behind the transducer array, diverging waves (DWs) are transmitted through multiple elements covering a broad imaging region. In DW imaging (DWI), a small number of DWs, usually 3–20, with different VS positions in the lateral direction are transmitted and coherently compounded for a single frame [15–17], thus leading to fast acquisition and less motion artifact. Using the broad coverage of DW, DWI is generally used for fast cardiac imaging with a phased array [16,18–21]. Because DWI offers dynamic transmit focusing with a small number of emissions, it can also be used to enhance the image quality of abdominal US, avoiding significant motion artifacts. Consequently, a recent paper suggested the use of DW for ultrafast abdominal imaging [22].

When the VSs are placed at infinity behind the array, plane waves (PWs) are transmitted. PW imaging (PWI) employs STF using the PWs steered at different angles [23]. It has been widely used as a means of ultrafast imaging for shear wave elastography [24–26] and blood flow imaging [27,28] or fast 3D scanning [29–31]. Although most PWI studies focus only on its ultrafast imaging capability, a few studies have reported its potential for high-resolution B-mode imaging [23,26,32]. In addition, PWI is more advantageous than DWI in terms of the spatial resolution [19,22] because PW STF can theoretically provide a uniform beam width over all depths [26,33]. However, PWI has not been widely used for large sectorial imaging because of the small coverage of PW compared to that of DW, which might be one of the reasons why PWI has not been implemented for abdominal imaging. To test the feasibility of PWI for this unexplored application, in our earlier study [34,35], we optimized PWI for the sectorial field-of-view and conducted simulation and phantom experiments. From that study, PWI was proven to enhance both the image quality and frame rate in convex array imaging, which results similar to that of linear array imaging when the PW angles, transmit aperture size, and synthesized PWs for each imaging point were properly selected.

The objective of this paper is to test the clinical applicability of the optimized PWI for abdominal ultrasonography by evaluating its in vivo image quality and comparing its quality with that of DWI and CF. To the best of our knowledge, this is the first in vivo study of PWI for abdominal ultrasonic imaging. Phantom images, in vivo images, and videos of the liver, kidney, and gallbladder of 30 healthy volunteers were obtained using PWI, DWI, and CF. First, the phantom images were used to measure the spatial resolution and image contrast for the quantitative evaluation. Then, in vivo abdominal US images were acquired from 30 healthy volunteers and assessed by three radiologists in terms of spatial resolution, image contrast, noise, and artifacts. In addition, in vivo video clips were also evaluated by radiologists to assess image quality under hand motion.

2. Materials and Methods

2.1. Imaging Techniques

Three imaging techniques were compared: CF, DWI, and PWI. The transmit configuration of each imaging technique is listed in Table 1. In CF imaging, traditional line-by-line scanning was performed using a focused beam with a focal depth of 10 cm and an F-number of 5.0. The focal depth and the F-number were optimized so that the spatial resolution was as constant with increasing depth as possible to obtain uniform spatial resolution in the entire field-of-view for the fair comparison with two-way dynamic focusing techniques (DWI and PWI). In addition, 128 focused beams were used for CF, and 32 VSs and 32 PWs were employed for DWI and PWI, respectively.

Table 1. Number of transmissions (N_{tx}), acquisition frame rate considering the N_{tx}, normalized amount of computations in the beamforming process, and Tx beam properties of conventional focusing (CF), diverging wave imaging (DWI), and plane wave imaging (PWI).

	N_{tx}	Acquisition Frame Rate (fps)	Amount of Computation (Normalized)	Tx Beam Properties
CF	128	31.9	1	Focal depth: 10 cm, F-number: 5.0
DWI	32	127.7	29.26	VS positions: x: from −2 cm to 2 cm with an interval of 4/31 cm, z: same as the vertex of the convex array
PWI	32	127.7	9.94	PW angles from −30° to 30° with an interval of 60/31°

Figure 1 illustrates the VSs used for DWI and PWs used for PWI. The red dots in Figure 1a show the 32 VSs, and the orange arcs represent the propagation of a DW originated from the first VS over time. The red lines in Figure 1b show the 32 steered PWs, and the orange lines illustrate the propagating PW. The blue arrows in Figure 1 show the directions of the DW and PW. The VS positions (Table 1) were chosen to employ the full aperture (yellow shaded area in Figure 1a) for each DW transmission. The outermost VS was positioned at $x = \pm 20$ mm (Figure S1) considering the 6 dB acceptance angle of the element of the convex array transducer used in this experiment. In PWI, the transmit aperture size was limited, indicated by a yellow shaded area in Figure 1b, such that it did not exceed the acceptance angle of the transducer element for the optimal PWI as proposed in our previous study [35].

In sector imaging, PW has a smaller beam propagation region (smaller coverage) compared with DW (orange shaded areas in Figure 1) because of the limited transmit aperture and a constant direction of propagation. Although this property of PW reduces the number of synthesized waves per imaging pixel, it does not diverge and maintains the wave intensity throughout the propagation, leading to a deeper penetration depth compared with that of DW. More importantly, the small propagation region reduces the amount of computations required in beamforming. Table 1 shows the normalized amount of computations for each imaging technique when a B-mode image with 1039 × 256 pixels was reconstructed. This value was obtained by normalizing the number of beamforming (channel-summation) operations required for a single compounded image using DWI or PWI by the value required for a single image using CF. As is well known, synthetic imaging (DWI and PWI) requires much more computations than traditional focusing (CF). Note that PWI with 32 PWs requires 2.9-times fewer computations compared with DWI with 32 VSs as shown in Table 1.

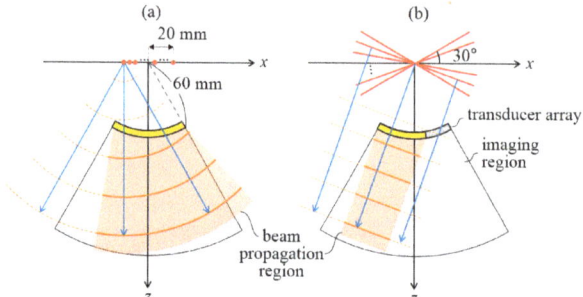

Figure 1. (a) Virtual sources (VSs) (red dots) and propagating DW (orange arcs) used for DWI and (b) steered PWs (solid red lines) and propagating PW (orange lines) used for PWI. Directions of DWs and PWs are indicated by blue arrows, and transmit apertures are marked with yellow shaded areas. Pixels in the beam propagation region (orange shaded area) for each DW or PW transmission should be reconstructed in the beamforming process. Difference in the size of beam propagation region between DWI and PWI shows the difference in the amount of computations required for DWI and PWI.

Given the imaging depth d, the sound speed c, and the time margin before the next emission τ, the acquisition rate for a single frame can be calculated by

$$FR_{max} = 1/\{(2d/c + \tau)\cdot N_{tx}\} \tag{1}$$

The acquisition frame rates of CF, DWI, and PWI are 31.9 fps, 127.7 fps, and 127.7 fps, respectively, when d = 15 cm, c = 1540 m/s, and τ = 50 μs (Table 1). However, during the in vivo data acquisition with the US system, the display frame rates (i.e., the frame rate at which the B-mode image was updated on the screen) were 26.6 fps, 10 fps, and 15 fps for CF, DWI, and PWI, respectively, which were lower than the acquisition rates due to the limited number of receive channels and computing power of the system. Although 2.9 times fewer computations were required for PWI than DWI (Table 1), the display frame rate of PWI supported by the system was only 1.5 higher than that of DWI due to the limited channel count of the system. The reasons for the low display frame rate will be further explained in discussion section. Note that the display frame rate could, however, be enhanced up to the acquisition frame rate by improving the computational algorithms in the beamforming process and upgrading the computational resources of the US system.

For all the imaging techniques, the transmit voltage was 80 V, the receive F-number was 1.0, and the 50% Tukey window was used for the receive apodization in the beamforming process. In DWI and PWI, only the imaging points reached by the DW or PW were calculated as the low-resolution image and compounded for the final image as described in [35].

2.2. System and Method for Data Acquisition

A research US system (E-cube 12R, Alpinion Medical Systems, Republic of Korea) with a convex array transducer (SC1-6, Alpinion Medical Systems, Republic of Korea) was used for data acquisition. The transducer has 128 elements and a center frequency of 3.6 MHz, and the system has a 128-channel transmit board and a 64-channel receive board. For DWI and PWI, which require full-channel reception, the same emission was repeated twice for the echo reception of the first and second sets of 64 channels. Beamforming and image processing were conducted on a graphics processing unit (GPU) (GeForce GTX 1080, NVIDIA, CA, USA) equipped in the system by using the CUDA computing platform. Thus, unfortunately, the acquisition frame rate of DWI and PWI in this study was twice the maximum acquisition frame rate (127.7/2 = 63.85 fps).

A commercial phantom (Model 539, ATS laboratories Inc., Bridgeport, CA, USA) was used for the phantom study. For the phantom images of CF, DWI, and PWI, radio-frequency (RF) data were acquired by fixing the transducer on the phantom. A cross-section of the phantom including point and cyst targets was selected, and three images of the same cross-section were reconstructed using the three imaging techniques (CF, DWI, and PWI).

In vivo abdominal ultrasonic images of the gallbladder, liver, and kidney were collected from 30 healthy male volunteers by a radiologist under institutional review board approval at Seoul Saint Mary's Hospital. Written informed consent was obtained from all volunteers. The radiologist obtained abdominal ultrasonic images and videos of each volunteer using CF, DWI, and PWI, sequentially, trying to obtain three images or videos (CF, DWI, and PWI) for the same cross-section as much as possible. Misaligned sets were excluded, and 52 image sets (a total of 156 images) were evaluated: 14 sets for the gallbladder, 18 sets for the liver, and 20 sets for the kidney. For the video evaluation, 22 video sets (a total of 66 clips) were chosen, and each video contains the real-time image of right hepatic lobe, gallbladder, and right kidney. Three video clips (CF, DWI, and PWI) of each set were synchronized to show the same cross-section at the same time point as much as possible. The time length of the synchronized videos was between 3 and 9 s.

2.3. Beamforming and Postprocessing of Image for the Evaluation

For the still images, the RF channel data were stored and beamforming and postprocessing were conducted offline. In the beamforming process, the RF data were demodulated to the base band, downsampled by a factor of 4, and then beamformed using the parameters shown in Section 2.1. To flatten the uneven brightness of the image across depths within an image and across different imaging techniques, automatic time-gain-compensation (TGC) was applied to all the images as in [8]. The imaging region was axially divided into 5 zones, and the 5 representative gain values were determined by the reciprocal of the median brightness of each zone. TGC was applied after the spline interpolation of the 5 gain values.

In the log compression, which highly affects the contrast of an image, the max value was automatically chosen to be 50 dB and 40 dB above the median brightness of the entire image for the phantom and in vivo images, respectively. The dynamic range was 80 dB and 57 dB for the phantom and in vivo images, respectively.

The RF channel data for in vivo videos could not be stored due to the limited storage capacity. The videos were obtained by recording displayed B-mode images on the screen. Because the automatic TGC was not implemented on the online reconstruction software in the system and the radiologist arbitrarily adjusted the gain during the acquisition, the brightness of the on-screen images among DWI, PWI, and CF was quite different. Thus, the automatic TGC was applied on a log scale to the recorded video clips. For this reason, unfortunately, the image contrast of video could not be evaluated because the brightness of the screen-captured video was already clipped with different ranges before the post TGC control.

2.4. Image and Video Evaluation

Three radiologists with 10 years, 8 years, and 5 years of abdominal ultrasonography experience assessed the phantom images and the in vivo images and videos of the human abdomen. The radiologists were asked to score each image or video on a 5-point Likert scale (1: very poor, 2: poor, 3: average, 4: good, and 5: very good) in terms of 4 evaluation items ('spatial resolution', 'contrast', 'noise', and 'artifacts'). The videos were not assessed in terms of 'contrast' because some grayscale values were saturated due to the unavailability of raw data as described in Section 2.3.

In the phantom study, 3 images (1 set) of a cross-section of the phantom were reconstructed using CF, DWI, and PWI. The 3 images were randomly ordered without labels and presented to evaluators. For the in vivo study, 156 images (52 sets) were randomly ordered and evaluated individually without any information about the patients and imaging techniques. For the assessment of in vivo videos (22 sets), the three synchronized videos of each set were played together side by side with random order. The radiologist could rewind and play back the videos freely during the assessment.

For the phantom study, the spatial resolution and contrast were also quantitatively measured. The spatial resolution was measured by the lateral length of the −6 dB contour of a point target [35]. The contrast ratio was calculated by $CR = \mu_b - \mu_c$, where μ_b and μ_c are the mean intensities of the background speckle and cyst regions, respectively [36].

2.5. Statistical Analysis

The Wilcoxon rank-sum test was used because it is known to be suitable for a Likert scale evaluation [37,38]. Because the absolute Likert scale values highly depend on the person's interpretation of the scale, the test was applied to each evaluator's scores. Three pairs of data (CF versus (vs.) DWI, CF vs. PWI, and DWI vs. PWI) were tested to statistically demonstrate that PWI offers a better image quality than does CF imaging and provides comparable performance to DWI. The mean score difference between two among three imaging techniques were obtained. For example, the mean score difference between PWI and DWI (P vs. D) was calculated as

$$d_{P \text{ vs. D}} = \frac{1}{N} \Sigma_n s_P(n) - s_D(n) \tag{2}$$

where $s_P(n)$ and $s_D(n)$ are scores of n-th image or video clip reconstructed by PWI and DWI, respectively.

3. Results

3.1. Phantom Study

Figure 2 shows the phantom images reconstructed by the CF, DWI, and PWI techniques. The −6 dB spatial resolutions of the point targets in Figure 2 are presented in Figure 3a, and magnified images of the point targets at 30, 80, and 120 mm are presented with −6 dB, −12 dB, and −20 dB contours in Figure 3b–d. In Figure 3a, the effective focal depth of CF seems slightly closer to the transducer than 100 mm because the point targets were vertically located 2.5 mm apart from the center scanline. For all imaging techniques, the spatial resolution deteriorates as the depth increases. Although the resolutions of the three techniques are similar at shallow depths, PWI and DWI clearly show a better spatial resolution than that of CF at depths ≥ 100 mm. In addition, PWI provides a slightly better resolution than that of DWI. This result might be explained by the nondiffraction property of PWs [26,33], and the superiority of PWI over DWI in terms of spatial resolution was also previously reported [19,22].

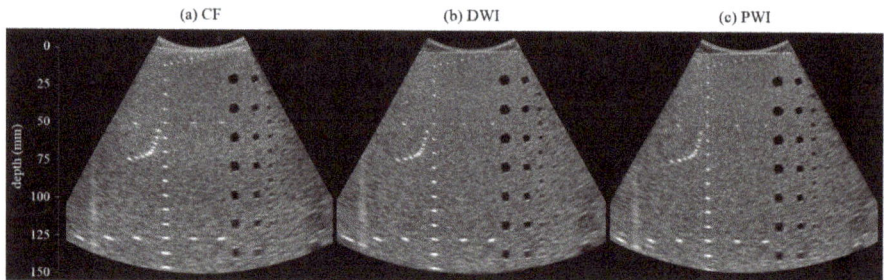

Figure 2. Phantom image set reconstructed by using (**a**) CF, (**b**) DWI, and (**c**) PWI. The enlarged images of point and cyst targets are shown in Figures 3 and 4.

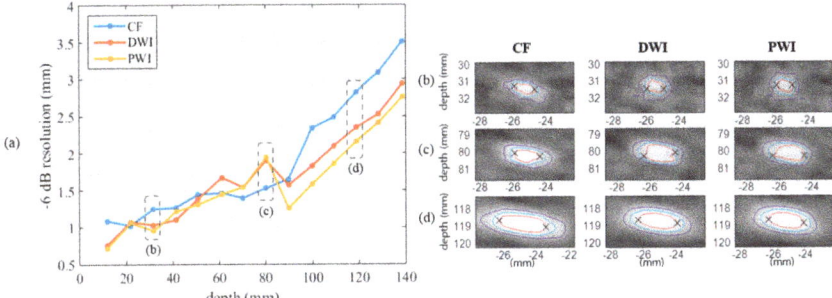

Figure 3. (**a**) Resolution (mm) over depths measured from point targets in the phantom images of CF, DWI, and PWI. DWI and PWI provided a better resolution at depths ≥ 100 mm. (**b–d**) Magnified point target images of Figure 2 at depths of (**b**) 30 mm, (**c**) 80 mm, and (**d**) 120 mm. Red, blue, and purple lines show −6 dB, −12 dB, and −20 dB contours, respectively, and the two crosses in each panel indicate where the maximum lateral distance is measured as the resolution value.

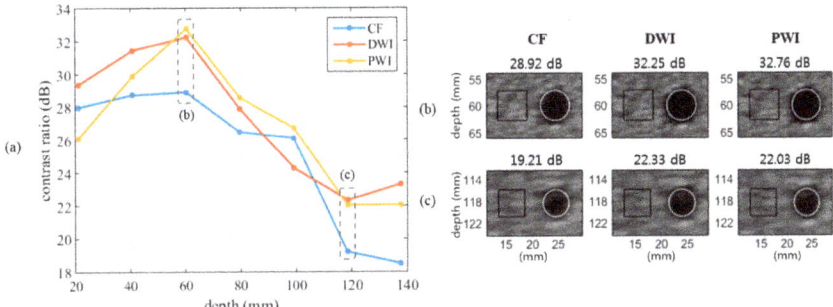

Figure 4. (**a**) Contrast ratio (dB) over all depths measured from cyst targets in the phantom images of CF, DWI, and PWI. The contrast ratio of CF is lower than those of DWI and PWI on average. (**b**, **c**) Magnified cyst target images of Figure 2 at depths of (**b**) 60 mm and (**c**) 120 mm. The black square and the white circle show the contrast measurement areas. The measured contrast ratio is presented above each panel.

The measured contrast of the cyst targets in Figure 2 is presented in Figure 4a, and the magnified cyst images are shown in Figure 4b,c. Except at a depth of 20 mm, PWI always offered a higher contrast ratio than CF. Compared with DWI, PWI provided a lower contrast at near depths but a similar contrast at mid and far depths. At near depths, DW had sufficient intensity before diverging further and the number of compounded DWs is greater than the number of compounded PWs (due to the broader coverage of a DW than a PW). This might lead to the higher contrast ratio of DWI than that of PWI at near depths ($z = 20$ and 40 mm). However, DWs lose the intensity more than PWs as it propagated. At mid and far depths ($z = 60$–140 mm), the contrast ratio of DWI and PWI became similar, although the number of compounded DWs is still greater than the number of compounded PWs. Both DWI and PWI showed the contrast degradation at near depths relative to the far depths. This might be due to the reverberation artifacts that appear more frequently in imaging techniques using broad beams than in conventional imaging. The mean contrast ratios of CF, DWI, and PWI were 25.15 dB, 27.27 dB, and 26.88 dB, respectively.

Table 2 lists the scores of phantom images of one set (Figure 2) reconstructed using CF, DWI, and PWI, and they were evaluated by three radiologists with consideration of the four items. PWI was scored higher than CF in all cases. All radiologists gave almost the same scores to DWI and PWI.

From this phantom study, PWI was proven to provide better image quality with a higher acquisition rate than CF and to yield a comparable performance to DWI with much lower computational costs.

Table 2. Evaluation results of phantom images (Figure 2) reconstructed by CF, DWI, and PWI on the Likert Scale (1–5) from three radiologists (Rad.). DWI and PWI were scored higher than CF in most cases.

Items		Spatial Resolution			Contrast			Noise			Artifacts		
Imaging		CF	DWI	PWI	CF	DWI	PWI	CF	DWI	PWI	CF	DWI	PWI
	Rad. 1	2	5	4	3	5	5	2	4	5	3	5	5
Score	Rad. 2	2	5	4	3	4	4	2	4	4	3	4	4
	Rad. 3	3	4	4	3	4	4	3	4	4	3	4	4

3.2. In Vivo Study

Figure 5 shows representative gallbladder, liver, and kidney images obtained using CF, DWI, and PWI. Although slightly different cross-sections were captured across the imaging techniques, the overall image quality, including the spatial resolution and image contrast, is better in DWI and PWI than in CF. Figure 6 shows the clinical evaluation results of 156 in vivo images (52 sets) by the three radiologists. A bar represents the mean of differences in scores between DWI and CF ($d_{D \text{ vs. } C}$), PWI and CF ($d_{P \text{ vs. } C}$), and PWI and DWI ($d_{P \text{ vs. } D}$) obtained by (2). The p-values of the statistical test are listed in Table 3, and the significant differences ($p < 0.05$) are shown in bold in Table 3 and marked by an asterisk in Figure 6. Both DWI and PWI show higher scores than CF for all evaluation items by all three radiologists. DWI and PWI received very similar scores. Although PWI shows slightly better spatial resolution and DWI presents slightly higher scores for other image qualities, significant differences are not observed.

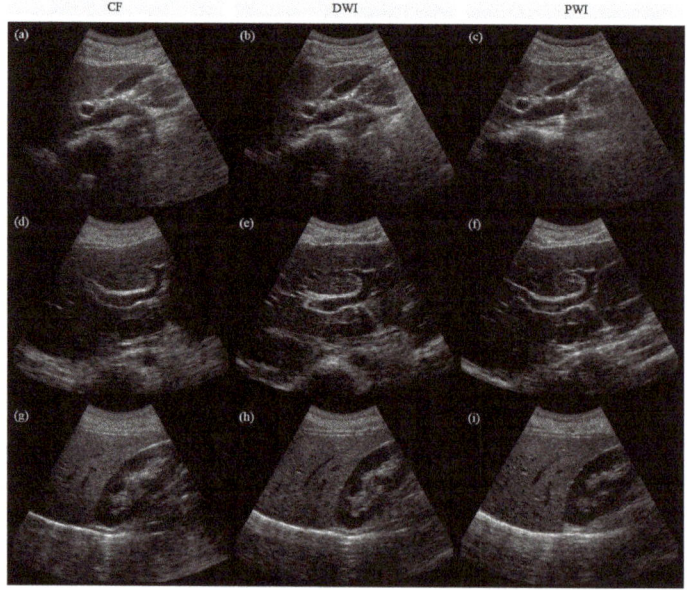

Figure 5. Representative in vivo images of (**a–c**) gallbladder, (**d–f**) liver, and (**g–i**) kidney, which were reconstructed by using CF (for (**a, d, g**)), DWI (for (**b, e, h**)), and PWI (for (**c, f, i**)). PWI showed better image quality with a 4-fold higher acquisition frame rate than CF and provided a comparable performance with a 2.9 times lower number of computations compared to DWI.

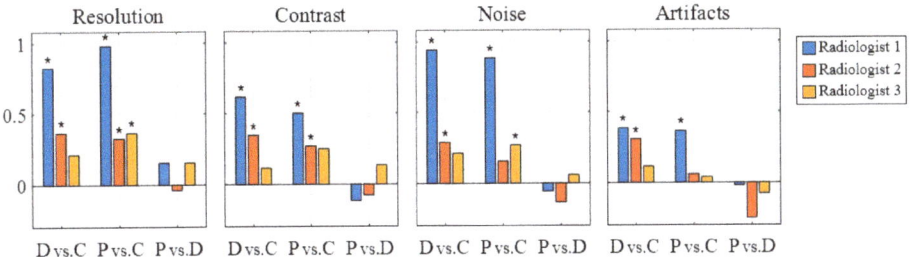

Figure 6. Mean score differences of in vivo images ($N = 52$) between DWI and CF ($\bar{d}_{D \text{ vs. C}}$), PWI and CF ($\bar{d}_{P \text{ vs. C}}$), and PWI and DWI ($\bar{d}_{P \text{ vs. D}}$). The asterisk indicates a significant difference ($p < 0.05$). DWI and PWI show higher scores than CF in all cases, with significant differences in some cases. PWI and DWI show a similar performance with no significant differences.

Table 3. P-values of the rank-sum test between each pair of CF, DWI, and PWI in the in vivo image evaluation ($N = 52$) (all P-values less than 0.05 (i.e., statistically significant differences) are shown in bold).

	Spatial Resolution			Contrast			Noise		
	D vs. C	**P vs. C**	**P vs. D**	**D vs. C**	**P vs. C**	**P vs. D**	**D vs. C**	**P vs. C**	**P vs. D**
Rad. 1	**0.000**	**0.000**	0.273	**0.002**	**0.012**	0.744	**0.000**	**0.000**	0.622
Rad. 2	**0.011**	**0.039**	0.691	**0.005**	**0.045**	0.829	**0.010**	0.111	0.887
Rad. 3	0.142	**0.024**	0.155	0.237	0.053	0.193	0.070	**0.041**	0.354

	Artifact		
	D vs. C	**P vs. C**	**P vs. D**
Rad. 1	**0.020**	**0.029**	0.499
Rad. 2	**0.027**	0.396	0.954
Rad. 3	0.194	0.350	0.666

Rad. = radiologist; D = DWI; P = PWI; C = CF.

Radiologist 1 found a highly significant enhancement ($p < 0.01$) of the 'spatial resolution', 'contrast', and 'noise' and a significant improvement ($p < 0.05$) in the 'unwanted artifacts' for the images obtained via DWI and PWI compared with those obtained via CF. Radiologist 2 also noted a significant enhancement ($p < 0.05$) of the 'resolution' and 'contrast' for DWI and PWI. Radiologist 3 indicated that PWI provides significantly better image quality ($p < 0.05$) based on the 'resolution' and 'noise' than CF. None of the radiologists found significant differences ($p > 0.1$) between DWI and PWI with respect to all the evaluation items.

Figure 7 shows the clinical evaluation results of the in vivo videos by the three radiologists. The representative video is available online as multimedia Video S1. The *p*-values of the rank-sum test of scores are listed in Table 4, and the significant differences ($p < 0.05$) are shown in bold in Table 4 and marked by an asterisk in Figure 7. As observed in the result of the still image assessment, PWI had higher average scores than CF in all cases and similar scores to that of DWI in the video evaluation. Radiologist 1 found that PWI significantly enhanced the image quality for all evaluation items ($p < 0.05$) compared to CF. Because noise is relatively easier to recognize from videos than from still images, two of the three radiologists indicated that the DWI and PWI videos showed a significant enhancement with respect to 'noise' compared with the CF videos. Significant differences were not observed between PWI and DWI.

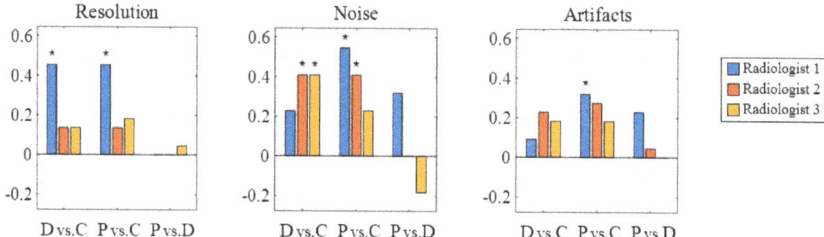

Figure 7. Mean of the score differences of in vivo videos ($N = 22$) between DWI and CF ($\bar{d}_{\text{D vs. C}}$), between PWI and CF ($\bar{d}_{\text{P vs. C}}$), and between PWI and DWI ($\bar{d}_{\text{P vs. D}}$). The asterisk indicates a significant difference ($p < 0.05$). DWI and PWI show higher scores on average than those of CF in all cases. PWI and DWI (P vs. D) show similar performance with no significant differences.

Table 4. P-values of the rank-sum test between each pair of CF, DWI, and PWI in the in vivo video evaluation ($N = 22$) (all P-values less than 0.05 (i.e., statistically significant differences) are shown in bold).

	Spatial Resolution			Noise			Artifact		
	D vs. C	P vs. C	P vs. D	D vs. C	P vs. C	P vs. D	D vs. C	P vs. C	P vs. D
Rad. 1	**0.025**	**0.035**	0.531	0.121	**0.013**	0.093	0.224	**0.021**	0.102
Rad. 2	0.259	0.283	0.552	**0.016**	**0.021**	0.479	0.136	0.092	0.408
Rad. 3	0.261	0.155	0.359	**0.041**	0.106	0.807	0.146	0.146	0.506

4. Discussion

In this paper, we demonstrated that PWI 1) provides significantly enhanced image quality with a 4-fold higher acquisition rate compared to line-by-line CF and 2) provides a comparable performance with a 2.9 times lower number of computations compared to DWI, based on quantitative and qualitative evaluations of phantom and in vivo images. In the phantom study, the spatial resolution at depths ≥ 100 mm was enhanced (~0.5 mm) and the contrast of cyst targets was improved (~2 dB higher on average) when using DWI and PWI compared with CF (Figures 3 and 4, Table 2). In the in vivo study, the radiologists assessed the still images of 52 sets and the video clips of 22 sets, including liver, gallbladder, and kidney.

Comparing PWI and CF, in the image evaluation (Figures 5 and 6), radiologist 1 rated PWI significantly higher than CF for all evaluation items and radiologists 2 and 3 recognized the significantly improved image quality of PWI in terms of 'resolution', 'contrast', and 'noise' items ($p < 0.05$). In the video evaluation (Video S1 and Figure 7), radiologist 1 found a significant enhancement in PWI in terms of 'resolution', 'contrast', and 'noise', while radiologist 2 found significant enhancements in terms of 'noise' compared to CF.

In addition to enhanced image quality, the fast acquisition rate is another advantage of PWI compared to CF. As the numbers of transmissions of PWI are 4-times lower than that of CF (Table 1), the acquisition rates under the physical speed of US in tissues are 4-fold higher than that of CF. This advantage of using a small number of emissions reduces the likelihood of motion artifacts, such as blurring and distortion, which are major issues in synthetic imaging.

A comparison between PWI and DWI showed that PWI had slightly better spatial resolution and DWI had slightly better contrast and reduced noise and artifacts (Figures 6 and 7). Similar results were reported by Tong et al. [19] and Kang et al. [22]. However, the score differences between PWI and DWI were quite small and none were significant ($p > 0.1$, Tables 3 and 4). Therefore, these findings imply that PWI is able to provide a comparable image quality to DWI in sector imaging.

More importantly, PWI required an approximately 3-times lower amount of computations (Table 1) relative to DWI. For sector imaging, DW is usually chosen to achieve dynamic transmit focusing, which might be related to the larger field-of-view of sector imaging compared with linear-scan imaging and the broader coverage region (beam propagation region in Figure 1) of DW compared with PW. In this paper, however, we found that PWI can provide comparable image quality with a much lower amount of computations compared to that of DWI when the PW angles and transmit aperture size are carefully selected as in [35].

4.1. Dependence on Evaluators

From the statistical analysis of the in vivo images and videos (Tables 3 and 4), significant differences were found most often in the assessment of radiologist 1, while the least significant differences among the three radiologists were found for the evaluation results of radiologist 3. This outcome might be associated with the evaluators' clinical experience. Radiologists 1, 2, and 3 had 10 years, 8 years, and 5 years of experience, respectively, and the most experienced radiologist gave the scores with the largest variance (variance in the image evaluation scores was 1.16, 0.79, and 0.79 for radiologists 1, 2, and 3, respectively). The more experienced radiologists might have assessed the images with greater confidence, resulting in more significant differences in many items.

4.2. Real-time Realization

Similar to other STF imaging techniques, DWI and PWI require massive computations because dozens of scanlines should be reconstructed per single transmission and reception event, while CF requires a one- or two-scanline reconstruction per event (Table 1). Thus, this computational load makes the real-time implementation of STF imaging challenging, although both DWI and PWI have a high acquisition frame rate. In this case, the lower number of computations of PWI compared with DWI can be beneficial.

Parallel processors can be successfully utilized for STF imaging to accelerate the reconstruction process because beamforming intrinsically performs the same operation on multiple data points. Software-based beamformers based on GPUs have been widely employed for STF imaging [13,39,40] as well as for conventional B-mode imaging, functional imaging, or three-dimensional imaging [29,30,41,42]. We also utilized a GPU for fast reconstruction of DWI and PWI. Although the display frame rate (real-time frame rate) of the system used in this study fell short of the acquisition frame rate, the process could be accelerated if the system supports a full channel reception and the online B-mode reconstruction software is further optimized, such as by using concurrent data copy and kernel execution. Indeed, using GeForce GTX 1080, it took 41.3 ms and 14.6 ms to compute a single synthesized (i.e., compounded) frame from channel data for DWI with 32 VSs and PWI with 32 PWs, respectively. Considering that parallel computing and data transfer technology is rapidly advancing, PWI with at least a 60-fps frame rate will soon be achievable.

4.3. Limitation of this Study

Despite the fast acquisition rates of PWI and DWI (Table 1), the display frame rates of PWI and DWI were lower than that of CF (26.6 fps, 10 fps, and 15 fps for CF, DWI, and PWI, respectively) in this study due to the lack of channel count and computing power of the system. The low display frame rates of PWI and DWI were mainly because they (1) need the full-aperture reception (128 channels) and (2) require 10–30 times more computations (Table 1) than CF. In CF, 64 channels were sufficient to receive echoes of a focused US beam from a straight scan line. However, DWI and PWI required a full 128-channel reception to collect echoes of a wide US beam reflected from a broad region. Unfortunately, the system supports only 64 reception channels and thus two times more transmit-receive sequences were performed to obtain 128-channel data with the 64-channel system. In addition, despite the use of a GPU for beamforming, the data transfer time and image reconstruction time for PWI and DWI was longer than the US echo acquisition time, which further decreases the display frame rates of PWI and DWI.

For the still image evaluation, the B-mode image was reconstructed offline from RF channel data stored and thus the frame rate of image was only affected by the limited number of receive channels. Hence, each still image of CF, PWI, and DWI was acquired at the rate of 31.9 fps, 63.85 fps, and 63.85 fps, respectively. For the video evaluation, the screen-captured videos were used, and thus the frame rate of video was the same as the display frame rate (26.6 fps, 10 fps, and 15 fps for CF, DWI, and PWI, respectively). Those limited frame rates of images and videos might have affected the evaluation results. Note that despite this unfavorable condition (lower frame rate than possible), DWI and PWI received better scores than CF. If the 128-channel acquisition is available, the motion artifacts in DWI and PWI would be further reduced. In addition, if the real-time reconstruction is realized and the reconstruction frame rate is close to the acquisition frame rate, the system noise presented in the B-mode image would also be reduced by frame averaging because more frames could be averaged within a fixed averaging time period the for image persistence.

Although we optimized parameters for each imaging (the focal depth and F-number of CF for a uniform resolution over depths, the VS positions of DWI for full-aperture transmission, and the PW angles and aperture size of PWI according to our previous study [35]), only a single set of parameters for each imaging technique was used to evaluate the image quality in this study. More exhaustive comparisons with changes in various parameters might be needed because the number and directions (or angles) of synthesized waves are major determinants of image quality in PWI and DWI.

5. Conclusions

We evaluated PWI against line-by-line CF imaging and another dynamic transmit focusing technique, DWI, through phantom and in vivo experiments. The phantom images and in vivo images and videos of the liver, kidney, and gallbladder of 30 healthy volunteers were assessed by three radiologists. PWI showed a significant enhancement ($p < 0.05$) of the spatial resolution, contrast, and noise and artifact reduction and presented a 4-fold higher acquisition rate compared to CF. PWI and DWI showed similar performance in in vivo images and video evaluations, although PWI showed a slightly better spatial resolution and DWI presented a slightly higher scores for other image qualities; however, significant differences were not found. With comparable performance to DWI, PWI can considerably lower the number of computations (approximately by 3 times in this study), which is the most challenging aspect for the realization of synthetic imaging. Therefore, we concluded that PWI represents a promising tool for abdominal ultrasonography by enhancing the spatial resolution and contrast from shallow to deep depths and realizing a higher acquisition rate.

Supplementary Materials: The following are available online at http://www.mdpi.com/1424-8220/20/19/5675/s1, Figure S1: We placed the outermost VSs ± 20 mm from the center so that the full aperture can be used for the DW transmission within the acceptance angle of the transducer element. Video S1: Representative in vivo abdominal US videos reconstructed by CF, DWI, and PWI (from left to right).

Author Contributions: Conceptualization, S.B., M.H.C. and T.-K.S.; methodology, S.B. and T.-K.S.; software, J.J. and S.B.; validation, S.B., J.J. and M.H.C.; formal analysis, S.B.; investigation, M.H.C., J.J.; resources, M.H.C. and T.-K.S.; data curation, S.B., J.J. and M.H.C.; writing—original draft preparation, S.B.; writing—review and editing, M.H.C. and T.-K.S.; visualization, S.B.; supervision, T.-K.S. and M.C; funding acquisition, T.-K.S. and M.H.C. All authors have read and agreed to the published version of the manuscript.

Funding: This work was supported by the R&D program (20007335, Development of Four Innovative Core Technologies to Commercialize Smart Endorectal Ultrasound Imaging System for General Use) funded by the Ministry of Trade, Industry & Energy (MOTIE, Korea).

Conflicts of Interest: The authors declare no conflict of interest. The funders had no role in the design of the study; in the collection, analyses, or interpretation of data; in the writing of the manuscript, or in the decision to publish the results.

References

1. Block, B. *Abdominal Ultrasound: Step by Step*, 2nd ed.; Thieme Medical Publishers: Stuttgart, Germany, 2012; ISBN 9783131383624.
2. Thomenius, K.E. Evolution of ultrasound beamformers. In Proceedings of the 1996 IEEE Ultrasonics Symposium Proceedings, San Antonio, TX, USA, 3–6 November 1996; pp. 1615–1622. [CrossRef]
3. Karaman, M.; O'Donnell, M. Synthetic aperture imaging for small scale systems. *IEEE Trans. Ultrason. Ferroelectr. Freq. Control.* **1995**, *42*, 429–442. [CrossRef]

4. Frazier, C.H.; O'Brien, W.D. Synthetic aperture techniques with a virtual source element. *IEEE Trans. Ultrason. Ferroelectr. Freq. Control.* **1998**, *45*, 196–207. [CrossRef] [PubMed]

5. Bae, M.-H.; Jeong, M.-K. A study of synthetic-aperture imaging with virtual source elements in B-mode ultrasound imaging systems. *IEEE Trans. Ultrason. Ferroelectr. Freq. Control.* **2000**, *47*, 1510–1519. [CrossRef] [PubMed]

6. Nagai, K. A new synthetic-aperture focusing method for ultrasonic B-scan imaging by the Fourier transform. *IEEE Trans. Sonics Ultrason.* **1985**, *SU-32*, 531–536. [CrossRef]

7. O'Donnell, M.; Thomas, L.J. Efficient synthetic aperture imaging from a circular aperture with possible application to catheter-based imaging. *IEEE Trans. Ultrason. Ferroelectr. Freq. Control.* **1992**, *39*, 366–380. [CrossRef]

8. Pedersen, M.H.; Gammelmark, K.L.; Jensen, J.A. In-vivo evaluation of convex array synthetic aperture imaging. *Ultrasound Med. Biol.* **2007**, *33*, 37–47. [CrossRef]

9. Gammelmark, K.L.; Jensen, J.A. Multielement synthetic transmit aperture imaging using temporal encoding. *IEEE Trans. Med. Imaging* **2003**, *22*, 552–563. [CrossRef]

10. Kim, C.; Yoon, C.; Park, J.-H.; Lee, Y.; Kim, W.H.; Chang, J.M.; Choi, B.I.; Song, T.-K.; Yoo, Y.-M. Evaluation of ultrasound synthetic aperture imaging using bidirectional pixel-based focusing: Preliminary phantom and in vivo breast study. *IEEE Trans. Biomed. Eng.* **2013**, *60*, 2716–2724. [CrossRef]

11. Kortbek, J.; Jensen, J.A.; Gammelmark, K.L. Sequential beamforming for synthetic aperture imaging. *Ultrasonics* **2013**, *53*, 1–16. [CrossRef]

12. Exploring n SIGHT Imaging—A Totally New Architecture for Premium Ultrasound, Philips White Paper. Available online: https://www.usa.philips.com/healthcare/resources/feature-detail/nsight (accessed on 15 September 2020).

13. cSound 3.0—Taking the Processing Power and Patient Care Advantages to a New Level with Artificial Intelligence, GE White Paper. Available online: https://landing1.gehealthcare.com/EU-EU-US19Q1-CVUS-Patient-Care-Elevated.html (accessed on 15 September 2020).

14. Jensen, J.A.; Nikolov, S.I.; Gammelmark, K.L.; Pedersen, M.H. Synthetic aperture ultrasound imaging. *Ultrasonics* **2006**, *44*, e5–e15. [CrossRef]

15. Papadacci, C.; Pernot, M.; Couade, M.; Fink, M.; Tanter, M. High-contrast ultrafast imaging of the heart. *IEEE Trans. Ultrason. Ferroelectr. Freq. Control.* **2014**, *61*, 288–301. [CrossRef] [PubMed]

16. Zhao, F.; Tong, L.; He, Q.; Luo, J. Coded excitation for diverging wave cardiac imaging: A feasibility study. *Phys. Med. Biol.* **2017**, *62*, 1565–1584. [CrossRef]

17. Zhang, M.; Varray, F.; Besson, A.; Carrillo, R.E.; Viallon, M.; Garcia, D.; Thiran, J.P.; Friboulet, D.; Liebgott, H.; Bernard, O. Extension of Fourier-based techniques for ultrafast imaging in ultrasound with diverging waves. *IEEE Trans. Ultrason. Ferroelectr. Freq. Control.* **2016**, *63*, 2125–2137. [CrossRef] [PubMed]

18. Hasegawa, H.; Kanai, H. High-frame-rate echocardiography using diverging transmit beams and parallel receive beamforming. *J. Med. Ultrason.* **2011**, *38*, 129–140. [CrossRef] [PubMed]

19. Tong, L.; Gao, H.; Choi, H.F.; D'hooge, J. Comparison of conventional parallel beamforming with plane wave and diverging wave imaging for cardiac applications: A simulation study. *IEEE Trans. Ultrason. Ferroelectr. Freq. Control.* **2012**, *59*, 1654–1663. [CrossRef] [PubMed]

20. Poree, J.; Posada, D.; Hodzic, A.; Tournoux, F.; Cloutier, G.; Garcia, D. High-frame-rate echocardiography using coherent compounding with Doppler-based motion-compensation. *IEEE Trans. Med. Imaging* **2016**, *35*, 1647–1657. [CrossRef] [PubMed]

21. Provost, J.; Papadacci, C.; Arango, J.E.; Imbault, M.; Fink, M.; Gennisson, J.-L.; Tanter, M.; Pernot, M. 3D ultrafast ultrasound imaging in vivo. *Phys. Med. Biol.* **2014**, *59*, L1–L13. [CrossRef]

22. Kang, J.; Go, D.; Song, I.; Yoo, Y. Wide field-of-view ultrafast curved array imaging using diverging waves. *IEEE Trans. Biomed. Eng.* **2020**, *67*, 1638–1649. [CrossRef]

23. Chang, J.H.; Song, T.-K. A new synthetic aperture focusing method to suppress the diffraction of ultrasound. *IEEE Trans. Ultrason. Ferroelectr. Freq. Control.* **2011**, *58*, 327–337. [CrossRef]

24. Zhao, H.; Song, P.; Urban, M.W.; Greenleaf, J.F.; Chen, S. Shear wave speed measurement using an unfocused ultrasound beam. *Ultrasound Med. Biol.* **2012**, *38*, 1646–1655. [CrossRef]

25. Bercoff, J.; Tanter, M.; Fink, M. Supersonic shear imaging: A new technique for soft tissue elasticity mapping. *IEEE Trans. Ultrason. Ferroelectr. Freq. Control.* **2004**, *51*, 396–409. [CrossRef] [PubMed]

26. Montaldo, G.; Tanter, M.; Bercoff, J.; Benech, N.; Fink, M. Coherent plane-wave compounding for very high frame rate ultrasonography and transient elastography. *IEEE Trans. Ultrason. Ferroelectr. Freq. Control.* **2009**, *56*, 489–506. [CrossRef] [PubMed]

27. Udesen, J.; Gran, F.; Hansen, K.L.; Jensen, J.A.; Thomsen, C.; Nielsen, M.B. High frame-rate blood vector velocity imaging using plane waves: Simulations and preliminary experiments. *IEEE Trans. Ultrason. Ferroelectr. Freq. Control.* **2008**, *55*, 1729–1743. [CrossRef] [PubMed]

28. Bercoff, J.; Montaldo, G.; Loupas, T.; Savery, D.; Mézière, F.; Fink, M.; Tanter, M. Ultrafast compound Doppler imaging: Providing full blood flow characterization. *IEEE Trans. Ultrason. Ferroelectr. Freq. Control.* **2011**, *58*, 134–147. [CrossRef] [PubMed]

29. Bae, S.; Park, J.; Song, T. Contrast and volume rate enhancement of 3D ultrasound imaging using aperiodic plane wave angles: A simulation study. *IEEE Trans. Ultrason. Ferroelectr. Freq. Control.* **2019**, *66*, 1731–1748. [CrossRef]

30. Flesch, M.; Pernot, M.; Provost, J.; Ferin, G.; Nguyen-Dinh, A.; Tanter, M.; Deffieux, T. 4D in-vivo ultrafast ultrasound imaging using a row-column addressed matrix and coherently-compounded orthogonal plane waves. *Phys. Med. Biol.* **2017**, *62*, aa63d9. [CrossRef]

31. Yang, M.; Sampson, R.; Wei, S.; Wenisch, T.F.; Fowlkes, B.; Kripfgans, O.; Chakrabarti, C. *High Volume Rate, High Resolution 3D Plane Wave Imaging*; IEEE International Ultrasonics Symposium: Chicago, IL, USA, 2014; pp. 1253–1256.

32. Jensen, J.; Stuart, M.; Jensen, J. Optimized plane wave imaging for fast and high quality ultrasound imaging. *IEEE Trans. Ultrason. Ferroelectr. Freq. Control.* **2016**, *63*, 1922–1934. [CrossRef]

33. Bae, S.; Song, T.-K. Methods for grating lobe suppression in ultrasound plane wave imaging. *Appl. Sci.* **2018**, *8*, 1–8. [CrossRef]

34. Bae, S.; Kim, P.; Kang, J.; Song, T. An optimized plane wave synthetic focusing imaging for high-resolution convex array imaging. In Proceedings of the 2015 IEEE International Ultrasonics Symposium (IUS), Taipei, Taiwan, 21–24 October 2015.

35. Bae, S.; Kim, P.; Song, T. Ultrasonic sector imaging using plane wave synthetic focusing with a convex array transducer. *J. Acoust. Soc. Am.* **2018**, *144*, 2627–2644. [CrossRef]

36. Zhao, J.; Wang, Y.; Zeng, X.; Yu, J.; Yiu, B.Y.S.; Yu, A.C.H. Plane wave compounding based on a joint transmitting-receiving adaptive beamformer. *IEEE Trans. Ultrason. Ferroelectr. Freq. Control.* **2015**, *62*, 1440–1452. [CrossRef]

37. Guerra, A.L.; Gidel, T.; Vezzetti, E. Toward a common procedure using likert and likert-type scales in small groups comparative design observations. In Proceedings of the International Design Conference, Dubrovnik, Croatia, 2016; pp. 23–32.

38. Riffenburgh, R.H. Tests on Ranked Data. In *Statistics in Medicine*; Riffenburgh, R.H., Ed.; Academic Press: Cambridge, MA, USA, 2006; pp. 281–303, ISBN 978-0-12-088770-5.

39. Yiu, B.Y.S.; Tsang, I.K.H.; Yu, A.C.H. GPU-based beamformer: Fast realization of plane wave compounding and synthetic aperture imaging. *IEEE Trans. Ultrason. Ferroelectr. Freq. Control.* **2011**, *58*, 1698–1705. [CrossRef] [PubMed]

40. Li, Y.F.; Li, P.C. Software beamforming: Comparison between a phased array and synthetic transmit aperture. *Ultrason. Imaging* **2011**, *33*, 109–118. [CrossRef] [PubMed]

41. Lok, U.W.; Li, P.C. Transform-based channel-data compression to improve the performance of a real-time GPU-based software beamformer. *IEEE Trans. Ultrason. Ferroelectr. Freq. Control.* **2016**, *63*, 369–380. [CrossRef] [PubMed]

42. Huang, Q.; Zeng, Z. A review on real-time 3D ultrasound imaging technology. *BioMed Res. Int.* **2017**, *2017*, 6027029. [CrossRef]

Article

Feasibility Study of Precise Balloon Catheter Tracking and Visualization with Fast Photoacoustic Microscopy

Jahae Kim [1,2,†], **Thi Thao Mai** [2,†], **Jin Young Kim** [3], **Jung-Joon Min** [4], **Chulhong Kim** [3] and **Changho Lee** [2,4,*]

1   Department of Nuclear Medicine, Chonnam National University Medical School & hospital, 42 Jebong-ro, Dong-gu, Gwangju 61469, Korea; jhbt0607@hanmail.net
2   Department of Artificial Intelligence Convergence, Chonnam National University, 77 Yongbong-ro, Buk-gu, Gwangju 61186, Korea; 196286@jnu.ac.kr
3   Departments of Creative IT Engineering and Electrical Engineering, Pohang University of Science and Technology (POSTECH), 77 Cheongam-ro, Nam-gu, Pohang, Gyeongbuk-do 37673, Korea; ronsan@postech.ac.kr (J.Y.K.); chulhong@postech.edu (C.K.)
4   Department of Nuclear Medicine, Chonnam National University Medical School & Hwasun Hospital, 264, Seoyang-ro, Hwasun-eup, Hwasun-gun, Jeollanam-do 58128, Korea; jjmin@jnu.ac.kr
*   Correspondence: ch31037@jnu.ac.kr; Tel.: +82-61-379-2885
†   Equally contributed this works.

Received: 10 August 2020; Accepted: 28 September 2020; Published: 29 September 2020

Abstract: Correct guiding of the catheter is a critical issue in almost all balloon catheter applications, including arterial stenosis expansion, coronary arterial diseases, and gastrointestinal tracking. To achieve safe and precise guiding of the balloon catheter, a novel imaging method with high-resolution, sufficient depth of penetration, and real-time display is required. Here, we present a new balloon catheter guiding method using fast photoacoustic microscopy (PAM) technique for precise balloon catheter tracking and visualization as a feasibility study. We implemented ex vivo and in vivo experiments with three different medium conditions of balloon catheter: no air, air, and water. Acquired cross-sectional, maximum amplitude projection (MAP), and volumetric 3D PAM images demonstrated its capability as a new imaging guiding tool for balloon catheter tracking and visualization.

Keywords: balloon catheter; photoacoustic imaging; image guiding

---

1. Introduction

A balloon catheter is commonly used for opening the narrow or blocked area in the body. The balloon catheter technique has been mostly applied in the field of coronary artery diseases. In stenotic coronary vessels, the contracted balloon enters the affected coronary artery. The balloon is inflated to widen the artery, squashing plaques against the artery wall. It can improve myocardial blood flow in a localized stenotic lesion when the deflated balloon is removed [1]. Similarly, it has been applied to peripheral arterial disease [2], extracranial or intracranial cerebral vascular lesion [3], stricture of the gastrointestinal tract [4], eustachian tube dysfunction [5], and ureteropelvic junction obstruction [6]. Unlike its function as an expander of the narrowed area, the balloon catheter can also function as a blocker at the site of ruptured or torn vessels. If a large amount of bleeding occurs from the aorta, the inflation of resuscitative balloon catheters can lead to blocking blood leakage from the perforated site [7]. Such the balloon catheter procedure can increase blood flow to the heart and brain, thus preventing cardiovascular collapse [8]. Recently, given a less-invasiveness of catheterization, a new hybrid operation which is combining open surgery and vascular intervention has been used in many cases such as heart disease [9], liver injury [10], subclavian artery injury [11], and lower extremity vascular injury [12]. The hybrid operation leads to minimizing the extent, the duration, and the cost of the traditional operation. In the hybrid operation, the balloon catheter technique still works as the main tool for vasculature intervention.

Whether the purpose of a balloon catheter is to open a blockage or block a torn site, it is necessary to check that the balloon catheter enters the body and that the balloon is located at the correct place in the body. To confirm the balloon's location, X-ray fluoroscopy with contrast dye is regarded as a gold standard. However, X-ray fluoroscopy devices cause substantial radiation exposure to patients and operators [13]. Furthermore, iodine-containing contrast is essential for fluoroscopy-guided angiography. It has been shown the contrast agent has adverse clinical consequences due to its high osmolality [14,15]. In addition, the high viscosity of the contrast agent takes more time to complete the interventional procedure [16]. Due to its relatively poor spatial resolution, X-ray fluoroscopy is limited to show the precise location of the catheter. To overcome these limitations of X-ray fluoroscopy, there have been experimental studies of balloon catheters using interactive MRI [17], carbon dioxide digital subtraction angiography (CO_2-DSA), [18] or ultrasound imaging (USI) [19]. However, MRI is not a proper solution to use in a real-surgical environment due to its huge size and limitation in the use of metal surgical tools known to interfere with the magnetic field. In addition, it is relatively expensive. CO_2-DSA needs to use additional agents to visualize vasculatures. Although USI is a better option to use in urgent and surgical conditions, it has difficulties showing the catheter clearly, due to its angle-dependent artifact. Furthermore, these techniques do not have high enough resolution to visualize images below 50 µm. This is a bottleneck of precise catheter guiding. In particular, although many operating sites use intraoperative imaging, such as fluoroscopy for vascular intervention parts in hybrid operation, high resolution and rapid follow-up imaging techniques are still required to track the catheter guiding precisely.

In recent years, photoacoustic imaging (PAI) has been spotlighted as a promising bio-imaging technology with the development of high-performance PAI laser source and rapid computing technology. This technique is originated from the photoacoustic (PA) effect discovered by Alexander Graham Bell in 1880. When a nano-second pulsed laser shines on the targeting sample, it undergoes absorption of laser energy and thermal-elastic expansion. As a result of this process, broad acoustic waves appear. These waves can be captured by common ultrasound transducers. Finally, two or three-dimensional images can be obtained via reconstruction algorithms. Thanks to less scattering of ultrasound in biological tissue, PAI can achieve deep tissue imaging while maintaining ultrasonic resolution [20–24]. Furthermore, depending on aiming applications, PAI can provide multiscale images from several nanometers to several centimeters selectively by controlling systemic specifications between optical and ultrasonic [25]. Normally, PA microscopy (PAM) can provide superior spatial resolution and sensitivity in relatively shallowed regions with focusing illuminated laser beam or detecting ultrasonic beam [26–29]. In contrast, PA tomography (PAT) enables to achieve deep tissue imaging with multi-arrayed transducers and broad laser illumination without any mechanical scanning [30–32]. Additionally, based on intrinsic absorption composites in the body, PAI can be used to visualize not only structural parameters including vasculature networks, location of melanoma, the structure of tendons, plaques and so on, but also functional parameters including blood velocity, oxygen saturated level, and metabolism ratios [33,34]. Using these advantages, PAI has been widely utilized to improve basic/preclinical studies and clinical translation in various areas, such as oncology, dermatology, ophthalmology, neurology and so on [35–37].

PAI is also utilized as a powerful tool for a wide range of image-guided applications. For minimally invasive biopsy of sentinel lymph nodes (SLNs), a handheld array probe-based PAI system has demonstrated its performance of visualizing the ability of needle insertion and guiding to SLNs dyed methylene blue (MB) and indocyanine green (ICG) [38,39]. A similar clinical PAI system has been successfully developed to detect the catheter with higher sensitivity than USI [40]. These approaches have suppressed the drawbacks of USI guiding by clear visualization of the metal part of the needle and the catheter. PAI is also used to effectively delineate cancerous regions for photothermal therapy monitoring and drug delivery [41,42]. By showing the margin of cancer, PAI can enhance the therapeutic effect. Moreover, thanks to its real-time displaying capability, relatively deep penetration depth, and high spatial resolution, PAI has shown the feasibility of surgical imaging-guided applications such as a precise needle guiding to a single vessel, incision of melanoma tumor, drug injection into tumor, spinal fusing surgery and so on [43–47].

In this study, we conceptually demonstrated the capability of PAI to track and visualize the correct location of the balloon catheter in the blood vessel under open surgery conditions. Especially, we identify the location of the catheter in the procedure of the endovascular intervention part constituting hybrid operation and the location of the balloon without any contrast agents. Using the fast-MEMS based PAM (f-MEMS-PAM) system, three-dimensional and cross-sectional PAM images were successfully obtained in real-time. In particular, by setting up three injection medium conditions (i.e., no air, air and water) in the balloon, the balloon catheter was selectively tracked and visualized.

2. Materials and Methods

2.1. Balloon Catheter

Figure 1a shows the balloon catheter and the inflation pump used in this study. Figure 1b shows an enlarged balloon catheter tip. A 2.0 × 20-mm balloon dilatation catheter (Medtronic, Sprinter Legend®) for coronary balloon angioplasty was used. The balloon catheter consists of a balloon with an optically transparent polyether block amide and a core made by polyvinyl chloride (PVC) marked with platinum. The balloon was inflated to a pressure of 6 atm using an inflation device.

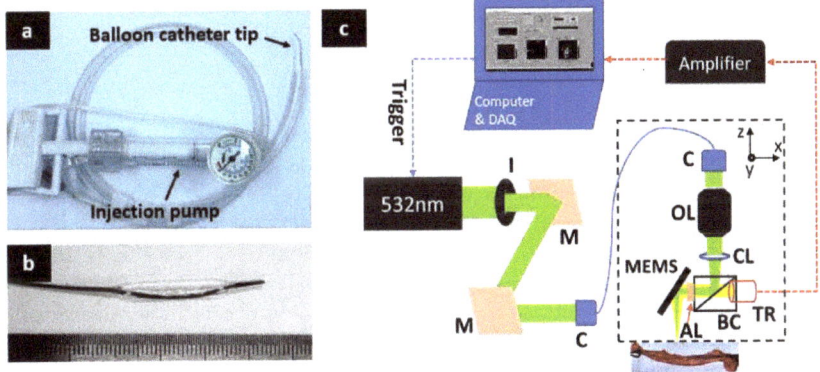

Figure 1. (**a**) Photography of the balloon catheter and the inflation pump, (**b**) Enlarged photograph of the tip of the balloon catheter, (**c**) Schematic of the fast-MEMS-based photoacoustic microscopy system for tracking and visualization of the balloon catheter. M, mirror; c, collimator; OL, objective lens, CL, correction lens, TR, transducer; BC, beam combiner; AL, acoustic lens; I, iris; MEMS, Micro Electro Mechanical Systems.

2.2. Fast Photoacoustic Microscopic Imaging

As shown in Figure 1c, the f-MEMS-PAM system was used to visualize the balloon catheter. In order to generate PA wave with a tiny beam size, the nano-second pulsed laser beam was emitted from a diode laser (SPOT-10-200-532, Elforlight, Daventry, UK) with a center wavelength of 532 nm, a pulsed width of 6 ns, and a repetition rate of 10 kHz. To achieve in vivo imaging, we used the laser energy around 10 mJ/cm² (Supplemental Material), which satisfied the ANSI limits (Max 20 mJ/cm² at visible light). The laser beam was then reshaped by an iris (SM1D12, Thorlabs, Newton, NJ, USA) which was held by a right-angle kinematic mount (KCB1E, Thorlabs, Newton, NJ, USA) that worked as a spatial filter. Two mirrors were inclined at a 45-degree angle of vertical and horizontal planes to adjust the direction of the beam to the collimator (F280APC-A, Thorlabs) before inserting into a single-mode optical fiber (P1-405BPM-FC-1, Thorlabs). The diverged beam from the optical fiber became the collimated beam after passing through the second collimator (F260APC-A, Thorlabs). The collimated beam was again focused by an objective lens (AC254-060-A, Thorlabs). It then penetrated through the beam combiner and projected into the sample. The beam combiner in front of the unfocused ultrasound transducer constituted by a normal prism and aluminum-coated prism could confocal align the laser beam with the ultrasound focus. A plano-concave lens (NT45-010, Edmund, Tucson, AZ, USA) as an acoustic lens was utilized to focus the acoustic beam used to support acoustic focusing on the transducer. Volumetric scanning was conducted with a 1-axis MEMS scanner (OpitchoMS-001, Opticho Inc., Ltd., Pohang, Korea) and a linear stepper motor stage (L-509-10SD00, PI) which achieved 25 Hz B-scan imaging controlled by a data acquisition (DAQ) board (PCIe-6321, NI instruments, Austin, TX, USA). The PA signal went back to the beam combiner. It was immediately detected by a high-frequency ultrasonic transducer (V214-BC-RM, 50 MHz, Olympus, Tokyo, JPN) and an RF-amplifier (ZX60-3018G-S+, Mini-Circuit, Brooklyn, NY, USA). The high-speed digitizer (ATS9371, AlazarTech Pointe-Claire, QC, Canada) digitalized PA signals with 12 bit and 1 GS/s sampling rates from the transducer. The whole system was controlled by a LabView program (NI instruments, Austin, TX, USA). Our system can achieve high-resolution imaging with 12 μm for lateral resolution and 45 μm for axial resolution [48]. Single B-scan and volumetric 3D PAM images were acquired at 0.25 and 12 s, respectively, which is a relatively fast imaging acquisition rate compared to other group reports [33,47–49]. Maximum projection amplitude (MAP) and cross-sectional PAM images were acquired and analyzed with MATLAB (R2016a, Mathworks, Natick, MA, USA). Furthermore, volumetric 3D PAM images and movies were reconstructed and produced with Amira program (Amira 6, FEI).

2.3. Automatic Surface Removing Algorithm

Due to the strong light absorption of blood at a wavelength of 532 nm, PAI can show blood vessel networks without any contrast agent [50]. Unfortunately, in the current systemic condition, the catheter was fully surrounded by blood vessels. These blood vessels usually generate the biggest signals. As a result, the MAP PAM image only shows the image of blood vessels. For delineating the correct location of the balloon catheter in blood vessels, these unexpected PA signals from blood vessels should be removed. Normally, the largest signals that appear on B-scan images are those for upper walls of blood vessels and catheters, in which the signal for blood vessels is larger. When the catheter is quite close to the walls of blood vessels, these signals will overlap. If the catheter's signal is too small, signals of upper and lower walls of the blood vessel are the two largest signals. Therefore, it is critical to identify these two signals from each A-scan for removing unwanted blood vessel signals. In order to distinguish between the upper blood vessel signal and the catheter signal, the depth of the signal should be considered. As shown in Figure 2b, the signal of the catheter is always deeper in all cases. Thus, in the two largest signals, the signal with a greater depth will be the catheter. In the case of the two largest signals belonging to blood vessels, a distance threshold is needed. The threshold should be smaller than the width of the blood vessel. If the distance between the two signals is less than the threshold, the second signal will belong to the catheter; otherwise, it will belong to the lower blood vessel wall. In this study, we used a threshold of 300 pixels. Random noise also should be considered. Noise may become the largest signal and appear on the MAP image. Interfering signals are often concentrated on the upper or lower edge of the MAP image. They can be eliminated by removing all large signals at the top and bottom edges of the image. The algorithm is summarized in Figure 2a. Figure 2b(i,ii) show the B-scan image and A-scan profile before applying the removing process, respectively. The biggest signal and the second biggest signal belong to blood vessel and catheter, respectively. The blood vessel signal is almost removed after adapting the surface removing process, as shown in Figure 2b(iii). In Figure 2b(iv), the signal of the blood vessel area becomes the constant line, and the biggest signal belongs to the catheter. Even if there is another signal next to the catheter signal, it will not affect the image reconstruction process because only the largest signal is selected, and the MAP image will only show the signal of the catheter. In this case, the signal from the surface and the catheter core has a sufficient distance. So, the removal process works well. Unfortunately, as shown in Figure 2c(i,ii), some regions show that the catheter is located very close to the surface of the blood vessel. In this case, there is trouble in removing the surface successfully. Although we enable to extract the only core region, we should consider the loss of the PA signals from the core as shown in Figure 2c(iii,iv).

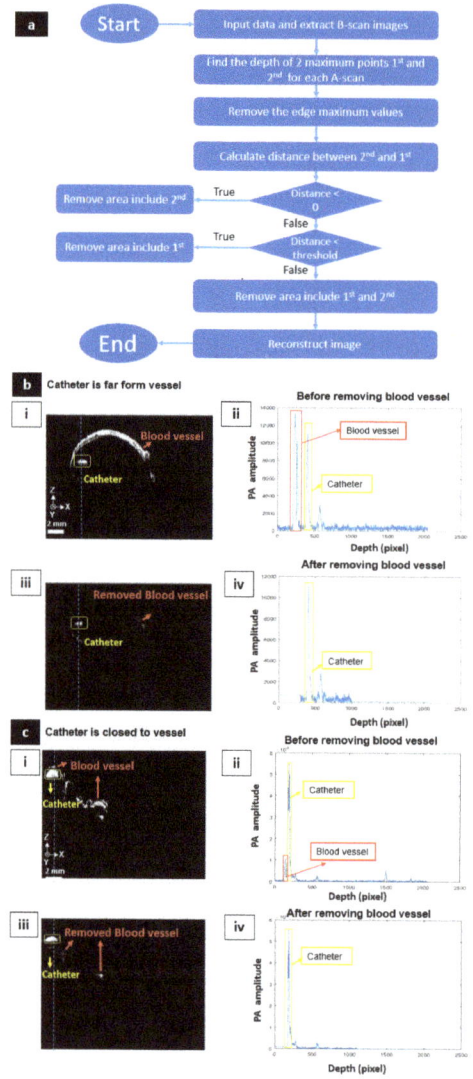

Figure 2. Surface removing process. (**a**) Flow chart showing the automatic surface removing algorithm. (**b**,**c**) Before/after applying the surface removing process when the catheter is far from the vessel and close to the vessel, respectively. (**i**,**ii**) Cross-sectional photoacoustic (PA) image and the selected depth-resolved A-scan profile before adapting the surface removing process, respectively. (**iii**,**iv**) Cross-sectional PA image and depth-resolved A-scan profile after adapting the surface removing process, respectively.

2.4. Animal Preparing

A 12-week-old male Sprague-Dawley rat weighing 380 g (Samtako, Korea) was used after one week of acclimation. The study protocol was approved by the Institutional Animal Ethical Committee of Chonnam National University Hospital. A rat was anesthetized with an intraperitoneal injection of Ketamine (80 mg/kg)/Xylazine (12 mg/kg). Skin and soft tissue incisions were made from the xiphoid process to both the femoral region. After exposing the abdominal aorta with around 1.1 mm diameter and both iliac arteries, proximal ligation at the highest abdominal aorta was obtained using a 4–0 silk suture. Distal ligations of both common iliac arteries approximately 1 cm from the iliac bifurcation were prepared in the same manner. Arteriotomy was performed through the upper half of the vessel circumference with surgical scissors at a location of 1–2 mm close to the distal ligation site. After removing arterial blood using gauzes, a guidewire was inserted through the arteriotomy and advanced to the proximal ligation site. The balloon catheter was then passed over a guidewire into the abdominal aorta. Beginning at the distal ligation part of the common iliac artery, the balloon was totally deflated. When the balloon was positioned at the abdominal aorta, PAM images were acquired in three medium conditions in the balloon: no air, air and water. After sacrificing the rat with overdose of the anesthetic mixture, we cut out of the aorta, and the inside of the aorta was washed three or more times with normal saline to confirm that there was no blood or thrombus in the blood vessels. After that, we conducted the same process of in vivo experiment with PAM.

3. Results

3.1. PA signal Characteristics of The Balloon Catheter

As shown in Figure 3a, PA signals of whole blood and the catheter core components, such as PVC and platinum, were measured under the same PAM setup at 532 nm wavelength. The used laser energy is approximately 3 mJ/cm^2. Compared to whole blood and platinum, PVC exhibited approximately 38% and 83% higher PA signals at the 532 nm wavelength. Thus, even though the use of 532 nm wavelength laser disturbed the detection of the catheter core directly because of the high PA signal from the blood vessel wall, we estimated that the catheter core also could be detectable due to the relatively high absorption of PVC. Figure 3b shows the PA signal ratio of the catheter core versus the blood vessel surface at the different distance between the catheter core and the surface of the blood vessel. We conducted ex vivo PA signal test with the extracted blood vessel and the inserted balloon catheter. By injecting water slowly into the balloon, we changed the distance between the catheter core and the blood vessel surface. The used laser energy is approximately 10 mJ/cm^2. As shown in Figure 3b, the maximum visible depth is about 1110 µm with a very low signal of the catheter core. The * mark indicates the invisible depth of the catheter core which is around 1370 µm. By increasing the distance between the catheter core and the blood vessel surface, the PA signal ratio was decreased exponentially.

3.2. The Balloon Catheter Visualization

As shown in Figure 4, MAP, cross-sectional and volumetric 3D PAM images of the catheter were obtained with different injection medium conditions for the balloon catheter (i.e., no air injection, air injection and water injection). When there was no air inside, the image of the catheter core was displayed as shown in Figure 4a(ii–iv) because of its black color. There were two locations (two junction points between the core of the catheter and the balloon due to air remaining) without signal on the PAM MAP, causing a loss of PA signal for the image. Air was then injected into the catheter as shown in Figure 4b(i). In Figure 4b(ii–iv), the entire PA signal of the catheter in the air area disappeared due to a strong loss of acoustic attenuation in air. The only non-balloon area was visualized. In Figure 4c(i), water was pumped into the balloon while still keeping a small air bubble (AB) inside the catheter. The core of the catheter was displayed in the water but completely lost in the air bubble area (Figure 4c(ii–iv)). The gold area (G) in the middle of the catheter core was also visualized in Figure 4a(ii,iii), Figure 4c(ii,iii) with relatively low PA signal level. The PA signal of the catheter core showed 231.96% higher than that in the G region for no air injection case and 248.25% for the water injection case. All these results were also shown in volumetric 3D images (Figure 4a(iv), Figure 4b(iv), and Figure 4b(iv)) and movies (Movie 1, Movie 2, and Movie 3). This simple experiment was performed to pre-check the ability of PAM to visualize the catheter at different injection medium conditions before conducting in-vivo and ex-vivo experiments. These obtained results demonstrated the feasibility of using the PAM system to monitor the balloon catheter with high-resolution in real-time.

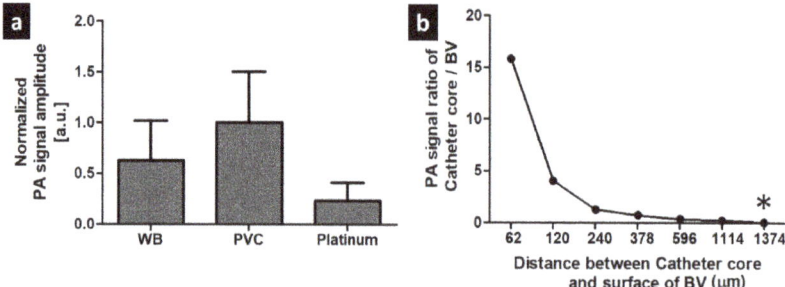

Figure 3. PA characteristics of the balloon catheter. (**a**) PA signals comparison among whole blood (WB), polyvinyl chloride (PVC), and platinum (**b**) PA signal ratio of the catheter core versus the blood vessel (BV) surface at the different distance between the catheter core and the BV surface. * mark indicates the invisible catheter core.

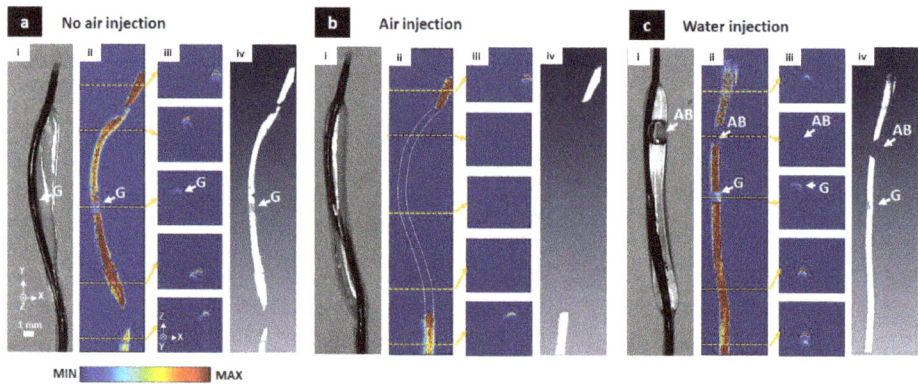

Figure 4. Maximum amplitude projection (MAP), cross-sectional and volumetric 3-dimentional photoacoustic microscopy (3D PAM) images of the balloon catheter with different injection conditions. (**a**) No air injection condition, (**b**) Air injection condition, (**c**) Water injection condition, (**i**) Photographs of the balloon catheter, (**ii**) Corresponding MAP PAM images, (**iii**) Five selected cross-sectional PAM images, (**iv**) Corresponding volumetric 3D PAM images (Movie 1, Movie 2, Movie 3). AB, air bubble; G, gold.

3.3. In Vivo Visualization of the Balloon Catheter

The catheter was inserted into the aorta of an anesthetized rat carefully. The results are shown in Figure 5. Under no air condition (Figure 5a), the whole blood vessel area was shown in the MAP PAM image (Figure 5a(ii)) without the surface removing process. In cross-sectional PAM images (Figure 5a(iii)), even the wall of the blood vessel showed the highest PA signal. Because the aorta receives blood through the small-sized microvessel network (called as vasa vasorum), the wall of blood vessel normally generates high PA signals regardless of the internal state of the aorta. The core of the catheter was also observed with a relatively low PA signal level. After applying the surface removing process, the core of the catheter was observed as shown in Figure 5a(iv). These results also were confirmed by volumetric 3D PAM images and a 3D rendering movie (Figure 5a(vi,vii), Movie 4). The signal marked with an oval white dot line (CR, contact region) was the most prominent part because it was adjacent to the blood vessel wall. Due to the actions of arterial muscles that changed the position of the inner catheter, this was easily identified when compared to the balloon catheter in Figure 4a. The lower part of the catheter core far away from the blood vessel wall was significantly reduced in the signal which was less 57.80% than the corresponding part of the blood vessel. The G region maintained the PA signal level because the core of the catheter was very close to the blood surface. Under air injection condition, the MAP image and corresponding cross-sectional images are shown in Figure 5b(ii,iii), respectively. In Figure 5b(iv), after removing the blood vessel, the signal again appeared in the CR area and the non-balloon area. The rest covered by the air medium did not show any signal of the catheter core. Only the background was shown. The same results also were confirmed in 3D rendering data (Figure 5b(vi,vii), Movie 5). Figure 5b(iv,v) show the MAP image and cross-sectional images under water injection condition, respectively. Water was the ideal medium for acoustic signal propagation. Thus, the signal of the catheter was fully collected in the case of water injection not only before removing the surface (Figure 5c(ii,iii)), but also after blood vessel removing (Figure 5c(iv,v)). We also observe the same results from volumetric 3D data (Figure 5c(vi,vii), Movie 6). The pumping of water, however, increased the distance of the scanner with the catheter. Thus, the received signal was quite weak compared to the biggest signals of no air injection case. The PA signal of the catheter core was less 78.66% than blood surface signals. The G region even had a signal 71.24% (no air injection) and 33.62% (water injection) lower than the black color of the core, although it could be easily visualized under no air condition and water injection.

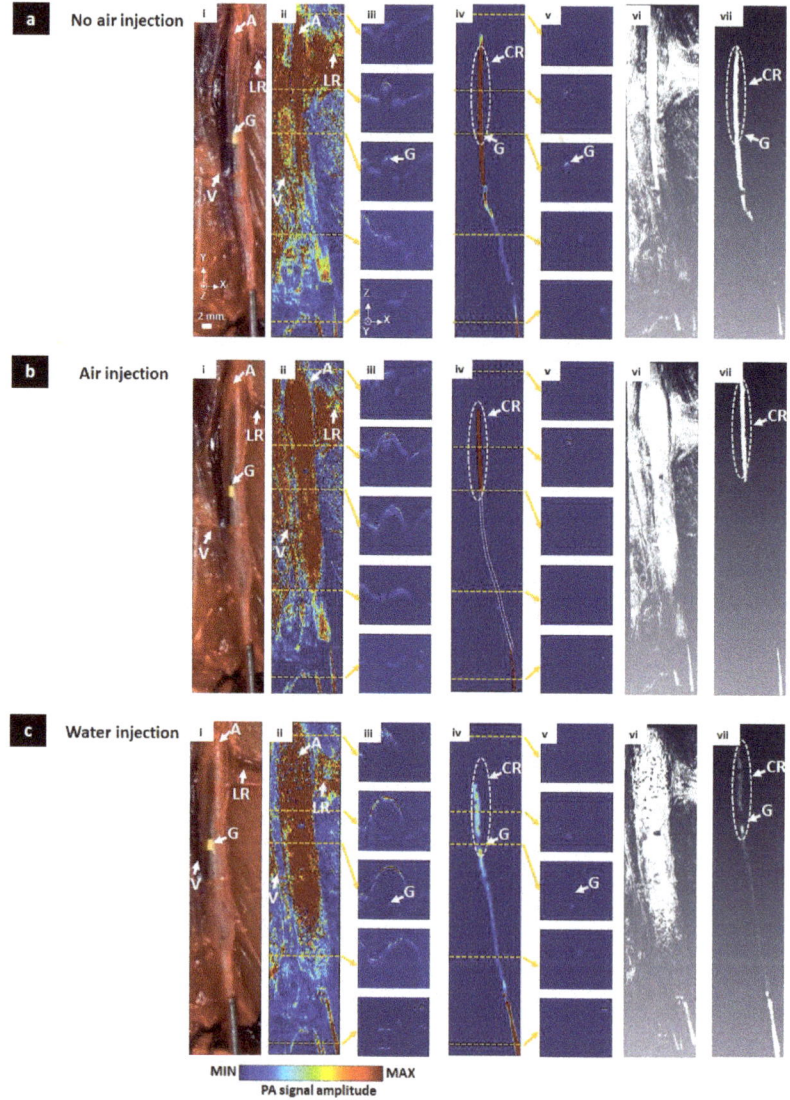

Figure 5. In vivo MAP, cross-sectional, volumetric 3D PAM images of the balloon catheter with different injection conditions. (**a**) No air injection condition, (**b**) Air injection condition, (**c**) Water injection condition, (**i**) Photographs of the balloon catheter, (**ii**) MAP PAM images without the surface removing process, (**iii**) Five selected cross-sectional PAM images without the surface removing process, (**iv**) MAP PAM images with the surface removing process, (**v**) Five selected cross-sectional PAM images with the surface removing process, (**vi,vii**) Volumetric 3D PAM images before and after the surface removing process (Movie 4, Movie 5, Movie 6). G, gold; CR, contacted region; A, aorta; LR, left renal artery; V, vein.

3.4. Ex Vivo Visualization of the Balloon Catheter

The aorta with the inner core catheter was cleverly separated from the body of the mouse. The position of the balloon catheter (Figure 6a(i)) was similar to that shown in Figure 4a. In Figure 6a(ii), the core catheter signal was outstanding compared to the remaining parts. After removing the surface, the catheter was fully shown in Figure 6a(iv). All cross-sectional images (Figure 6a(v)) included the signal. The PA signal level is almost maintained due to the short distance between the blood vessel and the catheter. Volumetric 3D data also showed the same results (Figure 6a(vi,vii), Movie 7). Results under air injection conditions, including MAP, cross-sectional, and volumetric 3D images before and after the surface removing process were shown in Figure 6b(ii–vii, Movie 8). Similar to the in vivo case, only the nearest surface area of the catheter core indicated by the CR area was detected regardless of medium condition. PA signal of another region was not detected. Finally, water injection was conducted as shown in Figure 6c(i). In Figure 6c(ii), the blood vessel signal after water injection overwhelmed the remaining signals, causing the signal of the core to be greatly reduced. However, the whole catheter including the G area could be monitored and visualized as shown in Figure 6c(iv). The PA signal of the catheter core was 60.64% less than blood surface signals. These results were also observed in Figure 6c(vi,vii) and Movie 9.

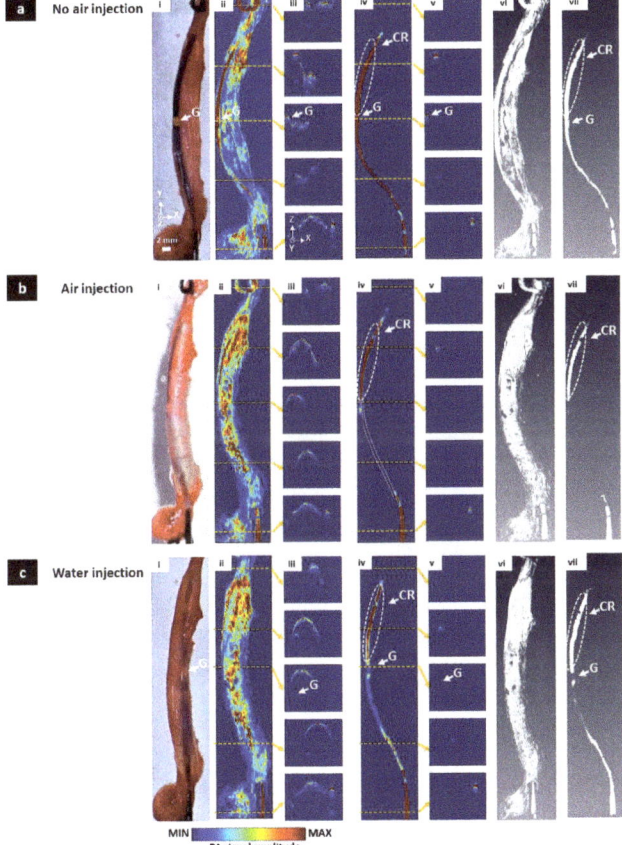

Figure 6. Ex vivo MAP, cross-sectional, and volumetric 3D PAM images of the balloon catheter with different injection conditions. (**a**) No air injection condition, (**b**) Air injection condition, (**c**) Water injection condition, (**i**) Photographs of the balloon catheter, (**ii**) MAP PAM images without the surface removing process, (**iii**) Five selected cross-sectional PAM images without the surface removing process, (**iv**) MAP PAM images with the surface removing process, (**v**) Five selected cross-sectional PAM images with the surface removing process, (**vi,vii**) Volumetric 3D PAM images before and after the surface removing process (Movie 7, Movie 8, Movie 9). G, gold; CR, contacted region.

4. Discussion

We demonstrated the feasibility of tracking and visualizing of the balloon catheter using the PAM technique under open surgery condition. Our approach successfully provided the precise tracking location of the catheter, as well as the inserted catheter shape in an unknown blood vessel. By providing high-resolution of cross-sectional and MAP PAM images, we could estimate whether the catheter was placed well in the aorta and whether the function of the balloon was working normally or not. Furthermore, at different injecting medium conditions in the balloon, we could acquire the black colored core of the catheter selectively without any PA contrast agent in ex/in vivo experiments. First, under no air injection condition, PA signals and the shape of the catheter core were almost shown in PAM images. Second, under air injection condition, all PA signals in the balloon disappeared because acoustic loss caused high acoustic attenuation in air. Third, under water injection condition, whole PA signals were detected again. Thus, we could successfully generate PAM images. Although direct contacted regions between the catheter core and the blood vessel wall showed high PA signals regardless of inserted medium conditions because of the intrinsic bending shape of the catheter, other contactless areas of the balloon catheter successfully worked despite uncontrollable in vivo experiments.

Unfortunately, although we carried out the tracking of the balloon catheter with PAM as the proof of concept, it still needs to improve its systemic performance for overcoming current limitations and further clinical study. First, the only use of a 532 nm pulsed laser beam with high blood absorption required an additional surface removing process that caused unexpected signal loss and post-processing time. As shown in/ex vivo MAP PAM images in Figures 4 and 5, there was a lot of signal loss due to unavoidable contracts between the catheter core and the blood vessel. Eventually, this lost information at the front end of the catheter interfered with correct catheter guidance. To overcome this issue, near-infrared region light (NIR, i.e., 680 nm–1800 nm) source and PA contrast agents such as MB and ICG can be solutions [51,52]. If NIR light can be used, more light energy will reach the catheter due to low light absorption and scattering in total hemoglobin [53,54]. NIR light provides an opportunity to avoid high PA signals in vascular and deep PA imaging. Furthermore, the balloon catheter filled with PA contrast agents will enhance the sensitivity of PA images. By choosing a specific wavelength laser such as 680 nm for MB and 800 nm for ICG, we can expect selective PA imaging of the balloon catheter.

Second, the current PAM imaging setup is not proper for clinical translation. Even though PAM provides high spatial resolution, its intrinsic imaging configuration which combination of the focused laser and acoustic beam makes PAM weak in visualizing deep tissue imaging. As shown in Figure 3b, in our experimental approach, we only enabled to show approximately 1 mm in a shallow depth. Therefore, for clinical implementation in the minimally invasive surgery such as hybrid operation, the handheld probe or laparoscopic probe should be updated [55]. To utilize PAI in the general catheter guiding, that works in several centimeter depths, a clinically based PAI system should be applied [56]. Furthermore, combining PA contrast agents and selecting the appropriate excitation light wavelength can be better identification of the balloon catheter in deep tissue. Additionally, the use of clinically approved PA agents is a better option than water in patient safety during the catheter guiding.

5. Conclusions

In this study, we conducted the tracking and visualization of the balloon catheter with PAM. Despite the limited single wavelength laser at 532 nm, we successfully demonstrated the potential of PAI technology for tracking the balloon catheter by applying the simple surface removing process and fast and high-resolution PAM scanning. These results were verified on in/ex vivo animal experiments. As a next plan, we will guide the balloon catheter in medium and large animals using clinical PAI systems, multi-wavelength lasers and PA contrast agents. We believe that the results of this study and advanced next trials will improve the possibility of clinical translation in the near future.

Sensors **2020**, *20*, 5585

Supplementary Materials: Supplementary materials can be found online at: http://www.mdpi.com/1424-8220/20/19/5585/s1.

Author Contributions: Conceptualization, J.K. and C.L.; methodology, J.K., J.-J.M. and C.L.; software, J.Y.K., C.K. and T.T.M.; validation, J.K., T.T.M. and C.L.; formal analysis, J.K., T.T.M. and C.L.; writing—original draft preparation, J.K., T.T.M. and C.L.; writing—review and editing, J.K., T.T.M. and C.L.; supervision, J.K. and C.L.; All authors have read and agreed to the published version of the manuscript.

Funding: National Research Foundation of Korea (NRF) grants funded by the Korean government (NRF-2015M3C1A3056407, NRF-2017M3A9E8023017, NRF-2017R1D1A1B03029556, NRF-2019R1F1A1062948), the Korea Institute for Advancement of Technology (KIAT) grant funded by the Korea Government (MOTIE) (P0008763, The Competency Development Program for Industry Specialist), and the Korea Health Technology R&D Project (HI18C0858) through the KHIDI (Korea Health Industry Development Institute) funded by the Ministry of Health & Welfare.

Acknowledgments: We thank the animal studies for this article from Munki Kim, DVM, (The Heart Research Center of Chonnam National University Hospital Designated by Korea Ministry of Health, Welfare and Family Affairs, Gwangju, Korea).

Conflicts of Interest: C.K. and J.Y.K. have financial interests in Opticho, which, however, did not support this work.

References

1. Kiemeneij, F.; Laarman, G.J.; de Melker, E. Transradial artery coronary angioplasty. *Am. Heart J.* **1995**, *129*, 1–7. [CrossRef]

2. Ouriel, K. Peripheral arterial disease. *Lancet* **2001**, *358*, 1257–1264. [CrossRef]

3. Higashida, R.T.; Tsai, F.Y.; Halbach, V.V.; Barnwell, S.L.; Dowd, C.F.; Hieshima, G.B. Interventional neurovascular techniques in the treatment of stroke—State-of-the-art therapy. *J. Intern. Med.* **1995**, *237*, 105–115. [CrossRef] [PubMed]

4. Lee, J.; Song, H.-Y.; Ko, H.K.; Park, J.-H.; Na, H.K.; Kim, Y.H.; Jung, H.-Y. Fluoroscopically guided balloon dilation or temporary stent placement for patients with gastric conduit strictures after esophagectomy with esophagogastrostomy. *Am. J. Roentgenol.* **2013**, *201*, 202–207. [CrossRef]

5. Kim, K.Y.; Tsauo, J.; Song, H.-Y.; Park, H.J.; Kang, W.S.; Park, J.-H.; Wang, Z. Fluoroscopy-guided balloon dilation in patients with Eustachian tube dysfunction. *Eur. Radiol.* **2018**, *28*, 910–919. [CrossRef] [PubMed]

6. Elabd, S.A.; Elbahnasy, A.M.; Farahat, Y.A.; Soliman, M.G.; Taha, M.R.; Elgarabawy, M.A.; Figenshau, R. Minimally-invasive correction of ureteropelvic junction obstruction: Do retrograde endo-incision techniques still have a role in the era of laparoscopic pyeloplasty? *Ther. Adv. Urol.* **2009**, *1*, 227–234. [CrossRef] [PubMed]

7. Stannard, A.; Eliason, J.L.; Rasmussen, T.E. Resuscitative endovascular balloon occlusion of the aorta (REBOA) as an adjunct for hemorrhagic shock. *J. Trauma Acute Care Surg.* **2011**, *71*, 1869–1872. [CrossRef] [PubMed]

8. Kim, D.H.; Chang, S.W.; Matsumoto, J. The utilization of resuscitative endovascular balloon occlusion of the aorta: Preparation, technique, and the implementation of a novel approach to stabilizing hemorrhage. *J. Thorac. Dis.* **2018**, *10*, 5550. [CrossRef]

9. Angelini, G.; Wilde, P.; Salerno, T.; Bosco, G.; Calafiore, A. Integrated left small thoracotomy and angioplasty for multivessel coronary artery revascularisation. *Lancet* **1996**, *347*, 757–758. [CrossRef]

10. Frevert, S.; Dahl, B.; Lönn, L. Update on the roles of angiography and embolisation in pelvic fracture. *Injury* **2008**, *39*, 1290–1294. [CrossRef]

11. Karkos, C.D.; Mair, R.; Markose, G.; Fishwick, G.; London, N.J.; Naylor, A.R. Hybrid procedures combining open and endovascular surgical techniques for the management of subclavian artery injuries. *J. Trauma Acute Care Surg.* **2007**, *63*, E107–E110. [CrossRef]

12. Tan, H.; Zhang, L.-Y.; Guo, Q.-S.; Yao, Y.-Z.; Sun, S.-J.; Wang, T.; Li, Y.-C.; Xiong, K.-L. "One-Stop Hybrid Procedure" in the Treatment of Vascular Injury of Lower Extremity. *Indian J. Surg.* **2015**, *77*, 75–78. [CrossRef]

13. Rehani, M.; Ciraj-Bjelac, O.; Vañó, E.; Miller, D.; Walsh, S.; Giordano, B.; Persliden, J. Radiological protection in fluoroscopically guided procedures performed outside the imaging department. *Ann. ICRP* **2010**, *40*, 1–102. [CrossRef] [PubMed]

14. Laird, J.R.; Katzen, B.T.; Scheinert, D.; Lammer, J.; Carpenter, J.; Buchbinder, M.; Dave, R.; Ansel, G.; Lansky, A.; Cristea, E. Nitinol stent implantation versus balloon angioplasty for lesions in the superficial femoral artery and proximal popliteal artery: Twelve-month results from the RESILIENT randomized trial. *Circ. Cardiovasc. Interv.* **2010**, *3*, 267–276. [CrossRef] [PubMed]

15. Reed, M.; Meier, P.; Tamhane, U.U.; Welch, K.B.; Moscucci, M.; Gurm, H.S. The relative renal safety of iodixanol compared with low-osmolar contrast media: A meta-analysis of randomized controlled trials. *Jacc Cardiovasc. Interv.* **2009**, *2*, 645–654. [PubMed]

16. Mogabgab, O.; Patel, V.G.; Michael, T.T.; Kotsia, A.; Christopoulos, G.; Banerjee, S.; Brilakis, E.S. Impact of contrast agent viscosity on coronary balloon deflation times: Bench testing results. *J. Interv. Cardiol.* **2014**, *27*, 177–181. [CrossRef]

17. Krueger, J.J.; Ewert, P.; Yilmaz, S.; Gelernter, D.; Peters, B.; Pietzner, K.; Bornstedt, A.; Schnackenburg, B.; Abdul-Khaliq, H.; Fleck, E. Magnetic resonance imaging–guided balloon angioplasty of coarctation of the aorta. *Circulation* **2006**, *113*, 1093–1100. [CrossRef]

18. Kusuyama, T.; Iida, H.; Mitsui, H. Intravascular ultrasound complements the diagnostic capability of carbon dioxide digital subtraction angiography for patients with allergies to iodinated contrast medium. *Catheter. Cardiovasc. Interv.* **2012**, *80*, E82–E86. [CrossRef]

19. Ewert, P.; Berger, F.; Daehnert, I.; Krings, G.; Dittrich, S.; Lange, P.E. Diagnostic Catheterization and Balloon Sizing of Atrial Septal Defects by Echocardiography Guidance Without Fluoroscopy. *Echocardiography* **2000**, *17*, 159–163. [CrossRef]

20. Kim, J.Y.; Lee, C.; Park, K.; Han, S.; Kim, C. High-speed and high-SNR photoacoustic microscopy based on a galvanometer mirror in non-conducting liquid. *Sci. Rep.* **2016**, *6*, 34803. [CrossRef]

21. Lee, C.; Han, S.; Kim, S.; Jeon, M.; Jeon, M.Y.; Kim, C.; Kim, J. Combined photoacoustic and optical coherence tomography using a single near-infrared supercontinuum laser source. *Appl. Opt.* **2013**, *52*, 1824–1828. [CrossRef] [PubMed]

22. Liu, W.; Yao, J. Photoacoustic microscopy: Principles and biomedical applications. *Biomed. Eng. Lett.* **2018**, *8*, 203–213. [CrossRef]

23. Steinberg, I.; Huland, D.M.; Vermesh, O.; Frostig, H.E.; Tummers, W.S.; Gambhir, S.S. Photoacoustic clinical imaging. *Photoacoustics* **2019**, *14*, 77–98. [CrossRef] [PubMed]

24. Wang, L.V.; Hu, S. Photoacoustic tomography: In vivo imaging from organelles to organs. *Science* **2012**, *335*, 1458–1462. [CrossRef] [PubMed]

25. Baik, J.W.; Kim, J.Y.; Cho, S.; Choi, S.; Kim, J.; Kim, C. Super Wide-field Photoacoustic Microscopy of Animals and Humans In Vivo. *IEEE Trans. Med. Imaging* **2019**, *39*, 975–984. [CrossRef]

26. Park, S.; Lee, C.; Kim, J.; Kim, C. Acoustic resolution photoacoustic microscopy. *Biomed. Eng. Lett.* **2014**, *4*, 213–222. [CrossRef]

27. Zhang, H.F.; Maslov, K.; Stoica, G.; Wang, L.V. Functional photoacoustic microscopy for high-resolution and noninvasive in vivo imaging. *Nat. Biotechnol.* **2006**, *24*, 848. [CrossRef]

28. Liu, C.; Liao, J.; Chen, L.; Chen, J.; Ding, R.; Gong, X.; Cui, C.; Pang, Z.; Zheng, W.; Song, L. The integrated high-resolution reflection-mode photoacoustic and fluorescence confocal microscopy. *Photoacoustics* **2019**, *14*, 12–18. [CrossRef]

29. Choi, W.; Park, E.-Y.; Jeon, S.; Kim, C. Clinical photoacoustic imaging platforms. *Biomed. Eng. Lett.* **2018**, *8*, 139–155. [CrossRef] [PubMed]

30. Lin, L.; Hu, P.; Shi, J.; Appleton, C.M.; Maslov, K.; Li, L.; Zhang, R.; Wang, L.V. Single-breath-hold photoacoustic computed tomography of the breast. *Nat. Commun.* **2018**, *9*, 2352. [CrossRef]

31. Lee, C.; Jeon, M.; Jeon, M.Y.; Kim, J.; Kim, C. In vitro photoacoustic measurement of hemoglobin oxygen saturation using a single pulsed broadband supercontinuum laser source. *Appl. Opt.* **2014**, *53*, 3884–3889. [CrossRef] [PubMed]

32. Cao, R.; Li, J.; Ning, B.; Sun, N.; Wang, T.; Zuo, Z.; Hu, S. Functional and oxygen-metabolic photoacoustic microscopy of the awake mouse brain. *Neuroimage* **2017**, *150*, 77–87. [CrossRef]

33. Jiao, S.; Jiang, M.; Hu, J.; Fawzi, A.; Zhou, Q.; Shung, K.K.; Puliafito, C.A.; Zhang, H.F. Photoacoustic ophthalmoscopy for in vivo retinal imaging. *Opt. Express* **2010**, *18*, 3967–3972. [CrossRef] [PubMed]

34. Hu, S.; Wang, L.V. Neurovascular photoacoustic tomography. *Front. Neuroenerg.* **2010**, *2*, 10. [CrossRef]

35. Zhang, C.; Wang, L.V.; Cheng, Y.-J.; Chen, J.; Wickline, S.A. Label-free photoacoustic microscopy of myocardial sheet architecture. *J. Biomed. Opt.* **2012**, *17*, 060506. [CrossRef] [PubMed]

36. Kim, C.; Song, K.H.; Gao, F.; Wang, L.V. Sentinel lymph nodes and lymphatic vessels: Noninvasive dual-modality in vivo mapping by using indocyanine green in rats—volumetric spectroscopic photoacoustic imaging and planar fluorescence imaging. *Radiology* **2010**, *255*, 442–450. [CrossRef] [PubMed]

37. Guo, X.; Tavakoli, B.; Kang, H.-J.; Kang, J.U.; Etienne-Cummings, R.; Boctor, E.M. Photoacoustic active ultrasound element for catheter tracking. In Proceedings of the Photons Plus Ultrasound: Imaging and Sensing, San Francisco, CA, USA, 3 March 2014; p. 89435M.

38. Mallidi, S.; Luke, G.P.; Emelianov, S. Photoacoustic imaging in cancer detection, diagnosis, and treatment guidance. *Trends Biotechnol.* **2011**, *29*, 213–221. [CrossRef]

39. Phan, T.T.V.; Bui, N.Q.; Cho, S.-W.; Bharathiraja, S.; Manivasagan, P.; Moorthy, M.S.; Mondal, S.; Kim, C.-S.; Oh, J. Photoacoustic Imaging-Guided Photothermal Therapy with Tumor-Targeting HA-FeOOH@ PPy Nanorods. *Sci. Rep.* **2018**, *8*, 8809. [CrossRef]

40. Francis, K.J.; Manohar, S. Photoacoustic imaging in percutaneous radiofrequency ablation: Device guidance and ablation visualization. *Phys. Med. Biol.* **2019**, *64*, 184001. [CrossRef]

41. Iskander-Rizk, S.; Kruizinga, P.; Beurskens, R.; Springeling, G.; Mastik, F.; de Groot, N.M.; Knops, P.; van der Steen, A.F.; van Soest, G. Real-time photoacoustic assessment of radiofrequency ablation lesion formation in the left atrium. *Photoacoustics* **2019**, *16*, 100150. [CrossRef]

42. Lee, D.; Lee, C.; Kim, S.; Zhou, Q.; Kim, J.; Kim, C. In vivo near infrared virtual intraoperative surgical photoacoustic optical coherence tomography. *Sci. Rep.* **2016**, *6*, 35176. [CrossRef] [PubMed]

43. Shubert, J.; Lediju Bell, M.A. Photoacoustic imaging of a human vertebra: Implications for guiding spinal fusion surgeries. *Phys. Med. Biol.* **2018**, *63*, 144001. [CrossRef] [PubMed]

44. Lediju Bell, M.A. Photoacoustic imaging for surgical guidance: Principles, applications, and outlook. *J. Appl. Phys.* **2020**, *128*, 060904. [CrossRef] [PubMed]

45. Moore, C.; Jokerst, J.V. Strategies for image-guided therapy, surgery, and drug delivery using photoacoustic imaging. *Theranostics* **2019**, *9*, 1550. [CrossRef]

46. Kim, J.Y.; Lee, C.; Park, K.; Lim, G.; Kim, C. Fast optical-resolution photoacoustic microscopy using a 2-axis water-proofing MEMS scanner. *Sci. Rep.* **2015**, *5*, 7932. [CrossRef]

47. Song, W.; Wei, Q.; Liu, T.; Kuai, D.; Zhang, H.F.; Burke, J.M.; Jiao, S. Integrating photoacoustic ophthalmoscopy with scanning laser ophthalmoscopy, optical coherence tomography, and fluorescein angiography for a multimodal retinal imaging platform. *J. Biomed. Opt.* **2012**, *17*, 061206. [CrossRef]

48. Liu, T.; Li, H.; Song, W.; Jiao, S.; Zhang, H.F. Fundus camera guided photoacoustic ophthalmoscopy. *Curr. Eye Res.* **2013**, *38*, 1229–1234. [CrossRef]

49. Nguyen, V.P.; Li, Y.; Zhang, W.; Wang, X.; Paulus, Y.M. High-resolution multimodal photoacoustic microscopy and optical coherence tomography image-guided laser induced branch retinal vein occlusion in living rabbits. *Sci. Rep.* **2019**, *9*, 1–14. [CrossRef]

50. Faber, D.J.; Aalders, M.C.G.; Mik, E.G.; Hooper, B.A.; van Gemert, M.J.C.; van Leeuwen, T.G. Oxygen Saturation-Dependent Absorption and Scattering of Blood. *Phys. Rev. Lett.* **2004**, *93*, 028102. [CrossRef]

51. Rajian, J.R.; Fabiilli, M.L.; Fowlkes, J.B.; Carson, P.L.; Wang, X. Drug delivery monitoring by photoacoustic tomography with an ICG encapsulated double emulsion. *Opt. Express* **2011**, *19*, 14335–14347. [CrossRef]

52. Ashkenazi, S. Photoacoustic lifetime imaging of dissolved oxygen using methylene blue. *J. Biomed. Opt.* **2010**, *15*, 040501. [CrossRef] [PubMed]

53. Sheng, Z.; Guo, B.; Hu, D.; Xu, S.; Wu, W.; Liew, W.H.; Yao, K.; Jiang, J.; Liu, C.; Zheng, H.; et al. Bright Aggregation-Induced-Emission Dots for Targeted Synergetic NIR-II Fluorescence and NIR-I Photoacoustic Imaging of Orthotopic Brain Tumors. *Adv. Mater.* **2018**, *30*, 1800766. [CrossRef] [PubMed]

54. Liu, X.; Lee, C.; Law, W.-C.; Zhu, D.; Liu, M.; Jeon, M.; Kim, J.; Prasad, P.N.; Kim, C.; Swihart, M.T. Au–Cu2–xSe Heterodimer Nanoparticles with Broad Localized Surface Plasmon Resonance as Contrast Agents for Deep Tissue Imaging. *Nano Lett.* **2013**, *13*, 4333–4339. [CrossRef]

55. Zhou, Y.; Xing, W.; Maslov, K.I.; Cornelius, L.A.; Wang, L.V. Handheld photoacoustic microscopy to detect melanoma depth in vivo. *Opt. Lett.* **2014**, *39*, 4731–4734. [CrossRef]

56. Kim, J.; Park, S.; Jung, Y.; Chang, S.; Park, J.; Zhang, Y.; Lovell, J.F.; Kim, C. Programmable real-time clinical photoacoustic and ultrasound imaging system. *Sci. Rep.* **2016**, *6*, 35137. [CrossRef] [PubMed]

Article

Investigation of the Effect of the Skull in Transcranial Photoacoustic Imaging: A Preliminary Ex Vivo Study

Rayyan Manwar [1,2], Karl Kratkiewicz [2] and Kamran Avanaki [1,2,3,]*

[1] Richard and Loan Hill Department of Bioengineering, University of Illinois at Chicago,
     Chicago, IL 60607, USA; r.manwar@wayne.edu
[2] Department of Biomedical Engineering, Wayne State University, Detroit, MI 48201, USA;
     karl.kratkiewicz@wayne.edu
[3] Department of Dermatology, University of Illinois at Chicago, Chicago, IL 60607, USA
*   Correspondence: ft5257@wayne.edu

Received: 2 June 2020; Accepted: 22 July 2020; Published: 28 July 2020

Abstract: Although transcranial photoacoustic imaging (TCPAI) has been used in small animal brain imaging, in animals with thicker skull bones or in humans both light illumination and ultrasound propagation paths are affected. Hence, the PA image is largely degraded and in some cases completely distorted. This study aims to investigate and determine the maximum thickness of the skull through which photoacoustic imaging is feasible in terms of retaining the imaging target structure without incorporating any post processing. We identify the effect of the skull on both the illumination path and acoustic propagation path separately and combined. In the experimental phase, the distorting effect of ex vivo sheep skull bones with thicknesses in the range of 0.7~1.3 mm are explored. We believe that the findings in this study facilitate the clinical translation of TCPAI.

Keywords: transcranial; skull bone; aberration; photoacoustic; distortion; brain imaging

---

1. Introduction

Transcranial imaging is considered as a significant milestone in the understanding of the underlying brain functionality. Transcranial Ultrasonography (TCUS) is a clinically approved non-invasive and rapid technique for the real-time measurement of cerebral blood flow characteristics in neonates [1–3]. TCUS is effective due to the very thin skull thickness in neonates. TCUS is the preferred modality to image the neonatal brain due to its portability, low cost, speed, and lack of ionizing radiation [4]. TCUS operates in low frequencies (0.5–2 MHz) to have sufficient skull penetration [5]. Among the pre-existing potential alternatives, intraoperative x-ray or CT may be used to navigate through bony anatomy [1,6,7]. Intraoperative magnetic resonance imaging (MRI) is another costly option [7]. X-ray, CT, and MRI all require sedation and exposure to ionizing radiation [6].

Photoacoustic imaging (PAI) has proved to be a promising tool for the diagnosis, prognosis, and treatment monitoring of neurological disorders in small and large animals [8–14]. PAI is a non-ionizing hybrid imaging modality based on the photoacoustic (PA) effect. PAI combines the high absorption contrast of optical imaging with the high spatial resolution of ultrasound imaging to visualize tissue chromophores in the optical quasi-diffusive or diffusive regime [15,16]. In PAI, the biological tissue is illuminated with a short-pulsed laser beam, generating acoustic waves via transient thermoelastic expansion [9,17–19]. The subsequent ultrasound waves propagating from within the tissue are then detected by an ultrasonic transducer array located outside the tissue. The ultrasound signals are used to form an image through a reconstruction algorithm [20]. Generated acoustic waves travel through the skull one way, unlike pulse-echo ultrasound. As a result, the waves are less susceptible to the attenuation that occurs when they encounter the skull–tissue interface [21,22].

One of the obstacles for PAI in transcranial imaging is the presence of the skull bone [23,24]. Skull bone represents a highly acoustical impedance mismatch and dispersive barrier for the propagation of acoustic waves [25]. The skull distorts the amplitude and phase of the received acoustic waves [26]. This distortion is contributed by four different phenomena: (i) the acoustic attenuation (i.e., the decrease in the acoustic signal amplitude) due to the absorption and scattering of the skull tissue [27–29]; (ii) the acoustic dispersion (i.e., the dependency of the speed of sound on frequency) modifies the phase of the acoustic wave [29]; (iii) the signal broadening, which is a frequency-dependent reduction in the acoustic wave amplitude [30]; and (iv) the temporal shift, where the significantly higher speed of sound in the bone (~2900 m/s [31]) as compared to the brain's soft tissue (~1500 m/s [32]) makes the acoustic waves travel faster through the skull and be detected earlier. The degree of attenuation, dispersion, broadening, and temporal shift are determined by the mechanical properties of the skull (i.e., bone type, density, porosity, and thickness), among which the tissue thickness has the most significant effect [33–35]. In transcranial photoacoustic imaging, there are two sources of signal attenuation: (1) acoustic, and (2) optical. Acoustic attenuation can be represented by $A = A_0 e^{-\alpha d}$, where A_0 is the signal amplitude before attenuation, d is the depth or thickness, and α is the attenuation coefficient. The attenuation coefficient is a function of frequency and is defined as: $\alpha = \omega^2 \eta / 2 c_p$, where ω is the angular frequency, η is the viscosity, and c_p is the phase velocity. Therefore, if the frequency or depth or both increase, the attenuation increases. Optical attenuation is studied based on the absorbing and scattering effects of the skull. The absorbing effect of the skull tissue can be represented by $A_{abs} = \varepsilon l C$, where ε is the molar absorptivity, l is the optical path length, and C is the concentration of the medium. The scattering effect of the tissue is a more complex event, and is modelled using the Extended Huygens–Fresnel (EHF) principle [36]. Studies have shown that the primary effect of scattering is a less steep slope of light intensity decay with depth than that predicted by the so-called single-scattering model that follows an exponential decay trend. A higher density causes increased optical and acoustic absorption, whereas a higher porosity causes more scattering [37].

The angle between the incident acoustic wave and the skull tissue affects the PA intensity. With increasing the incident angle, more shear waves are generated as compared to longitudinal waves, and hence the amplitude of the PA signal drops further. Yang and Wang et al. [38], evaluated the PA signal amplitude at two different frequencies (i.e., 1 and 2.25 MHz) as a function of the incident angle on the monkey's skull, and found that increasing the incident angle up to ~35° decreases the PA signal, whereas beyond that angle the PA signal amplitude starts increasing again. This phenomenon is applicable if the boundaries are part of a layered material (such as skull tissue), where the longitudinal waves first convert into shear waves at the tissue–skull interface and later the shear waves convert back to longitudinal waves (mode conversion) at the skull–tissue interface and vice-versa.

Due to the distorting effects of the skull, the PAI of a small animal brain (with semi optically and acoustically transparent skull) has been conducted [39]; however, there are only a few studies to validate the feasibility of photoacoustic technology for transcranial imaging in animals with thicker skulls [40,41]. Several PA signal/image enhancement algorithms were developed to improve the quality of the degraded images due to the presence of the skull [42]. Although some of the algorithms were effective, they were computationally expensive and were not run in real-time. Therefore, determining the maximum skull thickness that would allow the imaging target to be accurately reconstructed without any post-processing is essential.

In order to characterize the effect of the skull on the PAI, first we describe the skull bone structure and the corresponding physio-mechanical properties. The skull bone consists of three layers: the inner table, the middle diploe, and the outer table (see Figure 1a). The inner and outer table are cortical bones, whereas the middle diploe layer is the trabecular bone type [43]. The cortical and trabecular bones are anatomically different. At birth, the bones of the cranial vault are unilaminar tables (cortical type) and, thereafter, the intervening diploe (trabecular type) appears at about the fourth year [44]. With the age, the trabecular layer grows at a faster rate as compared to the cortical layers (26% volume per year turnover rate for trabecular and 3% for cortical bone) [45]. Cortical bone is a fairly solid (Figure 1b(i)) and dense material which consists of a minerals, organic parts, and water. The mineral ingredient is hydroxyapatite, and the organic parts are fibrous protein collagen and non-collagenous [46,47]. Trabecular bone primarily consists of lamellar, which are arranged in packets that make up an interconnected irregular array of plates and rods called trabeculae (Figure 1b(ii)). The trabecular bone of the central diploe is an energy absorbing lightweight meshed structure that provides cushioning, shear strength, and separation between the cortical plates in order to increase the inertial characteristics (bending strength) that allows the three-layered structure to endure mainly bending loads. Moreover, such a structure that makes the trabecular bone a highly porous, heterogeneous, and anisotropic material to absorb the external shock contains bone marrow and skull vasculatures. The density of cortical and trabecular bones range between 1.8 and 2.2 and 0.3 and 1.3 g/cm^3, respectively [48]. Since the density of the cortical bone is higher than that of the trabecular bone, energy is mostly absorbed by the cortical bone, whereas the porosity is higher in the trabecular bone and, therefore, scattering occurs within the trabecular diploe layers [49,50].

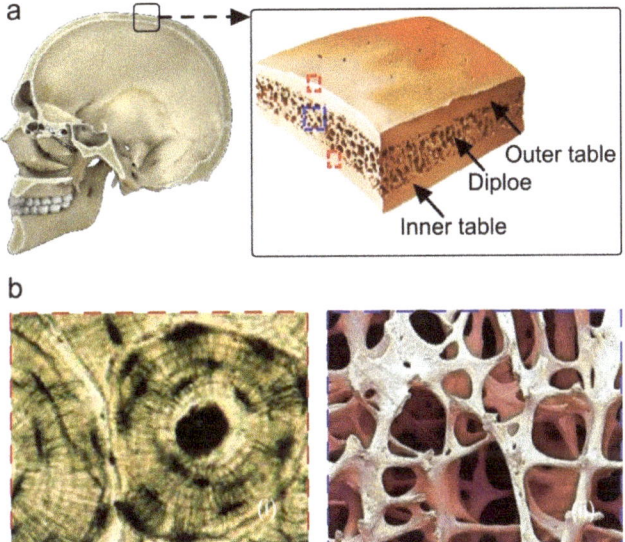

Figure 1. Skull bone structure. (**a**) Structural component of human skull bone, each layer magnified in (**bi**) the outer tables and (**bii**) diploe.

The mechanical and acoustic characteristics of the skull bone described in the literature can be summarized as follows: the characterization of the human skull in terms of the speed of sound and thickness were explored in several studies [25,51–54]; the longitudinal speed of sound and the acoustical attenuation coefficient of human calvaria were studied at frequencies ranging from 0.27 to 2.526 MHz [29]; the speed of sound in cortical bone is within the range of 2880–4220 m/s [55,56]; for the trabecular bones, the speed of sound is lower (2000–3000 m/s) and the attenuation is higher (15–30 dB/MHz/cm) as compared to those of the cortical bone [57]; in [58], the insertion loss and the elastic constants of the skull were measured; studies have also been performed to attain the optical properties of the skull bones [58–62]; the optical properties of human cranial bone were measured using the integrating sphere technique in [60]; the reduced scattering coefficient of the human skull follows $\mu_s'(\lambda) = 1533.02 \times \lambda^{-0.65}$ in the wavelength range, $\lambda = 800$–1000 nm [63]; in the near infrared region (600–900 nm), the reduced scattering and absorption coefficient of the human skull are in the range of 0.2–1.2 cm^{-1} and 20–25 cm^{-1}, respectively; the predominant ultrasound attenuation mechanism in the trabecular bone is scattering, while the absorption is considered to be a major attenuation mechanism in the cortical bone; the cortical bones exhibit higher optical scattering and absorption as compared to the trabecular bones [64]; a generalized pattern of acoustic transmittance and optical intensity decay as a function of time are shown in [64]—in this article, it is shown that due to a more prominent effect of optical scattering, the intensity decay has a slower rate in trabecular bones compared to acoustic transmittance. Although several studies have explored different skull properties—e.g., the geometry, scattering coefficient, speed of sound, insertion loss, and transmission dispersion [58,65,66]—the effect of the skull in the illumination path and the acoustic detection path, separately, has not been investigated quantitatively in a photoacoustic transcranial imaging experiment.

In this study, we investigate the feasibility of transcranial photoacoustic imaging by studying the effect of the skull in both the illumination path and the acoustic detection path, and determine the maximum skull thickness through which the accurate photoacoustic imaging of the structure and vasculature is feasible. Our investigation aims to explore and quantify the deterioration of PA images owing to the obstacle of the skull bone in three paths: (i) light illumination, (ii) acoustic propagation, and (iii) both light illumination and acoustic propagation.

2. Materials and Methods

2.1. PAI System

The PAI system used in this study composes of Phocus MOBILE, a 10 Hz Nd:YAG tunable laser (OPOTEK, Carlsbad, CA, USA) in the range 690 to 900 nm, that is controlled by an internal optical parametric oscillator (OPO). A silica fiber bundle consisting of 100 fibers with a total diameter of 1 cm has been used for light delivery. The average output energy at the fiber end was measured as ~20 mJ using an energy meter (QE12SP-H-MT-D0, Gentec-EO, Quebec, QC, Canada). The spot size was 8 mm on the skull piece, and the spot size on the target could not be measured since it was embedded in the phantom. Considering the numerical aperture of the optical fiber, the divergence of the light was ~30°.

Since the acoustic window near the temporal or occipital region is with a diameter of ~3 cm [67], phased array transducers are preferred. A phased array sensing surface has a smaller footprint area as compared to linear and curvilinear arrays. Moreover, a phased array provides a wider field of view and it has dynamic focusing capabilities, which increase the flexibility of scanning without or with a minimal mechanical movement of the array. We used a 64-element phased array P4-2 transducer probe (Philips Healthcare, Ville Platte, LA, USA) with a 2.5 MHz center frequency. The transducer was held in water inside an open top box using clamps. The clamps were attached to a two-axis mechanical stage for scanning. The probe was scanned in the *y*-axis to cover a distance of 2 cm with 48 total steps and a step size of 0.4 mm. The Vantage 128 imaging platform (Verasonics Inc., Kirkland, WA, USA) was used for the data acquisition and image processing.

2.2. Skull Tissue Preparation

There are structural differences between the sheep skull and human skull in terms of thickness, content, and architecture. Despite the differences, the structural components (diploe, outer, and inner table) in human and sheep skulls are similar. Here, we evaluate the effect of skull thickness on the reconstructed PA image; therefore, maintaining the skull thickness is important. To achieve similar thicknesses of the human skull at different ages, we have chosen the frontal skull bone of sheep head and mechanically configured the sheep skull to be flat and representative of human skull thicknesses. Three different skull thicknesses of 0.7, 1.0, and 1.3 mm were used. The skull samples were collected from ex vivo sheep heads. Using a Hole Dozer general purpose circular saw (Milwaukee Electric Tool, Brookfield, WI, USA), skull pieces with a diameter of 5 cm were cut. Later, 1.5 cm areas on the skull pieces were thinned down to the desired thickness using a drill bit. The thickness of each skull sample after preparation was measured using a H-2780 digital screw gauge (ULINE, Milton, ON, Canada) at 5 different points and averaged.

2.3. Phantom Experiments

We embedded an imaging target in a brain tissue-like mimicking phantom with optical properties similar to those of the brain tissue. To determine the light attenuating characteristics of the brain tissue, slices of sheep brain with a thicknesses between 0.5 and 1.75 cm were prepared. The experimental setup is shown in Figure 2. The ex vivo brain tissue was held on a metal plate with a circular hole. An energy meter (QE8SP-B-BL-INT-D0, Gentec-EO Inc., Quebec City, QC, Canada) was coaxially aligned to the optical fiber bundle through the hole; the surface of the sensor was protected with an optically transparent thin film. The brain tissue-mimicking phantom was realized by mixing gelatin (to represent the acoustic properties) and sugar-free psyllium hydrophilic mucilloid fiber (Metamucil, P&G, Cincinnati, OH, USA) (to represent the tissue optical attenuation and echogenicity). First, 8% gelatin was dissolved in water, followed by 4% fiber in a transparent one-side-open cubic acrylic box (Lanscoery, Monterey Park, CA, USA) [68]. Tissue-mimicking phantoms with thicknesses of 1, 2, 3, 4, and 5 cm were prepared.

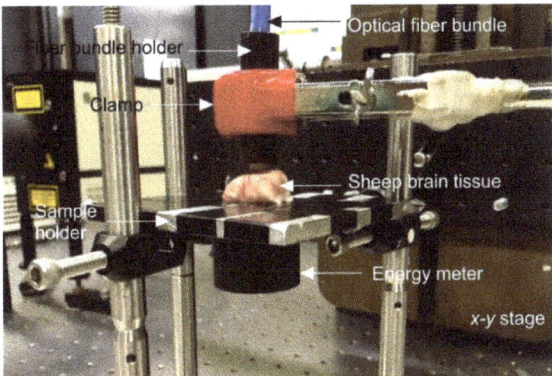

Figure 2. Experimental setup of the optical transmittance characterization to find the thickness of the tissue-mimicking phantom that is optically equivalent to brain tissue.

Next, we evaluated the effect of skull as a dispersive barrier in three paths: (i) light illumination, (ii) acoustic propagation, and (iii) both light illumination and acoustic propagation (see Figure 3a–c). The imaging target was a square loop of a 0.5 mm-thick copper rod covered with an insulating dark jacket, polyactic material (see Figure 4c), and held in the tissue-mimicking phantom mixture at a 20° angle versus the probe viewing plane at a 1 cm distance from the surface of the phantom. We initially tried to implement a 3D structure of blood vessels embedded within the gelatin phantom. Since the blood vessels were positioned in a three-dimensional coordinate, we could not inject blood evenly inside the blood vessels. This made the evaluation of the results difficult. We then used a thin plastic tube to represent the blood vessels. In addition to the fact that these tubes had additional absorption, the stationary blood within the tubes started forming sediment at the bottom of the tube. Therefore, we utilized a solid imaging target instead of the actual blood vasculature. According to the literature, the absorbance of polyactic material (with a normalized absorbance ~25% [69]) at the imaging wavelength of 690 nm is close to the absorbance of blood (with the normalized absorbance of ~20% [70]).

The ultrasound probe was horizontally scanned across the imaging target by manually rotating the x-axis knob of the x-y stage with a step size of 1 mm. Each 3D image was comprised of 50 2D B-scan images, which were later compiled into a 3D volume in Slicer 4.10 [71]. In the 3D slicer, the projection of the 3D volume was automatically adjusted to the default intensity range and, therefore, the intensity map had to be corrected. The process of correcting the intensity projection is as follows. Initially, we selected a B-scan frame for each configuration, where a specific portion of the square loop can be visualized without any post-processing; in this case, we have chosen the B-scan frames that correspond to the light green and orange dotted region of interests (ROIs), as shown in Figure 3. We then calculated the average intensity value of those specific frames and, later, applied them as a gain (intensity modifier) to the corresponding 3D volumes to project the corrected intensity map.

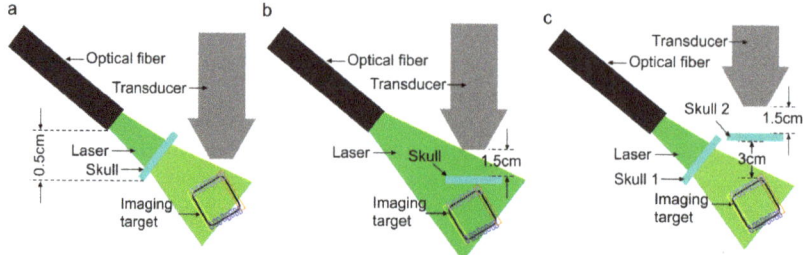

Figure 3. Schematic setups for imaging the square loop phantom with the skull as a barrier in: (**a**) the optical path, (**b**) the acoustic detection path, and (**c**) both the optical and acoustic detection paths. The optimized distances are calculated in Section 3.

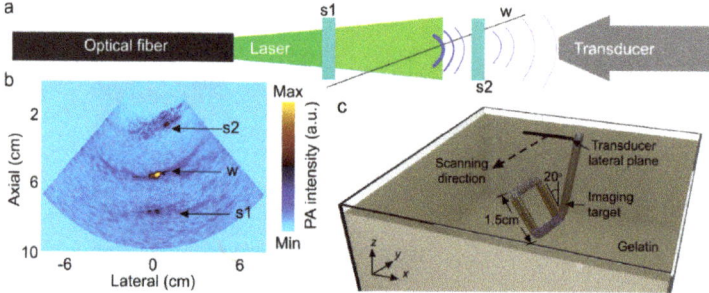

Figure 4. Experimental setup to find the optimum distance between the transducer and skull piece 2, and between the optical fiber and skull piece 1. (**a**) Schematic of the experimental setup, (**b**) photoacoustic (PA) image of the wire phantom in water with the optimum position of the transducer and optical fiber, and (**c**) a 3D model of the square loop imaging target in gelatin. Imaging target is slanted at 20° from the z axis. Scanning direction is along the x axis. Transducer lateral plane is along the y axis. s1: skull piece 1; s2: skull piece 2; w: polyactic wire. The distance between the transducer and s2 is <1.5 cm, the distance between the transducer and the imaging target is <3 cm, and the distance between the optical fiber bundle and s1 is 0.5 cm.

2.4. Quantitative Evaluation Parameters

We evaluated the PA images quantitatively in terms of the average intensity attenuation (AIA), smoothness (S), and image distortion (ID). These parameters were evaluated for the entire region on the square loop imaging phantom. AIA is defined as the averaged intensity in the specified area; S (can be viewed as lack of roughness) is extracted from the line profile across the specified ROI with and without the skull PA images, and is defined as the correlation between the peak values of the corresponding normalized line profiles. ID is defined as sum of the square of the regression (SSR) of the rising and falling edges of the line profile in the US image. Additionally, the image distortion in the PA images was defined as the difference between the contour profile of the PA image with and without the presence of the skull.

The processing protocol was as follows. The imaging target orientation was such that the light blue ROIs (indicated in both Figures 3 and 4c) were closest to the surface of the US probe (~4.5 cm) and the dark blue ROIs were the furthest away (~6 cm). We initially extracted several line profiles within the light green and orange ROIs (indicated in Figures 3 and 4c) and averaged them. The averaged line profile represents the signal intensity decay as a function of the distance between the imaging target and the transducer probe (i.e., 4.5 to 6 cm). Next, we extracted several parallel line profiles within each of the light and dark blue ROIs. The average and standard deviations of the extracted values 4.5, 5, 5.5, and 6 cm were presented in tables and figures. It is of note that the transducer probe was scanned in one direction. As a result, the light green and orange ROIs are located along the lateral plane of the transducer surface, and hence the entire ROI can be visualized in a 2D image, whereas the light and dark blue ROIs are seen as moving dots (due to the location of the ROIs in the cross-sectional imaging plane). Therefore, the light green and orange ROIs are represented by large rectangular boxes, whereas the light and dark blue ROIs are represented by small rectangular boxes.

3. Results and Discussion

Initially, to find out the optimum distance between the transducer and skull, we imaged a copper wire coated with polyactic jacket as the imaging target (see Figure 4a). By finding the optimum distances, any reflecting artifact overlapping with the signal coming from the imaging target were avoided. The phantom was held inside a transparent plastic container and fixed to the optical table. The transducer probe and skull pieces were held using optical rods and fixed to a customized *x-y* stage, made in the machine shop at Wayne State University. The experiment was performed in two stages: (1) First, the position of the sample was fixed with respect to the wire phantom and the transducer probe (5 cm away from the phantom) was moved towards the skull piece 2, using a mechanical stage in the *y*-axis with steps of 5 mm; this configuration was used to optimize the distance between the transducer and skull piece 2. (2) Once we determined the optimum position of the transducer with respect to skull piece 2, both the transducer and skull piece 2 were moved simultaneously from a distance of 5 cm towards the wire phantom, while the distance between the transducer and skull piece 1 was fixed; this configuration provided information regarding the US signal behavior generated from the phantom as a function of depth while the transducer was at a constant distance from skull piece 1. The optimum configuration was as follows: the P4-2 probe was at least 1.5 and 3 cm away from skull piece 2 and the imaging target, respectively; the optical fiber bundle was 0.5 cm away from skull piece 1 (see Figure 3a).

Next, using the experimental setup shown in Figure 2, we determined the thickness of the tissue-mimicking phantom that optically models the brain tissue. In this setup, the optical fiber bundle was placed right on top of the brain tissue or the brain-like tissue-mimicking phantom. With the laser energy measured (at the distal end of the fiber it was 30 mJ), we were able to measure the optical attenuation of different thicknesses of the brain tissue (i.e., T_b: 0.5 cm to 1.75 cm) and the brain tissue-mimicking phantom (i.e., T_m: 1 cm to 5 cm) through skull samples with thicknesses of 0.7, 1.0, and 1.3 mm; any thicknesses beyond the above thicknesses entirely blocked the light and thus were not considered in our experiments. The results shown in Figure 5 indicate that a 5 cm-thick tissue-mimicking phantom optically resembles the ~5 mm brain tissue.

We then evaluated the ultrasound images of the square loop phantom when the transducer was held 1.5 and 3 cm away from the skull piece and the square loop phantom, respectively; when the optical fiber was placed 0.5 cm away from the skull piece (see Figure 3); and when the orientation of the phantom was at an angle of 20° (Figure 4c) to the viewing plane of the probe. We imaged the target in both gelatin (see Figure 6b) and gelatin with fiber mixture (see Figure 6c). The US images without the skull clearly present the morphology of the square loop. In Figure 6d–f, a skull piece with thicknesses of 0.7, 1.0, or 1.3 mm were used when the square loop target was in gelatin with the fiber mixture. The ROIs on the squared loop phantom image were chosen to evaluate the effect of the skull aberration. The attenuating and aberrating effects of the skull are shown in Figure 6g. A summary of the quantitative evaluation is provided in Table 1.

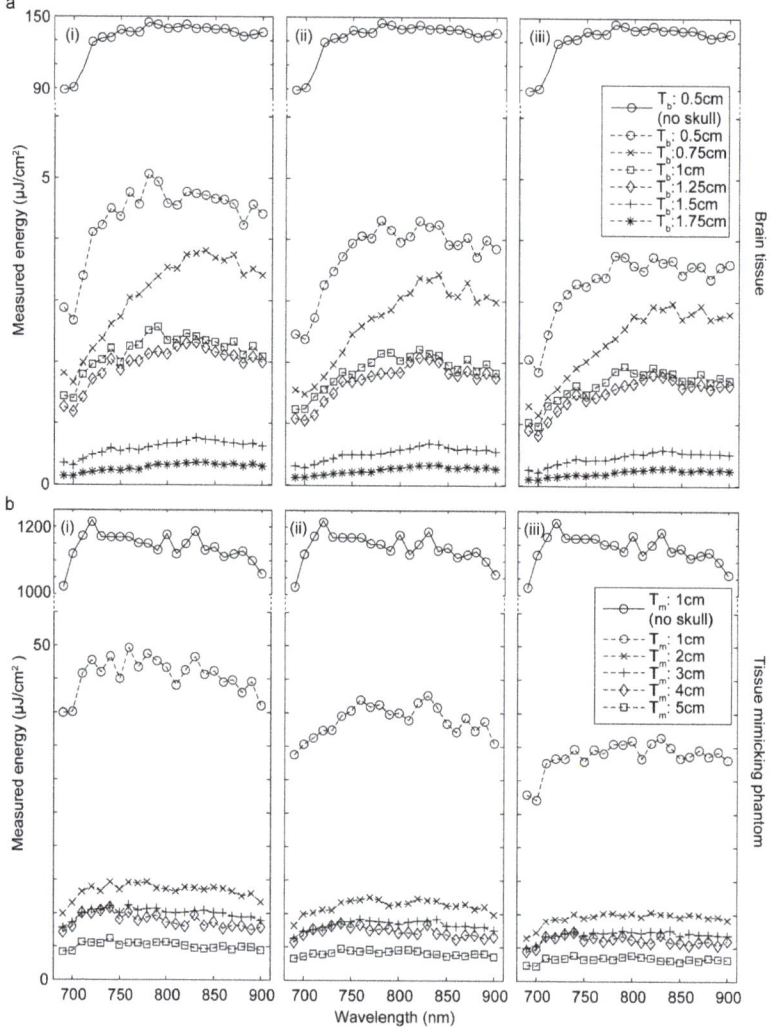

Figure 5. Optical energy measured with different thicknesses of (**a**) sheep brain tissue and (**b**) the brain tissue-mimicking phantom (gelatin + fiber) with a skull bone to block the illumination path with a thickness of (**i**) 0.7, (**ii**) 1.0, or (**iii**) 1.3 mm. Optical fiber bundle was placed on top of the brain tissue or the brain-like tissue-mimicking phantom. T_b: brain tissue thickness; T_m: tissue-mimicking phantom thickness.

Table 1. Summary of the quantitative evaluation of the US imaging through the skull in gelatin with a fiber mixture with different thicknesses. Please see the definition of the quantitative parameters in Section 2.4.

Skull Thickness (mm)	Image Average Intensity Attenuation (%)	Image Distortion (%)	Smoothness (%)
0.7	46.5 ± 1.30	32.5	50 ± 1.47
1.0	48.23 ± 3.74	35.6	10.7 ± 1.26
1.3	78.4 ± 4.23	56.28	1.8 ± 0.89

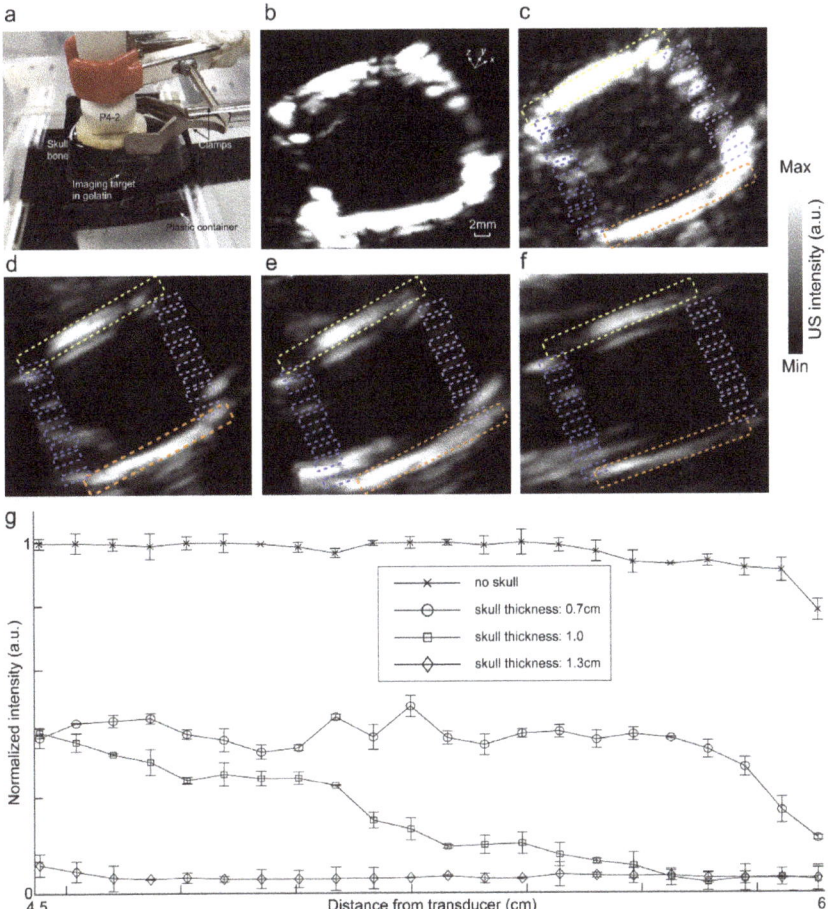

Figure 6. Image intensity attenuation and distortion analysis in a transcranial ultrasound imaging experiment. (**a**) Experimental setup. US image of the square imaging target (**b**) in gelatin and (**c**) in gelatin with the fiber mixture. Ultrasound images of the square imaging target in the gelatin with the fiber mixture through a skull piece with a thickness of (**d**) 0.7, (**e**) 1.0, and (**f**) 1.3 mm. (**g**) Average line profiles within the ROIs indicated with (i) light green, (ii) dark blue, (iii) light blue, and (iv) orange dotted boxes depicted in (**c**–**f**). Light green and orange ROIs are at the same depth from the transducer. This distance increases from the light to dark blue ROIs from 4.5 to 6 cm. The transducer was at the distance of 1.5 cm from the skull, and the skull was at the distance of 3 cm from the imaging target. ROI: region of interest.

Next, we studied the distorting effects of the skull on the PA images when it blocked only the acoustic detection path. The experimental setup as well as the PA images of the square loop imaging target with different thicknesses of the skull are shown in Figure 7a,b, respectively, and a summary of the quantitative evaluation is provided in Table 2. A bar chart to show the attenuating effect of skull as a function of the distance between the transducer and the imaging target for different skull thicknesses is provided in Figure 7c. Here, thicker skulls (1 and 1.3 mm skulls) impacted the PA signal intensity to decay abruptly at higher depths and, therefore, the average intensity decays faster along the light green and orange ROIs towards the dark blue ROIs. The contour profiles of the PA images to present the severity of the structural deformation with different thicknesses of the skull is provided in Figure 7d.

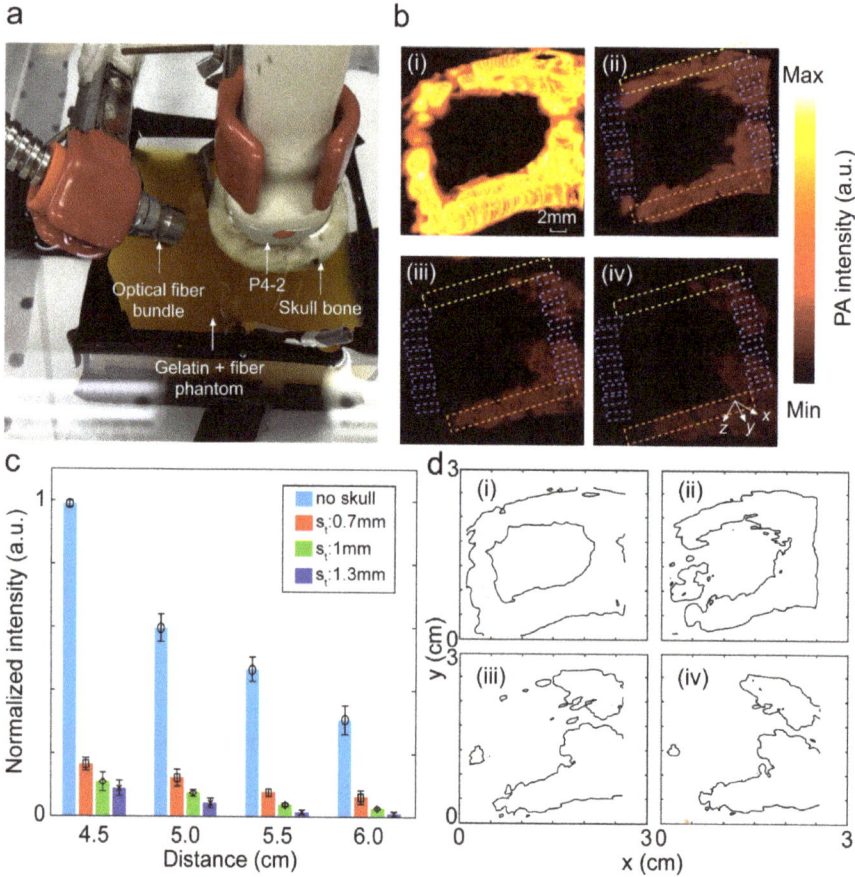

Figure 7. Image distortion analysis in photoacoustic imaging when the skull blocks the acoustic detection path. (**a**) Experimental setup. (**b**) PA image of the square loop imaging target (i) without skull, (ii) with a 0.7 mm skull, (iii) with a 1.0 mm skull, and (iv) with a 1.3 mm skull. (**c**) Average PA signal intensity within the ROIs depicted in (**bi–biv**) as a function of the distance between the transducer surface and the imaging target (4.5, 5, 5.5, and 6 cm). Light green and orange ROIs are at the same depth from the transducer. This distance increases from the light to dark blue ROIs from 4.5 to 6 cm. The transducer was at a distance of 1.5 cm from the skull, and the skull was at a distance of 3 cm from the square target. (**d**) Contour map of the PA images representing the skull-induced deformation (i) without the skull, (ii) with a 0.7 mm skull, (iii) with a 1.0 mm skull, and (iv) with a 1.3 mm skull. The transducer was at a distance of 1.5 cm from the skull, and the skull was at a distance of 3 cm from the square loop imaging target. S_t: skull thickness.

Table 2. Summary of the quantitative evaluation of PA imaging through the skull in gelatin with the fiber mixture when the acoustic propagation path is blocked with skull pieces with different thicknesses.

Skull Thickness (mm)	Image Average Intensity Attenuation (%)	Image Distortion (%)	Smoothness (%)
0.7	88.42 ± 1.12	32.12	5.74 ± 2.71
1	92.59 ± 0.27	72.45	4.41 ± 2.05
1.3	95.17 ± 0.21	79.73	2.46 ± 1.87

We then studied the distorting effects of the skull on the PA images when it blocked only the light illumination path. The experimental setup as well as the PA images of the square loop imaging target with different thicknesses of the skull are shown in Figure 8a,b, respectively, and a summary of the quantitative evaluation is provided in Table 3. A bar chart to show the attenuating effect of the skull as a function of the distance between the transducer and the imaging target for different skull thicknesses is provided in Figure 8c. Unlike the effect of the skull on the acoustic path, here the PA average intensity attenuation is higher; however, the intensity decay rate as a function of depth is comparatively lower. The contour profiles of the PA images to present the severity of structural deformation at different thicknesses of the skull is provided in Figure 8d.

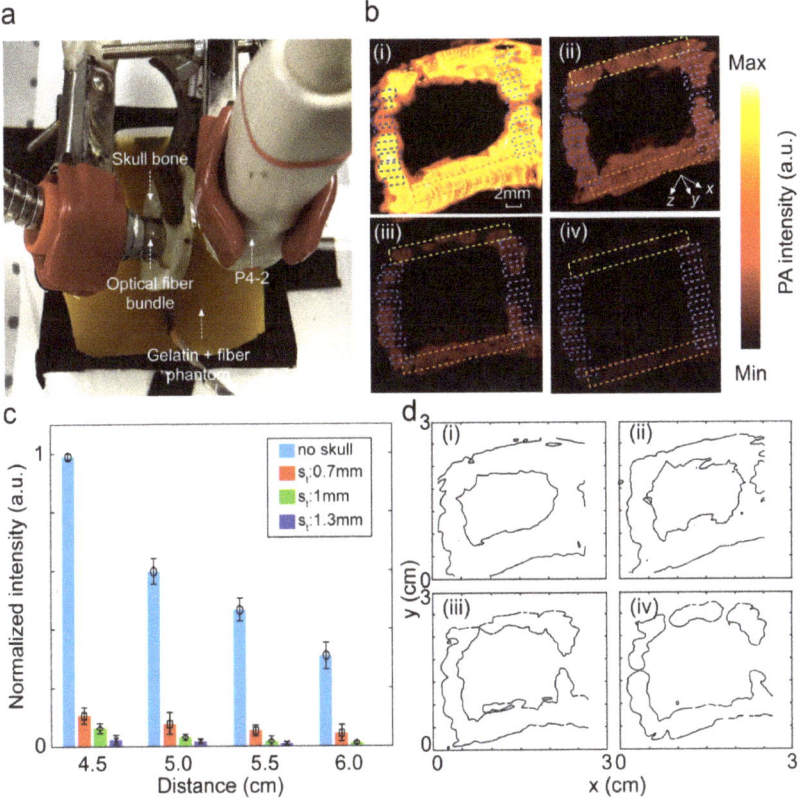

Figure 8. Image distortion analysis in photoacoustic imaging when the skull blocks the light illumination path. (**a**) Experimental setup. (**b**) PA image of the square loop imaging target (i) without the skull, (ii) with a 0.7 mm skull, (iii) with a 1.0 mm skull, and (iv) with a 1.3 mm skull. (**c**) Bar chart of the average PA signal intensity within the ROIs is depicted in (**bi–biv**) as a function of the distance between the transducer surface and imaging target (4.5, 5, 5.5, and 6 cm). (**d**) Contour map of PA images representing the skull-induced deformation (i) without the skull, (ii) with a 0.7 mm skull, (iii) with a 1.0 mm skull, and (iv) with a 1.3 mm skull. Light green and orange ROIs are at the same depth from the transducer. This distance increases from the light to dark blue ROIs from 4.5 to 6 cm. S_t: skull thickness.

Finally, we studied the distorting effects of the skull on the PA images when it blocked both the acoustic propagation and the light illumination paths. The experimental setup as well as the PA images of the square loop imaging target with different thicknesses of the skull are shown in Figure 9a,b, respectively, and the summary of the quantitative evaluation is provided in Table 4.

Table 3. Summary of the quantitative evaluation of the PA imaging through the skull in gelatin with the fiber mixture when the light illumination path is blocked with different thicknesses of skull.

Skull Thickness (mm)	Image Average Intensity Attenuation (%)	Image Distortion (%)	Smoothness (%)
0.7	91.10 ± 2.98	11.26	32.5 ± 1.56
1	95.07 ± 2.24	18.67	6.18 ± 1.06
1.3	97.03 ± 1.89	23.91	2.46 ± 0.74

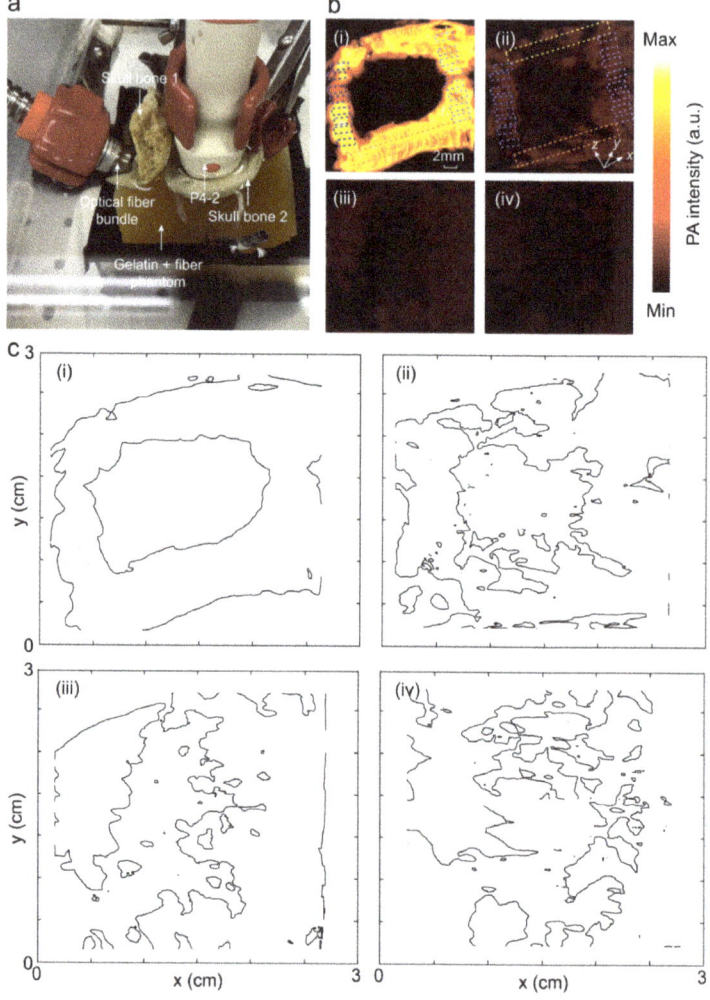

Figure 9. Image distortion analysis in photoacoustic imaging when the skull blocks both the light illumination and acoustic detection paths. (**a**) Experimental setup. (**b**) PA image of the square loop imaging target (i) without the skull, (ii) with a 0.7 mm skull, (iii) with a 1.0 mm skull, and (iv) with a 1.3 mm skull. (**c**) Contour map of the PA images representing the skull-induced deformation (i) without the skull, (ii) with a 0.7 mm skull, (iii) with a 1.0 mm skull, and (iv) with a 1.3 mm skull. The transducer was at a distance of 1.5 cm from the skull and the skull was at a distance of 3 cm from the square loop imaging target. Light green and orange ROIs are at the same depth from the transducer. This distance increases from the light to dark blue ROIs from 4.5 to 6 cm.

Table 4. Summary of the quantitative evaluation of the PA imaging through the skull in gelatin with the fiber mixture when both the light illumination and acoustic propagation paths are blocked by the skull.

Skull Thickness (mm)	Image Average Intensity Attenuation (%)	Image Distortion (%)	Smoothness (%)
0.7	92.3 ± 2.83	81.6	1.76 ± 1.38

The findings of this experiment were as follows: (i) The light was completely diffused inside the tissue-mimicking phantom after passing through the skull pieces, therefore a homogenous illumination of the target phantom was obtained. (ii) The horizontal sides of the phantom generated a higher PA signal amplitude compared to the vertical sides because of the transducer viewing plane. (iii) The only skull tissue that allowed seeing the structure of the imaging target accurately was the skull piece with a 0.7 mm thickness; with the 1.0 mm skull, the shape of the imaging target was almost visible (Figure 9c(iii)), and with the 1.3 mm skull (Figure 9c(iv)), the structure of the square loop target in the image was totally distorted and attenuated to such an extent that the target was not comprehensible.

The goal of this study was to evaluate the combined effect of the skull layers on the acoustic and optical attenuation. We used the architecture of the skull and its layer information, published in research articles, to explain the results. Furthermore, creating a ~1 cm-diameter flat cortical and trabecular layer tissues, thinned down to a millimeter thickness, requires sophisticated machinery, especially with the brittle nature of the skull layers, which was not available to us. A quantitative evaluation of the percent distribution of the acoustic and optical path blocked towards the overall evaluation parameters presented in Table 4 is shown in Figure 10. The individual contribution has been calculated based on their respective values presented in Tables 2 and 3.

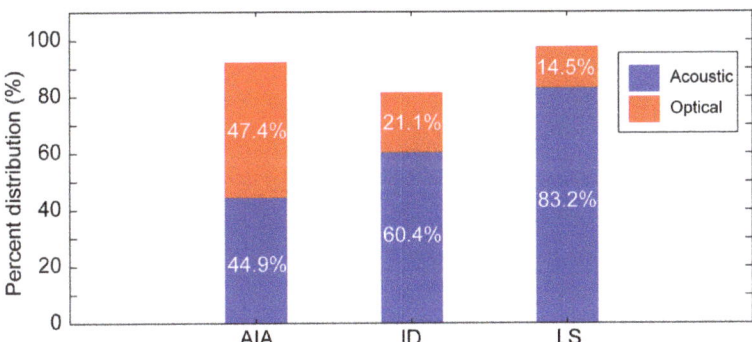

Figure 10. Percent distribution of the evaluating parameters from the acoustic and optical paths when both paths are blocked by the skull (thickness: 0.7 mm). AIA: average intensity attenuation; ID: image distortion; LS: lack of smoothness (1-S).

The quantitative evaluations in Figure 7, Figure 8, and Figure 10 show that blocking the acoustic propagation affects the PA signal more significantly as compared to blocking the illumination path. According to the literature, the optical illumination path is mainly affected by the cortical layers in the form of optical energy attenuation due to the higher absorption, scattering, and reflection of this layer [63,72–75], whereas the trabecular structure of diploe layer induces more acoustic scattering and insertion loss. Moreover, among the three quantitative measures, image distortion and the lack of smoothness are the major consequences of the diploe layer that constitutes the majority thickness of thick bones, whereas image average intensity attenuation is the consequence of the cortical layer [51,52,76–79]. In other words, blocking the optical path more significantly contributes to the amplitude decay, while blocking the acoustic propagation path contributes mainly to the distortion of the morphological map of the imaging target [29,52,80,81]. The combined significant acoustic distortion in the diploe layer and the optical attenuation in the cortical–diploe or cortical–tissue interfaces make the use of photoacoustic transcranial imaging challenging in animals with thicker skulls or in humans.

4. Conclusions

While photoacoustic imaging has shown great promise in the transcranial brain imaging of small animals, it is still underdeveloped for clinical us due to the presence of the skull. We studied the distorting effect of the skull when it is in the optical illumination path, acoustic detection path, and both simultaneously. We determined the maximum thickness of the skull through which PAI is feasible, such that the structure of the imaging target with no post processing is distinguishable; this thickness was ~0.7 mm. Utilizing sophisticated reconstruction algorithms as well as signal/image enhancement techniques [42,82–86], imaging through thicker skull tissues will be possible. Due to the complexity of creating flat cortical and trabecular layer tissues, we studied the combined effect of the skull layers on the acoustic distortion and optical attenuation. We concluded that the average intensity attenuation and distorting effect of the skull due to the blockage of the acoustic path is ~2.5% less and ~39.3% greater than those of the illumination path, respectively. These results can help in designing a more efficient photoacoustic imaging system suitable for transcranial brain imaging.

Author Contributions: Conceptualization, K.A.; methodology, R.M. and K.A.; software, R.M. and K.A.; validation, R.M., K.K., and K.A.; formal analysis, R.M., K.K., and K.A.; investigation, R.M. and K.A.; resources, K.A.; data curation, R.M. and K.A.; writing—original draft preparation, R.M. and K.A.; writing—review and editing, R.M., K.K., and K.A.; visualization, R.M. and K.A.; supervision, K.A. All authors have read and agreed to the published version of the manuscript.

Funding: We acknowledge funding support provided by National Institutes of Health (NIH) (R01 EB027769, R01 EB028661).

Conflicts of Interest: The authors declare no conflict of interest.

References

1. Kirsch, J.D.; Mathur, M.; Johnson, M.H.; Gowthaman, G.; Scoutt, L.M. Advances in transcranial Doppler US: Imaging ahead. *Radiographics* **2013**, *33*, E1–E14. [CrossRef]
2. Naqvi, J.; Yap, K.H.; Ahmad, G.; Ghosh, J. Transcranial Doppler ultrasound: A review of the physical principles and major applications in critical care. *Int. J. Vasc. Med.* **2013**, *2013*. [CrossRef] [PubMed]
3. Purkayastha, S.; Sorond, F. Transcranial Doppler ultrasound: Technique and application. *Semin. Neurol.* **2012**, *32*, 411–420. [CrossRef] [PubMed]
4. Van Wezel-Meijler, G.; Steggerda, S.J.; Leijser, L.M. Cranial ultrasonography in neonates: Role and limitations. *Semin. Neurol.* **2010**, *34*, 28–38. [CrossRef] [PubMed]
5. Chernyshev, O.Y.; Garami, Z.; Calleja, S.; Song, J.; Campbell, M.S.; Noser, E.A.; Shaltoni, H.; Chen, C.-I.; Iguchi, Y.; Grotta, J.C. Yield and accuracy of urgent combined carotid/transcranial ultrasound testing in acute cerebral ischemia. *Stroke* **2005**, *36*, 32–37. [CrossRef] [PubMed]
6. Bell, M.A.L.; Ostrowski, A.K.; Li, K.; Kazanzides, P.; Boctor, E.M. Localization of transcranial targets for photoacoustic-guided endonasal surgeries. *Photoacoustics* **2015**, *3*, 78–87. [CrossRef] [PubMed]
7. Schwartz, T.H.; Stieg, P.E.; Anand, V.K. Endoscopic transsphenoidal pituitary surgery with intraoperative magnetic resonance imaging. *Op. Neurosurg.* **2006**, *58*, ONS44–ONS51. [CrossRef]
8. Wang, L.V.; Hu, S. Photoacoustic tomography: In vivo imaging from organelles to organs. *Science* **2012**, *335*, 1458–1462. [CrossRef]
9. Nasiriavanaki, M.; Xia, J.; Wan, H.; Bauer, A.Q.; Culver, J.P.; Wang, L.V. High-resolution photoacoustic tomography of resting-state functional connectivity in the mouse brain. *Proc. Natl. Acad. Sci. USA* **2014**, *111*, 21–26. [CrossRef]
10. Hariri, A.; Fatima, A.; Mohammadian, N.; Bely, N.; Nasiriavanaki, M. Low cost photoacoustic spectroscopy system for evaluation of skin health. In Proceedings of the SPIE Optical Engineering+ Applications, San Diego, CA, USA, 28 August–1 September 2016.

11. Hariri, A.; Bely, N.; Chen, C.; Nasiriavanaki, M. Towards ultrahigh resting-state functional connectivity in the mouse brain using photoacoustic microscopy. In Proceedings of the SPIE BiOS, San Francisco, CA, USA, 13–14 February 2016; p. 97085A.

12. Manwar, R.; Li, X.; Mahmoodkalayeh, S.; Asano, E.; Zhu, D.; Avanaki, K. Deep Learning Protocol for Improved Photoacoustic Brain Imaging. *J. Biophotonics* **2020**. [CrossRef]

13. Mahmoodkalayeh, S.; Zarei, M.; Ansari, M.A.; Kratkiewicz, K.; Ranjbaran, M.; Manwar, R.; Avanaki, K. Improving vascular imaging with co-planar mutually guided photoacoustic and diffuse optical tomography: a simulation study. *Biomed. Optics Express* **2020**, *11*, 4333–4347. [CrossRef]

14. Liu, Y.; Liu, H.; Yan, H.; Liu, Y.; Zhang, J.; Shan, W.; Lai, P.; Li, H.; Ren, L.; Li, Z. Aggregation-Induced Absorption Enhancement for Deep Near-Infrared II Photoacoustic Imaging of Brain Gliomas In Vivo. *Adv. Sci.* **2019**, *6*, 1801615. [CrossRef] [PubMed]

15. Lindsey, B.D.; Light, E.D.; Nicoletto, H.A.; Bennett, E.R.; Laskowitz, D.T.; Smith, S.W. The ultrasound brain helmet: New transducers and volume registration for in vivo simultaneous multi-transducer 3-D transcranial imaging. *IEEE Trans. Ultrason. Ferroelectr. Freq. Control* **2011**, *58*, 1189–1202. [CrossRef] [PubMed]

16. Szabo, T.L.; Lewin, P.A. Ultrasound transducer selection in clinical imaging practice. *J. Ultrasound Med.* **2013**, *32*, 573–582. [CrossRef]

17. Wang, L.V. Tutorial on photoacoustic microscopy and computed tomography. *IEEE J. Sel. Top. Quantum Electron.* **2008**, *14*, 171–179. [CrossRef]

18. Zhou, Y.; Yao, J.; Wang, L.V. Tutorial on photoacoustic tomography. *J. Biomed. Opt.* **2016**, *21*, 61007. [CrossRef]

19. Mohammadi-Nejad, A.-R.; Mahmoudzadeh, M.; Hassanpour, M.S.; Wallois, F.; Muzik, O.; Papadelis, C.; Hansen, A.; Soltanian-Zadeh, H.; Gelovani, J.; Nasiriavanaki, M. Neonatal brain resting-state functional connectivity imaging modalities. *Photoacoustics* **2018**, *10*, 1–19. [CrossRef]

20. Fatima, A.; Kratkiewicz, K.; Manwar, R.; Zafar, M.; Zhang, R.; Huang, B.; Dadashzadeh, N.; Xia, J.; Avanaki, M. Review of Cost Reduction Methods in Photoacoustic Computed Tomography. *Photoacoustics* **2019**, *15*, 100137. [CrossRef]

21. Beard, P. Biomedical photoacoustic imaging. *Interface Focus* **2011**, *1*, 602–631. [CrossRef]

22. Kim, C.; Erpelding, T.N.; Jankovic, L.; Pashley, M.D.; Wang, L.V. Deeply penetrating in vivo photoacoustic imaging using a clinical ultrasound array system. *Biomed. Opt. Express* **2010**, *1*, 278–284. [CrossRef]

23. Mohammadi, L.; Behnam, H.; Tavakkoli, J.; Avanaki, M.R. Skull's Photoacoustic Attenuation and Dispersion Modeling with Deterministic Ray-Tracing: Towards Real-Time Aberration Correction. *Sensors* **2019**, *19*, 345. [CrossRef] [PubMed]

24. Mohammadi, L.; Manwar, R.; Behnam, H.; Tavakkoli, J.; Avanaki, M.R.N. Skull's aberration modeling: Towards photoacoustic human brain imaging. In Proceedings of the Photons Plus Ultrasound: Imaging and Sensing, San Francisco, CA, USA, 3–6 February 2019; p. 108785W.

25. Fry, F.J.; Barger, J.E. Acoustical properties of the human skull. *J. Acoust. Soc. Am.* **1978**, *63*, 1576–1590. [CrossRef]

26. White, P.J.; Clement, G.T.; Hynynen, K. Longitudinal and shear mode ultrasound propagation in human skull bone. *Ultrasound Med. Biol.* **2006**, *32*, 1085–1096. [CrossRef] [PubMed]

27. Treeby, B.E. Acoustic attenuation compensation in photoacoustic tomography using time-variant filtering. *J. Biomed. Opt.* **2013**, *18*, 36008. [CrossRef] [PubMed]

28. Treeby, B.E.; Cox, B.T. Modeling power law absorption and dispersion for acoustic propagation using the fractional Laplacian. *J. Acoust. Soc. Am.* **2010**, *127*, 2741–2748. [CrossRef]

29. Pichardo, S.; Sin, V.W.; Hynynen, K. Multi-frequency characterization of the speed of sound and attenuation coefficient for longitudinal transmission of freshly excised human skulls. *Phys. Med. Biol.* **2011**, *56*, 219–250. [CrossRef]

30. Dean-Ben, X.L.; Razansky, D.; Ntziachristos, V. The effects of acoustic attenuation in optoacoustic signals. *Phys. Med. Biol.* **2011**, *56*, 6129–6148. [CrossRef]

31. Fenster, A.; Lacefield, J.C. *Ultrasound Imaging and Therapy*; Taylor and Francis: Washington, DC, USA, 2015.

32. Nam, K.; Rosado-Mendez, I.M.; Rubert, N.C.; Madsen, E.L.; Zagzebski, J.A.; Hall, T.J. Ultrasound attenuation measurements using a reference phantom with sound speed mismatch. *Ultrason Imaging* **2011**, *33*, 251–263. [CrossRef]

33. Mohammadi, L.; Behnam, H.; Nasiriavanaki, M. Modeling skull's acoustic attenuation and dispersion on photoacoustic signal. In Proceedings of the Photons Plus Ultrasound: Imaging and Sensing, San Francisco, CA, USA, 29 January–1 February 2017; p. 100643U.

34. Volinski, B.; Hariri, A.; Fatima, A.; Xu, Q.; Nasiriavanaki, M. Photoacoustic investigation of a neonatal skull phantom. In Proceedings of the Photons Plus Ultrasound: Imaging and Sensing, San Francisco, CA, USA, 29 January–1 February 2017; p. 100643T.

35. Xu, Q.; Volinski, B.; Hariri, A.; Fatima, A.; Nasiriavanaki, M. Effect of small and large animal skull bone on photoacoustic signal. In Proceedings of the Photons Plus Ultrasound: Imaging and Sensing, San Francisco, CA, USA, 29 January–1 February 2017; p. 100643S.

36. Turani, Z.; Fatemizadeh, E.; Blumetti, T.; Daveluy, S.; Moraes, A.F.; Chen, W.; Mehregan, D.; Andersen, P.E.; Nasiriavanaki, M. Optical Radiomic Signatures Derived from Optical Coherence Tomography Images to Improve Identification of Melanoma. *Cancer Res.* **2019**, *79*, 2021–2030. [CrossRef]

37. Tournat, V.; Pagneux, V.; Lafarge, D.; Jaouen, L. Multiple scattering of acoustic waves and porous absorbing media. *Phys. Rev. E* **2004**, *70*, 26609. [CrossRef]

38. Yang, X.; Wang, L.V. Monkey brain cortex imaging by photoacoustic tomography. *J. Biomed. Opt.* **2008**, *13*, 44009. [CrossRef] [PubMed]

39. Yao, J.; Wang, L.V. Photoacoustic brain imaging: From microscopic to macroscopic scales. *Neurophotonics* **2014**, *1*, 11003. [CrossRef] [PubMed]

40. Zubiaurre-Elorza, L.; Soria-Pastor, S.; Junque, C.; Sala-Llonch, R.; Segarra, D.; Bargallo, N.; Macaya, A. Cortical thickness and behavior abnormalities in children born preterm. *PLoS ONE* **2012**, *7*, e42148. [CrossRef] [PubMed]

41. Ballardini, E.; Tarocco, A.; Baldan, A.; Antoniazzi, E.; Garani, G.; Borgna-Pignatti, C. Universal cranial ultrasound screening in preterm infants with gestational age 33–36 weeks: A retrospective analysis of 724 newborns. *Pediatr. Neurol.* **2014**, *51*, 790–794. [CrossRef]

42. Manwar, R.; Hosseinzadeh, M.; Hariri, A.; Kratkiewicz, K.; Noei, S.; N Avanaki, M. Photoacoustic Signal Enhancement: Towards Utilization of Low Energy Laser Diodes in Real-Time Photoacoustic Imaging. *Sensors* **2018**, *18*, 3498. [CrossRef]

43. Lynnerup, N.; Astrup, J.G.; Sejrsen, B. Thickness of the human cranial diploe in relation to age, sex and general body build. *Head Face Med.* **2005**, *1*, 13. [CrossRef]

44. Jin, S.-W.; Sim, K.-B.; Kim, S.-D. Development and growth of the normal cranial vault: An embryologic review. *J. Korean Neurosurg. Soc.* **2016**, *59*, 192. [CrossRef]

45. Webster, S.J.; Jee, S. *The Skeletal Tissues: Histology*; Elsevier Biomedical Press: Amsterdam, The Netherlands, 1983; pp. 200–254.

46. Abousleiman, Y.N.; Cheung, A.H.-D.; Ulm, F.-J. *Poromechanics III-Biot Centennial (1905–2005): Proceedings of the 3rd Biot Conference on Poromechanics, 24–27 May 2005, Norman, Oklahoma, USA*; CRC Press: Boca Raton, FL, USA, 2005.

47. Braidotti, P.; Branca, F.; Stagni, L. Scanning electron microscopy of human cortical bone failure surfaces. *J. Biomech.* **1997**, *30*, 155–162. [CrossRef]

48. Teoh, S.; Chui, C. Bone material properties and fracture analysis: Needle insertion for spinal surgery. *J. Mech. Behav. Biomed. Mater.* **2008**, *1*, 115–139. [CrossRef]

49. Njeh, C.F. *The Dependence of Ultrasound Velocity and Attenuation on the Material Properties of Cancellous Bone*; Sheffield Hallam University: Sheffield, UK, 1995.

50. Osterhoff, G.; Morgan, E.F.; Shefelbine, S.J.; Karim, L.; McNamara, L.M.; Augat, P. Bone mechanical properties and changes with osteoporosis. *Injury* **2016**, *47*, S11–S20. [CrossRef]

51. Wydra, A.; Malyarenko, E.; Shapoori, K.; Maev, R.G. Development of a practical ultrasonic approach for simultaneous measurement of the thickness and the sound speed in human skull bones: A laboratory phantom study. *Phys. Med. Biol.* **2013**, *58*, 1083. [CrossRef] [PubMed]

52. Geerits, T.W.; Kelder, O. Acoustic wave propagation through porous media: Theory and experiments. *J. Acoust. Soc. Am.* **1997**, *102*, 2495–2510. [CrossRef]

53. Manwar, R.; Kratkiewicz, K.; Avanaki, K. Overview of Ultrasound Detection Technologies for Photoacoustic Imaging. *Micromachines* **2020**, *11*, 692. [CrossRef]

54. Noguera, A.G. *Propagation of Ultrasound Through Freshly Excised Human Calvarium*; University of Nebraska-Lincoln: Lincoln, UK, 2012.

55. Han, S.; Rho, J.; Medige, J.; Ziv, I. Ultrasound velocity and broadband attenuation over a wide range of bone mineral density. *Osteoporos. Int.* **1996**, *6*, 291–296. [CrossRef]

56. Wang, S.; Chang, C.; Shih, C.; Teng, M. Evaluation of tibial cortical bone by ultrasound velocity in oriental females. *Br. J. Radiol.* **1997**, *70*, 1126–1130. [CrossRef]

57. Strelitzki, R.; Evans, J.; Clarke, A. The influence of porosity and pore size on the ultrasonic properties of bone investigated using a phantom material. *Osteoporos. Int.* **1997**, *7*, 370–375. [CrossRef]

58. Estrada, H.; Rebling, J.; Turner, J.; Razansky, D. Broadband acoustic properties of a murine skull. *Phys. Med. Biol.* **2016**, *61*, 1932. [CrossRef]

59. Bevilacqua, F.; Piguet, D.; Marquet, P.; Gross, J.D.; Tromberg, B.J.; Depeursinge, C. In vivo local determination of tissue optical properties: Applications to human brain. *Appl. Opt.* **1999**, *38*, 4939–4950. [CrossRef]

60. Antonio, A. Quantitative researches on the optical properties of human bone. *Nature* **1949**, *163*, 604. [CrossRef]

61. Firbank, M.; Hiraoka, M.; Essenpreis, M.; Delpy, D.T. Measurement of the optical properties of the skull in the wavelength range 650–950 nm. *Phys. Med. Biol.* **1993**, *38*, 503–510. [CrossRef]

62. Martin, B.; McElhaney, J.H. The acoustic properties of human skull bone. *J. Biomed. Mater. Res.* **1971**, *5*, 325–333. [CrossRef] [PubMed]

63. Bashkatov, A.N.; Genina, E.A.; Kochubey, V.I.; Tuchin, V.V. Optical properties of human cranial bone in the spectral range from 800 to 2000 nm. In Proceedings of the Saratov Fall Meeting 2005: Optical Technologies in Biophysics and Medicine VII, Saratov, Russia, 27–30 October 2005; pp. 616310–616311.

64. Takeuchi, A.; Araki, R.; Proskurin, S.; Takahashi, Y.; Yamada, Y.; Ishii, J.; Katayama, S.; Itabashi, A. A new method of bone tissue measurement based upon light scattering. *J. Bone Miner. Res.* **1997**, *12*, 261–266. [CrossRef] [PubMed]

65. White, D.; Curry, G.; Stevenson, R. The acoustic characteristics of the skull. *Ultrasound Med. Biol.* **1978**, *4*, 225241–239252. [CrossRef]

66. Wang, Q.; Reganti, N.; Yoshioka, Y.; Howell, M.; Clement, G.T. Comparison between diffuse infrared and acoustic transmission over the human skull. In Proceedings of the Meetings on Acoustics, Providence, RI, USA, 6–9 May 2014; p. 20002.

67. Lindsey, B.D.; Nicoletto, H.A.; Bennett, E.R.; Laskowitz, D.T.; Smith, S.W. Simultaneous bilateral real-time 3-D transcranial ultrasound imaging at 1 MHz through poor acoustic windows. *Ultrasound Med. Biol.* **2013**, *39*, 721–734. [CrossRef]

68. Bude, R.O.; Adler, R.S. An easily made, low-cost, tissue-like ultrasound phantom material. *J. Clin. Ultrasound* **1995**, *23*, 271–273. [CrossRef]

69. Tengsuthiwat, J.; Yorseng, K.; Siengchin, S.; Parameswaranpillai, J. Thermomechanical, water absorption, ultraviolet resistance and laser-assisted electroless plating behavior of Cu_2O and melamine–formaldehyde-coated sisal fiber-modified poly (lactic acid) composites. *Polym. Compos.* **2019**, *40*, 3264–3274. [CrossRef]

70. Karpienko, K.; Gnyba, M.; Milewska, D.; Wróbel, M.; Jędrzejewska-Szczerska, M. Blood equivalent phantom vs whole human blood, a comparative study. *J. Innov. Opt. Health Sci.* **2016**, *9*, 1650012. [CrossRef]

71. Fedorov, A.; Beichel, R.; Kalpathy-Cramer, J.; Finet, J.; Fillion-Robin, J.C.; Pujol, S.; Bauer, C.; Jennings, D.; Fennessy, F.; Sonka, M.; et al. 3D Slicer as an image computing platform for the Quantitative Imaging Network. *Magn. Reson. Imaging* **2012**, *30*, 1323–1341. [CrossRef]

72. Bell, M.A.L.; Ostrowski, A.K.; Li, K.; Kaanzides, P.; Boctor, E. Quantifying bone thickness, light transmission, and contrast interrelationships in transcranial photoacoustic imaging. In Proceedings of the Photons Plus Ultrasound: Imaging and Sensing, San Francisco, CA, USA, 3 March 2015; p. 93230C.

73. Boruah, S.; Paskoff, G.R.; Shender, B.S.; Subit, D.L.; Salzar, R.S.; Crandall, J.R. Variation of bone layer thicknesses and trabecular volume fraction in the adult male human calvarium. *Bone* **2015**, *77*, 120–134. [CrossRef]

74. Chaffaï, S.; Peyrin, F.; Nuzzo, S.; Porcher, R.; Berger, G.; Laugier, P. Ultrasonic characterization of human cancellous bone using transmission and backscatter measurements: Relationships to density and microstructure. *Bone* **2002**, *30*, 229–237. [CrossRef]

75. Dehghani, H.; Delpy, D.T. Near-infrared spectroscopy of the adult head: Effect of scattering and absorbing obstructions in the cerebrospinal fluid layer on light distribution in the tissue. *Appl. Opt.* **2000**, *39*, 4721–4729. [CrossRef] [PubMed]

76. Hughes, E.R.; Leighton, T.G.; Petley, G.W.; White, P.R. Ultrasonic propagation in cancellous bone: A new stratified model. *Ultrasound Med. Biol.* **1999**, *25*, 811–821. [CrossRef]
77. Lillie, E.M.; Urban, J.E.; Lynch, S.K.; Weaver, A.A.; Stitzel, J.D. Evaluation of skull cortical thickness changes with age and sex from computed tomography scans. *J. Bone Miner. Res.* **2016**, *31*, 299–307. [CrossRef] [PubMed]
78. McKelvie, M.; Palmer, S. The interaction of ultrasound with cancellous bone. *Phys. Med. Biol.* **1991**, *36*, 1331. [CrossRef]
79. Nicholson, P.; Strelitzki, R.; Cleveland, R.; Bouxsein, M. Scattering of ultrasound in cancellous bone: Predictions from a theoretical model. *J. Biomech.* **2000**, *33*, 503–506. [CrossRef]
80. Liu, J.; Lan, L.; Zhou, J.; Yang, Y. Influence of cancellous bone microstructure on ultrasonic attenuation: A theoretical prediction. *Biomed. Eng. Online* **2019**, *18*, 103. [CrossRef]
81. Samsudin, E.M.; Ismail, L.H.; Kadir, A.A.; Nasidi, I.N. Thickness, density and porosity relationship towards sound absorption performance of mixed palm oil fibers. In Proceedings of the 24th International Congress on Sound and Vibration (ICSV 24), London, UK, 23–27 July 2017.
82. Mahmoodkalayeh, S.; Jooya, H.Z.; Hariri, A.; Zhou, Y.; Xu, Q.; Ansari, M.A.; Avanaki, M.R. Low temperature-mediated enhancement of photoacoustic imaging depth. *Sci. Rep.* **2018**, *8*, 4873. [CrossRef]
83. Mozaffarzadeh, M.; Mahloojifar, A.; Orooji, M.; Adabi, S.; Nasiriavanaki, M. Double-Stage Delay Multiply and Sum Beamforming Algorithm: Application to Linear-Array Photoacoustic Imaging. *IEEE Trans. Biomed. Eng.* **2018**, *65*, 31–42. [CrossRef]
84. Estrada, H.C.; Huang, X.; Rebling, J.; Zwack, M.; Gottschalk, S.; Razansky, D. Virtual craniotomy for high-resolution optoacoustic brain microscopy. *Sci. Rep.* **2018**, *8*, 1459. [CrossRef]
85. Omidi, P.; Zafar, M.; Mozaffarzadeh, M.; Hariri, A.; Haung, X.; Orooji, M.; Nasiriavanaki, M. A novel dictionary-based image reconstruction for photoacoustic computed tomography. *Appl. Sci.* **2018**, *8*, 1570. [CrossRef]
86. Mozaffarzadeh, M.; Mahloojifar, A.; Orooji, M.; Kratkiewicz, K.; Adabi, S.; Nasiriavanaki, M. Linear-array photoacoustic imaging using minimum variance-based delay multiply and sum adaptive beamforming algorithm. *J. Biomed. Opt.* **2018**, *23*, 26002. [CrossRef] [PubMed]

Article

Analysis of Wave Patterns Under the Region of Macro-Fiber Composite Transducer to Improve the Analytical Modelling for Directivity Calculation in Isotropic Medium

Kumar Anubhav Tiwari [1,3,*], Renaldas Raisutis [1,2] and Liudas Mazeika [1]

[1] Ultrasound Research Institute, Kaunas University of Technology, K. Baršausko St. 59, LT-51423 Kaunas, Lithuania; renaldas.raisutis@ktu.lt (R.R.); liudas.mazeika@ktu.lt (L.M.)

[2] Department of Electrical Power Systems, Faculty of Electrical and Electronics Engineering, Kaunas University of Technology, Studentu g. 50, LT-51368 Kaunas, Lithuania

[3] Department of Multimedia Engineering, Kaunas University of Technology, Studentu g. 50, LT-51368 Kaunas, Lithuania

[*] Correspondence: k.tiwari@ktu.lt; Tel.: +370-64694913

Received: 9 March 2020; Accepted: 14 April 2020; Published: 17 April 2020

Abstract: Analytical modelling is an efficient approach to estimate the directivity of a transducer generating guided waves in the research field of ultrasonic non-destructive testing of the large and complex structures due to its short processing time as compared to the numerical modelling and experimental techniques. The wave patterns or the amplitude variations along the region of ultrasonic transducer itself depend on its behavior, excitation frequency, and the type of propagating wave mode. Depending on the wave-pattern of a propagating wave mode, the appropriate value of the amplitude correction factor must be multiplied to the amplitudes of the excitation signal for the accurate evaluation of directivity pattern of the ultrasonic transducers generating guided waves in analytical modelling. The objective of this work is to analyse the wave patterns under the region of macro-fiber composite (MFC) transducer to improve the accuracy of a previously developed analytical model for the prediction of directivity patterns. Firstly, the amplitude correction factor based on the wave patterns under the region of P1-type MFC (MFC-2814) transducer at two different frequencies (80 kHz, 3 periods and 220 kHz, 3 period) glued on 2 mm Al alloy plate has been estimated analytically in the case of an asymmetric (A0) guided Lamb wave. The validation of analytically estimated amplitude correction factor is performed by a proposed experimental method that allows analyzing the behaviour of MFC transducer under its region by gluing MFC on bottom surface and scanning the receiver on the top surface of the sample. Later on, the estimated amplitude correction factor is included in the previously developed 2D analytical model for the improvement in the directivity patterns of the A0 mode. The modified analytical model shows a significant improvement in the directivity pattern of the A0 wave mode in comparison to the results obtained by the previous model without considering the proper wave patterns. The results reveal that errors between the directivity estimated by the present modified 2D analytical model and experimental investigation are reduced by more than 58% in comparison to the previously developed analytical model.

Keywords: wave patterns; analytical model; directivity pattern; guided wave (GW); non-destructive testing (NDT); macro-fiber composite (MFC); transducer

---

1. Introduction

One of the key issues in the structural health monitoring (SHM) of various composite structures and components is to maintain the safety, reliability, and operational performance [1–3]. For the last few decades, ultrasonic guided waves (GWs) have been used for this purpose to detect and locate the defects in the structures. Among all the available non-destructive testing (NDT) techniques, ultrasonic guided wave (GW) testing has been the most promising due to its high sensitivity to the defects and wide coverage region [4,5]. Moreover, GW testing is fast, can cover up the defective regions to reasonable distances, and has the ability to detect defects underground, water, or a layer of insulation [6–8]. In comparison to guided wave testing, bulk wave testing is tedious and time-consuming, requires high-level training, uses the point-by-point scanning method, and needs a visible area and accessibility of the defective region [9,10]. Due to the high sensitivity of GWs to the variation in modulus of elasticity (E) of the material under testing and minimal amplitude damping of propagating wave modes, only a few measurements are required for the inspection of large infrastructures to detect internal and surface defects [1,10,11]. Researchers have successfully utilized GWs for inspecting defects/damages in metallic structures [12], concrete structures [13,14], pipes [15–17], and composite structures [18–24].

The Lamb wave is a specific type of guided wave that propagates in a plate-type structures and can be further categorized into the symmetric Lamb waves (S0, S1 ...) and asymmetric Lamb waves (A0, A1 ...) depending on the value of frequency-thickness product ($f \cdot t$), f is the excitation frequency of ultrasonic transducer and t is the thickness of propagating medium or structure under inspection. In the case of lower frequencies, only two fundamental guided Lamb modes (the S0 and A0) exist. Due to their high sensitivity in defective regions of structures, guided Lamb waves are widely used for the inspection of different types of defects such as delaminations, cracks and impact damages, etc. [3,25–27]. Many approaches and transducers are available for the generation of Lamb waves. Out of those, the interdigital transducers are gaining the most recognition [28–31].

Due to its small size, light weight, flat geometry, ability to work in actuation, transmission, and sensing mode, the macro fiber composite (MFC) transducer is one of the best interdigital transducers for NDT and SHM of composite structures [32–35]. The MFC transducer consists of rectangular shaped piezo ceramic rods. These rods are sandwiched between the layers of adhesive, electrodes, and polyimide film. The electrodes attached to the film form an interdigitated pattern. The electrodes transfer the applied electrical energy to/from the rods. In our research, MFC transducer of P1-type (M-2814-P1) with dimensions of 28 × 14 mm is used. The general parametric characteristics of the MFC-2814-P1 transducer are presented in Table 1 (32).

Table 1. General characteristics of MFC-P1-M2814 [32].

Features	Numerical Value
Active (length × width)	28 mm × 14 mm
Overall (length × width)	38 mm × 20 mm
Capacitance	0.61 nF
Free strain	1550 ppm
Blocking force	195 N
Operating voltage	−500 V to +1500 V
Operating bandwidth as a sensor	0 Hz to 1 MHz
Operating bandwidth as an actuator	0 Hz to 700 kHz
Maximum operational tensile strain	<4500 ppm
Linear-elastic tensile strain limit	1000 ppm

Guided Lamb waves (the A0 and S0) can be effectively transmitted and received by using an MFC transducer [34]. The S0 mode contains dominant in-plane whereas the A0 mode contains dominant out-of-plane components of the propagating waves. The inspection using MFC transducer can be easily combined with different contact and non-contact ultrasonic inspection methods for NDT and SHM of composite structures [36,37]. The MFC transducers can be easily glued or embedded within large and complex structures without damaging the surface [38]. In aerospace applications, the embedded MFCs are frequently used for generating and harvesting ultrasonic wave energy, SHM of a structure, and detecting defects and damages due to impact [35,39]. MFCs can control the twisting motion of aircraft wings as well as the airfoils' aerodynamic shaping [40,41]. Hence, it can increase the efficiency of an aircraft by improving its aerodynamic performance. In comparison to active fiber composite (AFC), MFC has a high fiber volume fraction which ensures its high stiffness and performance. Moreover, MFCs have better actuation performance compared to the most common piezoceramic actuators [35,40,41].

Although interdigital transducers are widely used for the transmission and reception of ultrasonic GWs, the dispersive nature and multi-modal behavior possessed by Lamb waves are the limiting factors for their adaptation and utilization in SHM. To ensure the effective application of a transducer for the inspection of a specific structure, the directivity of a transducer is one of the key parameters. Knowing the transducer directivity, the following amendments/adaptations can be performed [38]:

- The position of a transducer on the structure under inspection can be determined.
- The number/configuration of transducers can be decided.
- A specific wave mode (e.g., the S0, A0 and SH0 in LF ultrasonic) and excitation frequency can be selected for the inspection of defects.
- The best transducer for the specific application can be selected.

The analytical method is an efficient approach to calculate the directivity of transducers due to shorter processing time in comparison to the experimental or numerical analysis. An efficient 2D analytical model based on Huygens's principle was developed in our previous research for directivity estimation of the contact-type transducer at any distance and excitation frequency with known dispersive characteristics of propagation medium and behaviour of transducer [38]. The directivity patterns of the S0, A0, and fundamental shear-horizontal mode (SH0) for the P1-type MFC transducer glued on Al alloy plate were successfully estimated by this model and the obtained results showed a good compromise with the experimental results [38]. However, the correct wave patterns under the transducer region and their effect on directivity patterns at specific frequencies were not considered in our previous work [38]. In the previous model, the amplitude variations of the excitation signal were considered a fixed value for the directivity estimations at different frequencies. A similar assumption was considered by another researcher using numerical modelling [33].

The objective of this work is to analytically analyse the wave patterns under the region of an MFC transducer glued on isotropic medium and validate by the experimental investigation, which in turn improves the 2D analytical model for the estimation of directivity patterns. The P1-type MFC transducer with dimensions (28 × 14 mm) was glued on a 2 mm thick Al alloy plate. The wave patterns under the transducer region were analyzed analytically in order to improve the previously developed analytical model [38] for the accurate analysis of the directivity patterns. We showed that wave patterns of the excitation signal are different at different frequencies under the transducer region. Hence, the frequency-dependent amplitude correction factor is estimated and included in the model. We also propose a new experimental technique to validate the wave patterns and amplitude correction factors calculated analytically. In the proposed measurement technique, MFC transducer was glued on one side of Al plate and scanning was performed on the opposite side of the plate under the region of MFC transducer. The experiment was performed by using the low-frequency (LF) ultrasonic system developed by Ultrasound Institute, Kaunas University of Technology [36,38,42]. The point-type piezoceramic transducer operating in thickness mode was used in the experimental analysis for recording the Lamb waves. The receiving transducer was more sensitive to the out-of-plane wave components.

Hence, in this research, the improvement in the directivity of only A0 mode is discussed. The calculated amplitude correction factor based on the wave patterns under the region of MFC transducer was included in the analytical modeling. The analytical solution was verified by the experimental analysis, which clearly showed a significant improvement in the directivity pattern of the A0 mode as compared to the previously obtained results.

Section 2 of this article illustrates the detailed description of a problem. Section 3 presents the calculation of the modified amplitude factor based on the wave patterns along the region of MFC transducer. The verification of the calculated amplitude factor by the experimental analysis has been presented in Section 4. A comparison of the results obtained by an analytical model and experimental investigation has been performed in Section 5 followed by the conclusive remarks in Section 6.

2. Description of a Problem

According to the previously developed 2D analytical model based on Huygens's principle [38], the P1-type MFC transducer was considered as the number of line segments with distributed point sources along with its structure. The arbitrary points along the angles from 0° to 180° at a specified distance were considered as receiving elements. The schematic of the model is presented in Figure 1.

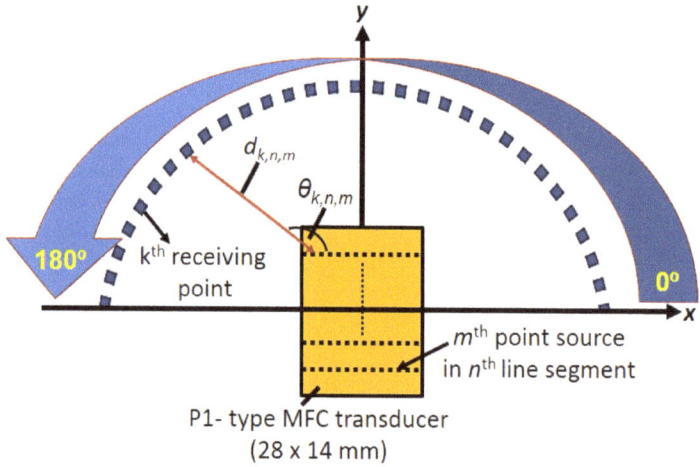

Figure 1. 2D analytical model schematic for estimation of the directivity of P1 MFC transducer [38].

At each receiving point, the signals propagating from all point sources were calculated and integrated in order to calculate the received signals along the angular region. The received signal spectrum can be expressed as [38]:

$$U_{R,k}(f, \theta_k) = \sum_{n=1}^{N} \sum_{m=1}^{M} U_{EC}(f) \cdot H_T\left(f, d_{k,n,m}, v_{ph}\right) \cdot \frac{1}{\sqrt{d_{k,n,m}}} \tag{1}$$

where k is the number of receiving element ($k = 1, 2, \ldots K$); θ_k is the angle between the kth receiving point and origin ($\theta_k = [(k-1) \cdot d\theta]$; $d\theta$ is the angular separation between receiving elements); n is line segment ($n = 1, 2, \ldots N$); m is point source ($m = 1, 2, \ldots M$); H ($f, d_{k,n,m}, v_{ph}$) is the transfer function [H ($f, d_{k,n,m}, v_{ph}$)= exp ($-a(f) \cdot d_{k,n,m}$)·exp ($-j2\pi f\, d_{k,n,m}/v_{ph}$ (f, h))]; $a(f)$ is the frequency-dependent attenuation coefficient; v_{ph} is phase dispersion velocity which depends on the thickness (h) of the plate and the frequency of excitation. $d_{k,n,m}$ is the distance from the mth point source to the kth receiving element; $U_{EC}(f)$ is FT of the input signal $u_{EC}(t)$; $U_{R,k}$ (f, θ_k) is the FT of the received signal and $1/\sqrt{d_{k,n,m}}$ is the diffraction factor corresponding to the distance.

The normalized amplitudes (A_{npp}) along the polar coordinates to plot the directivity pattern is expressed as:

$$A_{npp}(\theta_k) = \left[\frac{\max\left(FT^{-1}\left[U_{R,k}(f, \theta_k)\right]\right) - \min\left(FT^{-1}\left[U_{R,k}(f, \theta_k)\right]\right)}{\max\left(\max\left(FT^{-1}\left[U_{R,k}(f, \theta_k)\right]\right) - \min\left(FT^{-1}\left[U_{R,k}(f, \theta_k)\right]\right)\right)} \right] \tag{2}$$

The excitation signal was multiplied by the correction factor (A_F) corresponding to the particular Lamb wave mode (the S0, A0 or SH0) for the directivity estimation [38,43]. The approximated value of amplitude correction factor (A_F) in the model was considered depending on the behaviour of the P1-type MFC transducer for each of the wave modes. As P1 type MFC operates in elongation mode as shown in Figure 2a, A_F was considered as linearly increasing value from 0 at center up to 1(−1) at edges for upper/lower half-sections along the length (L_{MFC}) of MFC transducer for S0 mode (Figure 2b). On the other hand, A_F contained only two maximum labels with opposite polarities at the edges in the case of the A0 mode as presented in Figure 2c. It should be noted that 1 or −1 are the maximum labels and can be replaced by any numerical value.

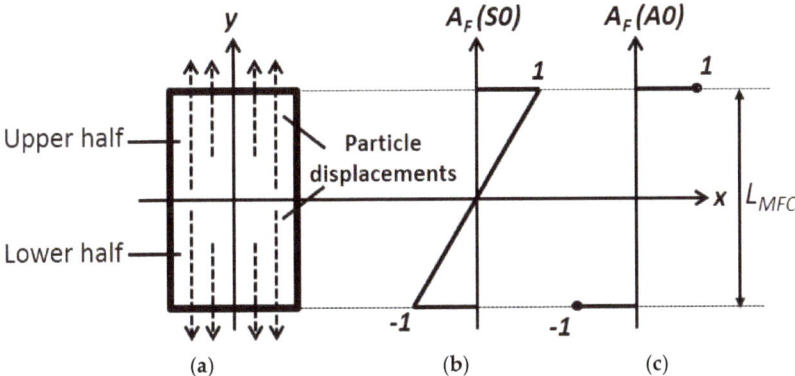

Figure 2. Particle displacements in P1-type MFC (28 × 14 mm) along with its length (**a**) and the amplitude correction factor (A_F) by previously developed model for the S0 (**b**) and the A0 mode (**c**).

Hence, the dependence of A_F on the excitation frequency and operative wavelength was not considered in the previous model. Although the directivity patterns of the A0 mode obtained by considering A_F using this approach showed significant similarity with the experimental results [38] at 80 kHz and 220 kHz excitation signals, there was still scope for improvement. The detailed specifications about the set-up and scanning procedure to estimate the directivity patterns by experimental analysis are described in [38].

The directivity patterns of the A0 mode calculated by a previous analytical model and the experiment at 300 mm from the center of the transducer with 80 kHz and 220 kHz excitations are shown in Figure 3 [38]. It can be clearly observed from Figure 3 that the number of side lobes in analytical results is equal to the experimental results but there is a significant difference in shapes. This is due to the fact that the signal distributions (amplitude correction factor (A_F)) along the structure of the MFC transducer were not considered correctly. The numerical model developed by Haig et al. for the estimation of directivity patterns of MFC on steel plate also had a similar limitation with the amplitude correction factor [33]. Therefore, this research aims to find the accurate amplitude correction factor for the P1-type MFC transducer according to the wave patterns in its structure along the length, which in turn could improve the analytical modelling for the prediction of directivity patterns. Only the A0 mode is considered in this work due to the very high sensitivity of receiving point-type piezoceramic transducer for out-of-plane radiations.

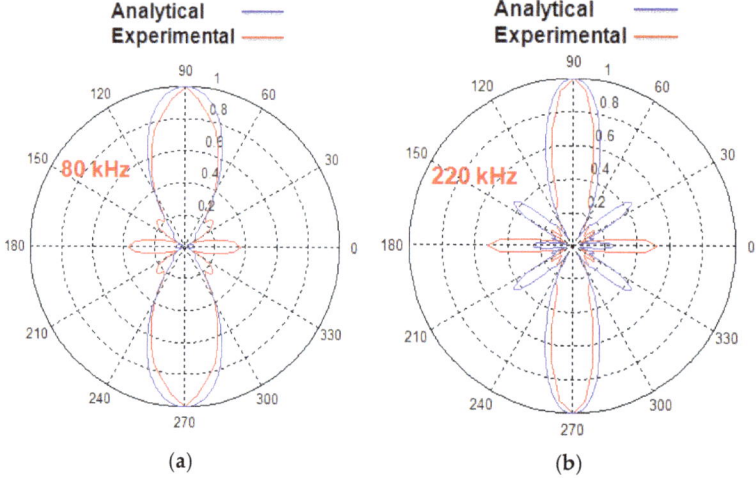

Figure 3. Directivity patterns of the A0 mode of P1-type MFC (28 × 14 mm) transducer at 300 mm distance from the center of a transducer without considering the appropriate wave patterns at 80 kHz (**a**) and 220 kHz (**b**).

3. Modified Amplitude Correction Factor (AF)

The amplitude correction factor is calculated by combining the information of wave patterns of the A0 and behavioral characteristics of P1-type MFC transducer. The wave patterns or signal distributions in the region of MFC transducer depend on the operating wavelength (λ) of the excitation signal along its length. The wavelength (λ) can be expressed by:

$$\lambda = \frac{v_{ph}}{f} \tag{3}$$

where v_{ph} is the phase velocity of propagation (Al alloy in our case of consideration) and f is the frequency of excitation signal.

The number of wavelengths (N_λ) along the length (L) of P1-type MFC (L = 28 mm) can be expressed by:

$$N_\lambda = \frac{L}{\lambda} \tag{4}$$

Due to the symmetry of the P1-type MFC structure and its operation in elongation (d33) mode, the approximated number of positive and negative peaks of the signals along the region of MFC transducer should be $2N_\lambda$. Therefore, rather than only two levels (Figure 2c), the profile of amplitude correction factor (A_F) for the A0 mode can be represented by D_λ discrete amplitude points in positive and negative directions alternatively and equispaced along the length of MFC.

A number of discrete levels must be an integer number, D_λ can be expressed as:

$$D_\lambda = \text{ceil} \lceil 2N_\lambda \rceil \tag{5}$$

where 'ceil' denotes a ceiling function that maps the real number to least integer greater than or equal to the number.

The spacing between the discrete values (Δ) of A_F will be given by:

$$\Delta = \frac{L}{D_\lambda - 1} \tag{6}$$

Hence, an amplitude correction factor of excitation signal in the analytical modelling can be expressed by two different mathematical functions depending on the number of discrete levels (i.e., even or odd) along the length of MFC transducer. The two cases are illustrated as follows:

$$A_F = \begin{cases} \delta(y) + \sum\limits_{p=1}^{\frac{D_\lambda-1}{2}} [\delta(y-2p\Delta) - \delta(y-(2p-1)\Delta)], & \text{if } D_\lambda \text{ is odd} \\ \sum\limits_{p=0}^{(\frac{D_\lambda}{2})-1} [\delta(y-2p\Delta) - \delta(y-(2p+1)\Delta)], & \text{if } D_\lambda \text{ is even} \end{cases} \tag{7}$$

where y is the longitudinal axis of MFC transducer; p is the discrete number depending on D_λ ($p = 1, 2$ ... $(D_\lambda-1)/2$ if D_λ is odd and $p = 1, 2$ ... $(D_\lambda/2)-1$ if D_λ is even); $\delta(y)$ denotes the unit impulse signal).

Two different excitation signals i.e., 80 kHz, 3-period and 220 kHz, 3-period with a Gaussian shape as shown in Figure 4a,b were considered in the analysis. The dispersion characteristics of the A0 mode in 2 mm Al plate were estimated using the computational package "Disperse" [44]. The phase velocity at 80 kHz and 220 kHz were observed as 1182 m/sec and 1795 m/sec, respectively, as shown in the dispersion curve (Figure 4c).

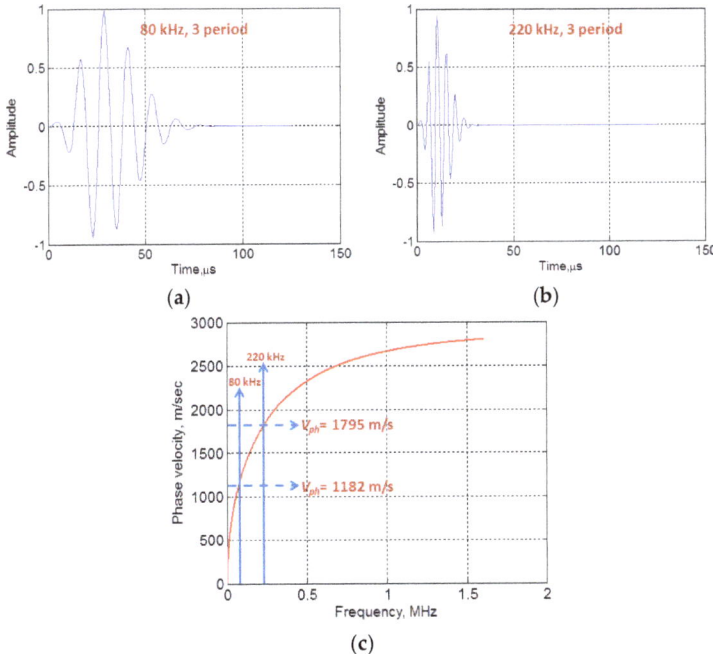

Figure 4. 80 kHz, 3 period (**a**) and 220 kHz, 3-period excitation signals (**b**) with Gaussian symmetry and the phase velocity dispersion curve of the A0 wave mode in 2 mm Al alloy plate (**c**).

Hence, the wavelength (λ_{80} and λ_{220}) and a number of wavelengths ($N_{\lambda 80}$ and $N_{\lambda 220}$) along the length and under the region of MFC transducer at 80 kHz and 220 kHz frequencies can be calculated by using Equations (3) and (4).

$$\lambda_{80} = 14.78 \text{ } mm; \lambda_{220} = 8.16 \text{ } mm; N_{\lambda 80} = 1.89; N_{\lambda 220} = 3.43 \tag{8}$$

Thus, the approximated number of discrete positive and negative amplitudes (D_λ) under the transducer region along its length can be calculated from Equation (5) as four and seven (corresponding to the excitation frequencies of 80 kHz and 220 kHz respectively. The schematic of modified amplitude correction factor A_F for the A0 mode is presented in Figure 5a,b in the case of 80 and 220 kHz excitation signals respectively. The A_F will have the following discrete values under the structure/region of MFC along its length:

- In the case of 80 kHz frequency, A_F will have the four discrete values (i.e., two with the same polarity and two with opposite polarity). The spatial separation (Δ) between the discrete values (Equation (7)) will be equal to 9.33 mm.
- Similarly, A_F will contain seven discrete values (i.e., four with the same polarity and three with opposite polarity) with the excitation frequency of 220 kHz. The spatial separation (Δ), in this case, will be 4.67 mm.

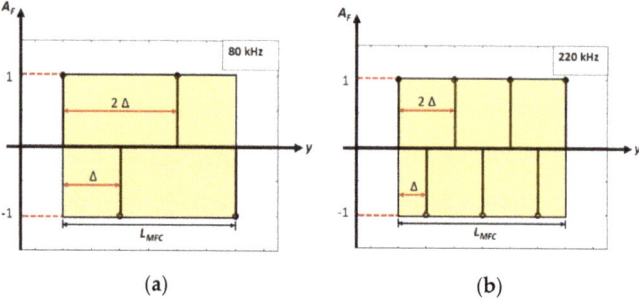

(a)   (b)

Figure 5. Amplitude correction factor A_F at 80 kHz (**a**) and 220 kHz (**b**) along the length of MFC transducer for the A0 mode.

After including the modified amplitude correction factor, the directivity pattern can be estimated by the analytical model [38].

4. Experimental Validation

The new measurement technique is proposed to experimentally analyze the behaviour of MFC transducer and wave patterns along with its structure for the verification of the estimated value of A_F in Section 3. The experiment was performed using the LF ultrasonic system ("Ultralab") developed by Ultrasound Research Institute of Kaunas University of Technology. The schematic of experimental investigation is presented in Figure 6a. The characteristics of the LF ultrasonic system are described in Table 2.

Table 2. Parameters of LF ultrasonic system [36,42,45].

Parameters	Numerical Value
No. of input channels	2
No. of bits of the analog-to-digital converter	10
Overall system gain (maximum)	113 dB
Ultrasonic system to computer interface	USB V.2
Frequency range	20 kHz–2 MHz

The P1-type MFC-2814 (28 × 14 mm) transducer was glued at the centre of the Al alloy plate with dimension (1000 × 1000 × 2 mm) on one side of a plate. The scanning with a 1 mm step was performed on the opposite side of plate under the cross-sectional area of (50 × 50 mm) which also covered the region of MFC transducer as described in Figure 6b. The experiment was repeated two times to record the data in the case of two different excitation signals, i.e., 80 kHz, 3-period and 220 kHz, 3-period with a Gaussian shape for exciting the MFC transducer as shown in Figure 3a,b. The sampling frequency was 100 MHz. The wideband contact-type ultrasonic transducer (maximum −6 dB bandwidth was equal to 300 kHz) was used to record the ultrasonic signals. Glycerol was used for effective acoustic contact between the transducer and Al alloy plate. All components including the ultrasonic system used in the experimental investigation were developed by Ultrasound Research Institute of the Kaunas University of Technology.

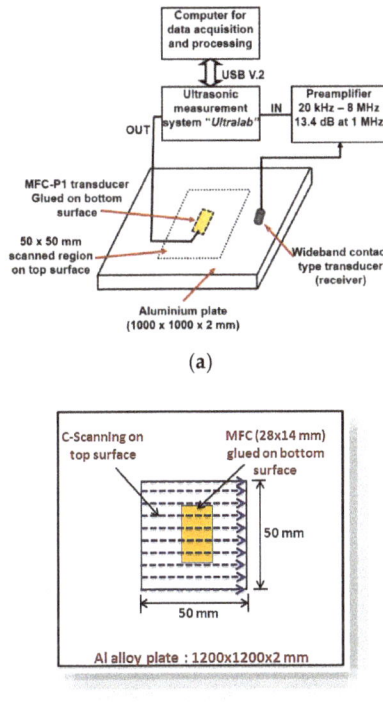

(a)

(b)

Figure 6. Schematic showing the experimental set-up (**a**) and C-scanning procedure in 50 × 50 mm region on the top surface with MFC glued on the bottom surface of Al alloy plate (**b**).

The B-scan images acquired along the length of MFC at 80 kHz and 220 kHz excitation frequencies are shown in Figure 7. It can be clearly observed from Figure 5a that approx. No. of discrete peaks D_λ is (3.5 ≈ 4) along the length of MFC at 80 kHz frequency. On the other hand, there is approx. No. of discrete peaks equal to 7 at 220 kHz frequency as shown in Figure 5b. Therefore, these results are similar to those obtained analytically and hence validate the calculation of amplitude correction factor (A_F) as described in Section 3. In order to view the two possible cases of signal peaks along the length of MFC transducer with more clear visibility, the C-scan images at 35 μs and 45 μs were acquired in the case of 80 kHz frequency. Similarly, the C-scan images at 20 μs and 24 μs were obtained for the excitation frequency of 220 kHz. The C-scan images are shown in Figure 8a–d. The time instants were chosen to show the wave patterns and number of signal peaks. The C-scan images provide a clearer visualization of the estimation of D_λ and hence, the A_F in the case of 80 kHz and 220 kHz frequency respectively.

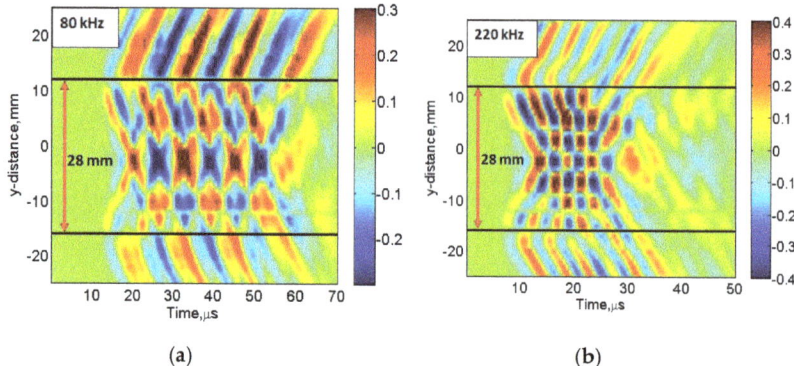

Figure 7. B-scan along the longitudinal axis of MFC transducer at 80 kHz (**a**) and at 220 kHz (**b**).

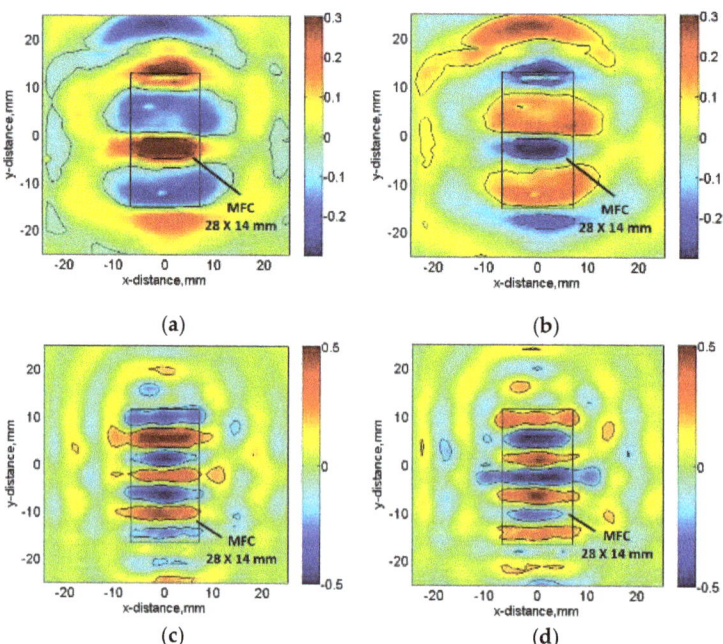

Figure 8. C-scan showing the wave patterns for the A0 mode: at 33 μs (**a**) and 40 μs (**b**) in the case of 80 kHz excitation; at 18.9 μs (**c**) and 21. 6 μs (**d**) in the case of 220 kHz excitation.

5. Results and Analysis

The estimation of amplitude correction factor analytically (Section 3) is validated by experimental analysis (Section 4). After including the modified values of amplitude correction factor A_F in the analytical model, the directivity patterns of MFC transducer at 80 kHz and 220 kHz at 300 mm distance from the center of the transducer are estimated in the case of the A0 mode. The experimental investigation to obtain the directivity patterns were already performed in the previous research [38]. The directivity patterns obtained by the modified analytical model are presented in Figure 9 with their corresponding experimental results.

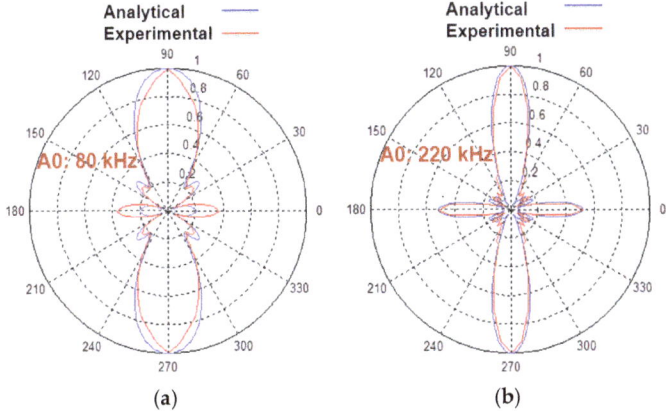

Figure 9. Comparison of directivity patterns of the A0 mode of P1-type MFC transducer with a modified analytical model at 80 kHz (**a**) and 220 kHz (**b**) and experimental analysis.

In comparison to the results obtained in the previously developed model [38] as presented in Figure 3, the directivity patterns obtained by the modified model show more similarities with experimental results. Therefore, the inclusion of spatial distribution of the amplitudes of excitation signal significantly improves the previously developed analytical model. This could also improve the numerical model developed by Haig et al. by resolving a similar limitation with the amplitude correction factor [33]. In order to quantitatively estimate the improvement in results as compared to the previous model, the error between the normalized amplitudes along the polar coordinates of experimental results with that obtained by previously developed model and the modified analytical model is compared. The MFC transducer is symmetric in construction. Thus, the directivity pattern along 0° to 90° with an angular separation of 5° is considered for the comparative analysis of the previous and new modified model. The absolute value of the difference between the normalized amplitudes of experimental results and the modelling results (amplitude error) along the polar axis (0° to 90°) is presented in the case of 80 and 220 kHz frequencies. The comparative results are presented in Figure 10a,b.

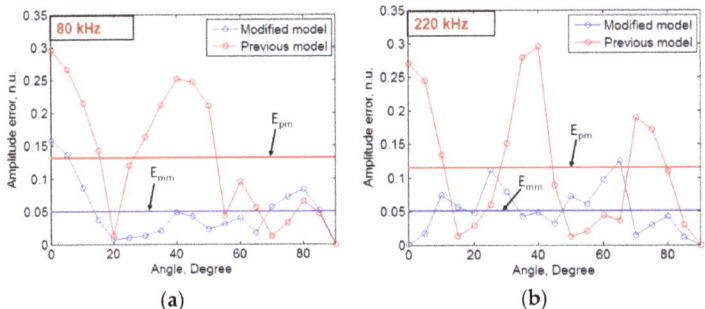

Figure 10. Comparison of amplitude errors in the results obtained by modified model and previously developed model in the case of 80 kHz (**a**) and 220 kHz (**b**) (E_{pm} -Mean error in the previous model, E_{mm} -Mean error in the modified model).

It is clearly observed from Figure 10a,b that the amplitude error is significantly reduced in the modified model compared to the previously developed model. At 80 kHz frequency (Figure 10a), the range of amplitude error was observed as (0–0.3) and (0–0.16) in the case of the previous model and new modified model respectively. The corresponding mean error (E_{pm} and E_{mm}) in this case was estimated as 0.13 (E_{pm}) and 0.05 (E_{mm}), respectively. Hence, at 80 kHz frequency, the relative error in the estimation of directivity patterns by new modified model is reduced by 61.54% as compared to the previous model. In the case of 220 kHz frequency, the amplitude error lies in the range of (0–0.3) for previously developed model and (0–0.13) for the modified model. The mean error (E_{pm} and E_{mm}) at 220 kHz was calculated as 0.12 and 0.05 in the case of a previously developed model and newly developed modified analytical model, respectively. The relative error at 220 kHz is reduced by 58.33% in the modified model in comparison to the previous model.

6. Conclusions

In this work, the accuracy of the previously developed 2D analytical model to predict and estimate the directivity pattern of the MFC transducer in the isotropic medium is increased by including the correct wave patterns of the excitation signal under the spatial region of the transducer. The wave patterns along the structure of MFC transducer are estimated analytically and validated by experimental analysis for 80 kHz and 220 kHz frequencies for the A0 mode. A new measurement technique is also proposed to analyse the spatial behaviour of the MFC transducer and wave patterns by gluing the MFC on one side of the sample and scanning on the opposite side under its region. The C-scan images under the MFC transducer and the B-scan images along the longitudinal axis of the MFC transducer were obtained at different frequencies.

In this way, we showed the dependency of amplitude correction factor on excitation frequency and included it in the present model. The P1-type MFC transducer and 2-mm thick Al alloy medium are used for a demonstration of modelling. It should be noted that dispersive phase velocity in the modelling is included by calculating the theoretical dispersion curves based on the thickness of propagating medium. In comparison to the previously developed analytical model, the error between the experimental and analytical results is reduced by 61.54% and 58.33% in the case of 80 kHz and 220 kHz, respectively. The model has significant flexibility by providing the option of selecting any isotropic propagation medium, frequency of excitation, and spatial dimensions (length and width) of transducer. In general, it is possible to include completely all distributions of the wave under the transducer. However, this leads to longer simulation time. The proposed method simplifies this task as the number of excitation points in modelling is essentially reduced compared to the case when the total spatial distribution of excitation amplitudes is taken into account.

Author Contributions: Conceptualization, K.A.T. and R.R.; methodology, K.A.T., R.R., and L.M.; software, K.A.T.; validation, K.A.T.; formal analysis, K.A.T. and R.R.; investigation, K.A.T. and L.M.; resources, K.A.T.; data curation, L.M., R.R., and K.A.T.; writing—original draft preparation, K.A.T.; writing—review and editing, K.A.T. and R.R; visualization, K.A.T. supervision, L.M. and R.R.; project administration, R.R.; funding acquisition, K.A.T., R.R., and L.M. All authors have read and agreed to the published version of the manuscript.

Funding: This research received no external funding.

Acknowledgments: This work was performed at Ultrasound Research Institute of the Kaunas University of Technology, Lithuania.

Conflicts of Interest: The authors declare no conflict of interest.

References

1. Cawley, P. Practical Guided Wave Inspection and Applications to Structural Health Monitoring. In Proceedings of the 5th Australasian Congress on Applied Mechanics, Brisbane, Australia, 10–12 December 2007; Martin, V., Faris, A., Daniel, B., Griffiths, J., Hargreaves, D., McAree, R., Meehan, P., Tan, A., Eds.; Engineers Australia: Brisbane, Australia, 2007; pp. 12–21.
2. Diamanti, K.; Soutis, C. Structural health monitoring techniques for aircraft composite structures. *Prog. Aerosp. Sci.* **2010**, *46*, 342–352. [CrossRef]
3. Yelve, N.P.; Mitra, M.; Mujumdar, P. Detection of delamination in composite laminates using Lamb wave based nonlinear method. *Compos. Struct.* **2017**, *159*, 257–266. [CrossRef]

4. Wilcox, P.; Konstantinidis, G.; Croxford, A.J.; Drinkwater, B.W. Strategies for Guided Wave Structural Health Monitoring. *AIP Conf. Proc.* **2007**, *894*, 1469–1476.

5. Raghavan, A.; Cesnik, C.E.S. Review of Guided-wave Structural Health Monitoring. *Shock. Vib. Dig.* **2007**, *39*, 91–114. [CrossRef]

6. Rose, J.L. Ultrasonic Guided Waves in Structural Health Monitoring. *Key Eng. Mater.* **2004**, *270*, 14–21. [CrossRef]

7. Michaels, J.E.; Dawson, A.J.; Michaels, T.E. *Massimo Ruzzene Approaches to Hybrid SHM and NDE of Composite Aerospace Structures, 9 March*; Kundu, T., Ed.; PIE: San Diego, CA, USA, 2014; pp. 9064-27–9064-35.

8. Ostiguy, P.-C.; Quaegebeur, N.; Masson, P. Non-destructive evaluation of coating thickness using guided waves. *NDT E Int.* **2015**, *76*, 17–25. [CrossRef]

9. Rose, J.L. Successes and Challenges in Ultrasonic Guided Waves for NDT and SHM. *Mater. Eval.* **2010**, *68*, 494–500.

10. Delrue, S.; Abeele, K.V.D. Detection of defect parameters using nonlinear air-coupled emission by ultrasonic guided waves at contact acoustic nonlinearities. *Ultrasonics* **2015**, *63*, 147–154. [CrossRef]

11. Clarke, T.; Cawley, P.; Wilcox, P.D.; Croxford, A.J. Evaluation of the damage detection capability of a sparse-array guided-wave SHM system applied to a complex structure under varying thermal conditions. *IEEE Trans. Ultrason. Ferroelectr. Freq. Control.* **2009**, *56*, 2666–2678. [CrossRef]

12. Rathod, V.; Mahapatra, D.R. Ultrasonic Lamb wave based monitoring of corrosion type of damage in plate using a circular array of piezoelectric transducers. *NDT E Int.* **2011**, *44*, 628–636. [CrossRef]

13. Sharma, A.; Sharma, S.; Sharma, S.; Mukherjee, A. Ultrasonic guided waves for monitoring corrosion of FRP wrapped concrete structures. *Constr. Build. Mater.* **2015**, *96*, 690–702. [CrossRef]

14. Lu, Y.; Li, J.; Ye, L.; Wang, N. Guided waves for damage detection in rebar-reinforced concrete beams. *Constr. Build. Mater.* **2013**, *47*, 370–378. [CrossRef]

15. Willey, C.L.; Simonetti, F.; Nagy, P.B.; Instanes, G. Guided wave tomography of pipes with high-order helical modes. *NDT E Int.* **2014**, *65*, 8–21. [CrossRef]

16. Løvstad, A.; Cawley, P. The reflection of the fundamental torsional guided wave from multiple circular holes in pipes. *NDT E Int.* **2011**, *44*, 553–562. [CrossRef]

17. Leinov, E.; Lowe, M.J.; Cawley, P. Investigation of guided wave propagation and attenuation in pipe buried in sand. *J. Sound Vib.* **2015**, *347*, 96–114. [CrossRef]

18. Mustapha, S.; Ye, L. Propagation behaviour of guided waves in tapered sandwich structures and debonding identification using time reversal. *Wave Motion* **2015**, *57*, 154–170. [CrossRef]

19. Putkis, O.; Dalton, R.; Croxford, A. The anisotropic propagation of ultrasonic guided waves in composite materials and implications for practical applications. *Ultrasonics* **2016**, *65*, 390–399. [CrossRef]

20. Castaings, M.; Singh, D.; Viot, P. Sizing of impact damages in composite materials using ultrasonic guided waves. *NDT E Int.* **2012**, *46*, 22–31. [CrossRef]

21. Raisutis, R.; Kazys, R.J.; Žukauskas, E.; Mažeika, L. Ultrasonic air-coupled testing of square-shape CFRP composite rods by means of guided waves. *NDT E Int.* **2011**, *44*, 645–654. [CrossRef]

22. Deng, Q.-T.; Yang, Z.-C. Propagation of guided waves in bonded composite structures with tapered adhesive layer. *Appl. Math. Model.* **2011**, *35*, 5369–5381. [CrossRef]

23. Masserey, B.; Raemy, C.; Fromme, P. High-frequency guided ultrasonic waves for hidden defect detection in multi-layered aircraft structures. *Ultrasonics* **2014**, *54*, 1720–1728. [CrossRef] [PubMed]

24. Puthillath, P.; Rose, J.L. Ultrasonic guided wave inspection of a titanium repair patch bonded to an aluminum aircraft skin. *Int. J. Adhes. Adhes.* **2010**, *30*, 566–573. [CrossRef]

25. Pieczonka, Ł.; Ambroziński, Ł.; Staszewski, W.J.; Barnoncel, D.; Pérès, P. Damage detection in composite panels based on mode-converted Lamb waves sensed using 3D laser scanning vibrometer. *Opt. Lasers Eng.* **2017**, *99*, 80–87. [CrossRef]

26. Ge, L.; Wang, X.; Jin, C. Numerical modeling of PZT-induced Lamb wave-based crack detection in plate-like structures. *Wave Motion* **2014**, *51*, 867–885. [CrossRef]

27. Ochôa, P.; Infante, V.; Silva, J.M.; Groves, R. Detection of multiple low-energy impact damage in composite plates using Lamb wave techniques. *Compos. Part B Eng.* **2015**, *80*, 291–298. [CrossRef]

28. Mamishev, A.V.; Sundara-Rajan, K.; Yang, F.; Du, Y.; Zahn, M. Interdigital sensors and transducers. *Proc. IEEE* **2004**, *92*, 808–845. [CrossRef]

29. Bellan, F.; Bulletti, A.; Capineri, L.; Masotti, L.; Yaralioglu, G.; Degertekin, F.L.; Khuri-Yakub, B.; Guasti, F.; Rosi, E. A new design and manufacturing process for embedded Lamb waves interdigital transducers based on piezopolymer film. *Sensors Actuators A Phys.* **2005**, *123*, 379–387. [CrossRef]

30. Na, J.K.; Blackshire, J.L.; Kuhr, S. Design, fabrication, and characterization of single-element interdigital transducers for NDT applications. *Sensors Actuators A Phys.* **2008**, *148*, 359–365. [CrossRef]

31. Mu, J.; Rose, J.L. Guided wave propagation and mode differentiation in hollow cylinders with viscoelastic coatings. *J. Acoust. Soc. Am.* **2008**, *124*, 866. [CrossRef]

32. MFC P1 Type. Available online: https://www.smart-material.com/MFC-product-P1.html (accessed on 23 April 2018).

33. Haig, A.G.; Sanderson, R.; Mudge, P.J.; Balachandran, W. Macro-fibre composite actuators for the transduction of Lamb and horizontal shear ultrasonic guided waves. *Insight Non-Destructive Test. Cond. Monit.* **2013**, *55*, 72–77. [CrossRef]

34. Ren, G.; Jhang, K.-Y. Application of Macrofiber Composite for Smart Transducer of Lamb Wave Inspection. *Adv. Mater. Sci. Eng.* **2013**, *2013*, 1–5. [CrossRef]

35. Tiwari, K.A.; Raisutis, R. Investigation of the 3D displacement characteristics for a macro-fiber composite transducer (MFC-P1). *Mater. Teh.* **2018**, *52*, 235–239. [CrossRef]

36. Tiwari, K.A.; Raisutis, R. Identification and characterization of defects in glass fiber reinforced plastic by refining the guided lamb waves. *Materials* **2018**, *11*, 1173. [CrossRef] [PubMed]

37. Tiwari, K.A.; Raisutis, R. Comparative Analysis of Non-Contact Ultrasonic Methods for Defect estimation of composites in remote areas. In Proceedings of the CBU International Conference Proceedings, Central Bohemia University, Prague, Czech Republic, 23–25 March 2016; Volume 4, pp. 846–851.

38. Tiwari, K.A.; Raisutis, R.; Mažeika, L.; Samaitis, V. 2D Analytical Model for the Directivity Prediction of Ultrasonic Contact Type Transducers in the Generation of Guided Waves. *Sensors* **2018**, *18*, 987. [CrossRef]

39. Pullin, R.; Eaton, M.; Pearson, M.; Featherston, C.; Lees, J.; Naylon, J.; Kural, A.; Simpson, D.J.; Holford, K. On the Development of a Damage Detection System using Macro-fibre Composite Sensors. *J. Physics Conf. Ser.* **2012**, *382*, 012049. [CrossRef]

40. Wang, X.; Zhou, W.; Xun, G.; Wu, Z. Dynamic shape control of piezocomposite-actuated morphing wings with vibration suppression. *J. Intell. Mater. Syst. Struct.* **2017**, *29*, 358–370. [CrossRef]

41. Debiasi, M.; Leong, C.W.; Bouremel, Y.; Yap, C. Application of macro-fiber-composite materials on UAV wings. In Proceedings of the Aerospace Technology Seminar (ATS), Singapore, 2013; pp. 1–20. Available online: https://scholar.google.com/scholar?hl=en&as_sdt=0%2C5&q=Application+of+macro-fiber-composite+materials+on+UAV+wings&btnG= (accessed on 8 March 2020).

42. Tiwari, K.A.; Raisutis, R. Post-processing of ultrasonic signals for the analysis of defects in wind turbine blade using guided waves. *J. Strain Anal. Eng. Des.* **2018**, *53*, 546–555. [CrossRef]

43. Tiwari, K.A.; Raisutis, R.; Mazeika, L.; Samaitis, V. Development of a 2D analytical model for the prediction of directivity pattern of transducers in the generation of guided wave modes. *Procedia Struct. Integr.* **2017**, *5*, 973–980. [CrossRef]

44. Pavlakovic, B.; Lowe, M.; Alleyne, D.; Cawley, P. Disperse: A General Purpose Program for Creating Dispersion Curves. *Rev. Prog. Quant. Nondestruct. Eval.* **1997**, *16A*, 185–192. [CrossRef]

45. Tiwari, K.A.; Raisutis, R.; Samaitis, V. Hybrid signal processing technique to improve the defect estimation in ultrasonic non-destructive testing of composite structures. *Sensors* **2017**, *17*, 2858. [CrossRef]

Article

The Contribution of Elastic Wave NDT to the Characterization of Modern Cementitious Media

Gerlinde Lefever [1,*], Didier Snoeck [1,2], Nele De Belie [2], Sandra Van Vlierberghe [3], Danny Van Hemelrijck [1] and Dimitrios G. Aggelis [1]

[1] Department Mechanics of Materials and Constructions, Vrije Universiteit Brussel (VUB), Pleinlaan 2, 1050 Brussels, Belgium; didier.snoeck@ugent.be (D.S.); Danny.Van.Hemelrijck@vub.be (D.V.H.); Dimitrios.Aggelis@vub.be (D.G.A.)

[2] Magnel-Vandepitte Laboratory for Structural Engineering and Building Materials, Department of Structural Engineering and Building Materials, Faculty of Engineering and Architecture, Ghent University, Tech Lane Ghent Science Park, Technologiepark Zwijnaarde 60, 9052 Ghent, Belgium; nele.debelie@ugent.be

[3] Polymer Chemistry & Biomaterials Research Group, Centre of Macromolecular Chemistry, Ghent University, Krijgslaan 281 S4-Bis, 9000 Ghent, Belgium; sandra.vanvlierberghe@ugent.be

* Correspondence: gerlinde.lefever@vub.be; Tel.: +32-(0)2-629-29-27

Received: 15 April 2020; Accepted: 19 May 2020; Published: 23 May 2020

Abstract: To mitigate autogenous shrinkage in cementitious materials and simultaneously preserve the material's mechanical performance, superabsorbent polymers and nanosilica are included in the mixture design. The use of the specific additives influences both the hydration process and the hardened microstructure, while autogenous healing of cracks can be stimulated. These three stages are monitored by means of non-destructive testing, showing the sensitivity of elastic waves to the occurring phenomena. Whereas the action of the superabsorbent polymers was evidenced by acoustic emission, the use of ultrasound revealed the differences in the developed microstructure and the self-healing of cracks by a comparison with more commonly performed mechanical tests. The ability of NDT to determine these various features renders it a promising measuring method for future characterization of innovative cementitious materials.

Keywords: acoustic emission; ultrasound; hydrogel; nanosilica

1. Introduction

Recently, there have been many developments in cementitious media, especially in the field of admixtures aiming to enhance mechanical properties, but mostly to extend durability and in doing so, to improve sustainability. In the process of developing innovative materials, monitoring techniques play an important role. Specifically, elastic wave methods allow non-invasive and non-destructive characterization of the mechanical properties (i.e., direct calculation of stiffness and correlation with the strength) as well as the monitoring of processes like setting and hydration of concrete with admixtures by means of active (ultrasound) or passive elastic waves (AE).

In the present paper, cementitious materials with different admixtures are tested. These admixtures are superabsorbent polymers (SAPs), nanosilica (NS) and a combination of both. SAPs are applied in concrete mainly to prevent shrinkage cracking by internal curing [1–5]. They function by initially absorbing (extra) water and releasing it to the cementitious matrix at a later stage, when the evaporation rate as well as the chemical hydration reaction reduce the amount of available water in the mixture and increase the capillary pressure in the system. In a normal situation, when the pressure becomes too high, air enters into the system, demonstrated by a sudden drop of capillary pressure and signifying a high risk for shrinkage cracking [6]. This is also escorted by AE bursts recorded within the same time frame of the pressure drop [7]. In case SAPs are present, the absorbed water inside the SAPs is released due to the increase of capillary pressure. This smoothens the effect of evaporation rate, ideally avoids the drop in internal relative humidity, and allows for continued hydration, enabling the cementitious material to resist the tensile forces leading to cracking. Practically, SAPs eliminate cracking as was revealed from the dedicated restrained ring tests in the time frame of the study (1 month) [8]. The contribution of the SAPs in controlling the shrinkage cracking is undeniable. However, they impose a certain reduction in mechanical properties due to the increase of the porosity. The dry SAP grains under study are normally 100 ± 21 µm in size while after water absorption, their size can reach up to 257 ± 55 µm, as they absorb approximately 26 times their mass in water when included in a cementitious mix. This value was obtained by measuring the flow of fresh mortars with and without SAPs using the flow table test [9]. The amount of additional water, necessary to obtain an identical flow of SAP mortars compared to the reference material, determines the absorption capacity of the SAP in the studied environment. After the water is drained back to the cementitious matrix, cracking is avoided [10] but the microstructure is affected by the remaining cavities which are a permanent part of the hardened microstructure. Recent results have shown a decrease of the order of 20% in compressive strength and flexural strength for mixtures with SAPs compared to the reference mix without SAPs [11,12]. To compensate for the reduction in strength, nanosilica (NS) particles are used in the mixtures. NS has shown the ability to increase the strength of a cementitious material due to its large surface area, which provides nucleation sites for the hydration of cement, early pozzolanic reaction and filler action. Recent results show that actually NS particles help to restore the mechanical properties in mixes with SAPs to the level of the reference material, while at the same time, the mixes benefit by the cracking mitigation action of SAPs [8].

SAPs are not only used as an admixture to mitigate autogenous shrinkage [13], they are also interesting materials to obtain sealing and healing characteristics [14,15]. Upon crack formation, the SAPs absorb moisture and/or fluids and this can be provided to the cementitious matrix to stimulate further hydration, pozzolanic activity and calcium carbonate crystallization, up to minimally 8 years of age [16]. The further hydration is promoted by nearly 40% compared to the reference cementitious material [17], and can even be repeated for a second healing cycle [18]. The healing characteristic is an interesting feature and requires high amounts (1 m% of binder) of SAPs to be added, although detrimental for the mechanical properties [19,20], which can be counteracted by the addition of NS [21].

The present paper discusses the non-destructive techniques (NDT) used to monitor the material in three phases. First, the fresh stage is evaluated, where the material is curing with simultaneous monitoring by AE. Secondly, ultrasound is used to check the elastic properties of the media in the hardened stage, as the microstructure is significantly modified. Finally, the self-healing stage is monitored, when the specimens are subjected to wet-dry curing cycles to check the potential for crack closure and restoration of mechanical properties due to the action of SAPs that can maintain water and lead to a second stage of hydration and promote the precipitation of $CaCO_3$ inside the crack. For the first time in literature, AE is shown sensitive to the activity of SAPs, allowing to monitor the whole duration of internal curing, which so far was only possible with expensive and cumbersome neutron tomography and nuclear magnetic resonance (NMR) testing [22]. In addition, it is the first time that elastic wave measurements indicate the mechanical healing due to the wet-dry cyclic curing, which is later on confirmed by mechanical reloading. In the following section, a brief introduction of elastic wave NDT for the specific applications is provided.

2. Elastic Wave NDT in Cementitious Media

2.1. Acoustic Emission in Fresh Concrete

Acoustic emission (AE) monitoring has been applied mainly in the last two decades for monitoring of fresh cement paste. Sources that have been targeted include grain settlement, water mobility, hydration reaction and cracking. Some studies indicated start or peak of AE activity during the calorimetric temperature peak that implies relation to the hydration reaction [23–25]. However, other studies [26,27] showed the large majority of recorded hits occurring earlier than this peak, leading to the conclusion that significant processes (possibly of lower intensity and thus more difficult to register) occur from as early as the mixing time, much before the chemical reaction of hydration initiates and any heat is developed. Differences in the acquisition equipment (including sensor frequency range and sensitivity), the coupling (with or without waveguide) and the specimen size do not allow for robust conclusions relatively to the original sources. Recent studies showed that individual physical mechanisms like bubble creation in the fresh cementitious matrix and aggregate impacts can be recorded as AE events [28], while it was verified that most of the AE activity during the first 2 h after mixing originates from cement grains settlement [7]. Due to its sensitivity down to the attoJ (10^{-18} J) level, AE is influenced by the size of the grains (fly-ash suspension with mean grain size of 57 µm induces lower frequencies and higher energies of AE than normal cement with average grain size of 12 µm) during settlement (up to 2–3 h after mixing). Furthermore, AE energy exhibits peaks close to the moment of capillary pressure breakdown, giving a good indication when the risk of plastic cracking increases. Therefore, the reception of high energy AE bursts during this stage indicates the starting of the detrimental action of cracking and allows external curing treatment to mitigate it [7]. A recent review on this topic is composed by Aggelis et al. [29].

2.2. Ultrasonic Assessment of Hardened Cementitious Media

Elastic waves have been more widely used for characterization of hardened cementitious media resulting in a vast literature on the subject. Indicatively, apart from the well-known general correlations of pulse velocity to strength [30,31], phase velocity has shown sensitivity to frequency and mix parameters like the water and aggregate content [32]. In addition, the amount of heterogeneity in the form of actual or simulated damage alters the wave characteristics decreasing the wave velocity and amplitude [33–36].

In addition, elastic waves have been used in certain cases to evaluate the repair effectiveness in concrete materials and structures [37–39]. The wave velocity and amplitude are restored, while this has also been applied in concrete with a self-healing vascular network, showing restoration of wave parameters for healed cracks of width up to 500 µm [40]. In the following section, the aforementioned elastic wave techniques are used for monitoring of all stages of the materials' life. Focus is given on the NDT aspect of the study, while preliminary results concerning the material properties have been recently published [21].

From the monitoring point of view, measurements were conducted in three stages:

(1) During the hydration of mortar specimens in order to check the AE activity of the modified and reference mixes and specifically monitor the action of SAPs for the first time in literature;
(2) elastic wave measurements on the sound material after 28 days to check the effect of the microstructure on the elastic properties;
(3) elastic wave measurements during the healing cycles to examine in a simple way if the mechanical properties are restored.

3. Experimental Details

3.1. Materials and Mechanical Testing

Four mortar mixtures were made: a reference mixture, a mixture holding SAPs, a mixture holding nanosilica and finally a mixture combining both SAPs and NS. The cement used for all mixtures is a high-strength ordinary Portland cement, CEM I 52.5 Strong (Holcim, Nivelles, Belgium). To obtain the reference mortar mixture, cement, river sand and tap water were added in a proportion of 1:2:0.35. To allow for an easier compaction, a superplasticizer was included at an amount of 0.4% by weight of the binder.

The superabsorbent polymer used in this research is a copolymer of acrylamide and sodium acrylate, produced by bulk polymerization. The SAP presents the ideal characteristics for internal curing purpose with a particle size equal to 100 ± 21.5 µm [4]. The swelling capacity of the SAP is equal to 305.0 ± 3.7 g/g SAP in demineralized water and 61.0 ± 1.0 g/g SAP in cement filtrate [22], measured following the RILEM recommendations [41]. The necessary amount of SAPs for efficient mitigation of autogenous shrinkage can be calculated by means of Powers' hydration model [13] to obtain the highest possible degree of hydration. In case of the reference mixture under study, an amount of 0.24% by mass of the binder should be added together with 26 g of water per gram of SAP, leading to an entrained additional amount $(w/c)_e$ of 0.063. However, it was chosen to lower the amount of SAP included to the mortar mixtures to an amount of 0.2% by mass of the binder, to partially mitigate autogenous shrinkage and limit the reduction in mechanical properties. Compared to the reference mixture, the amount of superplasticizer was kept constant in the SAP mixture and the workability was the same in all mixtures (flow value of 138 ± 1 mm).

To counteract the decrease in compressive strength, caused by the formation of macropores after water release from the SAPs, a nano-reinforcement was introduced. The nanomaterial used was a colloidal nanosilica, containing 40% of synthetic amorphous silica in a water solution. The nanosilica particles have a nominal diameter of approximately 12 nm and a specific surface area between 18 and 258 m²/g. Cement was in this case replaced by nanosilica in an amount of 2% by mass of cement, so that a constant mass of binder was maintained. Also, the amount of superplasticizer added was increased to 0.76% with respect to the total weight of the binder material to account for the decrease in flowability of the fresh mortar caused by the nanoparticles. All mixtures showed the same workability. Table 1 summarizes the mixture proportions of the mortar blends used throughout this study.

Table 1. Ratios of mixture components with respect to the binder content.

	Cement	Water	Sand	Superplasticizer	SAP	Dry NS
Reference	1	0.35	2	0.004		
0.2% SAP	1	0.402	2	0.004	0.002	
2% NS	0.98	0.35	2	0.076		0.02
0.2% SAP + 2% NS	0.98	0.402	2	0.076	0.002	0.02

To obtain the mechanical properties of the various mixtures, three prism specimens measuring 40 mm × 40 mm × 160 mm were cast per mixture and cured in plastic foil at 20 ± 1 °C. Their compressive strength was measured according to ASTM C349-18 [42]. The average densities after 28 days of curing and the compressive strengths are summarized in Table 2, along with the standard deviations. It can be seen from the results that the addition of SAPs indeed has a strong influence on the mechanical performance, decreasing the compressive strength, while the use of NS restores the compressive strength.

Table 2. Density (g/m³) and compressive strength (MPa) of the four mixtures under study, measured at 28 days of curing.

	Density (g/cm³)	Compressive Strength (MPa)
Reference	2.16 ± 0.01	77.29 ± 1.17
0.2% SAP	2.17 ± 0.05	72.36 ± 2.55
2% NS	2.20 ± 0.01	88.60 ± 1.44
0.2% SAP + 2% NS	2.16 ± 0.02	78.39 ± 2.37

3.2. Acoustic Emission Monitoring

To monitor the hydration process, a metallic mold equipped with three piezoelectric sensors was used. The sensors were of type R15α and had an operating frequency between 50 and 400 kHz and a resonance frequency at 150 kHz. The three sensors were placed along the sides of a prism specimen of 40 mm × 40 mm × 160 mm: two of them were oppositely attached to the longitudinal faces of the beam mold, while the third was placed on the bottom surface. The fresh mortar specimens were monitored for a period of three days in sealed conditions. The set-up is shown in Figure 1.

(a)   (b)

Figure 1. Set-up for acoustic emission monitoring: (**a**) top view showing two sensors with magnetic holders at opposite sides of the beam and (**b**) a side view revealing the bottom sensor.

3.3. Surface Wave Measurements

In order to conduct elastic wave measurements, two pico sensors were placed on the top of the specimen. Pico sensors have their sensitivity peak at 450 kHz but they are broadband sensors, which operate between 50 to 800 kHz. The two sensors were located at a distance of either 50 mm or 30 mm, depending on the type of specimen used (plain for sound property determination and with steel rebar for mechanical loading and reloading purposes, respectively), and the excitation took place through a pencil lead break at a distance of approximately 1 cm from the first sensor as shown in Figure 2a. Typical signals received by the two sensors on sound material are depicted in Figure 2b. The signal in the 2nd sensor arrives later and is much lower compared to the 1st, due to the extra distance. Considering the delay between the onset of the two waveforms, the longitudinal wave velocity could be calculated. In addition, by identifying the dominant Rayleigh cycle in both waveforms, the Rayleigh wave velocity was also calculated, as the ratio of the sensor distance over the time delay between the characteristic points (Figure 2b). Apart from the surface measurements, ultrasonic measurements were conducted with a commercial high-power device through the longitudinal axis as well at a resonant frequency of 54 kHz on the specimens without rebar.

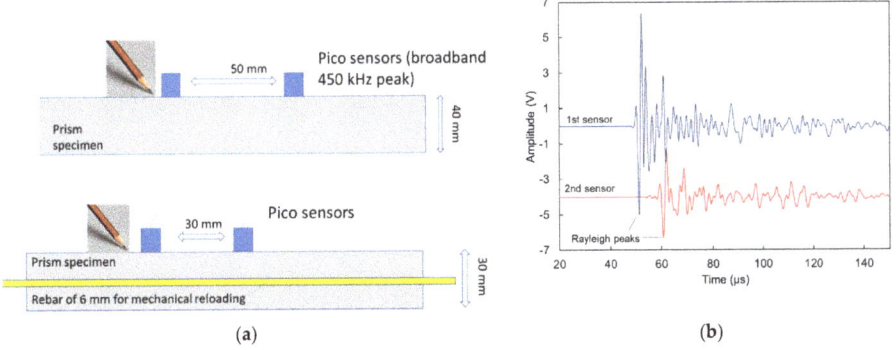

(a)   (b)

Figure 2. (**a**) Experimental set-up for surface wave measurements; (**b**) shows typical waveforms after pencil lead break in front of the first sensors.

The surface wave measurements on sound material were conducted on prism specimens of dimensions 40 mm × 40 mm × 160 mm, six per mixture, after 28 days of curing. After measurement of the sound material, these prisms were cracked by means of a three-point bending test. A carbon fiber reinforced polymer (CFRP) laminate allowed the two halves of the prism specimens to be kept together (at the position of the CFRP laminate). A metal framework was then placed around the specimen, following the procedure described in [43]. By means of restraining with the metal framework, the crack width opening was decreased to approximately 150 μm for all specimens. The determination of the crack width was done by microscopic measurements, using a Leica S8 APO optical microscope equipped with a DFC 295 camera. Along the crack mouth opening, three positions were chosen and a micrograph was taken. In each of these pictures, five measurements of the crack opening were conducted, leading to a total of 15 measurements per specimen.

Afterwards, wet-dry healing cycles were applied on five out of six specimens and this for all mixtures, for a period of 28 days. These cycles consist of 1 h submersion in water at 20 ± 1 °C and 23 h of dry conditions at a relative humidity of 60 ± 5% and temperature of 20 ± 1 °C for a period of 28 days. The remaining specimen was kept in dry conditions, identical to the environmental conditions of the dry period during the healing cycles. Measurements of the crack width were repeated in the exact same 15 locations as described above after 3, 7, 14 and 28 days of wet-dry curing. By means of these microscopic measurements, visual crack closure, implying possible healing of cracks, can be seen. A side and bottom view of the specimens with metal framework can be seen in Figure 3a,b, respectively. A more detailed explanation on the experimental testing procedure and the results can be found in [21].

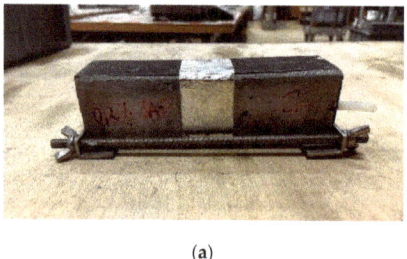

(a)   (b)

Figure 3. Prism specimens cracked in three-point bending with metal framework for restrained crack opening: (**a**) side view and (**b**) bottom view showing the crack in the center of the specimen.

For the further examination of healing by means of mechanical loading and reloading, although the sensors and excitation remained the same, the specimens' geometry was slightly modified (cross-section of 30 mm × 30 mm, and length of 360 mm) and a thin steel rebar of 6 mm diameter and a length of 700 mm was embedded at casting [44]. Tensile loading and reloading were performed by clamping the reinforcement bar of the test specimens into an Instron 5982 Floor Model Testing System (Instron GmbH, Darmstadt, Germany). The capacity of the load cell is 100 kN and a uniaxial tensile load was applied at a speed of 0.01 mm/s. In the loading stage, the tensile load-displacement response presents initially a linear increase, characterized by the stiffness of the composite beam. When cracking occurs, a sudden drop in the load is noticed. This occurs for every additional crack, until no new cracks were formed and the final part of the load-displacement curve shows the capacity of the steel rebar only. The displacement was increased further on, until a certain opening of the cracks could be maintained after release of the applied load. The opening of the initial cracks was then also measured by means of microscopy and this in five locations on each of the four sides of the mortar specimen, leading to a total of 20 measurements per crack. Upon reloading of a cracked specimen, the response is identical to the final part measured in the loading stage. However, when healing of cracks has taken place, following the same healing procedure as described above, the load-displacement curve could show a regain in stiffness as well as the occurrence of new cracks. Figure 4a shows typical specimens of the latter case, while Figure 4b shows the location of the sensors in either side of a crack after mechanical loading. Three specimens were cast per mixture. The test was performed after 28 days of curing in plastic foil, at a room temperature of 20 ± 1 °C.

(a)   (b)

Figure 4. Specimens used during wet-dry healing cycles: (**a**) geometry of the specimens with central rebar and (**b**) positioning of the sensors around the crack opening.

During the aforementioned mechanical loading, multiple cracks initiated in the mortar matrix and measured between 50 and 500 μm. At this moment, one or two cracks per specimen were arbitrarily chosen to be followed up during the wet-dry healing cycles. After choosing the cracks to be monitored, the sensors were placed around these chosen cracks, at 30 mm apart. Several surface wave measurements, consisting of a pencil lead break test as explained in Section 3.3, were then conducted and repeated after 3, 7, 14 and 28 days of wet-dry curing, close to the end of the dry period. After this 28-day period, mechanical reloading was performed to investigate whether a regain in mechanical properties could be obtained.

4. Results

4.1. Acoustic Emission Monitoring During Hydration

Results of the cumulative AE activity are seen in Figure 5a, where various curves of reference mortar and mortar with SAPs are included for a monitoring period of approximately three days. It is obvious that the SAPs' modified mixtures exhibit much higher activity that starts to evolve at approximately 11 h after mixing. According to previous studies, this is practically the time when SAPs start to release their water back to the mixture [22]. AE monitors the whole period of SAP contribution, showing that the phenomenon comes to completion after 40 h, again in correlation with literature [22]. The AE activity may come from the water flow in the porosity of cement as well as from the detachment of the SAPs from their cavity as they shrink. While this is still under consideration, it is the first time that AE is used to monitor the phenomenon, which so far could be traced only by cumbersome and expensive neutron tomography and NMR [22].

Figure 5b focuses on the first 15 h of AE, where the nearly vertical increase due to the higher rate of SAPs activity is clearly seen. Earlier, most mixes exhibit similar AE rates from the start of the monitoring, while at approximately 2 h the AE evolves to a lower rate. This initial period of high activity before 2 h coincides well with the measured settlement in cement, showing once again the sensitivity of the AE sensors to the micro-level processes [7].

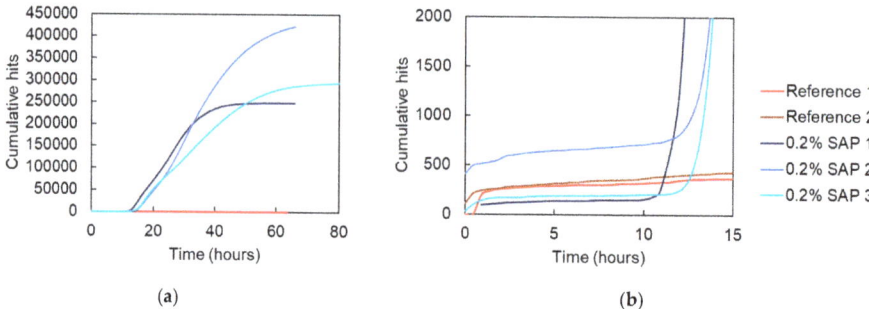

Figure 5. Cumulative hits versus time of various reference (2 replicates) and SAP-containing mortar prisms (3 replicates) in (**a**) and a zoom during the first 15 h in (**b**).

Looking at the cumulative AE activity, a nearly constant rate is depicted for several hours during the activation of SAPs (i.e., at least between 12 and 30 h). However, more detail is offered by AE parameters like the amplitude and duration. In Figure 6a, it is clear that from the moment of the onset of the phenomenon (approximately 11 h as aforementioned), a rapid increase in the amplitudes is noted, reaching values of even 70 dB at 17–18 h. This level is maintained until approximately 26 h, also illustrated by the moving average red line of 250 points included in the graph when a gradual decrease starts to occur and continues until the end of monitoring at 85 h. Similar conclusions are provided in Figure 6b, where the AE duration is depicted. There, the average line starts at approximately 17 μs at 11 h, reaches a plateau of 100 μs until 26 h of curing and then gradually decreases to the initial level throughout the rest of the monitoring period. The results are in agreement with NMR data that show that these specific SAPs release water from final setting, at approximately 11 h after mixing, and most entrained water is released in between 22 h to 30 h, and then levelling down to slower pace when studied in sealed conditions [22]. This is also an indication that despite the inherently large scatter of AE data, quantitative information can still be drawn to accurately characterize microstructural processes and to determine the time frame for internal curing by the SAPs.

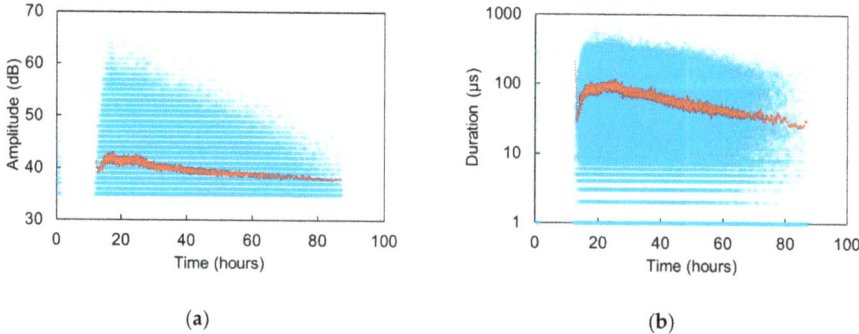

Figure 6. Amplitude (**a**) and duration (**b**) of AE waveforms versus time for a typical mortar specimen with SAPs. Each point stands for the amplitude of one AE signal and the red line stands for the moving average of 250 points.

It is also noticed that different processes can be discriminated based on the AE characteristics. For example, the activity received during the period dominated by settlement (roughly first two hours), shows 50 to 400 times higher average energy (ranging between 500 and 10,000 attoJ) than the activity during the steady state of SAPs action (at approximately 25 h, ranging between 10 and 50 attoJ). In accordance, the typical duration of settlement AE signals (270–370 μs) is 3 to 4 times longer than SAPs activity signals (70–130 µs). Therefore, something that years ago seemed impossible (characterization of sources in fresh cement) and caused a lot of confusion to researchers, now starts to become substantiated and offers unique insights in the hidden processes within fresh cementitious media.

Selecting 'representative' AE waveforms is not straightforward due to the inherent experimental scatter of the parameters. However, it is always important to have a look at the raw data on which the analysis is based. Figure 7a,b show three AE signals from the period of intense SAP action (at 25 h), and the settlement (first 2 h) respectively. The waveform shapes do not fundamentally differ in shape, apart from the longer average duration of settlement signals in Figure 7b. Figure 7c,d show the corresponding FFT of the same waveforms. The main content is in any case in the band 50 to 200 kHz, which is expected reasonable due to the resonance of the sensors, while occasionally the magnitude of settlement signals Figure 7d reaches higher values than of SAPs action.

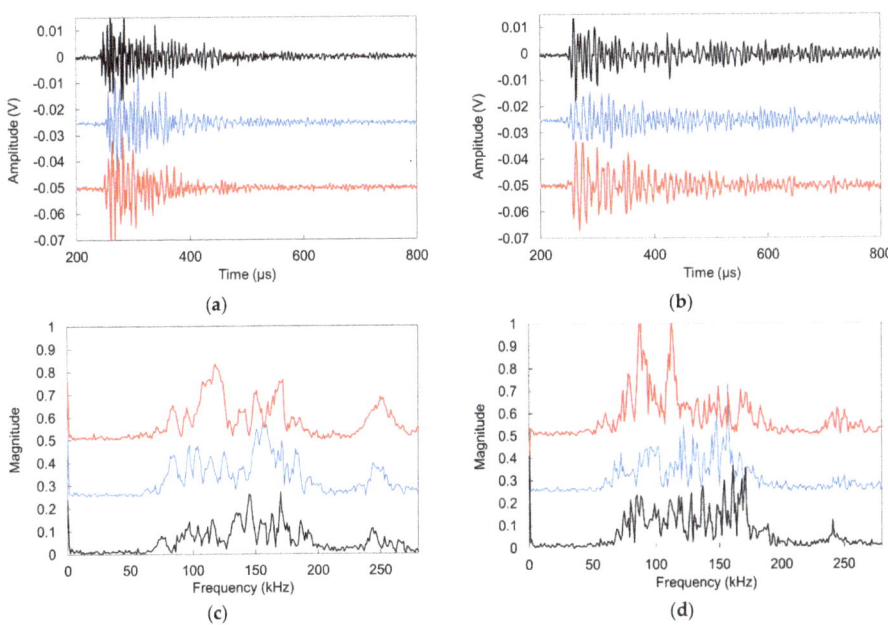

Figure 7. Typical AE waveforms: (**a**) during SAP activity (25 h after mixing) and (**b**) during settlement (<1 h after mixing). Typical FFT magnitudes are shown in (**c**) for SAP activity signals and in (**d**) for settlement signals (several signals are translated on the vertical axis for readability).

4.2. Ultrasonic Measurements on the Hardened Material

Figure 8a,b show the longitudinal and Rayleigh wave velocities respectively as measured by the pico sensors on the surface after pencil lead break. The results confirm the effect of the cavities created by the SAPs as the longitudinal wave velocity drops for the specific mix by more than 10%, while the Rayleigh wave velocity decreases by 4%. In addition, the beneficial effect of NS is also evident since the velocity of the mix with SAPs and NS is restored to the same level as the reference mix (above 5500 m/s). Mortar containing only NS exhibits even higher Rayleigh velocity values than the reference mix Figure 8b), due to the reinforcement by the nanomaterial. Considering the densities of all mixes, the Young's moduli of the materials range between 46 GPa for SAPs mix and 55 to 57 GPa for the other mixes. A point that should be highlighted is the larger influence on the wave velocity when NS is added to SAP samples, compared to the addition to reference mixtures. This may be explained by the formation of products, caused by the pozzolanic nature of the nanosilica, within the macropores created by the emptying of the SAPs. Further research is however necessary to substantiate this assumption.

The influence of heterogeneity in the form of cavities is not only demonstrated by the lower velocity values but also by the experimental scatter they exhibit. The more heterogeneous the material, the more random it becomes, which is depicted in the coefficient of variation (COV) values, calculated as the standard deviation over the average and shown again in Figure 8a,b. Indicatively, while the COV for longitudinal waves of reference mortar is 5%, it increases to more than 11% for material with SAPs. Concerning the typical measurement error and taking into account the sampling rate of 10 MHz (time step 0.1 µs), this is calculated at an average of 0.98% for the longitudinal and at 0.56% for the Rayleigh wave velocities.

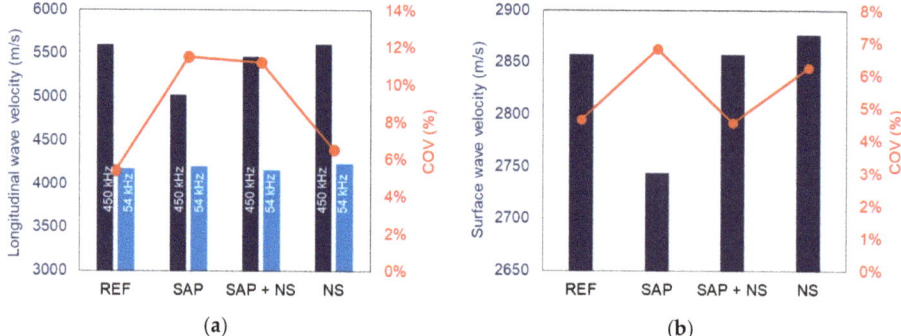

Figure 8. Elastic wave velocity and coefficient of variation of various mixes: (a) longitudinal waves and (b) surface or Rayleigh waves. Results of 450 kHz refer to the measurements with pico sensors after pencil lead excitation and 54 kHz refers to the experiments conducted with the commercial high-power ultrasonic device.

As aforementioned, apart from the surface measurements, ultrasonic tests took place through the longitudinal axis with a frequency of 54 kHz. The results are also seen in Figure 8a. The velocity values are lower than the higher frequency ones, something normal due to the well-known dispersion exhibited by cementitious media [33,45]. In addition, it is seen that lower frequencies and therefore, longer wavelengths do not help much to characterize between the various mixes in this scale, as the results are all within a range of 80 m/s (4160 m/s to 4240 m/s), without strong characterization power over the mixes. Indeed, 54 kHz results in wavelength λ of approximately 70–80 mm. Considering the dimensionless parameter $\alpha = \pi D/\lambda$, where D is the inclusion diameter (SAP cavities have a maximum diameter size of 300 µm), and λ, the wavelength, it results in a value around 0.025, much lower than 1. This clearly indicates that the phenomena fall into the "long wavelength" regime [45], where limited interaction between the heterogeneity and the wavelength is expected. On the other hand, concerning the surface measurements mentioned above, Figure 9 shows typical spectra after pencil lead break excitation as received by the Pico sensors used for wave measurements on the surface. The main peaks come at approximately 400 kHz resulting in a representative Rayleigh wavelength of 7 mm. For this wavelength, the corresponding value of parameter α is 0.26, one order of magnitude higher than for the 54 kHz measurements. Therefore, although still lower than 1, the surface wave measurements after pencil lead excitation start to deviate from the long wavelength regime and the microstructure starts influencing more critically the results through scattering. In addition, there is substantial content even at higher frequencies up to 600 kHz, which would result in even smaller wavelengths, higher α values and stronger interaction.

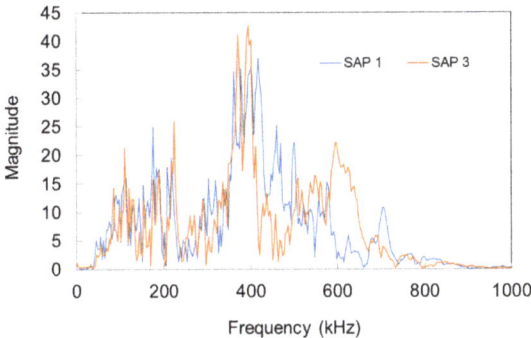

Figure 9. Typical FFT spectra from waveforms received on the surface of mortars with SAPs.

One correlation that is also worth mentioning is the one between wave velocity and strength. While this is well known for cementitious materials, this is the first time that it is confirmed for this type of admixtures. Mortar with SAPs indeed exhibits simultaneously the lower strength and the lower Rayleigh wave velocity, while mortar with NS has the highest values exhibiting 22% higher strength and 5% higher velocity than the SAPs mixes Figure 10.

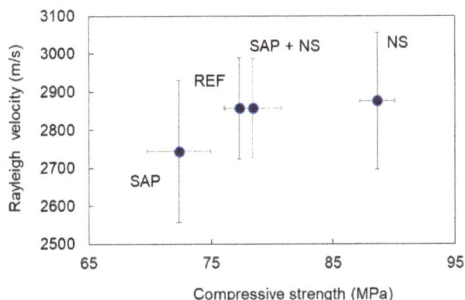

Figure 10. Correlation between average Rayleigh wave velocity and compressive strength of mortars.

The above results demonstrate clearly that NDT based on elastic waves can be used to characterize advanced mixes as wave velocity correlates well with the expected microstructure and final mechanical properties, extending the knowledge from conventional materials to innovative cementitious mixes.

4.3. Drying-Wetting Cycles

As aforementioned, healing cycles were conducted to check the capacity of crack closure and possible mechanical restoration. Initially, the specimens with rebars were loaded until no new cracks initiated. Upon reloading, the specimens were cured in wet-dry cycles, as explained in the experimental section. At specific ages, surface wave measurements were conducted with the two sensors placed at either side of a crack to check the effect of wet-dry curing on the signal transmission through possible sealing or healing. Specifically, the samples were studied six times (sound and cracked condition at 0 days and later at 3, 7, 14, 28 days during wet-dry cycles). The waveforms in Figure 11a correspond to the 2nd receiver on a reference sample. It is seen that after the crack occurrence, the waveform (2nd from top) loses much of its amplitude compared to the "sound" one (top waveform), while the Rayleigh cycle cannot be identified any longer. Throughout the wet-dry cycles, there seems to be an increase of the energy of the waveform, without however, being able to clearly detect the Rayleigh cycle similarly to before cracking. The increase can be due to the closure of the crack from late hydration

Sensors **2020**, *20*, 2959

products and calcium carbonate precipitation. This result is comparable to monitoring the healing capability on impacted plates with and without SAPs, by means of resonance analysis using a tap hammer [46]. Figure 11b shows a typical case for a SAP + NS specimen. The initial waveforms show the same tendency compared to the reference specimen, since after cracking, the transmission is seriously decreased, as seen by the reduction of amplitude. At later times however, the waveform starts to restore its content, signifying that more energy passes through the volume of the crack, while at 28 days, the Rayleigh peak becomes visible again, although not as clear as the one before cracking. This restoration of wave energy was noticed in most of the SAP + NS specimens while other mixes showed much weaker restoration. This is the first time that surface wave amplitude is used to monitor the crack closure effect of stimulated autogenous healing while in the past, it has been used for crack closure after epoxy repair [37].

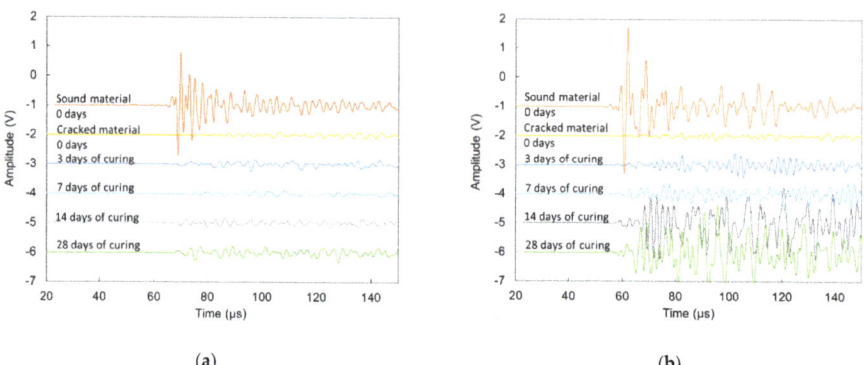

(a)   (b)

Figure 11. Waveforms received at successive stages of curing for (**a**) reference mortar sample and (**b**) SAP + NS mortar sample.

Considering all the monitored locations (four different cracks for each mix during 28 days of wet-dry cycles), an average value for the attenuation coefficient can be calculated. It is measured by the ratio of the maximum amplitude of the waveform of the 2nd receiver over the maximum amplitude of the 1st "reference" receiver (close to excitation, receiving the signal before passing through the crack), divided over the sensor to sensor distance of 30 mm and expressed in dB. The low values below 0.4 dB/mm at 0 days, as shown in Figure 12a, correspond to the attenuation of the sound media before cracking. Just after cracking, the attenuation strongly increased, to values around 1 dB/mm showing the influence of the discontinuity on the wave path. As the wet-dry cycles are performed, the attenuation of all mixes shows a decreasing trend, evident of the fact that cracks are closing due to further hydration products that are formed between the crack sides and the deposition of calcium carbonate. In addition to the general decreasing trend, it is obvious that the attenuation of SAP + NS mortars exhibit much lower values than the other mixes signifying much better transmission conditions through the volume of the crack.

Wave attenuation can be discussed in relation to the microscopy results on the same mixes that were cracked in bending and followed the same wet-dry cycles, as explained earlier. Figure 12b shows the average crack width for the four considered mixtures: reference, SAP, NS and SAP + NS.

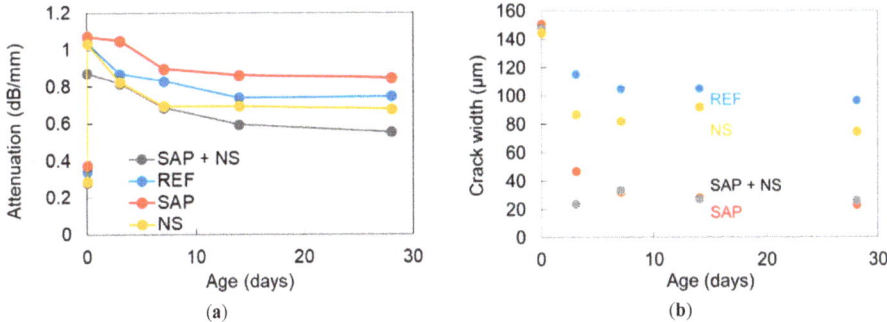

Figure 12. (a) Surface wave attenuation and (b) crack width for various mortar mixes vs. curing time in wet-dry cycles.

The initial crack is very similar as this is imposed by the metal frame where the specimens are fit into after the mechanical cracking, thus the value of 150 μm is the starting point to check the healing or sealing potential of the different compositions. From the second measurement at 3 days, the average readings of the cracks were reduced, even for the mixtures without SAPs showing a value below 120 μm and 90 μm for reference and NS samples respectively. For SAP and SAP + NS compositions, the average crack width exhibited a much stronger reduction being close to 40 μm. Wet-dry cycle curing until 28 days has a small additional effect especially for the SAP specimens, which exhibit an average final crack width of 24 μm while for SAP + NS the final value was 34 μm. Therefore, microscopy results confirm that the addition of SAPs contributes to the closure of cracks, either as standalone admixture or in combination with NS. In general, the trend of decreasing attenuation is in agreement with the closing trend of cracks, initially exhibiting stronger rate and later being saturated. However, there is one point that needs to be highlighted. While SAP and SAP + NS mixes exhibit similar crack closure at 28 days in Figure 12b the attenuation shows much lower values for SAP + NS mixes than SAPs alone. The reason behind these differences is likely to be caused by the variation on the initial crack widths that exists for the specimens with rebars, studied for the attenuation profile. An average value, considering all initial crack width measurements, was equal to 89 ± 33 μm for SAP + NS samples, 139 ± 64 μm and 139 ± 48 μm for reference and NS, respectively, while for SAP specimens a mean crack width of 202 ± 102 μm was found. The significantly larger average crack width in SAP samples can lead to a limited total healing, as the total amount of healing products necessary to fully close the cracks is higher compared to the SAP + NS series. This trend, indicated by the lower attenuation of SAP + NS, was tested by mechanical loading, where the same specimens used for surface wave measurements were reloaded in tension after 28 days of wet-dry cycles. The average regain in equivalent stiffness of all mixtures is shown in Figure 13. The equivalent stiffness, measured by the slope of the load-displacement curve, during reloading is compared to the one of the loading stage. A regain of only 10% was seen for the reference samples, while for the SAP specimens this regain was increased up to 22%. This means that, even though the crack widths in the SAP specimens were on average wider compared to the reference material, the healing ratio is still higher for SAP inclusion. This is due to the promotion of further hydration by the SAPs by nearly 40%, as confirmed earlier by NMR measurements [17] and visualized by means of X-ray tomography [47]. When comparing the SAP and SAP + NS samples, better healing conditions are given for the latter series. This is due to the stronger initial cementitious matrix, resulting in smaller crack widths compared to cracks in the reference and NS specimens. The included SAPs also improved further hydration in this case. Moreover, the addition of NS had a positive influence on the healing capacity when compared to the reference series. In this case, the average crack openings were comparable. A possible explanation could be the formation of other healing products, caused by the pozzolanic nature of the nanomaterial, promoting the healing capability. The latter phenomenon would confirm the restoration of stiffness for SAP + NS, being much stronger than for other mixes, like the restoration of signal transmission implied through the decrease of attenuation. This is subject for further research.

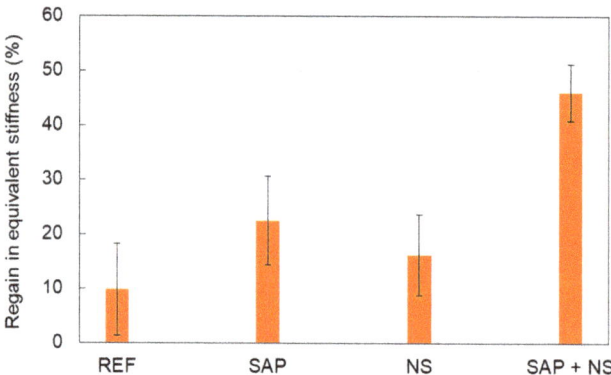

Figure 13. Regain in equivalent stiffness (slope of load-displacement curves) of all mixture series.

For completeness, it is mentioned that the assessment after cracking is based on the amplitude and not on the wave velocity for two basic reasons. The first is that the wave amplitude (or inversely attenuation) is much more sensitive to the cracks and heterogeneity in general, as well known from the literature [33–36] and also shown in this study. In the examples of Figure 11a,b above, the peak amplitude drops by 95% after the crack (compare "sound material" and "cracked material" waveforms). At the same time, judging from the onset of these waveforms, the velocity drops by about 50% (from approximately 5500 m/s to approximately 2800 m/s in both cases). Therefore, amplitude shows much stronger sensitivity and characterization capacity for the same cracking than the velocity. Furthermore, an important point is that after the development of the crack, the reliability of picking the onset of the waveform is compromised due to the low amplitude of the opening cycles of the waveform. These peaks are quite low and similar to the noise level, reducing the reliability of a specific pick for the onset. This of course does not hold for the attenuation, which is measured by the peak amplitude of the waveform and can be clearly depicted in all cases.

5. Conclusions

This paper studied the use of elastic wave NDT as a promising method to monitor the various processes occurring in cementitious materials and to characterize their inner microstructure. The mixtures under study contained different additives, being SAPs to mitigate autogenous shrinkage and NS to counteract the reduction in strength caused by SAP inclusion. The effects of these components on the hydration process, the final microstructure and the self-healing efficiency were measured by wave methods and a comparison with the results of more common experimental procedures was made.

Acoustic emission monitoring of reference and SAP mortars revealed the action of the SAPs during hydration. A steep increase in received AE hits was noticed between approximately 11 h and 40 h of curing in case of SAP samples, whereas this was not seen for the reference mortars. The increase in hits therefore can be linked to the release of water by the SAPs as the desiccation of the mortar and internal curing initiate. Moreover, by analysis of the received waveforms, the action of the SAPs could be distinguished from the settlement, occurring within the first hours of curing.

Secondly, using ultrasonic measurements performed by pencil lead break tests, differences in microstructure between the four mixtures were exposed by variations in longitudinal and surface wave velocities measured. The creation of cavities after water release by the SAPs lowered both wave velocities and increased the scatter on the results, while the inclusion of NS increased the wave velocity due to the reinforcement of the matrix.

Finally, it was seen that the addition of SAPs and NS improved the self-healing capacity of the mortar specimens. Tensile tests were performed to obtain multiple cracking and after 28 days of healing in wet-dry cycles, the tensile reloading showed that a partial regain in stiffness could be obtained for the SAP + NS mixtures. During these wet-dry cycles, ultrasonic tests were conducted next to specific crack openings to receive information on waves travelling across the crack opening. By examination of the attenuation, decreasing over time as curing in wet-dry cycles was performed, the increased healing capability of the SAP + NS mixtures compared to other mixtures was confirmed.

The improved healing capacity of SAP + NS mixtures, determined by means of wave measurements and regain in mechanical stiffness and confirmed by the results of water permeability tests and visual crack closure [21] is an interesting feature of the newly obtained cementitious material. Together with the mitigation of autogenous shrinkage through the inclusion of SAPs, without having a negative influence on the compressive strength, the combination of SAPs and nanosilica shows to be a promising addition to cementitious mixtures, meeting the continuously increasing requirements regarding the performance of construction materials.

Author Contributions: Conceptualization, D.S., N.D.B., S.V.V., D.V.H. and D.G.A.; formal analysis, G.L., D.G.A.; funding acquisition, D.S., N.D.B., D.V.H. and D.G.A.; investigation, G.L.; methodology, G.L., D.S., N.D.B., D.V.H. and D.G.A.; resources, S.V.V.; supervision, D.S., N.D.B., D.V.H. and D.G.A.; validation, D.S., N.D.B., D.V.H. and D.G.A.; writing—original draft, G.L., D.G.A.; writing—review and editing, D.S., N.D.B., S.V.V., D.V.H. and D.G.A. All authors have read and agreed to the published version of the manuscript.

Funding: This research is funded by the Research Foundation—Flanders (FWO), grant number G.0A28.16.6. The financial support is therefore gratefully acknowledged by all authors. As a Postdoctoral Research Fellow of the Research Foundation-Flanders (FWO-Vlaanderen), D. Snoeck would like to thank the foundation for the financial support (12J3620N).

Acknowledgments: We would like to thank Alexander Assmann (BASF) for providing the SAP and Yves De Vreese (W. R. Grace & Co.-Conn.) for providing the nanosilica under study.

Conflicts of Interest: The authors declare no conflict of interest.

References

1. Craeye, B.; Geirnaert, M.; de Schutter, M. Super absorbing polymers as an internal curing agent for mitigation of early-age cracking of high-performance concrete bridge decks. *Constr. Build. Mater.* **2011**, *25*, 1–13. [CrossRef]
2. Schröfl, C.; Mechtcherine, V.; Gorges, M. Relation between the molecular structure and the efficiency of superabsorbent polymers (SAP) as concrete admixture to mitigate autogenous shrinkage. *Cem. Concr. Res.* **2012**, *42*, 865–873. [CrossRef]
3. Mechtherine, V.; Gorges, M.; Schröfl, C.; Assmann, A.; Brameshuber, W.; Ribeiro, A.; Cusson, D.; Custodio, J.; da Silva, E.F.; Ichimiya, K.; et al. Effect of internal curing by using superabsorbent polymers (SAP) on autogenous shrinkage and other properties of a high-performance fine-grained concrete: Results of a RILEM round-robin test. *Mater. Struct.* **2014**, *47*, 541–562. [CrossRef]
4. Snoeck, D.; Jensen, O.; de Belie, N. The influence of superabsorbent polymers on the autogenous shrinkage properties of cement pastes with supplementary cementitious materials. *Cem. Concr. Res.* **2015**, *74*, 59–67. [CrossRef]
5. Shen, D.; Wang, X.; Cheng, D.; Zhang, J.; Jiang, G. Effect of internal curing with super absorbent polymers on autogenous shrinkage of concrete at early age. *Constr. Build. Mater.* **2016**, *106*, 512–522. [CrossRef]
6. Slowik, V.; Schmidt, M.; Fritzsch, R. Capillary pressure in fresh cement-based materials and identification of the air entry value. *Cem. Concr. Compos.* **2008**, *30*, 557–565. [CrossRef]
7. Dzaye, E.; de Schutter, G.; Aggelis, D. Monitoring early-age acoustic emission of cement paste and fly ash paste. *Cem. Concr. Res.* **2020**, *129*, 105964. [CrossRef]
8. Lefever, G.; Tsangouri, E.; Snoeck, D.; Aggelis, D.; de Belie, N.; van Vlierberghe, S.; van Hemelrijck, D. Combined use of superabsorbent polymers and nanosilica for reduction of restrained shrinkage and strength compensation in cementitious mortars. *Constr. Build. Mater.* **2020**, *251*, 118966. [CrossRef]

9. Belgisch Instituut Voor Normalisatie (BIN). *Methods of Test for Mortar Masonry—Part 3: Determination of Consistence of Fresh Mortar (by Flow Table)*; European Committee for Standardization (CEN): Brussels, Belgium, 1999.

10. Snoeck, D.; Pel, L.; de Belie, N. Superabsorbent polumers to mitigate plastic drying shrinkage in a cement paste as studied by NMR. *Cem. Concr. Compos.* **2018**, *93*, 54–62. [CrossRef]

11. Mignon, A.; Snoeck, D.; D'Halluin, K.; Balcaen, L.; Vanhaecke, F.; Dubruel, P.; van Vlierberghe, S.; de Belie, N. Alginate biopolymers: Counteracting the impact of superabsorbent polymers on mortar strength. *Constr. Build. Mater.* **2016**, *110*, 169–174. [CrossRef]

12. Wehbe, Y.; Ghahremaninezhad, A. Combined effect of shrinkage reducing admixtures (SRA) and superabsorbent polymers (SAP) on the autogenous shrinkage, hydration and properties of cementitious materials. *Constr. Build. Mater.* **2017**, *138*, 151–162. [CrossRef]

13. Jensen, O.; Hansen, P. Water-entrained cement-based materials: I. Principles and theoretical background. *Cem. Concr. Res.* **2001**, *31*, 647–654. [CrossRef]

14. Snoeck, D.; van Tittelboom, K.; Steuperaert, S.; Dubruel, P.; de Belie, N. Self-healing cementitious materials by the combination of microfibres and superabsorbent polymers. *J. Intell. Mater. Syst. Struct.* **2014**, *25*, 13–24. [CrossRef]

15. Snoeck, D. Superabsorbent polymers to seal and heal cracks in cementitious materials. *RILEM Tech. Lett.* **2018**, *3*, 32–38. [CrossRef]

16. Snoeck, D.; de Belie, N. Autogenous healing in strain-hardening cementitious materials with and without superabsorbent polymers: An 8-year study. *Front. Mater.* **2019**, *6*, 1–12. [CrossRef]

17. Snoeck, D.; Pel, L.; de Belie, N. Autogenous healing in cementitious materials with superabsorbent polymers quantified by means of NMR. *Sci. Rep.* **2020**, *10*, 642. [CrossRef]

18. Snoeck, D.; de Belie, N. Repeated autogenous healing in strain-hardening cementitious composites by using superabsorbent polymers. *J. Mater. Civ. Eng.* **2015**, *28*, 04015086. [CrossRef]

19. Hasholt, M.; Jensen, O.; Kovler, K.; Zhutovsky, S. Can superabsorbent polymers mitigate autogenous shrinkage of internally cured concrete without compromising the strength? *Constr. Build. Mater.* **2012**, *31*, 226–230. [CrossRef]

20. Snoeck, D.; Schaubroeck, D.; Dubruel, P.; de Belie, N. Effect of high amounts of superabsorbent polymers and additional water on the workability, microstructure and strength of mortars with a water-to-cement ratio of 0.50. *Constr. Build. Mater.* **2014**, *72*, 148–157. [CrossRef]

21. Lefever, G.; Snoeck, D.; Aggelis, D.; de Belie, N.; van Vlierberghe, S.; van Hemelrijck, D. Evaluation of the self-healing ability of mortar mixtures containing superabsorbent polymers and nanosilica. *Materials* **2020**, *13*, 380. [CrossRef]

22. Snoeck, D.; Pel, L.; de Belie, N. The water kinetics of superabsorbent polymers during cement hydration and internal curing visualized and studied by NMR. *Sci. Rep.* **2017**, *7*, 1–14. [CrossRef] [PubMed]

23. Van den Abeele, K.; Desadeleer, W.; de Schutter, G.; Wevers, M. Active and passive monitoring of the early hydration process in concrete using linear and nonlinear acoustics. *Cem. Concr. Res.* **2009**, *39*, 426–432. [CrossRef]

24. Chotard, T.; Barthelemy, J.; Smith, A.; Gimet-Breart, N.; Huger, M.; Fargeot, D.; Gault, C. Acoustic emission monitoring of calcium aluminate cement setting at the early age. *J. Mater. Sci. Lett.* **2001**, *20*, 667–669. [CrossRef]

25. Assi, L.; Soltangharaei, V.; Anay, R.; Ziehl, P.; Matta, F. Unsupervised and supervised pattern recognition of acoustic emission signals during early hydration of Portland cement paste. *Cem. Concr. Res.* **2018**, *103*, 216–226. [CrossRef]

26. Iliopoulos, S.; el Khattabi, Y.; Aggelis, D. Towards the Establishment of a Continuous Nondestructive Monitoring Technique for Fresh Concrete. *J. Nondestruct. Eval.* **2016**, *35*, 37. [CrossRef]

27. Topolar, L.; Pazdera, L.; Kucharczykova, B.; Smutny, J.; Mikulasek, K. Using Acoustic Emission Methods to Monitor Cement Composites during Setting and Hardening. *Appl. Sci.* **2017**, *7*, 451. [CrossRef]

28. Dzaye, E.; de Schutter, G.; Aggelis, D. Study on mechanical acoustic emission sources in fresh concrete. *Arch. Civ. Mech. Eng.* **2018**, *18*, 742–754. [CrossRef]

29. Aggelis, D.; Grosse, C.; Shiotani, T. Acoustic Emission Characterization of Fresh Cement-Based Materials. In *Advanced Techniques for Testing of Cement-Based Materials*; Serdar, M., Gabrijel, I., Schlicke, D., Staquet, S., Azenha, M., Eds.; Springer Nature Switzerland AG: Basel, Switzerland, 2020; pp. 1–22.

30. Komlos, K.; Popovics, S.; Nürnbergerova, T.; Babal, B.; Popovics, J. Ultrasonic pulse velocity test of concrete properties as specified in various standards. *Cem. Concr. Compos.* **1996**, *18*, 357–364. [CrossRef]

31. Kaplan, M. The effects of age and water/cement ratio upon the relation between ultrasonic pulse velocity and compressive strength of concrete. *Mag. Concr. Res.* **1959**, *11*, 85–92. [CrossRef]

32. Philippidis, T.; Aggelis, D. Experimental study of wave dispersion and attenuation in concrete. *Ultrasonics* **2005**, *43*, 584–595. [CrossRef]

33. Chaix, J.-F.; Garnier, V.; Corneloup, G. Ultrasonic wave propagation in heterogeneous solid media: Theoretical analysis and experimental validation. *Ultrasonics* **2006**, *44*, 200–210. [CrossRef] [PubMed]

34. Ju, T.; Achenbach, J.; Jacobs, L.; Guimaraes, M.; Qu, J. Ultrasonic nondestructive evaluation of alkali-silica reaction damage in concrete prism samples. *Mater. Struct.* **2017**, *50*, 60. [CrossRef]

35. Aggelis, D.; Shiotani, T. Exprimental study of surface wave propagation in strongly heterogeneous media. *J. Acoust. Soc. Am.* **2007**, *122*, EL151–EL157. [CrossRef] [PubMed]

36. Selleck, S.; Landis, E.; Peterson, M.; Shah, S.; Achenbach, J. Ultrasonic investigation of concrete with distributed damage. *ACI Mater. J.* **1998**, *95*, 27–36.

37. Aggelis, D.; Shiotani, T.; Polyzos, D. Characterization of surface crack depth and repair evaluation using Rayleigh waves. *Cem. Concr. Compos.* **2009**, *31*, 77–83. [CrossRef]

38. Shiotani, T.; Momoki, S.; Chai, H.; Aggelis, D. Elastic wave validation of large concrete structures repaired by means of cement grouting. *Constr. Build. Mater.* **2009**, *23*, 2647–2652. [CrossRef]

39. Benmeddour, F.; Villain, G.; Abraham, O.; Choinska, M. Development of an ultrasonic experimental device to characterise concrete for structural repair. *Constr. Build. Mater.* **2012**, *37*, 934–942. [CrossRef]

40. Tsangouri, E.; Lelon, J.; Minnebo, P.; Asaue, H.; Shiotani, T.; van Tittelboom, K.; de Belie, N.; Aggelis, D.; van Hemelrijck, D. Feasibility study on real-scale, self-healing concrete slab by developing a smart capsules network and assessed by a plethora of advanced monitoring techniques. *Constr. Build. Mater.* **2019**, *228*, 116780. [CrossRef]

41. Snoeck, D.; Schröfl, C.; Mechtcherine, V. Recommendation of RILEM TC 260-RSC: Testing sorption by superabsorbent polymers (SAP) prior to implementation in cement-based material. *Mater. Struct.* **2018**, *51*, 116. [CrossRef]

42. ASTM International. *ASTM Standard C 349-18: Standard Test Method for Compressive Strength of Hydraulic-Cement Mortars (Using Portions of Prisms Broken in Flexure)*; ASTM Standards: Conshohocken, PA, USA, 2018.

43. Van Mullem, T.; Gruyaert, E.; Debbaut, B.; Caspeele, R.; de Belie, N. Novel active crack width control technique to reduce the variation on water permeability results for self-healing concrete. *Constr. Build. Mater.* **2019**, *203*, 541–551. [CrossRef]

44. Wang, J.; Snoeck, D.; van Vlierberghe, S.; Verstraete, W.; de Belie, N. Application of hydrogel encapsulated carbonate precipitating bacteria for approaching a realistic self-healing in concrete. *Constr. Build. Mater.* **2014**, *68*, 110–119. [CrossRef]

45. Iliopoulos, S.; Malm, F.; Grosse, C.; Aggelis, D.; Polyzos, D. Concrete wave dispersion interpretation through Mindlin's strain gradient elastic theory. *J. Acoust. Soc. Am.* **2017**, *142*, EL89–EL94. [CrossRef] [PubMed]

46. Snoeck, D.; de Schryver, T.; de Belie, N. Enhanced impact energy absorption in self-healing strain-hardening cementitious materials with superabsorbent polymers. *Constr. Build. Mater.* **2018**, *191*, 13–22. [CrossRef]

47. Snoeck, D.; Dewanckele, J.; Cnudde, V.; de Belie, N. X-ray computed microtomography to study autogenous healing of cementitious material promoted by superabsorbent polymers. *Cem. Concr. Compos.* **2016**, *65*, 83–93. [CrossRef]

Article

Improved Depth-of-Field Photoacoustic Microscopy with a Multifocal Point Transducer for Biomedical Imaging

Thanh Phuoc Nguyen [1,*], Van Tu Nguyen [2], Sudip Mondal [3], Van Hiep Pham [2], Dinh Dat Vu [2], Byung-Gak Kim [4] and Junghwan Oh [2,3,*]

1   Department of Mechatronics, Cao Thang Technical College, Ho Chi Minh City 700000, Vietnam
2   Interdisciplinary Program of Biomedical Mechanical and Electrical Engineering, Pukyong National University, Busan 48513, Korea; nguyen.vantu91@gmail.com (V.T.N.); pvhiep.mta.hut@gmail.com (V.H.P.); dinhdatvn96@gmail.com (D.D.V.)
3   Center for Marine-Integrated Biomedical Technology, Pukyong National University, Busan 48513, Korea; mailsudipmondal@gmail.com
4   College of Future Convergence, Pukyong National University, Busan 48513, Korea; bgkim@pknu.ac.kr
*   Correspondence: nguyenthanhphuoc@caothang.edu.vn (T.P.N.); jungoh@pknu.ac.kr (J.O.); Tel.: +82-51-629-5771 (J.O.)

Received: 5 February 2020; Accepted: 2 April 2020; Published: 3 April 2020

Abstract: In this study, a photoacoustic microscopy (PAM) system based on a multifocal point (MFP) transducer was fabricated to produce a large depth-of-field tissue image. The customized MFP transducer has seven focal points, distributed along with the transducer's axis, fabricated by separate spherically-focused surfaces. These surfaces generate distinct focal zones that are overlapped to extend the depth-of-field. This design allows extending the focal zone of 10 mm for the 11 MHz MFP transducer, which is a great improvement over the 0.48 mm focal zone of the 11 MHz single focal point (SFP) transducer. The PAM image penetration depths of a chicken-hemoglobin phantom using SFP and MFP transducers were measured as 5 mm and 8 mm, respectively. The significant increase in the PAM image-based penetration depth of the chicken-hemoglobin phantom was a result of using the customized MFP transducer.

Keywords: ultrasound; photoacoustic imaging; photoacoustic microscopy; biomedical imaging; multifocal point transducer

1. Introduction

Of late, optical techniques have been used widely in biomedical imaging, which has improved the performance of in vivo diagnosis with high optical contrast [1,2]. This approach produces strong light scattering effects and a low spatial resolution. The optical microscopy penetration depth is limited to approximately 1 mm. Photoacoustic (PA) imaging (PAI) is a biomedical imaging modality based on the PA effect [3–7]. When short-pulsed laser light is directed on tissues, chromophores absorb some of the light energy that is converted into acoustic waves due to rapid thermal expansion. Photoacoustic microscopy (PAM) is also widely used to image tissues through optical absorption [8–13]. PAM can image optical contrast beyond the existing depth limit for high-resolution optical imaging [14]. The spatial resolution depends on the performance of the ultrasonic transducer. Many studies developed photoacoustic tomography by employing a short pulse to generate ultrasound waves in biological tissues; such techniques are used for in vivo biomedical imaging [1,15].

Ultrasonic transducers play an important role in PAI systems [1,13,16,17]. The transducer's parameters have a significant effect on image quality [18]. The main parameters of ultrasonic transducers are center frequency, bandwidth, focal length, focal zone, aperture size, and lateral and axial resolutions. Single focal point (SFP) transducers have a limited focal zone so as to acquire a deep image. Ultrasonic array transducers can control focal depths through dynamic focusing algorithms to capture the target image [19–22]. However, the fabrication process of this type of transducer is highly complicated. Multifocal point (MFP) transducers, which are developed to be used in imaging systems, are proposed to increase the image depth in ultrasound imaging [23]. The newly designed MFP transducers show a significant increase in focal zone compared with SFP transducers. Combined with PA systems, the penetration depth in deep imaging can be increased.

In PAI systems, the depth of penetration depends on the frequency of the ultrasonic wave. Higher frequencies have a small depth of penetration, whereas lower frequencies have a greater depth of penetration. To obtain the correct depth of an image, PAM requires an appropriate high frequency with a short wavelength and an enhanced resolution. Previous studies reported the use of phased array transducers for the PAI systems with complex dynamic focusing algorithms to acquire images [5,21,24–29]. Chulhong et al. [5] imaged biological tissues using a combined handheld PA microimaging probe and a transducer array. The lymph nodes containing methylene blue at a depth of 4.5 cm in tissue were visible in the PA images. Vogt et al. [21] used four clinical ultrasound transducer arrays in a PAI system with frequencies of 2.5, 8, 8.7, and 12.4 MHz. The largest penetration depths in the PAI system were obtained from the transducer with the lowest frequency, i.e., 2.5 MHz. A real-time 512-channel PA system with a 5 MHz transducer array imaged mouse brain vasculature at a depth of 4.5 mm with a lateral resolution of 200 μm [25]. However, a combination of two systems poses complications for setup and operation.

In addition, for array ultrasound signals, the complex algorithm of the synthetic-aperture focusing technique (SAFT) had been applied to acquire qualified final images [21,25–27,29]. This study was conducted with the motive to design a custom-made MFP transducer with seven focal points and a long focal zone for deeper imaging applications, which does not require the application of SAFT for reconstruction or any post-processing to obtain the image. Because MFP transducers are designed using a single piezoelectric element, it is easy to acquire the image using only a single-channel ultrasonic pulser/receiver.

The contributions of this study are presented in three aspects. First, two types of focused transducer (SFP and MFP) were designed and fabricated, both made of the same type of 28 μm polyvinylidene fluoride (PVDF) film and producing the same center frequency. In the case of the SFP transducer, with only one focal point, a large image depth cannot be obtained owing to the limited size of the focal zone. The main objective of this study was to expand the MFP transducer's focal zone, which may create many focal points at different depths with the use of a multi-spherical pattern (MSP) model. Second, the proposed design of the MFP transducer can be driven by a one-channel ultrasonic pulser/receiver system because of its single element function. This enables simple operation and data processing to acquire images without applying the complex SAFT as the transducer array system. Third, the significant difference between the focal zone and penetration depth of images from two transducers was distinguished in ultrasound imaging of the wire phantom, PAI of the needle, and hemoglobin (Hb) embedded in chicken tissues.

2. Materials and Methods

2.1. Transducer Design Materials

The piezoelectric element is the most important element of an ultrasonic transducer, which is made of piezoelectric polymers. PVDF membrane (Piezotech S.A.S, France) is the preferred polymer and has been extensively studied for many decades in the manufacture of high-frequency transducers [23,30–33]. PVDF is a special material used in high-purity applications, as well as in solvents, acids, and hydrocarbon resistance. Polymerization, stretching, and polling processes for a 28 μm PVDF element can be applied for developing transducers. In this study, PVDF was selected to fabricate SFP and MFP transducers due to its advanced properties. Table 1 details the properties of

PVDF used in this study. Although PVDF film's acoustic impedance (~4 MRayl) is lower than that of piezoceramics and crystal materials, PVDF shows a good mechanical versatility, which makes it easy to press the film into a spherical shape. The transducer developed from this PVDF has a normal broad bandwidth. In addition, PVDF film has a small dielectric constant appropriate for electrical impedance matching [34].

Table 1. Properties of PVDF[1] material.

Property	Value
Electromechanical coupling coefficient (K_t)	0.15
Relative clamped dielectric constant $\varepsilon^S / \varepsilon$	11
Mechanical quality factor (Q_m)	~20
Density (kg/m^3)	1780
Longitudinal wave velocity (m/s)	2110
Acoustic impedance (MRayl)	3.9
Curie temperature (°C)	100
Melting temperature (°C)	160~180

[1] Data reported by Piezo film sensor, AMP Inc, Valley Forge, PA.

2.2. Transducer Design

SFP transducers have a functional limitation in the focal zone and penetration depth. To extend the length of the focal zone, the multifocal point transducer was designed with a focal zone of 11 mm. In this study, the structure of the MFP transducer was designed similarly to the design reported in a previous study [23]. Figure 1 shows the profile and focal zones distribution of the developed transducer. The surface of the MFP transducer was designed by connecting seven parts with same areas in order to create the same level of intensity in their focused areas. The "Ri" is the radius of part "i" ($i = 1$–7). The distance between two focal points $b = R_j - R_{j-1} = 1.5$ mm ($j = 2$–7) The parameters of the seven-focal-point transducer were designed as shown in Table 2.

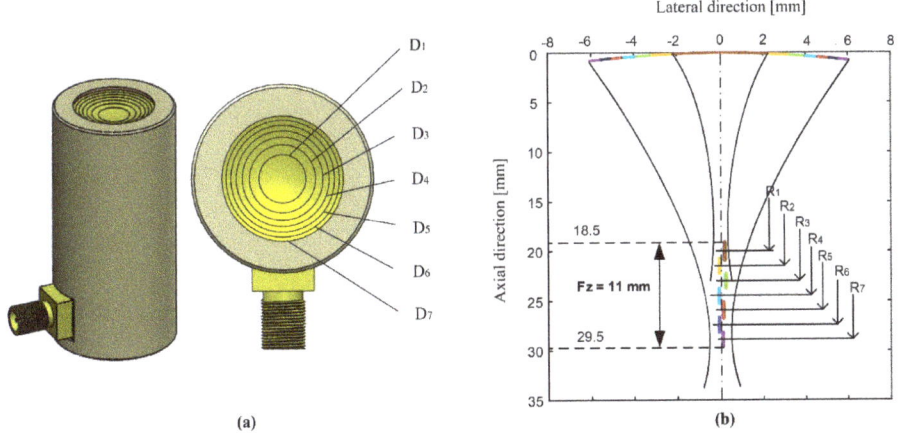

(a)　　　　　　　　　　　　　　　(b)

Figure 1. The (**a**) profile and (**b**) distribution of the focal zones of the multifocal point (MFP) transducer.

Table 2. Parameters of the MFP transducer.

Part Number i ($i = 1$–7)	Focal Length R_i (mm)	Aperture Diameter D_i (mm)	F-Number (R_i/D_i)	Active Area (mm^2)
1	20.00	4.60	4.3	1.256
2	21.49	6.49	3.3	1.256
3	22.97	7.94	2.8	1.256
4	24.45	9.16	2.6	1.256
5	25.92	10.24	2.5	1.256
6	27.40	11.21	2.4	1.256
7	28.87	12.10	2.3	1.256

The most important feature in Table 2 shows that a longer focal length obtains a smaller F-number. To obtain the best axial resolution, the system needs the smallest F-number for the best quality image (Axial resolution = speed of sound x F-number/center frequency). The seventh part of the MFP transducer has the smallest F-number of 2.3 and the longest focal length of 28.87 mm, which can obtain better axial resolutions at deeper depths.

Figure 2 shows the comparison of the transducers' focal zones. For the SFP transducer, the front face was formed by a steel ball bearing of 12.7 mm in radius. The SFP transducer has only one focal point at the focus depth of 12.7 mm and only one focal zone of 0.48 mm. The front surface's parameter of the MFP transducer was designed and simulated using the Matlab (version 2013a, Mathworks, Natick, MA) software. The MFP transducer has seven focal points, which created seven focal zones.

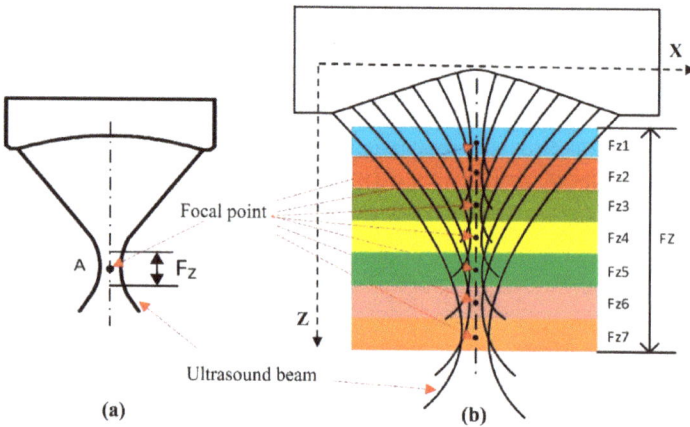

Figure 2. The distribution of the focal points and focal zones of (**a**) the single focal point (SFP) and (**b**) MFP transducer.

The parameters of the MFP transducer were carefully calculated to obtain tightly focused areas from the first to the last focal zone. Table 3 shows the overlap length between the two closed focal zones, which confirms that the focal depths are continued on the completely focused areas. Based on the distribution of the focal zones, the MFP transducer can capture a good target image as the object is placed in the area of the focal zone between 18.50 mm and 29.50 mm.

Table 3. The distribution of the focal zones of the MFP transducer.

Focal Zone	Range of Focus (mm)	Length of Focus (mm)	Overlap Length (mm)
F_{Z1}	18.50 – 22.20	3.70	–
F_{Z2}	20.18 – 22.80	2.61	2.02
F_{Z3}	21.97 – 23.98	2.01	0.83
F_{Z4}	23.59 – 25.31	1.72	0.39
F_{Z5}	25.15 – 26.71	1.55	0.16
F_{Z6}	26.67 – 28.19	1.45	0.04
F_{Z7}	28.18 – 29.50	1.38	0.01

2.3. Transducer Fabrication

To acquire a large penetration depth image, a combination of the light absorption and acoustic detector was used. Two types of ring-shaped transducer (SFP and MFP) were developed to compare the different performances in PAM imaging. Figure 3 shows the device used for forming the multi-spherical surfaces of the MFP transducer. The press-fit system (Figure 3a) was fabricated with five aluminum plates and four stainless steel rods with screws. The parameters of the MSP (Figure 3b) with seven spherically-focused surfaces were simulated using Matlab, and the original computer numerical control (CNC) machine fabricated the transducer, as shown in Figure 3c. The components of the press-fit system were designed using Matlab and made by using the CNC machine, as shown in Figure 3d. The MFP transducer was designed using Solidworks, as shown in Figure 3e. Figure 3f shows a photograph of the fabricated MFP transducer with seven focal points.

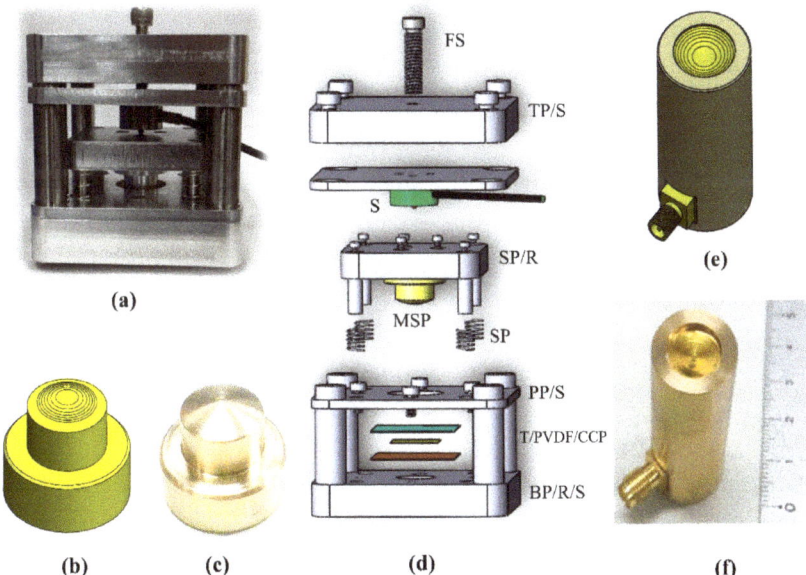

Figure 3. (**a**) A photograph of the press-fit structure; (**b**) the designed multi-spherical pattern (MSP); (**c**) a photograph of the MSP; (**d**) the press-fit's components: base plate/rod/screw (BP/R/S), Teflon/PVDF/copper-clad polyimide (T/PVDF/CCP), pressure plate/screw (PP/S), spring (SP), MSP, slide plate/rod (SP/R), sensor of force (S), top plate/screw (TP/S), and force screw (FS); (**e**) the profile of MFP transducer; and (**f**) a photograph of the MFP transducer.

For the MFP transducer, the fabrication process proceeded in two phases, as shown in Figure 4. The press-fit system was used in the first phase to form a multi-spherical profile of the active membrane. The copper-clad polyimide (CCP; Hanwha Corp., FCCL, Korea), 28 µm PVDF membrane (Piezotech S.A.S, France), and Teflon films (4 × 4 cm) were prepared. For bonding the two films of CCP and PVDF, the epoxy (EPO TEK 301, Epoxy Technology, Billerica, MA, USA) was applied. To avoid tearing the films, springs were placed to minimize the vibration in the manufacturing process. The force sensor was controlled to display the tension value of the active film surface. The top and bottom plates were fixed using screws through rods. To ensure a uniform pressure on the surface of the MSP transducer, the forcing screw was rotated by a hexagonal bar wrench.

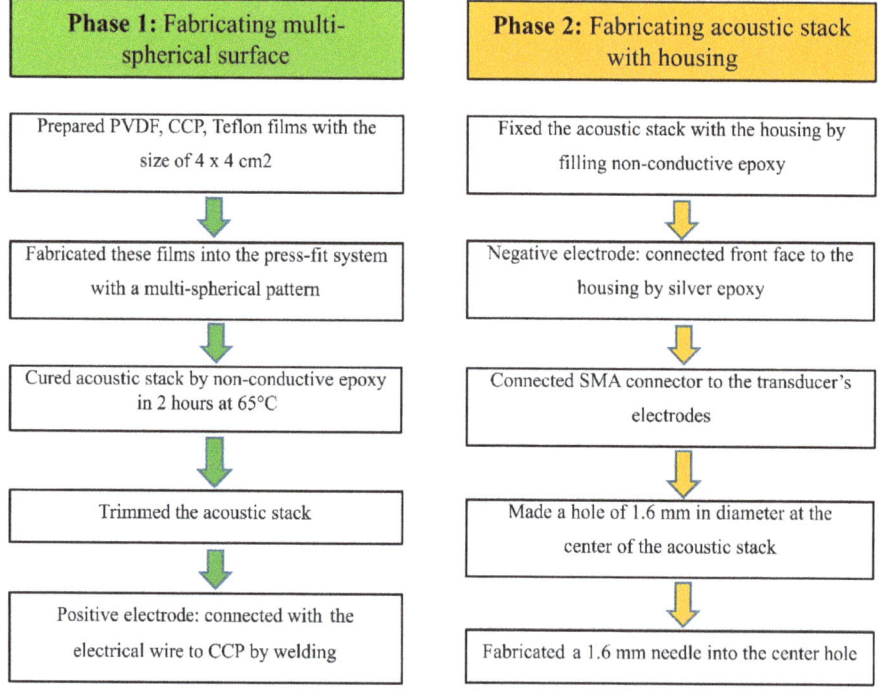

Figure 4. The fabrication process of the ring-shaped MFP transducer.

The press-fit system was inverted after these films were inserted into the base plate's center hole. The Teflon tube was filled with the nonconductive epoxy to keep the spherical profile of the PVDF film after curing. The fabricated system was heated at 65 °C for 2 hours. After disassembling the press-fit system, the acoustic stack was taken out with an epoxy plug connected to it. The CCP and PVDF were trimmed close to the epoxy plug. The pin of the SMA connector (Mouser Electronics, TX, USA) was soldered to a small CCP line through an electrical wire.

In the second phase, the acoustic stack was fabricated to a transducer housing. The acoustic stack was concentrically attached to the transducer housing. An open space was filled with a nonconductive epoxy inside the housing to keep the transducer's long-term electrical and mechanical stability. Following the epoxy curing, the transducer housing was connected to the connector. A piece of silver epoxy (H20 epoxy, Epoxy Technology, Inc., USA) was cast between a piece of PVDF and the housing to create a ground path. A drill was used to form a hole of 1.6 mm at the center of the transducer's surface, and a 14G needle (Syringe needle, Anhui, China) was then inserted into the hole with a thin layer of UV adhesive (Norland products, Inc, Cranbury, NJ, USA) to maintain the spherical form of the

PVDF film. The laser cable was inserted inside the needle and adjusted at the transducer's focal point to obtain the best resolution PA image.

For the SFP transducer, a steel ball bearing (Hecto, Jiangsu, China) of 25.4 mm in diameter was used to form a single spherical surface, producing a focal point at 12.7 mm and one focal zone of 0.48 mm. The same 28 µm PVDF film was used to fabricate an SFP transducer with an aperture diameter of 12 mm. Figure 5 shows the transducers' cross-sectional view and photographs of the ring-shaped transducers.

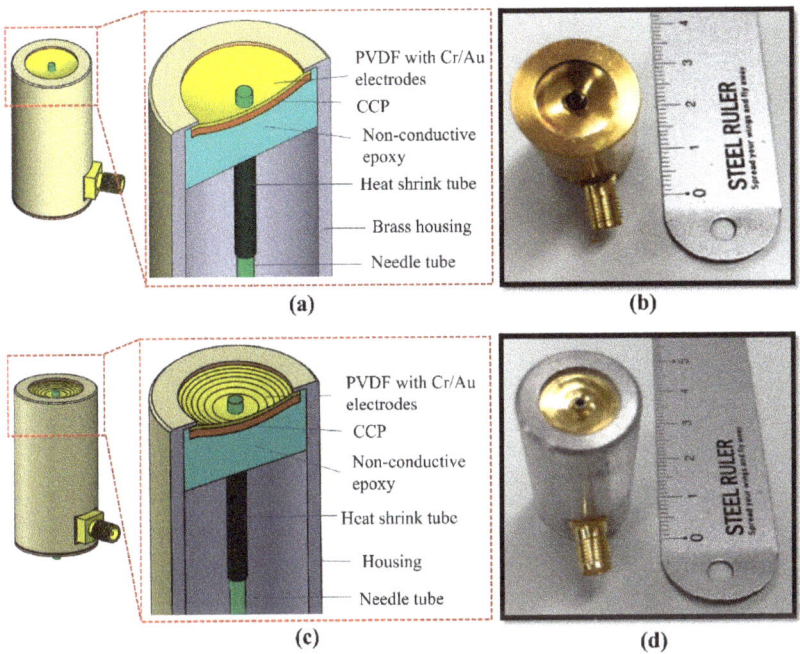

Figure 5. (**a**) A cross sectional view and (**b**) SFP ring-shaped transducer photograph. (**c**) A cross sectional view and (**d**) MFP ring-shaped transducer photograph.

2.4. Phantom Fabrication

2.4.1. Wire Phantom

Sixteen phantom wires (25-µm) were placed diagonally, with an equal distance of 1 mm in the vertical axis and horizontal axis (Figure 6a). The transducer was moved along the X-axis to scan the wires image, which placed at the transducer's focused position in degassed water. The reflected pulse-echo signal from the wire was used to figure the beam shape in the lateral direction. The final image was obtained by image processing, importing data into Matlab-based (Version. 2013a, Mathworks, Natick, MA, USA) software.

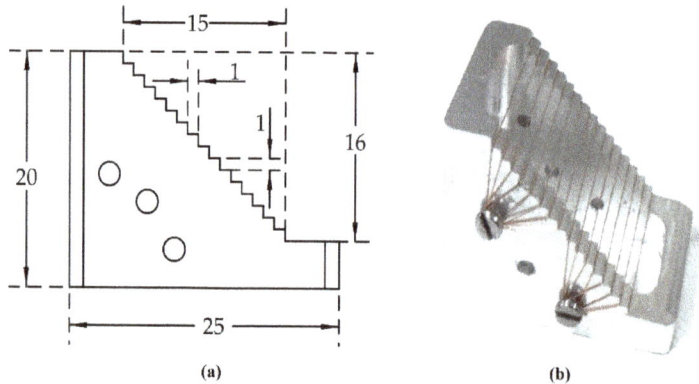

Figure 6. (a) The structure of the wire phantom. (b) A photograph of the wire phantom.

2.4.2. Chicken-Needle Phantom

Figure 7 shows the structure (Figure 7a) and photograph of a chicken-needle (CN) phantom. The chicken meat was placed into an acrylic mold, and 20G (syringe needle, Anhui, China) stainless steel needles (outer diameter, 0.79 mm) were inserted at different depths of the acrylic mold (Figure 7b). The needle phantom was composed of seven needles, which were separated by 1.5 mm in the Z-axis and 2.5 mm in the X-axis. The top surface of the chicken meat sample was flattened in the same plane as that of the top plane of the acrylic mold.

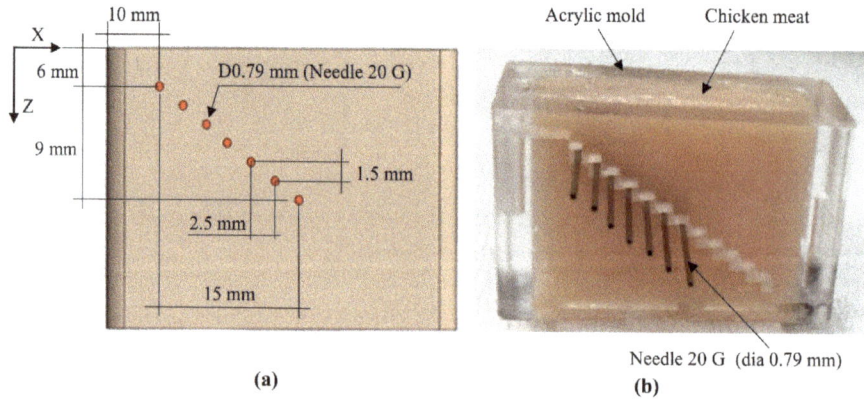

Figure 7. (a) The structure and (b) a photograph of the chicken-needle (CN) phantom.

2.4.3. Chicken-Hemoglobin Phantom

Figure 8 shows the structure (Figure 8a) and photograph of a chicken-hemoglobin (CHb) phantom, in which transparent polytetrafluoroethylene (PTFE) tubes (Zeus, Orangeburg, USA) containing Hb (Sigma-Aldrich, Merck, Seoul, South Korea) were embedded. The hemoglobin concentration of 13.6 g/dL is suitable for optical absorption coefficient of 4.0 cm^{-1} at 800 nm. The blood hemoglobin (Hb) concentration test is one of the most commonly performed tests; the normal Hb level in the human body ranges from 12 to 16 g/dL [21]. Hb is the iron-containing metalloprotein involved in the transport of oxygen that is found in nearly all vertebrates' red blood cells, as well as in some invertebrates' tissue. Hb in the blood transports oxygen to the whole body from the lungs or gills. Hb samples were injected into transparent tubes with a 1.6 mm outer diameter and 25 mm length. Hb-filled tubes were embedded at different depths of the chicken tissue in the acrylic mold (Figure 8b). To produce imaging targets at different depths, the CHb phantom comprised six Hb-filled tubes, which were positioned diagonally by 1.5 mm in the Z-axis and 2.5 mm in the X-axis.

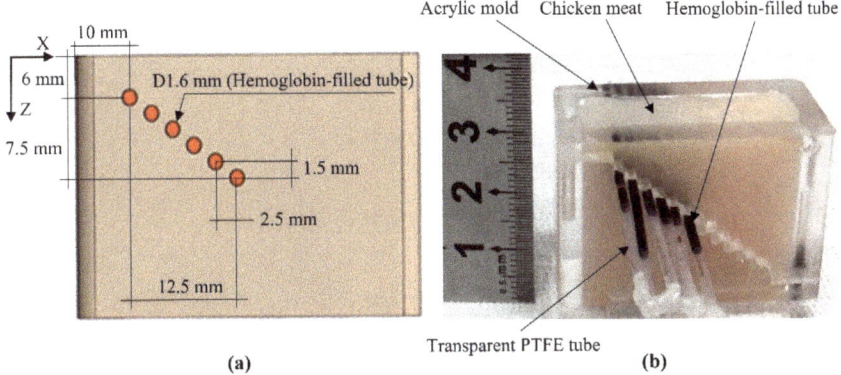

Figure 8. (a) The structure and (b) a photograph of the chicken-hemoglobin (CHb) phantom. The polytetrafluoroethylene (PTFE) tube with an inner diameter of 1.6 mm, wall thickness of 0.038 mm, and transparent color.

2.5. Experimental Setup

2.5.1. Ultrasound Imaging System

Figure 9 shows a schematic diagram of the experimental process. A computer-controlled remote (DPR 500, JSR Ultrasonics, Pittsford, NY, USA) pulser/receiver and the transducer was connected to excite an electrical impulse at a 200 Hz repetition rate at 50 Ω damping with 3 μJ energy per pulse. To measure the pulse-echo and frequency spectra of the transducer, a glass plate was positioned at the focused position as a target. The reflected signal was received using a 500 MHz bandwidth receiver with a high pass filter of 5 MHz and a low pass filter of 500 MHz. The attained raw data were digitized at a sampling frequency of 500 megasamples/s. An 8-bit digitizer (NI PCI-5153EX, National Instruments, Austin, TX, USA) was used to digitize echoes.

A stepper motor (UE63PP, Newport Corporation, CA, USA) was used to control the movement of the transducer, and a universal motion controller/driver (ESP300, Newport Corporation, CA, USA) was used to drive the motors' motion. A LabView (LabView 2014, National Instrument, Austin, TX, USA) program was built to control all the processes mentioned above. A computer-controlled scanning stage was moved along the X-axis to acquire a B-scan image.

An Agilent Keysight 4396B impedance analyzer (Agilent Technologies, Santa Clara, CA, USA) was used to measure the electrical impedance (magnitude and phase) of the fabricated SFP and MFP transducers.

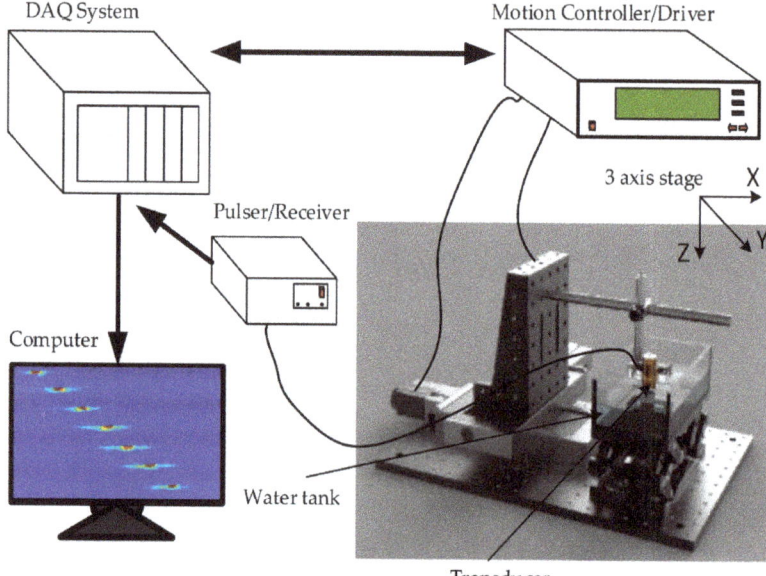

Figure 9. A schematic of the experimental ultrasound system.

2.5.2. Photoacoustic Microscopy System

The ex vivo experiments were conducted on the tissue phantom models. Figure 10 shows a schematic diagram of the experimental setup for the PAM system. Briefly, a tunable OPO laser (Surelite OPO Plus, Continuum, CA, USA) pumped by an Nd:YAG laser (Surelite III, San Jose, CA, USA) was applied as a light source with a 6 ns pulse width, 10 Hz repetition rate, and 650–1064 nm wavelength. A multimode fiber with a diameter of 1 mm and a light divergence of 30 degrees (NA = 0.5) was used to deliver the 800 nm pulsed laser beams with a laser energy of 0.18–1.98 mJ/pulse, which is well below the safety limit (20 mJ/cm^2) of the American National Standards Institute. An 800 nm wavelength was employed to acquire the PA images. The input optical fiber was connected to a plano-convex (focal length: 50 mm; Thorlabs, Newton, NJ, USA). The fiber's output end was connected to the custom transducers and aligned to the center of the illuminated area. To obtain the largest penetration depth in the PAM image, the fiber's output end was adjusted to ensure a guaranteed maximum overlap area between the laser beam and the focal zone. The signals were then digitized and stored in coordination with a laser system, using a data acquisition (DAQ) system to capture the PA signals. The LabView program (Version 2012, National Instruments, Austin, TX, USA) was used to control the scanning procedure. Finally, through Hilbert's transformation, the detected PA signals were transformed into PA images.

To evaluate the capability of transducers in PAM imaging, ex vivo experiments were conducted with both SFP and MFP transducers. Two types of tissue phantom were fabricated for the scanning system to obtain the PAM images, which demonstrated the differential penetration depths from two types of custom transducer. A tissue phantom was positioned inside a water tank through a thin transparent plastic membrane. An ultrasound gel was cast on the top plane of the tissue under the plastic membrane for acoustic coupling. The transducer was immersed in a tank of water during the experimental procedure. The B-scan mode was performed by linearly scanning the specimen along the transverse direction, which demonstrated the in-depth structure of the target. The large focal zone of the MFP transducer combined with the laser energy which can capture a deeper depth image with a high resolution. This is due to the seventh part of the MFP transducer which has the smallest F-number and the longest focal length related to the axial resolution.

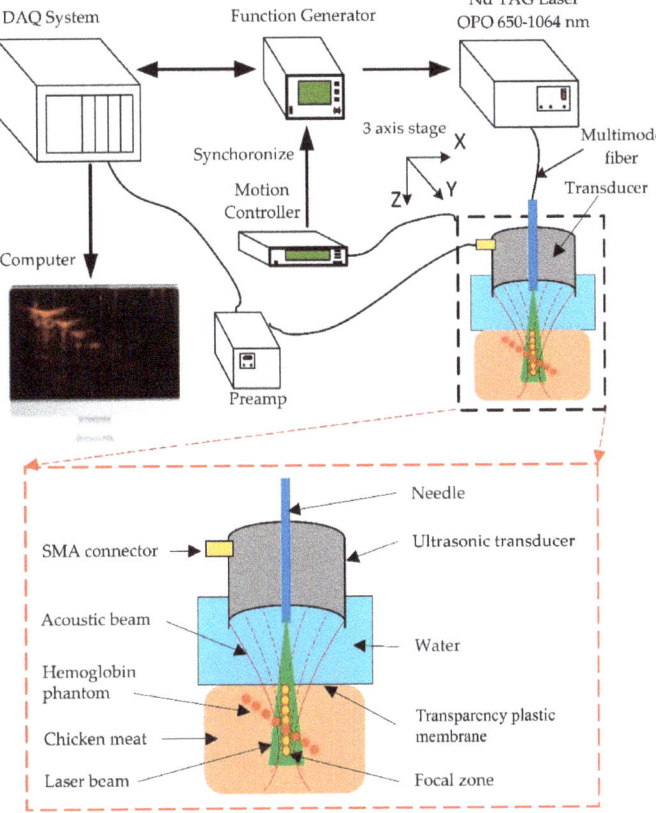

Figure 10. A schematic of the photoacoustic microscopy (PAM) experimental system. Laser OPO wavelengths: 650–1064 nm. Multimode fiber diameter: 1 mm, NA = 0.5.

3. Performance Evaluation

3.1. Ultrasound Characterization

Figure 11 shows the measured pulse-echo response and the transducers' frequency spectrum. According to the ultrasound pulse-echo test, the SFP and MFP transducers had the same center frequency of 11 MHz, and the −6 dB bandwidths of the SFP and MFP transducers were 91% and 109%.

Figure 12 shows the measured electrical impedance (magnitude and phase) of the fabricated SFP and MFP transducers. In the SFP transducer, the electrical impedance was measured at a magnitude of 36 Ω and a phase angle of 46° at 11 MHz. In the MFP transducer, the electrical impedance was measured at a magnitude of 32 Ω and a phase angle of 15° at 11 MHz.

Figure 13 shows B-scan images of the wire phantom attained using the ring-shaped transducers. The wires were positioned in the transducers' focal zones and produced the bright points in the image; otherwise, they appeared as blurred points in the images. In the ultrasound image of the 11 MHz SFP transducer (Figure 13a), only a single bright point at a depth of 12.7 mm was displayed, because the focal length of the SFP transducer was 12.7 mm. Using the seven-focal-point transducer, the ultrasound image of the wire phantom at 11 MHz was captured, as shown in Figure 13b. We observed that the

MFP transducer displayed eleven bright points, showing a large focal zone (10 mm) for deeper images, which is significantly higher than the SFP transducer's focal zone (0.48 mm).

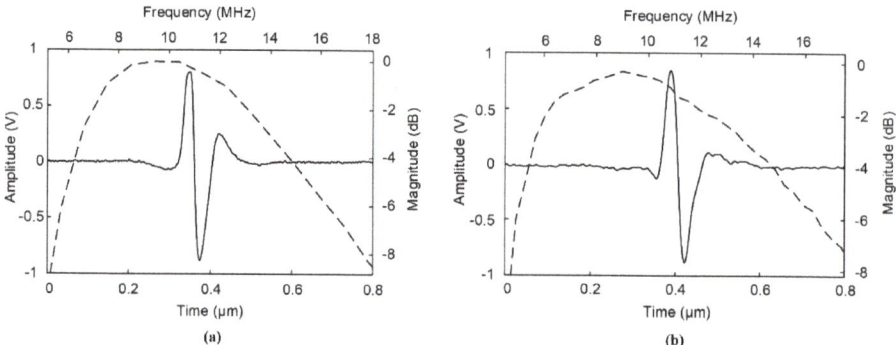

Figure 11. The measured pulse-echo and frequency spectra of transducers after forming a hole of (**a**) the single focal point transducer and (**b**) the seven-focal point transducer.

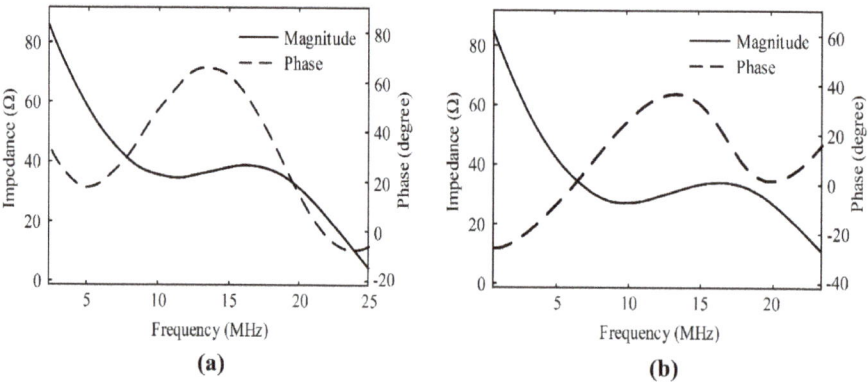

Figure 12. The measured electrical impedance magnitude and phase of (**a**) the single focal point transducer and (**b**) a seven-focal-point transducer.

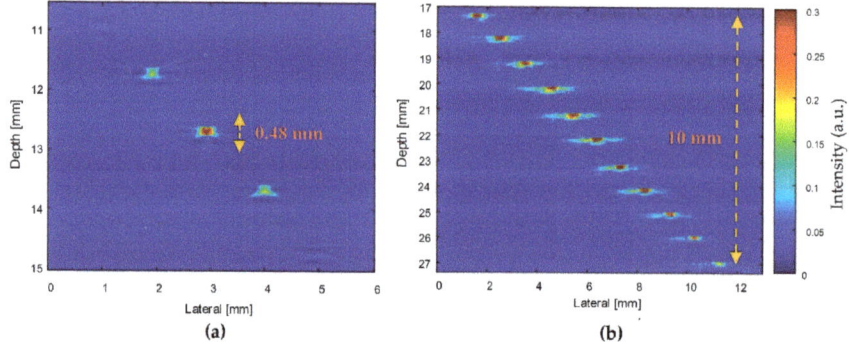

Figure 13. Ultrasound images of the wire phantoms obtained using (**a**) the SFP transducer and (**b**) the MFP transducer.

Figure 14 shows the lateral and axial resolutions of the SFP and MFP transducers. Using the wire phantom target, the spatial resolution of transducers was determined at full width at half maximum. At a depth of 12.7 mm, the SFP transducer had lateral and axial resolutions of 200 μm and 90 μm, respectively. In contrast, the MFP transducer can capture wire images at depths from 17.4 mm to 27.4 mm. The measured lateral resolutions at depths from 17.4 mm to 24.4 mm had a similar value of 360 μm. Lateral resolutions decreased linearly to 200 μm at a depth of 27.4 mm (Figure 14a). The measured axial resolution of the MFP transducer had a similar value of 140 μm at depths from 17.4 mm to 25.4 mm (Figure 14b).

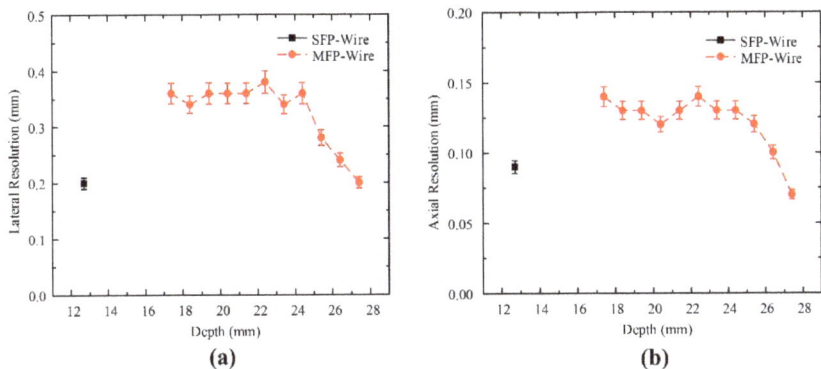

Figure 14. Ultrasound (**a**) lateral and (**b**) axial resolution measurements versus depth in the wire phantoms of the SFP (solid line) and MFP (dashed line) transducers. Error bars denote 95% confidence.

Although the MFP transducer has weak lateral and axial resolutions compared to SFP transducer, the MFP transducer can maintain good image quality at different depths due to its large focal zone. In case of the SFP transducer, the focal length and aperture diameter are 12.7 mm and 12 mm, from which a good F-number (focal length/aperture diameter: 12.7/12 = 1.05) could be acquired. Meanwhile, the MFP transducer has seven F-numbers from 4.4 mm to 2.3 mm, as shown in Table 2. Axial resolution is affected by the F-number, speed of sound, and center frequency of the transducer (Axial resolution = speed of sound x F-number/center frequency). A smaller F-number contributes to the improvement of the axial resolution.

3.2. Penetration Depth in Photoacoustic Microscopy Images

Figure 15 shows the PAM images of the CN phantom obtained using two transducer types. The SFP transducer's PAM image displayed only one bright point at a depth of 6 mm (Figure 15a). In this case, a penetration depth of 6 mm was measured from the top surface of chicken meat to the image depth, in which the first needle was placed in the chicken tissue. The deeper needles' images were not displayed clearly. The MFP transducer captured the PAM image of the first four needles at depths from 6 mm to 10 mm (Figure 15b). The brightest point is the first needle's image at a depth of 6 mm. At a depth of 10 mm, the image of the fourth needle can be seen clearly. The image depth in chicken tissue was measured from the chicken meat's surface to the farthest image point (0–10 mm). Therefore, the penetration depth of the MFP transducer in this CN phantom was measured to be 10 mm. However, blood vessels in the chicken phantom can absorb laser energy, which generates ultrasound waves and reflected signals to the transducer for forming images. Figure 15b shows the PAM image of the blood vessels at the top surface of the chicken meat, at the depth of 3 mm above the first needle, and at the right side from a depth of 0 to 13.5 mm. In this case, the penetration depth was measured to be 13.5 mm. Therefore, in this phantom, the MFP transducer's penetration depth for the stainless steel needles is 10 mm and the penetration depth for chicken blood vessels is 13.5 mm.

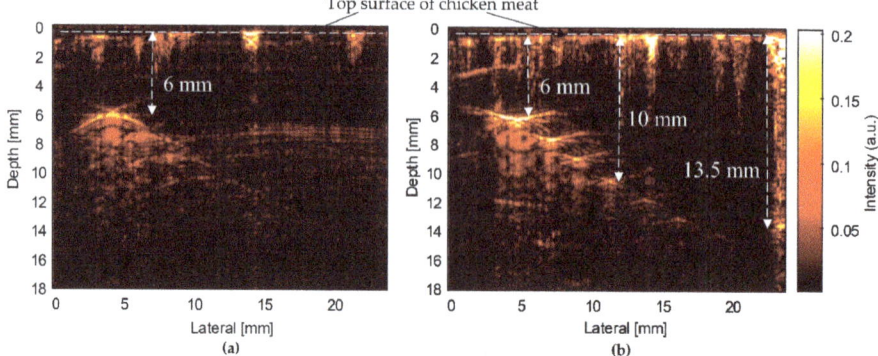

Figure 15. PAM images of the CN phantom acquired using (**a**) SFP and (**b**) MFP transducers. The penetration depths of the SFP and MFP transducers are 6 mm and 10 mm.

Figure 16 shows the PAM images of the CHb phantom obtained using two fabricated transducers. The SFP transducer's PAM image displayed only a single bright point at a depth of 5 mm (Figure 16a). Therefore, the penetration depth of the SFP transducer in this phantom was measured to be 5 mm. The MFP transducer's PAM image displayed three bright points at depths from 6.5 mm to 10 mm (Figure 16b). Therefore, the penetration depth of the MFP transducer in the CHb phantom is 8 mm, which was measured from the chicken meat's surface to the third image point.

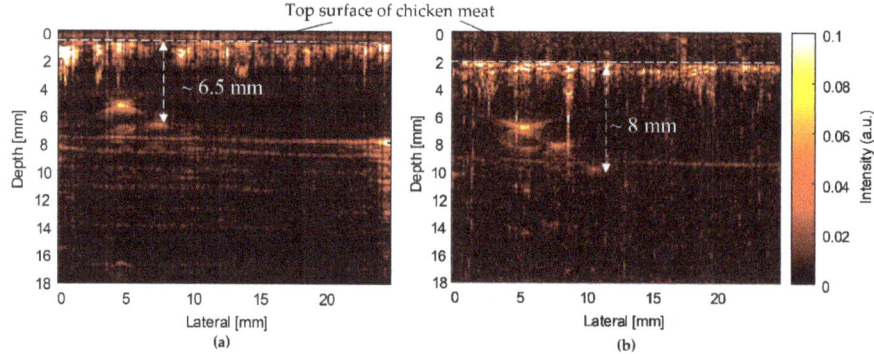

Figure 16. PAM images of the CHb phantom acquired using (**a**) SFP and (**b**) MFP transducers.

Table 4 summarizes the transducers' parameters used in imaging systems. One type of piezoelectric material was designed for the same values of center frequency and different focal zones for two different transducers.

Table 4. Operating parameters of ultrasonic transducers.

Parameter	SFP Transducer	MFP Transducer
Center frequency (MHz)	11	11
-6 dB Bandwidth (%)	91	109
SNR (dB)	38.7	34
Focal length (mm)	12.7	18.5 - 29.5
Axial resolution (μm)	90	140 - 70
Lateral resolution (μm)	200	360 - 200
Focal zone (mm)	0.48	10

Figure 17 shows the mean intensity depth profiles for two transducers in the wire phantom, CN phantom, and CHb phantom. For the ultrasound image of the wire phantom shown in Figure 13, the intensity was considerably high at the focus depth of the transducers, and many more ultrasound waves reflected to the transducers (Figure 17a). The SFP transducer has only one focal point; therefore, the highest intensity was concentrated at a focal depth of 12.7 mm, whereas other depths had lower intensities. In case of the MFP transducer, a seven-focal-point transducer created a larger focal zone, which showed similar intensities at depths from 18.4 mm to 24.4 mm. For the PAM images of the CN and CHb phantoms shown in Figures 15 and 16, the intensity decreased proportionally with depth in Figure 17b. However, the intensities of the PAM image from the MFP transducer were generally lower in the CHb phantom, because light absorbed by Hb is weaker than that absorbed by a needle in a chicken tissue. Owing to light diffusion in the chicken tissue, the intensity at a depth of 10 mm in the CHb phantom was 10 dB, whereas that in the wire phantom was 28 dB at the same depth.

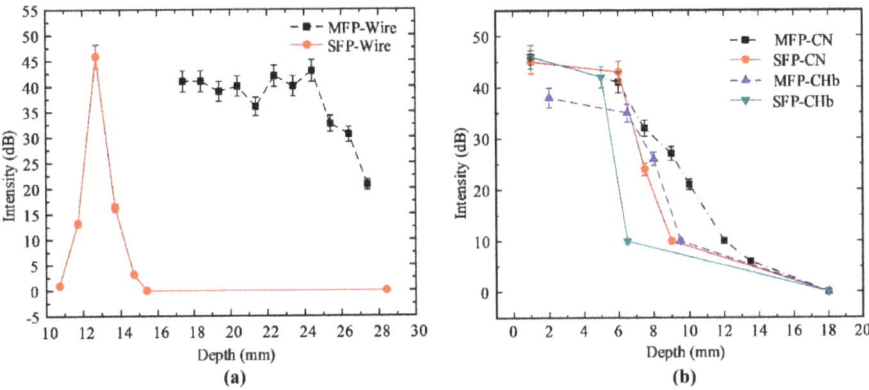

Figure 17. Target intensity versus depth for log-compressed (**a**) ultrasound images in the wire phantom and (**b**) photoacoustic images in the CN phantom and CHb phantom of the SFP (solid line) and MFP (dashed line) transducers. Error bars denote 95% confidence.

4. Discussion

The feasibility of extending the depth-of-field in PAM imaging using a custom MFP transducer was investigated in this study. The goal of this study was to harness the advantages of a single piezoelectric element for designing an MFP transducer for large-depth image applications. The SFP and MFP transducers were designed to compare the different features through their structures. The conventionally focused transducer created only one focal point with a short focal zone. To extend the length of the focal zone, the front face structure of the MFP transducer was designed with many spherically-focused surfaces, which produced many focal points and different focal zones. For the MFP transducer, the B-scan image was obtained by scanning a transducer along the X-axis over the wire phantom-based motion control system. The ultrasound data were collected from a single-channel ultrasonic pulser/receiver and plotted using the Matlab software without applying SAFT processing as with multi-element systems.

For ultrasound and PA applications, a spherically focused transducer can enhance the sensitivity to detect small defects within objects or living tissues. At the same center frequency of 11 MHz, the MFP transducer had a focal zone of 10 mm and the SFP transducer had a focal zone of only 0.48 mm, which was demonstrated by ultrasound images of the wire phantom.

For biomedical imaging applications, the PAM images of the CN phantoms in Figure 15 were obtained by two types of transducers. For the SFP transducer, the measured penetration depth and intensity were 6 mm (Figure 15a) and 43 dB (Figure 17b), respectively. Meanwhile, the measured penetration depth of the MFP transducer's CN phantom image was 10 mm (Figure 15b) and the intensities were decreased from the phantom surface to the farthest image point (46–10 dB), as represented in Figure 17b. In the case of the Hb-filled tubes embedded in chicken tissue (Figure 16), the penetration depths of the SFP and MFP transducers were detected at 6.5 mm and 8 mm, respectively. The measured intensity at the focused depth of the CHb phantom image using the SFP transducer was 43 dB (Figure 17b), whereas in the MFP transducer, it decreased from the phantom surface to the farthest image point (35–10 dB), as represented in Figure 17b. In the same chicken tissue, needles absorbed much more laser light than Hb; therefore, at a deeper depth (10 mm) in the chicken tissue, the ultrasound waves still reflect to the MFP transducer to construct images. From the structure of the MFP transducer, the focal zone length can be extended by increasing the number of focal points and the aperture size.

Using phantoms (wire phantom, CN phantom, and CHb phantom), the performance of the MFP transducer was demonstrated to be suitable. However, B-scan images produced by a fixed focal point transducer can still benefit from SAFT, as the non-focal point scatterers were displayed as arcs due to the finite width of the transducer field away from the focus. As evident from Figures 13, 15 and 16, the developed MFP transducer is not free from this artifact. Instead of the arcs produced by the SFP transducer in the wire phantom image (Figure 13a), the MFP transducer produced horizontal spreading (Figure 13b), which has a much better appearance than the arcs, but is still an artifact of the same nature as the arcs. For the B-scan images of the chicken phantoms (Figures 15 and 16), the MFP transducer produces arcs similar to those produced by the SFP transducer. The effects of horizontal spreading in water vs. arcs in the chicken phantoms are probably because of the difference in the sound speeds between water and the chicken phantom. Both the arcs and the horizontal spreading can be reduced using SAFT, which will be more difficult to implement for the MFP transducer.

Table 4 shows the measured spatial resolution of transducers. However, the axial resolution of the MFP transducer can be improved at greater depths. This is because the axial resolution has an effect on the transducer's parameters such as the center frequency and F-number. The MFP transducer has seven spherically-focused surfaces, which generate various focus depths with different F-numbers. The bigger aperture diameter has a smaller F-number (focal length/aperture diameter). The smaller F-number and the bigger center frequency results in a better axial resolution (Axial resolution = speed of sound x F-number/center frequency). The seventh part of the MFP transducer has the smallest F-number of 2.3 and the longest focal length of 28.87 mm. Therefore, the measured axial resolution of this part was 70 μm, which was the better axial resolution at deeper depths (Figure 17b).

5. Conclusions

This study described a novel design and evaluation of an MFP transducer with a significantly increased focal zone (10 mm) versus the SFP transducer (0.48 mm). The image of eleven phantom wires revealed the proposed MFP transducer's extended focal zone. Additionally, the capacity to extend the focal zone for a larger-sized target was demonstrated, thereby enabling the imaging without the need for depth scans or any complex SAFTs. In ex vivo imaging, the penetration depth of CN and CHb phantoms increased to 10 mm and 8 mm, respectively. The intensity of the CN image decreased from the phantom surface to the farthest image point (4610 dB), whereas that in the CHb decreased from 35 to 10 dB. Specifically, the proposed seven-focal-point transducer is capable of generating seven focal zones along the axial direction simultaneously. Therefore, for large-depth imaging applications, MFP transducers have great potential.

Author Contributions: Conceptualization, Supervision, and Funding Acquisition, J.O.; Methodology, T.P.N., V.T.N.; Software, D.D.V., V.H.P., and T.P.N.; Validation, T.P.N., and J.O.; Writing—Original Draft Preparation, T.P.N., V.T.N., S.M. and B.-G.K.; Writing—Review and Editing, all authors. All authors have read and agreed to the published version of the manuscript.

Funding: This work was supported by the Technology Development Program (S2829803) funded by the Ministry of SMEs and Startups (MSS, Korea).

Conflicts of Interest: The authors declare no conflict of interest.

References

1. Wang, L.V. Multiscale photoacoustic microscopy and computed tomography. *Nat. Photonics* **2009**, *3*, 503. [CrossRef] [PubMed]

2. Balas, C. A novel optical imaging method for the early detection, quantitative grading, and mapping of cancerous and precancerous lesions of cervix. *IEEE Trans. Biomed. Eng.* **2001**, *48*, 96–104. [CrossRef] [PubMed]

3. Bremer, C.; Bredow, S.; Mahmood, U.; Weissleder, R.; Tung, C.-H. Optical imaging of matrix metalloproteinase–2 activity in tumors: Feasibility study in a mouse model. *Radiology* **2001**, *221*, 523–529. [CrossRef] [PubMed]

4. Lao, Y.; Xing, D.; Yang, S.; Xiang, L. Noninvasive photoacoustic imaging of the developing vasculature during early tumor growth. *Phys. Med. Biol.* **2008**, *53*, 4203. [CrossRef]

5. Kim, C.; Erpelding, T.N.; Jankovic, L.; Pashley, M.D.; Wang, L.V. Deeply penetrating in vivo photoacoustic imaging using a clinical ultrasound array system. *Biomed. Opt. Express* **2010**, *1*, 278–284. [CrossRef]

6. Xu, M.; Wang, L.V. Photoacoustic imaging in biomedicine. *Rev. Sci. Instrum.* **2006**, *77*, 041101. [CrossRef]

7. Li, X.; Wei, W.; Zhou, Q.; Shung, K.K.; Chen, Z. Intravascular photoacoustic imaging at 35 and 80 MHz. *J. Biomed. Opt.* **2012**, *17*, 106005. [CrossRef]

8. Lu, H.; Shao, P.; Ranasinghesagara, J.; DeWolf, T.; Harrison, T.; Gibson, W.; Zemp, R.J. In improved depth-of-field photoacoustic microscopy with a custom high-frequency annular array transducer, in photons plus ultrasound: Imaging and sensing 2011. *Int. Soc. Opt. Photonics* **2011**, *7899*, 78993R.

9. Ning, B.; Kennedy, M.J.; Dixon, A.J.; Sun, N.; Cao, R.; Soetikno, B.T.; Chen, R.; Zhou, Q.; Shung, K.K.; Hossack, J.A.; et al. Simultaneous photoacoustic microscopy of microvascular anatomy, oxygen saturation, and blood flow. *Opt. Lett.* **2015**, *40*, 910–913. [CrossRef]

10. Tian, C.; Xie, Z.; Fabiilli, M.L.; Wang, X. Imaging and sensing based on dual-pulse nonlinear photoacoustic contrast: A preliminary study on fatty liver. *Opt. Lett.* **2015**, *40*, 2253–2256. [CrossRef]

11. Sun, Y.; O'Neill, B. Imaging high-intensity focused ultrasound-induced tissue denaturation by multispectral photoacoustic method: An ex vivo study. *Appl. Opt.* **2013**, *52*, 1764–1770. [CrossRef] [PubMed]

12. Yang, Z.; Chen, J.; Yao, J.; Lin, R.; Meng, J.; Liu, C.; Yang, J.; Li, X.; Wang, L.; Song, L.; et al. Multi-parametric quantitative microvascular imaging with optical-resolution photoacoustic microscopy in vivo. *Opt. Express* **2014**, *22*, 1500–1511. [CrossRef] [PubMed]

13. Wang, L.V.; Hu, S. Photoacoustic tomography: In vivo imaging from organelles to organs. *Science* **2012**, *335*, 1458–1462. [CrossRef] [PubMed]

14. Wang, L.V. Tutorial on photoacoustic microscopy and computed tomography. *IEEE J. Sel. Top. Quantum Electron.* **2008**, *14*, 171–179. [CrossRef]

15. Ku, G.; Wang, X.; Stoica, G.; Wang, L.V. Multiple-bandwidth photoacoustic tomography. *Phys. Med. Biol.* **2004**, *49*, 1329. [CrossRef]

16. Wang, X.; Pang, Y.; Ku, G.; Xie, X.; Stoica, G.; Wang, L.V. Noninvasive laser-induced photoacoustic tomography for structural and functional in vivo imaging of the brain. *Nat. Biotechnol.* **2003**, *21*, 803. [CrossRef]

17. Ku, G.; Wang, L.V. Deeply penetrating photoacoustic tomography in biological tissues enhanced with an optical contrast agent. *Opt. Lett.* **2005**, *30*, 507–509. [CrossRef]

18. Li, L.; Zemp, R.J.; Lungu, G.F.; Stoica, G.; Wang, L.V. Photoacoustic imaging of lacZ gene expression in vivo. *J. Biomed. Opt.* **2007**, *12*, 020504. [CrossRef]

19. Ding, Q.; Tao, C.; Liu, X. Photoacoustics and speed-of-sound dual mode imaging with a long depth-of-field by using annular ultrasound array. *Opt. Express* **2017**, *25*, 6141–6150. [CrossRef]

20. Li, R.; Phillips, E.; Wang, P.; Goergen, C.J.; Cheng, J.X. Label-free in vivo imaging of peripheral nerve by multispectral photoacoustic tomography. *J. Biophotonics* **2016**, *9*, 124–128. [CrossRef]

21. Vogt, W.C.; Jia, C.; Wear, K.A.; Garra, B.S.; Pfefer, T.J. Phantom-based image quality test methods for photoacoustic imaging systems. *J. Biomed. Opt.* **2017**, *22*, 095002.

22. Jansen, C.H.; Brangsch, J.; Reimann, C.; Adams, L.; Hamm, B.; Botnar, R.M.; Makowski, M.R. In vivo high-frequency ultrasound for the characterization of thrombi associated with aortic aneurysms in an experimental mouse model. *Ultrasound Med. Boil.* **2017**, *43*, 2882–2890. [CrossRef] [PubMed]

23. Nguyen, T.P.; Truong, N.T.P.; Bui, N.Q.; Nguyen, V.T.; Hoang, G.; Choi, J.; Phan, T.T.V.; Pham, V.H.; Kim, B.-G.; Oh, J.; et al. Design, fabrication, and evaluation of multifocal point transducer for high-frequency ultrasound applications. *Sensors* **2019**, *19*, 609. [CrossRef] [PubMed]

24. Zemp, R.J.; Song, L.; Bitton, R.; Shung, K.K.; Wang, L.V. Realtime photoacoustic microscopy in vivo with a 30-MHz ultrasound array transducer. *Opt. Express* **2008**, *16*, 7915–7928. [CrossRef] [PubMed]

25. Gamelin, J.; Maurudis, A.; Aguirre, A.; Huang, F.; Guo, P.; Wang, L.V.; Zhu, Q. A real-time photoacoustic tomography system for small animals. *Opt. Express* **2009**, *17*, 10489–10498. [CrossRef] [PubMed]

26. Song, L.; Maslov, K.; Wang, L.V. Section-illumination photoacoustic microscopy for dynamic 3D imaging of microcirculation in vivo. *Opt. Lett.* **2010**, *35*, 1482–1484. [CrossRef]

27. Niederhauser, J.J.; Jaeger, M.; Lemor, R.; Weber, P.; Frenz, M. Combined ultrasound and optoacoustic system for real-time high-contrast vascular imaging in vivo. *IEEE Trans. Med. Imaging* **2005**, *24*, 436–440. [CrossRef]

28. Zhou, Y.; Li, G.; Zhu, L.; Li, C.; Cornelius, L.A.; Wang, L.V. Handheld photoacoustic probe to detect both melanoma depth and volume at high speed in vivo. *J. Biophotonics* **2015**, *8*, 961–967. [CrossRef]

29. Ilovitsh, A.; Ilovitsh, T.; Foiret, J.; Stephens, D.N.; Ferrara, K.W. Simultaneous axial multifocal imaging using a single acoustical transmission: A practical implementation. *IEEE Trans. Ultrason. Ferroelectr. Freq. Control* **2018**, *66*, 273–284. [CrossRef]

30. Ketterling, J.A.; Lizzi, F.L.; Aristizábal, O.; Turnbull, D.H. Design and fabrication of a 40-MHz annular array transducer. *IEEE Trans. Ultrason. Ferroelectr. Freq. Control* **2005**, *52*, 672. [CrossRef]

31. Kim, J.; Lindsey, B.D.; Li, S.; Dayton, P.A.; Jiang, X. In dual-frequency transducer with a wideband PVDF receiver for contrast-enhanced, adjustable harmonic imaging, health monitoring of structural and biological systems 2017. *Int. Soc. Opt. Photonics* **2017**, *10170*, 101700T.

32. Ketterling, J.A.; Aristizábal, O.; Turnbull, D.H. High-frequency piezopolymer transducers with a copper-clad polyimide backing layer. *IEEE Trans. Ultrason. Ferroelectr. Freq. Control* **2006**, *53*, 1376–1380. [CrossRef] [PubMed]

33. Snook, K.A.; Jian-Zhong, Z.; Alves, C.H.F.; Cannata, J.M.; Wo-Hsing, C.; Meyer, R.J.; Ritter, T.A.; Shung, K.K. Design, fabrication, and evaluation of high frequency, single-element transducers incorporating different materials. *IEEE Trans. Ultrason. Ferroelectr. Freq. Control* **2002**, *49*, 169–176. [CrossRef] [PubMed]

34. Cannata, J.M.; Ritter, T.A.; Wo-Hsing, C.; Silverman, R.H.; Shung, K.K. Design of efficient, broadband single-element (20-80 MHz) ultrasonic transducers for medical imaging applications. *IEEE Trans. Ultrason. Ferroelectr. Freq. Control* **2003**, *50*, 1548–1557. [CrossRef] [PubMed]

Article

3D Measurement of Human Chest and Abdomen Surface Based on 3D Fourier Transform and Time Phase Unwrapping

Haibin Wu, Shuang Yu * and Xiaoyang Yu

The Higher Educational Key Laboratory for Measuring & Control Technology and Instrumentations of Heilongjiang Province, Harbin University of Science and Technology, Harbin 150080, China; woo@hrbust.edu.cn (H.W.); yuxiaoyang@hrbust.edu.cn (X.Y.)
* Correspondence: yushuang@hrbust.edu.cn; Tel.: +86-451-8639-5333

Received: 13 January 2020; Accepted: 16 February 2020; Published: 17 February 2020

Abstract: Monitoring respiratory movements is an effective way to improve radiotherapy treatments of thoracic and abdominal tumors, but the current approach is limited to measuring specific points in the chest and abdomen. In this paper, a dynamic three-dimensional (3D) measurement approach of the human chest and abdomen surface is proposed, which can infer tumor movement more accurately, so the radiotherapy damage to the human body can be reduced. Firstly, color stripe patterns in the RGB color model are projected, then after color correction, the collected stripe image sequences are separated into the three RGB primary color stripe image sequences. Secondly, a fringe projection approach is used to extract the folded phase combined 3D Fourier transform with 3D Gaussian filtering. By the relationship between adjacent fringe images in the time sequence, Gaussian filter parameters with individual characteristics are designed and optimized to improve the accuracy of wrapped phase extraction. In addition, based on the difference between the fractional parts of the folded phase error, one remainder equation can be determined, which is used for time-phase unwrapping. The simulation model and human experiments show that the proposed approach can obtain the 3D image sequences of the chest and abdomen surface in respiratory motion effectively and accurately with strong anti-interference ability.

Keywords: 3D measurement; fringe projection; 3D Fourier transform; phase unwrapping; phase measurement

---

1. Introduction

During radiotherapy, respiratory movements can cause tumors and normal tissues of the chest and abdomen to move at a certain frequency and amplitude. Sometimes, respiratory movements may affect the radiotherapy effect and even cause radiotherapy damage to the human body. In order to solve this respiratory motion problem, now the most effective real-time tracking method is to monitor extracorporeal respiratory movement. Based on extracorporeal respiratory movement, the respiratory movement of the tumor can be inferred, and then, the relative position of the target area and the field can be controlled by the radiotherapy system [1].

The fact that the respiratory movement of the tumor can be deduced from extracorporeal respiratory movement has been proven to be effective [2]. Based on this premise, dynamic 3D measurement of the chest and abdomen surface can be used to infer tumor movement more accurately [3].

Currently, optical measurement has become the most practical method to solve the problem of dynamic 3D measurement of the chest and abdomen surface. The optical method contains the following three types, point imaging, line imaging, and surface imaging, of which the surface imaging method is the best choice for dynamic 3D measurement. One image or many images can be captured by the surface imaging method. Single image acquisition methods include binocular vision [4], spatially encoded light [5], Fourier transform profile [6], etc. Multiple image acquisition methods include the phase-shift profile [7], modulation measurement profile [8], etc. Due to the fact that the multiple image acquisition method needs to capture many images, it has low efficiency and is not suitable for dynamic measurements. Thus, the one image acquisition method is a good solution for dynamic three-dimensional measurement.

Among these single image acquisition methods, the binocular vision method needs complex stereo matching and has low accuracy [9]; and the spatially encoded light method needs to be coded and decoded by the neighboring pixels, which makes the measurement resolution limited and may cause measurement failure in the case of surface height jump or shade [10]. Fluoroscopic real-time tumor tracking radiotherapy following 4D treatment planning was developed and shown to be feasible to improve the accuracy of the radiotherapy for mobile tumors [11]. In radiation therapy, the projection patterns need to be simple and continuous changes. Because the Fourier transform profile method has significant advantages in noise suppression and full field measurement and the sinusoidal fringe patterns projected by this method meet the requirements of radiotherapy, the Fourier transform profile method is suitable for measuring the 3D motion of the thoracic and abdominal surfaces. However, when only one stripe pattern is projected to measure the entire chest and abdomen surface, the difference in light intensity between adjacent pixels is small due to the large measurement range, and the anti-interference ability is low [12].

In this paper, a dynamic 3D measurement approach of the human chest and abdomen surface during respiration is proposed, which provides a basis for inferring and tracking tumor respiration movement during radiotherapy. This approach adopts a single color stripe pattern with three periods. Through combining one coded pattern with the three RGB primary colors, the sinusoidal stripe pattern with three different periods can be formed. During measurement, the projection pattern does not change, and the deformed stripe image of the chest and abdomen surface is collected in real time. Then, after color coupling correction and color separation, the single color deformation fringe images with three different periods can be formed. The proposed approach can obtain three deformed fringe images by one unchanged pattern. Taking each image sequence of single color fringe as a whole, the 3D Fourier fringe analysis (3D-FFA) method is used to extract the folded phase. This method has higher anti-interference ability. The three-frequency time phase unwrapping method is adopted. The absolute phase is obtained by the folding phase of three monochromatic fringes. This method has a large unwrapping range and strong anti-interference ability. According to the principle of triangulation, the 3D coordinates of the chest and abdomen surface are obtained from the absolute phase [13].

Section 2 introduces the proposed 3D measurement system. Section 3 describes the folding phase extraction method. Section 4 derives a three-frequency time phase unwrapping method. Section 5 shows and analyzes the experimental results, and the conclusions are given in Section 6.

2. 3D Measurement System Description

Figure 1 is a schematic diagram of the 3D measurement system for the human chest and abdomen surface. The method mainly includes the following five parts.

(1) Pattern projection and image acquisition. The computer generates different periods of three RGB primary color cosine stripe patterns, and these three parts are combined into a composite color stripe pattern, then this pattern is projected to the chest and abdomen surface of the human. The camera captures stripe images of the chest and abdomen surface, which change with breathing movements at regular intervals to get a composite color stripe image sequence. The projection pattern in the proposed approach does not change, which can reduce the time of projection pattern conversion and setup. In addition, the measurement system only collects one composite color stripe pattern, which can decrease the image acquisition time. All these advantages can lay the foundation of the dynamic 3D measurement for the human chest and abdomen surface.

(2) Image color correction and separation. For compound color stripe image sequences, color coupling correction and color separation should be made based on the correction matrix of each pixel, and the three RGB primary color fringe image sequences of different periods can be separated. Because the color coupling phenomenon exists at the coincident intersection of the three color channel spectral response curves in the 3CCD industrial camera [14], the color calibration based on hardware equipment should be completed before the measurement. That is to say, the projector projects four patterns of full red, full green, full blue, and full black to the chest and abdomen surface, and the four images will be captured by the camera. Using these four images, the correction matrix of each pixel is obtained according to the Casti illumination model [15].

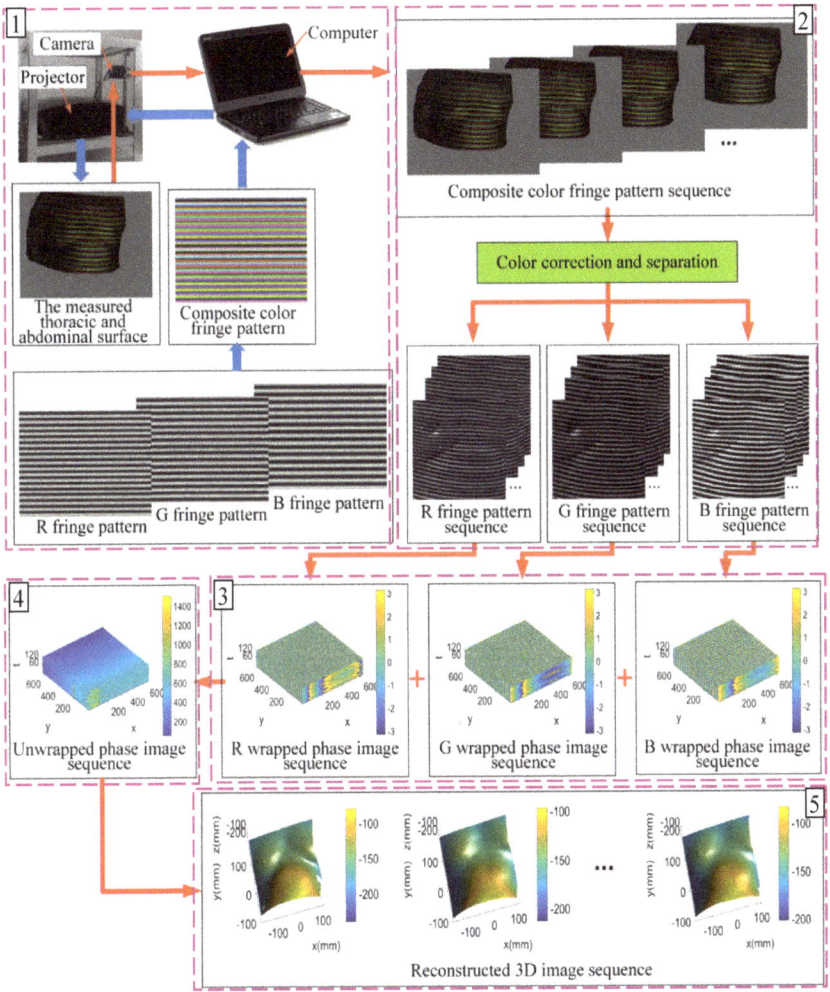

Figure 1. Schematic diagram of the 3D measurement principle and system of the human chest and abdomen surface. (**1**) Pattern projection and image acquisition. (**2**) Image color correction and separation. (**3**) Folded phase extraction. (**4**) Folded phase unwrapping. (**5**) Three-dimensional image sequence acquisition.

(3) Folded phase extraction. For each image in the RGB stripe image sequence, three-dimensional Fourier fringe analysis (3D-FFA) is used to extract the folding phase of each pixel and get the folded phase map of each image, and then, RGB folding phase map sequences can be formed. Fourier fringe analysis (FFA) is extended from one-dimensional Fourier fringe analysis (1D-FFA) to two-dimensional Fourier fringe analysis (2D-FFA) by using the properties of 2D fringe images. After this process, useful signals and interference can be separated better. This method becomes an effective measurement for 3D measurement of flat surfaces [16]. In this paper, the image sequences of the chest and abdomen surface are taken as a 3D one, which is analyzed by 3D Fourier transformation. Useful signals and interference can be separated further by increasing the time dimension, so as to reduce the influence of interference and improve the accuracy of measurement.

(4) Folded phase unwrapping. According to the RGB folding phase diagram at the same time, with the proposed method of three-frequency time phase unwrapping in this paper, the folded phase is expanded into a continuous absolute phase, and the absolute phase diagram at that moment is obtained; thus, an absolute phase sequence can be formed. In the phase unwrapping method of this paper, the unwrapping operation depends on the difference of the decimal part of the measured folded phase, which can ensure that the absolute phase error does not exceed the folded phase error under certain conditions. In addition, we can judge whether there is any big error based on the absolute phase value, which can eliminate or reduce the effect of large absolute phase error by eliminating or interpolating operation. The phase unwrapping is achieved by solving the remainder equation set in the maximum range.

(5) Three-dimensional image sequence acquisition. According to the absolute phase diagram sequence, the 3D coordinates are calculated to form a 3D image sequence of the human chest and abdomen surface based on the triangulation principle. The sequence expresses the 3D shape of the human chest and abdomen surface at each sampling moment during respiratory movement.

3. Folded Phase Extraction Method

In this paper, 3D-FFA combines 3D Fourier transform with 3D Gauss filtering in the frequency domain to achieve folded phase extraction.

3.1. Folded Phase Extraction Principle

Taking an image in the R fringe image sequence as an example, the principle of folded phase extraction is as follows. Firstly, the intensity of fringe image sequences $i_r(x,y,t)$ at different times t can be described as:

$$i_r(x, y, t) = a_r(x, y, t) + b_r(x, y, t) \cos[2\pi(f_{x0}x + f_{y0}y + f_{t0}t) + \varphi_r(x, y, t)] \tag{1}$$

where x represents the row coordinate of stripe images, y denotes the column coordinate, $a_r(x, y, t)$ is the background light intensity, $b_r(x, y, t)$ is the modulation of fringes, f_{x0}, f_{y0}, and f_{t0} are the carrier frequencies in the direction of x, y, and t, and $\varphi_r(x, y, t)$ is the phase distribution function. Equation (1) can be further expressed as:

$$i_r(x, y, t) = a_r(x, y, t) + d_r(x, y, t) \exp[j2\pi(f_{x0}x + f_{y0}y + f_{t0}t)] + \\ d_r^*(x, y, t) \exp[-j2\pi(f_{x0}x + f_{y0}y + f_{t0}t)] \tag{2}$$

where:

$$d_r(x, y, t) = \frac{1}{2}b_r(x, y, t) \exp[j\varphi_r(x, y, t)] \tag{3}$$

$$d_r^*(x, y, t) = \frac{1}{2}b_r(x, y, t) \exp[-j\varphi_r(x, y, t)] \tag{4}$$

After 3D Fourier transform of Equation (2), we can obtain:

$$I_r(f_x, f_y, f_t) = A_r(f_x, f_y, f_t) + D_r(f_x - f_{x0}, f_y - f_{y0}, f_t - f_{t0}) + D_r^*(f_x + f_{x0}, f_y + f_{y0}, f_t + f_{t0}) \quad (5)$$

where f_{x0}, f_{y0}, and f_{t0} are the frequency domain variables in the direction of axes x, y, and t, respectively, $A_r(f_x, f_y, f_t)$ is the background light spectrum, and $D_r(f_x - f_{x0}, f_y - f_{y0}, f_t - f_{t0})$ and $D_r^*(f_x + f_{x0}, f_y + f_{y0}, f_t + f_{t0})$ are the spectra of deformed fringes.

In addition, a 3D filter is used to separate the first level spectrum of $D_r(f_x - f_{x0}, f_y - f_{y0}, f_t - f_{t0})$ and move it to the origin of the frequency domain. After obtaining $D_r(f_x, f_y, f_t)$, the 3D inverse Fourier transform is performed, and the phase distribution function is as follows,

$$\varphi_r(x, y, t) = \tan^{-1} \frac{\text{Im}\{d_r(x, y, t)\}}{\text{Re}\{d_r(x, y, t)\}} \quad (6)$$

where $\text{Im}\{d_r(x, y, t)\}$ denotes the imaginary part of $d_r(x, y, t)$ and $\text{Re}\{d_r(x, y, t)\}$ is the real part. Similarly, $\varphi_g(x, y, t)$ and $\varphi_b(x, y, t)$ can be obtained.

3.2. Three-Dimensional Gauss Filter

When extracting the positive first-order spectrum of fringe image sequences, the 3D filter must also have the function of filtering interference, which is important for 3D Fourier analysis. Because of the interference existing in the environment, the measured object and the measurement system, and the spectrum leakage led by signal truncation, the folding phase error will happen, which will cause 3D measurement errors. Currently, 3D-FFA mainly uses two types of 3D filters [17]: one is the 3D rectangular filter, and the other is the 3D Butterworth filter [18,19]. The former has truncation problems and large leakage errors. For phase unwrapping, the latter has many problems such as a complex algorithm, accumulated error, and unreliability. By comparison, the Gauss filter has the advantages of a small ringing effect and a good effect of eliminating spectrum leakage. For the determined chest and abdomen surface of the measured human, the center frequency and the width of filter in the 3D direction are determined by experiments, which can make it have good adaptability and filter performance. Moreover, previous studies proved that the effect of the 2D Gauss filter is better than the Hanning window and rectangular window [20]. In this paper, the 3D Gauss filter is used as follows,

$$H(f_x, f_y, f_t) = e^{-[\frac{(f_x - f_{0x})^2}{2\sigma_x^2} + \frac{(f_y - f_{0y})^2}{2\sigma_y^2} + \frac{(f_t - f_{0t})^2}{2\sigma_t^2}]} \quad (7)$$

where f_{x0}, f_{y0}, and f_{t0} denote the center frequency of the direction of the x, y, and t axis, respectively, and σ_x, σ_y, and σ_t represent the filter widths in the three directions, respectively. Figure 2 gives the schematic diagram of filtering in the frequency domain with 1D-FFA, 2D-FFA, and 3D-FFA. As shown in Figure 2, the cut-off frequency in the x-axis, y-axis, and z-axis directions are $f_{x1} = f_{x0} - \sigma_x/2$ and $f_{x2} = f_{x0} + \sigma_x/2$, $f_{y1} = f_{y0} - \sigma_y/2$ and $f_{y2} = f_{y0} + \sigma_y/2$, and $f_{t1} = f_{t0} - \sigma_t/2$ and $f_{t2} = f_{t0} + \sigma_t/2$, respectively. The cut-off frequency value is determined experimentally for each measured object, so that the measured signal passes through as much as possible, and the interference signal passes as little as possible.

In principle, 3D-FFA filters have stronger anti-interference ability than 1D-FFA and 2D-FFA. The 3D shock interference is taken as an example to explain this theory. The spectrum amplitude of the shock interference $\delta(x, y, t)$ obtained by 3D Fourier transform is one, and its frequency components cover the entire 3D frequency domain. 1D-FFA can only be filtered along the f_x axis in the one-dimensional frequency domain, and its pass band is $f_{x1} < f_x < f_{x2}$. As shown in Figure 2a, it can only filter out interference signals in the one-dimensional frequency domain. 2D-FFA filters along the f_x axis and f_y axis in the 2D frequency domain, the pass bands are $f_{x1} < f_x < f_{x2}$ and $f_{y1} < f_y < f_{y2}$. As shown in Figure 2b, it can filter out interference signals in the 2D frequency domain, which can further significantly weaken the interference signal. 3D-FFA filters along the f_x, f_y, and f_t axis in the 3D frequency domain, and the pass bands are $f_{x1} < f_x < f_{x2}$, $f_{y1} < f_y < f_{y2}$, and $f_{t1} < f_t < f_{t2}$, respectively. As shown in Figure 2c, it can filter out interference signals in the 3D frequency domain, which can weaken the interference signal once again.

Due to adding the relationship between adjacent fringe images in time sequence, we can get the Gaussian filter parameters with individual characteristics by designing and optimizing the filter parameters to reduce the ringing effect and spectrum leakage; this can improve the accuracy of wrapped phase extraction. Moreover, 3D-FFA processes all the images at the same time, so it has high efficiency and is suitable for dynamic measurement.

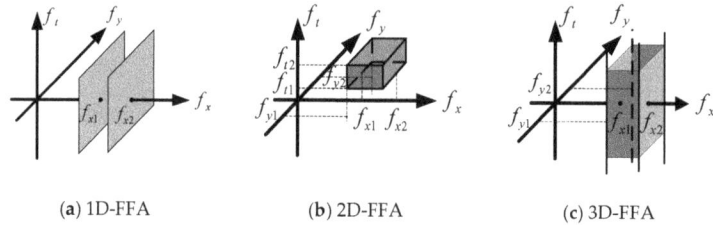

(a) 1D-FFA   (b) 2D-FFA   (c) 3D-FFA

Figure 2. Schematic diagram of filtering in the frequency domain with three methods. FFA, Fourier fringe analysis.

4. Tri-Frequency Time Phase Unwrapping Method

Firstly, positive integers P_r, P_g, and P_b ($P_r < P_g < P_b$) are used to represent the fringe period of the R, G, and B stripe patterns; then r_r, r_g, and r_b are used to denote the folding phases of the R, G, and B stripe images, and they are the solutions of φ_r, φ_g, and φ_b, respectively in the main value intervals, whose value range is $[0, P_r]$, $[0, P_g]$, and $[0, P_b]$. Therefore, there are the following equations.

$$\varphi_r \times P_r / (2 \times \pi) = r_r (\text{mod} P_r) \tag{8}$$

$$\varphi_g \times P_g / (2 \times \pi) = r_g (\text{mod} P_g) \tag{9}$$

$$\varphi_b \times P_b / (2 \times \pi) = r_b (\text{mod} P_b) \tag{10}$$

In these equations, P_r, P_g, and P_b are the three modulus values, and r_r, r_g, and r_b are the corresponding three remainders.

In addition, N represents the distance between the measured pixel point and the phase origin, which also denotes the absolute phase after the unwrapping processing. Let $N_r = [N/P_r]$, $N_g = [N/P_g]$, $N_b = [N/P_b]$, where $[\,]$ represents a rounding down operation, then N can be written as follows:

$$N = N_r P_r + r_r = N_g P_g + r_g = N_b P_b + r_b \tag{11}$$

If P_r, P_g, and P_b have the greatest common divisor P, there are $\Omega_r = P_r / P$, $\Omega_g = P_g / P$, $\Omega_b = P_b / P$, $\Omega = \Omega_r \times \Omega_g \times \Omega_b$, $\lambda_r = \Omega / \Omega_r$, $\lambda_g = \Omega / \Omega_g$, and $\lambda_b = \Omega / \Omega_b$. Assuming that Ω_r, Ω_g, and Ω_b are mutually prime, then λ_r and Ω_r are relatively prime, that is the modulus inverse $\overline{\lambda}_r$ of λ_r exists for Ω_r, and there is $\overline{\lambda}_r \lambda_r \equiv 1 (\text{mod} \Omega_r)$. For the same reason, there are $\overline{\lambda}_g \lambda_g \equiv 1 (\text{mod} \Omega_g)$ and $\overline{\lambda}_b \lambda_b \equiv 1 (\text{mod} \Omega_b)$.

Let:

$$q_r = [r_r / P], q_g = [r_g / P], q_b = [r_b / P] \tag{12}$$

Then, there is:

$$r_r = q_r P + r^c, r_g = q_g P + r^c, r_b = q_b P + r^c \tag{13}$$

where $N \equiv r^c (\mathrm{mod}P)$ and $N_0 = [N/P]$, then:

$$\overline{\lambda}_r \lambda_r q_r + \overline{\lambda}_g \lambda_g q_g + \overline{\lambda}_b \lambda_b q_b = N_0 (\mathrm{mod}\Omega) \tag{14}$$

so:

$$N = PN_0 + r^c \tag{15}$$

It is obvious that the folding phases of the R, G, and B fringe images, i.e., the remainders r_r, r_g, and r_b, can be used to unwrap the phase and obtain the absolute phase by Equations (12)–(15).

However, the above discussion does not consider the effect of the residual measurement errors. According to Equation (14), if there are measurement errors Δr_r, Δr_g, and Δr_b in the folding phases r_r, r_g, and r_b, the errors Δq_r, Δq_g, and Δq_b belonging to q_r, q_g, and q_b, respectively will be produced, i.e., $|\Delta q_r| \geq 1$, $|\Delta q_g| \geq 1$, and $|\Delta q_b| \geq 1$. In the meantime, $(\overline{\lambda}_r \lambda_r q_r + \overline{\lambda}_g \lambda_g q_g + \overline{\lambda}_b \lambda_b q_b)$ will also produce coarse errors, i.e., $|\Delta q_r \overline{\lambda}_r \lambda_r| \geq \overline{\lambda}_r \times \Omega_g \times \Omega_b$, $|\Delta q_g \overline{\lambda}_g \lambda_g| \geq \overline{\lambda}_g \times \Omega_r \times \Omega_b$, and $|\Delta q_b \overline{\lambda}_b \lambda_b| \geq \overline{\lambda}_b \times \Omega_r \times \Omega_g$. All these can lead the N_0 and N to produce coarse errors $\Delta N_{MAX} \approx \Delta N_{0MAX}$. Under the condition of $\overline{\lambda}_r \geq 3$, $\overline{\lambda}_g \geq 3$, and $\overline{\lambda}_b \geq 3$, there is $|\Delta N_{MAX}| \geq 3 \times \Omega_r \times \Omega_g$. Consider that the pixel resolution of the existing device is usually limited within $\Omega_b \leq 9$, i.e., $|\Delta N_{MAX}| \geq \Omega/3|$. ΔN_{MAX} may further cause $(\overline{\lambda}_r \lambda_r q_r + \overline{\lambda}_g \lambda_g q_g + \overline{\lambda}_b \lambda_b q_b)$ to make errors in the remainder operation with the modulo $\Omega = \Omega_r \times \Omega_g \times \Omega_b$, which may result in a larger absolute phase error. To sum up, ΔN_{MAX} can cause a large error in the measurement or even lead the measurement to fail. Therefore, some measures should be adopted to avoid the appearance of ΔN_{MAX} or reduce its influence.

To avoid the appearance of ΔN_{MAX}, assume that the remainder measurement error satisfies Condition A,

$$|\Delta r_r / P| < 0.25, |\Delta r_g / P| < 0.25, |\Delta r_b / P| < 0.25 \tag{16}$$

The measured values $\hat{r}_r$, $\hat{r}_g$, and $\hat{r}_b$ of the remainders r_r, r_g, and r_b can be expressed as:

$$\hat{r}_r / P = [\hat{r}_r / P] + \{\hat{r}_r / P\}, \hat{r}_g / P = [\hat{r}_g / P] + \{\hat{r}_g / P\}, \hat{r}_b / P = [\hat{r}_b / P] + \{\hat{r}_b / P\} \tag{17}$$

where $\{\}$ is a fractional operation. Then, if the difference between the fractional part of the remainder measurement satisfies Condition B, then:

$$\begin{aligned} |\Delta r_{rg} / P| &= \left| \{\hat{r}_r / P\} - \{\hat{r}_g / P\} \right| < 0.5 \\ |\Delta r_{rb} / P| &= |\{\hat{r}_r / P\} - \{\hat{r}_b / P\}| < 0.5 \\ |\Delta r_{gb} / P| &= \left| \{\hat{r}_g / P\} - \{\hat{r}_b / P\} \right| < 0.5 \end{aligned} \tag{18}$$

Then, according to the following equation, q_r, q_g, and q_b can be obtained as follows:

$$q_r = [\hat{r}_r / P], q_g = [\hat{r}_g / P], q_b = [\hat{r}_b / P] \tag{19}$$

Otherwise, according to the following formula, q_r, q_g, and q_b can be calculated:

$$q_r = [\hat{r}_r / P + 0.5], q_g = [\hat{r}_g / P + 0.5], q_b = [\hat{r}_b / P + 0.5] \tag{20}$$

This ensures that $\Delta q_r = 0$, $\Delta q_g = 0$, $\Delta q_b = 0$, then $\Delta N_0 = 0$, which can eliminate the coarse error ΔN_{MAX}.

According to Equation (14), N_0 is obtained from q_r, q_g, and q_b, then the absolute phase can be obtained by the following equation:

$$N = PN_0 + \frac{\{\hat{r}_r / P\} + \{\hat{r}_g / P\} + \{\hat{r}_b / P\}}{3} P \tag{21}$$

According to the above equation, the absolute phase error $\Delta N = \frac{\Delta r_r + \Delta r_g + \Delta r_b}{3}$ does not exceed the folding phase errors Δr_r, Δr_g, and Δr_b.

If the residual measurement error does not satisfy Condition A, ΔN_{MAX} may occur. In order to reduce the influence of Δ_{MAX}, the proposed method in this paper increases the part of judging and processing the absolute phase. After the phase unwrapping is completed, the difference $\Delta N_k = N_k - N_{k-1}$ between the absolute phase measurement value N_k of each pixel k and the absolute phase measurement value N_{k-1} of its neighboring pixel is calculated. If it meets the following Condition C,

$$|\Delta N_k| < 3 \times \Omega_r \times \Omega_g \tag{22}$$

We regarded N_i as the valid measurement value, and the absolute phase error does not exceed the folding phase errors Δr_r, Δr_g, and Δr_b; otherwise, $|\Delta N_k| \geq 3 \times \Omega_r \times \Omega_g|$, which means the spatial distance between adjacent pixels is no less than 1/3 of the range. This is obviously unreasonable for a relatively flat surface such as the chest and abdomen surface of the human body. If N_i is invalid, the pixel can be rejected as an immeasurable point.

If necessary, the interpolation method can be used to obtain the absolute phase based on the surrounding pixels of the absolute phase. Elimination or interpolation usually has little effect on the measurement because the sample points in the image are large and dense. Conversely, if a larger absolute phase error cannot be identified and eliminated, it will affect the measurement result seriously or even make the measurement result unusable. This is also a challenging problem in the 3D measurement of Fourier fringe analysis [21]. To sum up, under Conditions of A, B, and C, Equations (19)–(21) are combined to form our proposed 3D measurement method.

In addition, compared with the three-frequency differential method, the proposed method has a larger unwrapping range. Let $P_r = P_0 - W_r$, $P_g = P_0$, $P_b = P_0 + W_b$, where W_r and W_b are positive integers, and the phase unwrapping range of the proposed method is $P_{rgbO} = P_r P_g P_b$. When phase unwrapping is performed by the three-frequency differential method, the light stripes R and G are used for phase unwrapping to form a synthetic light stripe RG with the phase unwrapping range $P_{rg} = P_r P_g/(P_g - P_r)$; then, use the light stripes G and B for phase unwrapping to form a synthetic light stripe GB with a phase unwrapping range $P_{gb} = P_g P_b/(P_b - P_g)$; moreover, the phase unwrapping is further performed by using the synthesized light stripes RG and GB. When $P_{gb} > P_{rg}$, its phase unwrapping range is $P_{rgbH} = P_r P_g P_b/(2W_r W_b + (W_r - W_b)P_0)$. When $W_r \geq W_b$, $2W_r W_b + (W_r - W_b)P_0 \geq 2$, then $P_{rgbO} \geq 2P_{rgbH}$. Only when $W_r < W_b$, $2W_r W_b + (W_r - W_b)P_0 = 1$, that is $P_{rgbO} = P_{rgbH}$ is possible. According to the pixel resolution of the currently available digital pattern projection device and digital image acquisition device, the phase unwrapping range should be 300 to 10,000 pixels. Only in these two cases, one being $P_r = 9$ pixel, $P_g = 11$ pixel, $P_b = 14$ pixel, $P_{rgbH} = 1386$ pixels and the other being $P_r = 12$ pixel, $P_g = 17$ pixel, $P_b = 29$ pixel, $P_{rgbH} = 5916$ pixels, there will be $P_{rgbO} = P_{rgbH}$. Under these situations, it is difficult to achieve comprehensive optimization of the measurement range, resolution, and anti-interference ability by flexible selection of P_r, P_g, and P_b.

5. Experimental Results and Analysis

5.1. Simulation Experiments of Folding Phase Extraction

Ethical approval to undertake this project was examined by the Human Research Ethics Committee for Non-Clinical Faculties, School of Measurement-Control Technology and Communication Engineering, Harbin University of Science and Technology on 1 March 2019. The title of the project is "Projection on Patient Body Surface in Invasive Surgeries (National Natural Science Foundation of China, 61671190)". Informed consent form was obtained from the subject. Simulation experiments were conducted by using a planar cosine light image sequence with a size of 768×768 pixels and a period P_0 of 35 pixels. To simulate human respiratory movement that approximates periodic motion, let the measured plane do relative paralleled movement to the image plane of the camera 2mm each time and perform a periodic reciprocating motion with a period of 20mm, then collect the plane cosine light images from 120 positions.

$$i_m(x,y,t) = 128 + 75 \times \cos[2\pi x/P_0 + 2\pi(t-1)/20] \ m = 1,2,3,\cdots,120. \tag{23}$$

The 3D representation of a cosine fringe image sequence with one period in the t-axis direction is shown in Figure 3a. Extract the folding phases of the sixth frame image by 1D-FFA, 2D-FFA, and 3D-FFA, respectively. Then, these extracted folded phases are subtracted from the folded phase, respectively. The folding phase errors of the three methods are shown respectively in Figure 3a–c.

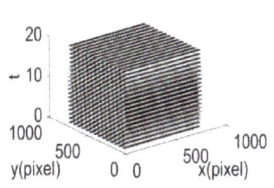

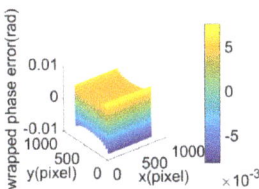

(**a**) Cosine fringe image sequence of one period   (**b**) Folding phase error using 1D-FFA

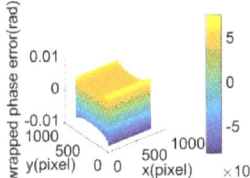

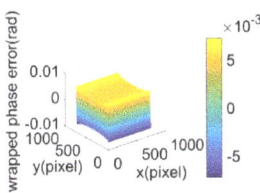

(**c**) Folding phase error using2D-FFA   (**d**) Folding phase error using 3D-FFA

Figure 3. Cosine light image sequence and folding image error of one image.

According to Figure 3, in the extracted folded phase by the three methods, there were errors caused by truncation and sampling. The calculated folding phase value of measured points and their standard values were used to calculate the RMS error, and the difference between the maximum folding phase value and the minimum folding phase value was defined as peak-to-valley (PV) error. As shown in Table 1, the PV and RMS of the folded phase error were approximately equal, and they were so small that could be neglected. It can be seen that these three methods could extract the folding phase effectively.

Table 1. Folding phase error of the stripe image (unit: rad).

Methods	1D-FFA	2D-FFA	3D-FFA
PV	0.0156	0.0156	0.0145
RMS	0.0040	0.0040	0.0044

To evaluate the anti-interference ability of 3D-FFA, the following interference signals were added to the cosine fringe image sequence:

$$\gamma = 75 \times I_p \times [2 \times \text{rand}(768,768,120) - 1] \tag{24}$$

where rand() is a function of generating random numbers in $[0,1]$ and I_p is the ratio percentage of the interference signal amplitude to the cosine modulation.

By analyzing the fringe images collected in the experiments, the results showed that the noise mainly consisted of salt and pepper noise and Gaussian noise, whose probability distribution curves were superimposed and integrated to form a uniform noise probability curve. Therefore, the uniformly distributed random noise could be used in the simulation experiment, which could fully simulate the effect of noise and was better than the direct superposition of Gaussian noise and salt and pepper noise.

Taking $I_p = 60\%$ as an example, the folding phase error of the sixth frame image is shown in Figure 4. It can be seen that the folding phase error extracted by 3D-FFA was obviously smaller than that of the other two methods, which showed that the anti-interference ability of the 3D-FFA was the strongest.

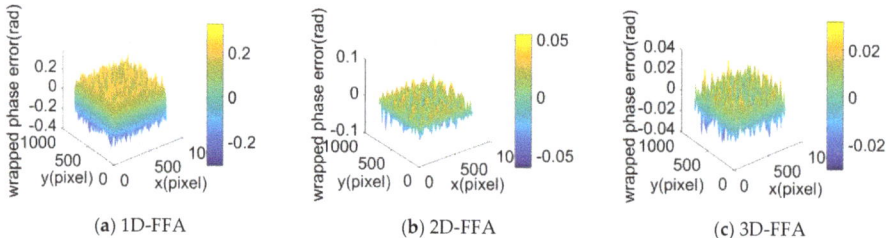

(a) 1D-FFA   (b) 2D-FFA   (c) 3D-FFA

Figure 4. Folding phase error of a stripe image after adding interference.

When I_p was 1%, 2%, 3%, 5%, 10%, 20%, 40%, 60%, 80%, and 100%, respectively, the folded phase error curves of the three methods are shown in Figure 5. As far as the RMS error and peak-valley error of folded phase were concerned, one was that they increased with the increase of interference; the other was that the results obtained by the 2D-FFA method were significantly lower than the 1D-FFA method; the third was that the results obtained by 3D-FFA method were significantly lower than the 2D-FFA method; the fourth was that the 3D-FFA method and the 2D-FFA method were almost invariant and approximate to the interference when the interference ratio was less than 10%.

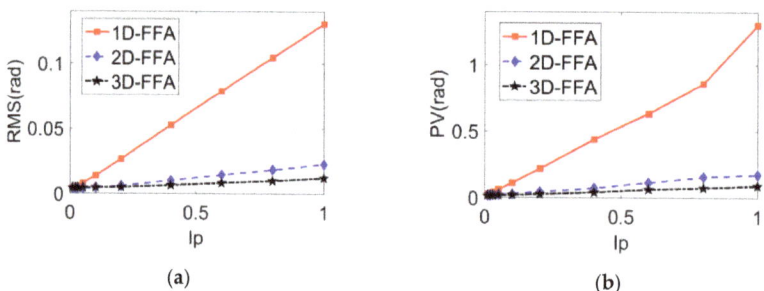

(a)   (b)

Figure 5. The folded phase error curves of the sixth frame stripe images with different interference percentages. (**a**) Comparison of the root mean square of the folded phase error; (**b**) comparison of peak and valley values of folded phase errors.

In order to compare the anti-interference ability of the three methods quantitatively, the ratio of the folding phase errors of the three methods with different interference percentages is shown in Figure 6. In Figure 6a, the red line is the RMS error ratio of 1D-FFA and 2D-FFA, which was between four and six; the blue line is the RMS error ratio of 2D-FFA and 3D-FFA, which was between one and two. In Figure 6b, the red line is the PV error ratio of 1D-FFA and 2D-FFA, which was between four and eight; the blue line is the PV error ratio of 2D-FFA and 3D-FFA, which was between one and two. Obviously, the anti-interference capability of 2D-FFA method was much better than that of the 1D-FFA method, and the one of the 3D-FFA method was much better than that of the 2D-FFA method; the bigger the interference, the more obvious the superiority of the anti-interference capability. When the interference reached 40%, the anti-interference ability of the 3D-FFA method was about twice that of the 2D-FFA method.

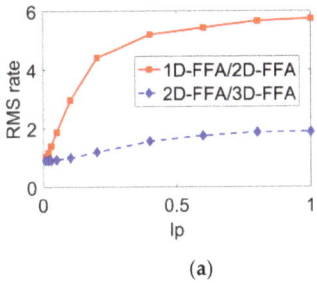
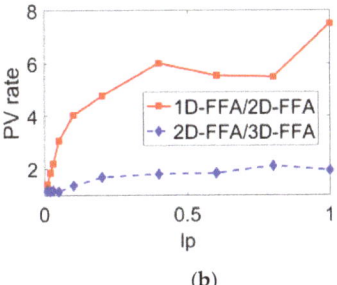

(a)   (b)

Figure 6. The ratio curves between the folded phase errors of the three methods with different interference ratios of the sixth frame. (**a**) RMS ratio curve of folded phase error; (**b**) PV ratio curve of folded phase error.

5.2. Chest Model Measurement Experiments

Based on the proposed method in this paper, a 3D experimental apparatus was constructed for human chest and abdomen surface measurement. The device used a projector (InFocus IN82, InFocus Corporation, Wilsonville, OH, USA) to project a color stripe pattern with a resolution of 1024 × 768 pixels. The pattern parameter was $P_r = 25$ pixel, $P_g = 30$ pixel, $P_b = 35$ pixel, and the measured surface stripe images with a resolution of 1624 × 1236 pixels were collected by using a 3CCD industrial camera (AT-200GE, JAI Ltd., Copenhagen, Denmark).

In the measurement experiments, the chest model simulated respiratory movement and reciprocating motion on the guide rail for 20 mm. For each mobile 2 mm, we collected an image as shown in Figure 7. A total of 120 images was collected and sent to the computer to form a sequence of stripe images. Intercept the stripe image sequence with the size of 768 × 768 pixels from the fringe image sequence, and then get the 3D image sequence of the tested area by using the 3D measurement method. Figure 8a–c shows the sixth frame of 3D images formed by taking folded phases using 1D-FFA, 2D-FFA, and 3D-FFA, respectively. All of them could reproduce the 3D surface of the measured area correctly, which verified the 3D measurement method presented in this paper. The visual effect of the three methods was basically the same, because the measurement errors of the three were basically the same when the interference could be ignored in the darkroom.

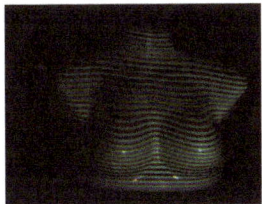

Figure 7. Image of the chest model after the projected stripe pattern.

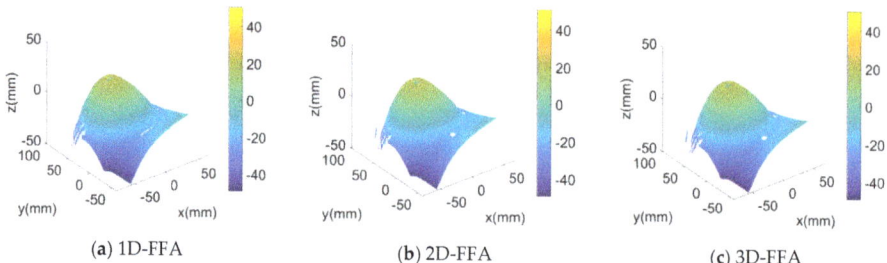

Figure 8. Measurement results of the sixth frame stripe image of the chest model.

In order to verify and compare the anti-interference ability of the 3D measurement method in this paper, we added the interference γ of different I_p to the stripe sequence of the chest model. Take the sixth frame stripe image as an example, the measurement results are shown in Figure 9. When the interference was 5%, the error of the three measurements was similar.

Beginning with $I_p = 10\%$, the error of 1D-FFA measurement result increased rapidly, and the errors of the other two measurements also increased. The error of measurement based on 2D-FFA was larger than 3D-FFA in the range of 10%–80% for I_p, although it was difficult to observe visually. From the beginning of $I_p = 80\%$, the error of measurement based on 2D-FFA was larger than that based on 3D-FFA. It showed that the proposed method had the strongest anti-interference ability, and the greater the interference, the more obvious the advantage.

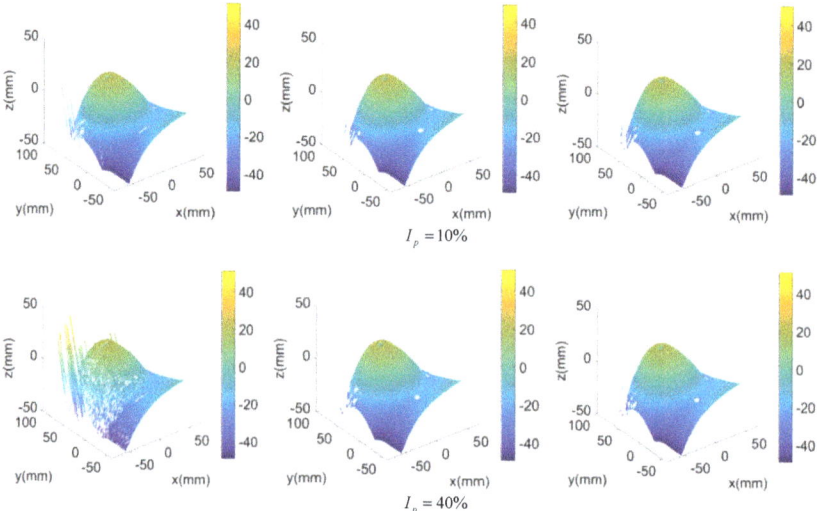

Figure 9. *Cont.*

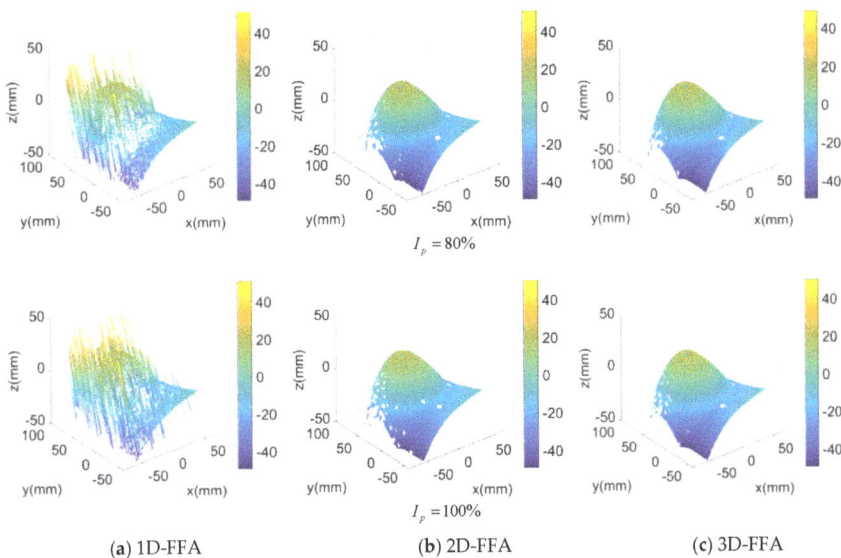

(a) 1D-FFA   (b) 2D-FFA   (c) 3D-FFA

Figure 9. Measurement results of the sixth stripe image of the chest model with different interference.

5.3. Measurement Experiments of the Human Chest and Abdomen Surface

With the proposed 3D measurement method, a 3D measurement experiment of the chest and abdomen surface was conducted during its respiratory movement, as shown in Figure 10a, and the 3D surface depth image sequence of the measured area is shown in Figure 10b. The sixth frame in the 3D image sequence is shown in Figure 10c. That is to say that, the proposed method could reconstruct the 3D surface of the human chest and abdomen correctly. In Figure 10c, the enlarged part of the square area on the left is shown on the right, which indicated that there was maximum measurement error in the indoor light environment, which formed an undetectable blank spot.

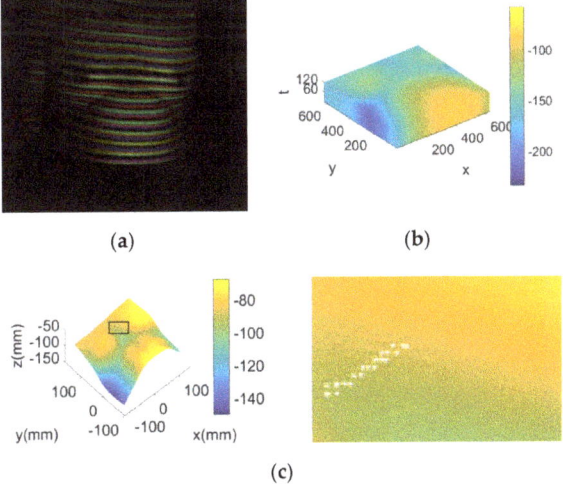

Figure 10. *Cont.*

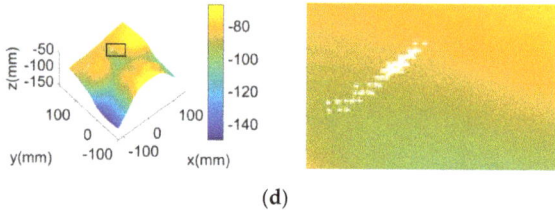

(d)

Figure 10. Surface of the human chest and abdomen and its measurement results. (**a**) Measured surface of the chest and abdomen; (**b**) 3D surface depth image sequence of the measured area; (**c**) measurement results of the sixth stripe image and the local magnification map when ignoring the interference; (**d**) measurement results of the sixth stripe image and the local magnification map after adding 30% interference.

In the measurement process, interference was added artificially. When the interference was small, the measurement results were basically unchanged. When the interference reached 30%, the measurement results are shown in Figure 10d, which indicated that the undetectable points increased significantly, and the measurement results started to become significantly worse. If necessary, the measured value of the white area could be obtained by the difference between the effective measurement values of its peripheral adjacent points.

However, for a relatively flat surface such as the surface of the human chest and abdomen, eliminating the isolated white spot or obtaining the blank area by the difference had little effect on the measurement results. As shown in Figure 11, a fixed point near the diaphragm was selected, and the height (z) of the point in the image sequence changed with time (t) to form the height (z)-time (t) respiratory movement curve of the point. The measurement results of two typical respiratory movement states, deep breathing and rapid breathing, are given in Figure 11a,b, respectively.

The measured results showed that the proposed method in this paper could recover the 3D shape of the surface of the chest and abdomen at different times correctly, realize the dynamic 3D measurement of the surface, and obtain the respiratory motion trajectory of the surface points. This method had strong anti-interference ability and could eliminate the maximum measurement error or its influence basically.

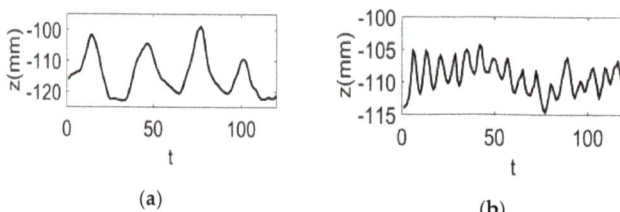

(a) (b)

Figure 11. The height (z)-time (t) respiratory motion curve at a point near the diaphragm. (**a**) Curve of the point's motion in deep breathing; (**b**) curve of the point's motion in rapid breathing.

6. Conclusions

In this paper, a 3D dynamic surface measurement method was proposed for the human chest and abdomen. The RGB trichromatic cosine stripe pattern was synthesized into a color stripe projection pattern. The projection pattern was invariable, and one measurement could be realized by collecting one stripe image. It had less pattern projection and image acquisition time, which provided the foundation for dynamic measurement. Color correction and color separation for the color stripe image sequence were used to form the RGB monochromatic deformation stripe image sequence with a different period. The 3D Fourier transform and the 3D Gaussian filter were combined to carry out 3D-FFA to extract the folding phase. The interference effect could be reduced, and the measurement accuracy could be improved by increasing the time dimension frequency domain filter. The phase unwrapping operation depended on the difference between the fractional parts of the folded phase error, which could ensure that the absolute phase error did not exceed the measurement error of the remainder. Removal or interpolation based on absolute phase measurement could eliminate or reduce the effect of gross absolute phase error. The phase unwrapping based on the remainder equation had a larger phase unwrapping range and preferred space.

In view of this proposed method, the theoretical analysis and experimental results showed that the method presented in this paper could obtain the 3D image sequence of the human chest and abdomen surface and the respiratory motion curve of the chest and abdomen surface points in the respiratory movement effectively and accurately. It had the characteristics of strong anti-interference ability and a wide range of development and could eliminate gross absolute phase error. It should be pointed out that although the method in this paper achieved the analysis of five frames of 3D images per second and respiratory motion in about four seconds, its operation time should be reduced from both the perspective of improving the analysis effect and engineering demand. Therefore, we plan to adopt multithread optimization measures to improve the operation speed.

Author Contributions: This article was completed by all authors. H.W. designed and implemented the classification algorithm. S.Y. made the experimental analysis of the algorithm. X.Y. participated in the writing of the paper. All authors have read and agreed to the published version of the manuscript.

Funding: This work was supported by the National Natural Science Foundation of China (61671190, 61571168), the University Nursing Program for Young Scholars with Creative Talents in Heilongjiang Province (UNPYSCT-2017086), and Fundamental Research Foundation for Universities of Heilongjiang Province (LGYC2018JQ014).

Conflicts of Interest: The authors declare no conflict of interest.

References

1. Fassi, A.; Schaerer, J.; Fernandes, M.; Riboldi, M.; Sarrut, D.; Baroni, G. Tumor tracking method based on a deformable 4D CT breathing motion model driven by an external surface surrogate. *Int. J. Radiat. Oncol. Biol. Phys.* **2014**, *88*, 182–188. [CrossRef] [PubMed]

2. Fayad, H.; Pan, T.; Pradier, O.; Visvikis, D. Patient specific respiratory motion modeling using a 3D patient's external surface. *Med. Phys.* **2012**, *39*, 3386–3395. [CrossRef] [PubMed]

3. Povsic, K.; Jezersek, M.; Mozina, J. Real-time 3D visualization of the thoraco-abdominal surface during breathing with body movement and deformation extraction. *Physiol. Meas.* **2015**, *36*, 1497–1516. [CrossRef] [PubMed]

4. Xiao, H.; Hou, L.; You, Y. A Measurement Method of Multi-line Structure light Stereo vision based on Image Fusion. *J. Sichuan Univ. (Eng. Sci. Ed.)* **2015**, *47*, 154–158.

5. Sam, V.D.J.; Dirckx, J.J.J. Real-time structured light profilometry: A review. *Opt. Lasers Eng.* **2016**, *87*, 18–31.

6. Zuo, C.; Tao, T.; Feng, S.; Huang, L.; Asundi, A.; Chen, Q. Micro Fourier Transform Profilometry (mu FTP): 3D shape measurement at 10,000 frames per second. *Opt. Lasers Eng.* **2018**, *102*, 70–91. [CrossRef]

7. Lu, L.; Ding, Y.; Luan, Y.; Luan, Y.; Yin, Y.; Liu, Q.; Xi, J. Automated approach for the surface profile measurement of moving objects based on PSP. *Opt. Express* **2017**, *25*, 32120–32131. [CrossRef]

8. Zhong, M.; Su, X.Y.; Chen, W.J.; You, Z.S.; Lu, M.T.; Jing, H.L. Modulation measuring profilometry with auto-synchronous phase shifting and vertical scanning. *Opt. Express* **2014**, *22*, 31620–31634. [CrossRef]

9. El-Haddad, M.T.; Tao, Y.K.K. Automated stereo vision instrument tracking for intraoperative OCT guided anterior segment ophthalmic surgical maneuvers. *Biomed. Opt. Express* **2015**, *6*, 3014–3031. [CrossRef]

10. Su, X.Y.; Zhang, Q.C. Dynamic 3-D shape measurement method: A review. *Opt. Lasers Eng.* **2010**, *48*, 191–204. [CrossRef]

11. Shirato, H.; Shimizu, S.; Kitamura, K.; Nishioka, T. Four-dimensional treatment planning and fluoroscopic real-time tumor tracking radiotherapy for moving tumor. *Int. J. Radiat. Oncol. Biol. Phys.* **2000**, *48*, 435–442. [CrossRef]

12. Cao, S.P.; Cao, Y.P.; Zhang, Q.C. Fourier transform profilometry of a single-field fringe for dynamic objects using an interlaced scanning camera. *Opt. Commun.* **2016**, *367*, 130–136. [CrossRef]

13. Geng, J. Structured-light 3D surface imaging: A tutorial. *Adv. Opt. Photonics* **2011**, *3*, 128–160. [CrossRef]

14. Padilla, M.; Servin, M.; Garnica, G. Fourier analysis of RGB fringe-projection profilometry and robust phase-demodulation methods against crosstalk distortion. *Opt. Express* **2016**, *24*, 15417–15428. [CrossRef] [PubMed]

15. Caspi, D.; Kiryati, N.; Shamir, J. Range imaging with adaptive color structured light. *IEEE Trans. Pattern Anal. Mach. Intell.* **1998**, *20*, 470–480. [CrossRef]

16. Hu, Y.; Chen, Q.; Zhang, Y.Z.; Feng, S.J.; Tao, T.Y.; Li, H.; Yin, W.; Zuo, C. Dynamic microscopic 3D shape measurement based on marker-embedded Fourier transform profilometry. *Appl. Opt.* **2018**, *57*, 772–780. [CrossRef]

17. Shi, H.J.; Zhu, F.P.; He, X.Y. Low-Frequency Vibration Measurement Based on Spatiotemporal Analysis of Shadow Moire. *Acta Opt. Sin.* **2011**, *31*, 120–124.

18. Abdul-Rahman, H.S.; Gdeisat, M.A.; Burton, D.R.; Lalor, M.J.; Lilley, F.; Abid, A. Three-dimensional Fourier fringe analysis. *Opt. Lasers Eng.* **2008**, *46*, 446–455. [CrossRef]

19. Zhang, Q.C.; Hou, Z.L.; Su, X.Y. 3D fringe analysis and phase calculation for the dynamic 3D measurement. *AIP Conf. Proc.* **2010**, *1236*, 395–400.

20. Bu, P.; Chen, W.J.; Su, X.Y. Analysis on measuring accuracy of Fourier transform profilometry due to different filtering window. *Laser J.* **2003**, *24*, 43–45.

21. Zheng, D.L.; Da, F.P.; Kemao, Q.; Seah, H.S. Phase-shifting profilometry combined with Gray-code patterns projection: Unwrapping error removal by an adaptive median filter. *Opt. Express* **2017**, *25*, 4700–4713. [CrossRef] [PubMed]

Article

Active 3D Imaging of Vegetation Based on Multi-Wavelength Fluorescence LiDAR

Xingmin Zhao [1], Shuo Shi [1,*], Jian Yang [2], Wei Gong [1], Jia Sun [2], Biwu Chen [1], Kuanghui Guo [1] and Bowen Chen [1]

[1] State Key Laboratory of Information Engineering in Surveying, Mapping and Remote Sensing, Wuhan University, Wuhan 430079, China; zhaoxingmin@whu.edu.cn (X.Z.); weigong@whu.edu.cn (W.G.); cbw_think@whu.edu.cn (B.C.); kuanghuiguo@whu.edu.cn (K.G.); chenbowen1204@whu.edu.cn (B.C.)

[2] Faculty of Information Engineering, China University of Geosciences, Wuhan 430074, China; yangjian@cug.edu.cn (J.Y.); sunjia@cug.edu.cn (J.S.)

* Correspondence: shishuo@whu.edu.cn; Tel.: +86-1399-552-5676

Received: 17 January 2020; Accepted: 9 February 2020; Published: 10 February 2020

Abstract: Comprehensive and accurate vegetation monitoring is required in forestry and agricultural applications. The optical remote sensing method could be a solution. However, the traditional light detection and ranging (LiDAR) scans a surface to create point clouds and provide only 3D-state information. Active laser-induced fluorescence (LIF) only measures the photosynthesis and biochemical status of vegetation and lacks information about spatial structures. In this work, we present a new Multi-Wavelength Fluorescence LiDAR (MWFL) system. The system extended the multi-channel fluorescence detection of LIF on the basis of the LiDAR scanning and ranging mechanism. Based on the principle prototype of the MWFL system, we carried out vegetation-monitoring experiments in the laboratory. The results showed that MWFL simultaneously acquires the 3D spatial structure and physiological states for precision vegetation monitoring. Laboratory experiments on interior scenes verified the system's performance. Fluorescence point cloud classification results were evaluated at four wavelengths and by comparing them with normal vectors, to assess the MWFL system capabilities. The overall classification accuracy and Kappa coefficient increased from 70.7% and 0.17 at the single wavelength to 88.9% and 0.75 at four wavelengths. The overall classification accuracy and Kappa coefficient improved from 76.2% and 0.29 at the normal vectors to 92.5% and 0.84 at the normal vectors with four wavelengths. The study demonstrated that active 3D fluorescence imaging of vegetation based on the MWFL system has a great application potential in the field of remote sensing detection and vegetation monitoring.

Keywords: fluorescence LiDAR; laser-induced fluorescence; vegetation monitoring; classification discrimination

1. Introduction

Plants play a considerable role in the carbon and water cycles of the global ecosystem [1,2]. A prompt and effective monitoring of vegetation is of great significance for ecological environmental monitoring and agricultural guidance. Researchers have regarded the optical remote sensing monitoring method as an ideal and feasible way, owing to its several advantages, such as quickness, accuracy, and non-destruction of plants [3]. Many optical remote sensing imaging techniques have been applied to vegetation detection in recent decades. Passive hyperspectral reflection imaging, a commonly used form of optical imaging, can provide abundant biochemical components of plants. However, such a method lacks the spatial expression in 3D space and can be affected by various factors, such as the external environment, including weather conditions, and measurement time [4,5]. The LiDAR technology, an active detection sensor with several technical advantages, such as high temporal–spatial resolution and non-destruction, has received great attention from researchers [6]. Such technology has obtained a wide range of applications on vegetation structural parameter inversion [7,8]. However, this technology typically utilizes a single band of near-infrared laser for the detection of spatial location [9], and it lacks spectral information associated with the materialized components. Therefore, the signal acquisition of multi-wavelength channels can be extended on the basis of the single-wavelength LiDAR, to expand the detection capability of vegetation biochemistry and growth state.

The fluorescence spectral properties of vegetation provide an available indicator for its detection. Fluorescence is treated as the radiation appearance of the energy loss by the oscillating motion of electrons during electromagnetic radiation [10]. The present research means for the vegetation fluorescence emission include stimulation chlorophyll fluorescence by passive solar energy and active artificial light source-laser for induction. Sun-induced fluorescence (SIF)-based vegetation remote sensing monitoring has also achieved development in recent years. SIF can provide global-scale chlorophyll fluorescence detection through on-board data and an indicator for studying the vegetation's ecological environment [11]. However, SIF only provides the fluorescence bands (685 and 740 nm) associated with the chlorophyll in vegetation, due to the extraction method limitation [12–14]. This passive fluorescence detection provides fluorescence spatial distribution in 2D images only and lacks the 3D structural detection capacity of LiDAR.

Laser-induced fluorescence (LIF) is an active means of generating fluorescence by using a single short-wavelength laser as the excitation light. Vegetation absorbs the light energy of a given wavelength, and a part of it is dissipated by light emission at long wavelengths within a short time [15,16]. Vegetation has typical characteristic signals and spectral shapes in the case of being produced by laser stimulation. With the discovery and exploration of the LIF technology, monitoring vegetation by using this mechanism has become possible. Chappelle et al. [16] used an ultraviolet (UV) laser to stimulate fluorescence signals on leaves and proposed the utilization of fluorescence to distinguish vegetation types. They further explored the ability of fluorescence as a probe to resolve plant species and stress states [17]. The production of chlorophyll fluorescence peaks in vivo was also explained [18]. Researchers studied the utilization of fluorescence characteristic peaks to develop a series of correlation studies on vegetation growth status [19], biochemical content, and environmental stress factors [20]. Fluorescence characteristics have a strong indication of leaf nitrogen content [21], water deficiency [22], and fungal infections [23]. Spectral properties of LIF demonstrate a powerful ability to monitor the vegetation status as the reflection [24,25]. Artificial-light-source-induced fluorescence signals have more comprehensive spectral characteristics than SIF and are excited with only a single-wavelength light economically. Accordingly, we supposed that the LiDAR laser source is used to simultaneously achieve laser scanning ranging and fluorescence induction forming multi-wavelength reception for implementing a multi-wavelength fluorescence LiDAR. The system expands the ability to monitor the physiological states of vegetation by adding several channels for receiving fluorescence signals.

Some existing fluorescence LiDAR systems have achieved a good ability of detecting marine oil spills [26,27] and terrestrial water bodies [28]. A number of fluorescence imaging systems had been previously constructed to express the distribution characteristics of fluorescence signals [29,30]. These proactive vegetation fluorescence imaging systems can monitor growth and stress status by using the LIF technology from an imaging perspective [31]. However, these fluorescence imaging systems express the spatial distribution of fluorescence emission signals in the form of 2D images. The imaging spatial scale is extremely small to be directly applied to the remote sensing of vegetation.

Simultaneous monitoring of the external appearance and internal biochemical status has a comprehensive perception for vegetation remote sensing. We stimulate the vegetation fluorescence reception for multi-wavelength channels on the basis of the scanning ranging function of single-wavelength LiDAR. In this way, 3D spatial structural and growth state information of vegetation can be simultaneously acquired. However, constructing such LiDAR with a multi-wavelength reception of fluorescence manifests several problems. First, several wavelengths need to be selected for the fluorescence emission detection. Vegetation emits fluorescence in the form of a continuous spectrum through UV laser excitation. The receiving wavelength design can represent biochemical information and reduce the hardware cost of the system. Second, the enhancement of multi-channel fluorescence data in the background signal is also a problem, since there is a large number of non-vegetation signals as non-interested targets in the system data. Third, the data of these multiple different system units must be organized and visualized. The data output comes from several units of the system. These data require an integrated expression of the 3D spatial structure and fluorescence emissions of the vegetation.

In this study, (1) the MWFL system was proposed, created, and integrated, to perform experimental verification in the laboratory. The system design of the four wavelengths corresponding to the characteristics of the vegetation fluorescence, system components, and system-based data form were introduced. (2) The system experimented fluorescence signal imaging and scanned canopy distribution of the vegetation to verify the 3D imaging ability of the system. (3) Three-dimensional fluorescence imaging based on spectral enhancement pretreatment was adopted and achieved a good effect on the experimental scenes. (4) System evaluation based on point cloud classification was applied to classify 3D fluorescence point cloud data on vegetation, to further quantitatively explain the system advantages. The ability of the MWFL system to monitor spatial and biochemical status of vegetation through 3D fluorescence imaging was demonstrated. The feasibility of the MWFL system and the efficiency of 3D fluorescence imaging for vegetation detection were assessed.

2. Materials and Methods

2.1. System Description

2.1.1. Selection of Fluorescence Wavelengths

MWFL, an active remote sensing monitoring device, adds several channels to receive the vegetation fluorescence compared with the ranging LiDAR. When excited by the short-wavelength light source, the energy of vegetation fluorescence is emitted in a longer continuous wavelength range. The fluorescence-receiving wavelengths of the system design must be optimally selected to represent vegetation fluorescence characteristics and to be as few as possible, considering the system cost.

In this study, two leaves in different physiological states were picked. The continuous fluorescence spectra of the points measured through ICCD (Intensified Charge Coupled Device) excited with a UV laser in the laboratory were recorded. Figure 1a shows that the two leaves had different physiological states. The upper right corner of the right leaf had turned brown. Three points, namely A, B, and C, in the two leaves were located in the fresh green, yellow, and brown areas, respectively. The color characterization of the exterior leaves reflected the concentration distribution of the internal pigment. Figure 1b shows the continuous fluorescence spectral shape of points A, B, and C (wavelength range of 360–800 nm).

Vegetation has typical fluorescence spectral emission waveforms during UV laser induction and exhibits characteristic peaks, namely F460, F525, F685, and F740. The summit of F525 is sometimes less evident, or merely a slight rise on the fluorescence spectrum is observed [32]. The characteristic peaks of F685 and F740 are closely related to the chlorophyll content of leaves [33]. F460 is mainly caused by water-soluble compound NADPH, vitamin K, and beta-carotene; the prime contributor to the characteristic peak of F525 is riboflavin [32]. The measured points of the selected leaves show typical fluorescence spectral curves, but they differ from each other. Points A, B, and C represent the process of leaves turning from green to yellow and are eventually withered. Figure 1b demonstrates that the fluorescence spectrum reflects the changes in biochemical substances inside the leaves during this process. In the green leaf, chlorophyll closely related to photosynthesis reactions actively works, as indicated by F685 and F740 on the spectrum of point A. The spectrum of point B shows that the intensity of F740 first decreases, and that of F685 slightly increases when the leaf turns yellow. By contrast, F460 related to lutein and carotene can have a relatively large increase in strength. The chlorophyll content decrease is accompanied by a decrease in the F740 intensity; an increase in F685 may be due to the weak resorption effect [34]. As illustrated in Figure 1, the fluorescence spectrum of point C indicates that the strength of F460 and F525 is low when the leaves are withered, given that the corresponding biochemical substances are decreasing. Simultaneously, chlorophyll is almost exhausted, and the corresponding strength of F685 and F740 has become low, although not obvious. The change process of the fluorescence spectrum demonstrates that the intensity variation of the four characteristic wavelengths, namely F460, F525, F685, and F740, can represent the degree of yellowing in the leaves.

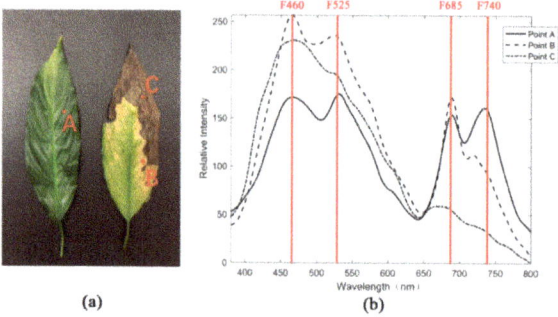

(a) (b)

Figure 1. Induced continuous fluorescence spectrum of leaves in different physiological states excited with an UV laser. (**a**) Two leaves of different physiological states (Scene 1), and three points, namely A, B, and C, were located in fresh green, yellow, and brown areas, respectively; (**b**) continuous fluorescence spectrum of points A, B, and C excited with a UV laser (355 nm) was detected by the wavelength of 380–800 nm.

The developed active laser fluorescence imaging system predecessor was applied to the vegetation detection. The wavelengths of fluorescence imaging system Lichtenthaler et al. studied were blue, green, red, and far-red, corresponding to fluorescence emission [35]. Langsdorf et al. developed multicolor fluorescence imaging to determine whether the nitrogen-stress state of leaves is related to these four wavelengths [29]. The range of detection bands of fluorescence imaging systems for vegetation nutrition stress and disease diagnosis detection has been focusing on the four wavelengths, namely F460, F525, F685, and F740 [31,36,37], in recent years, in spite of a slight offset in the wavelength position. Considering the point measurement results of the vegetation fluorescence spectrum in the laboratory and receiving bands of the previous fluorescence imaging system, 460, 525, 685, and 740 nm were selected as the four receiving wavelength centers of the MWFL system.

2.1.2. System Components

The MWFL system design aims to simultaneously obtain the 3D spatial structure and four wavelengths of the fluorescence characteristics of the vegetation target. This system can implement two detection mechanisms, namely reflection ranging and laser-induced fluorescence. In addition to the scanning and ranging functions of the single-wavelength LiDAR, the system also has the module for fluorescence detection and reception. The MWFL system includes system components of laser emission, scanning, ranging, receiving detection, and data processing. Figure 2 shows the block diagram of the MWFL system.

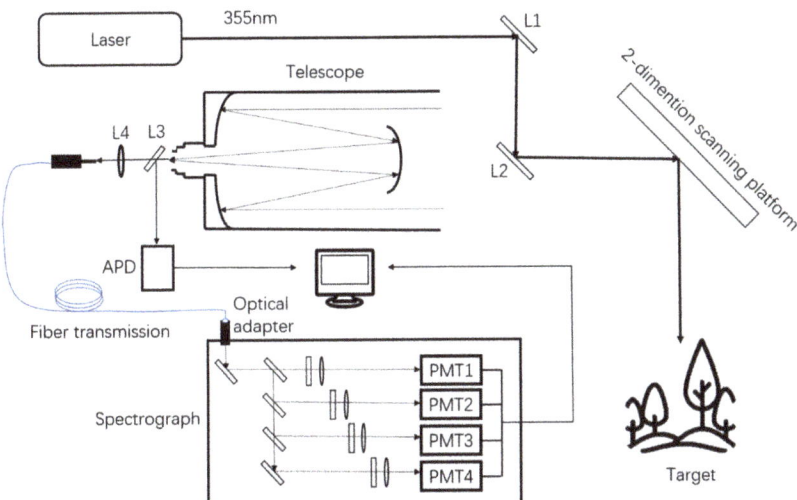

Figure 2. Block diagram of the multi-wavelength fluorescence LiDAR.

In the MWFL system, the laser source uses a 355 nm UV laser as a laser-emitting unit considering excitation efficiency. The UV laser is not only the excitation source of vegetation fluorescence, but its reflective signal is the system's distance-measuring source. The parameters of the laser source are set to meet the requirements of laser pulse ranging and vegetation fluorescence induction. The L1 mirror has high reflectivity to the UV-wavelength laser, which acts as a reflection and filter, for optimal design. The L2 reflective mirror reflects the laser light to the center of the 2D scanning platform of a scanning unit. The beam can be scanned in the x and y directions on vegetation canopy target as the platform rotates. The system echo signals, including reflective UV laser and vegetation fluorescence signals, are received through the view field of Schmidt–Cassegrain telescope. The connection between the center of the scanning platform and the center of the L2 mirror is collinear with the central axis of the telescope to form an optical coaxial design. Such a setup is the requirement for ranging and spectral detection in the single point and is beneficial to improve the detection signal to noise ratio.

The receiving detection unit mainly includes objects, such as telescope, spectrometer, and transmission fiber. The L3 mirror can reflect the UV band and transmit the long band, which can separate the UV reflection and fluorescence signal from vegetation. The reflective signal is recorded by an APD (Avalanche Photo Diode) of the ranging unit, which is compared with the time of the initial pulse by means of TOF (Time of Flight), to obtain the distance value of the single point through the pulse method. After focusing through the L4 convex lens and coupling, fluorescence signals are transmitted through the fiber to the spectrometer. Inside the spectrometer are a four-wavelength splitter module and corresponding photodetectors. With regard to the spectroscopic module, the continuum signal of the vegetation fluorescence introduced into the spectrometer is separated from each other by dichroic filters passing through narrow-band filters and into the four photomultiplier tube arrays with single-wavelength response centers on 460, 525, 685, and 740 nm. The four-channel photoelectric signal is converted by analog–digital transformation. The fluorescence intensity is acquired by integration and transmitted to the data-processing unit of the system. Table 1 shows the technical parameters of MWFL system.

Table 1. Technical parameters of MWFL system design.

Multi-Wavelength Fluorescence LiDAR	
Laser wavelength	355 nm
Repetition rate	7 kHz
Pulse width	3~5 ns
Pulse energy	18 µJ
Beam divergence	<1 mrad
Telescope aperture	200 mm
Spatial resolution	Distance: 10 mm Scanning: 2 mm @20m

Compared with the existing vegetation-monitoring fluorescence LiDAR [38,39], the MWFL system has a combination of scanning ranging and LIF to achieve 3D fluorescence imaging of vegetation targets. This imaging method can form an integrated monitoring of the vegetation's external growth and internal biochemical components.

2.1.3. Data Description

The form of the MWFL system data is spatially presented in a point cloud format. Each point has a fluorescence spectral property. The MWFL system breaks through the limitations of traditional single-wavelength ranging LiDAR only for 3D space detection, given its fluorescence spectral features and expanded ability to detect vegetation. The type of system data is divided into two parts: 3D point cloud data and four-wavelength data of fluorescence signal. Figure 3 illustrates the formation process of the MWFL system data form.

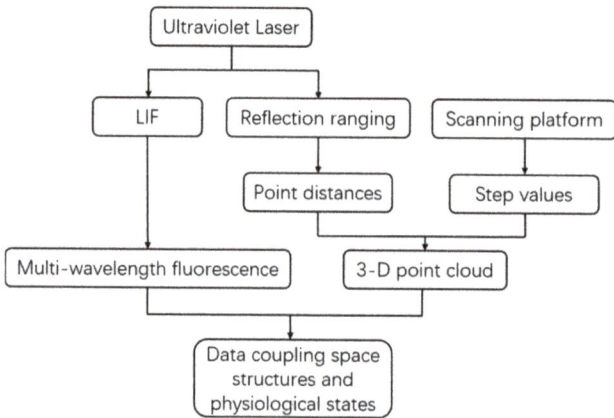

Figure 3. Data formation process of multi-wavelength fluorescence LiDAR.

Figure 3 shows that the UV laser generates the multi-wavelength fluorescence signal for a single point via LIF, and its reflection is used for ranging. The design method for single-point measurement was described in Section 2.1.2. The data-processing unit of the MWFL system records the distance values at the single-point position and the fluorescence intensity values of the four channels. The system scanning platform can be rotated in two directions, to perform 2D scanning detection on vegetation targets. The rotary step values are simultaneously recorded. The data that are saved and transferred to the data-processing unit consisted of three parts: the distance values of points, the signal intensity of four channels, and the step values of the platform scanning. Among these parts, the distances and step values between points constitute the 3D point cloud spatial distribution in the form of spherical coordinates. These coordinates can be converted into a spatial Cartesian coordinate system. The vegetation fluorescence intensity in the four channels constitutes multi-wavelength fluorescence characteristic data. The data outputted by the system include the XYZ coordinates and the fluorescence intensity values of F460, F525, F685, and F740 of the vegetation target. The system generates a remote-sensing data form for vegetation targets. This new form of data couples the 3D distribution and fluorescence spectra of vegetation detection. As a result, the integrated monitoring of spatial and physiological status of vegetation target is enabled. We hope that this new data format based on the MWFL system can be applied to the remote-sensing monitoring of vegetation for improving the accuracy of qualitative and quantitative detections of vegetation.

2.2. Sample Materials

Two scenes were presented as samples to demonstrate the ability of MWFL system's 3D fluorescence imaging in characterizing the vegetation states and the ability to couple spatial and physiological states. The two leaves mentioned in Section 2.1.1 for wavelength selection were recommended as Scene 1 to implement 3D fluorescence imaging on the basis of point cloud. This task was carried out to study the spectral-imaging differences in the green, yellow, and brown areas of the leaf.

A scanning experiment of the potted vegetation was conducted, to prove the detection advantage of the system on the 3D canopy as Scene 2 in an experimental scene (Figure 4). The leaves in this potted vegetation were spatially distributed at different angles and positions. Moreover, the leaves represent their different physiological states. Such a featured scene can be used as an observation sample with spatially complex states and physiological differences. For two scenes arranged in the laboratory, the ability of MWFL system to effectively monitor vegetation can be verified. Scene 1 expresses the spectral detection performance of the system for fluorescence emission at the leaf level. Scene 2 shows the fluorescence point cloud imaging capability of vegetation with 3D morphology.

Figure 4. Scene 2 for 3D point cloud imaging of the MWFL scanning experiment.

2.3. Methods

2.3.1. D Fluorescence Imaging Based on Spectral Enhancement

The spectral signal of the MWFL system comes from the photoelectric conversion of four channels. However, the fluorescence spectral information of vegetation from the target of interest is often insufficiently prominent, due to the ground-scene background. During the 3D imaging of vegetation fluorescence, appropriate methods should be adopted to highlight the fluorescence characteristics of vegetation for adapting to the perception of human eyes. The method of processing remote-sensing hyperspectral image uses hyperspectral enhancement application and obtains exceptional analytical results [40,41]. Histogram equalization (HE) is a commonly used image spectral enhancement method that redistributes the spectra intensity by histogram distribution [42]. In this work, the raw spectral data obtained by the system were processed by the HE method. The fluorescence characteristics of the vegetation point cloud after treatment in this way were significantly and visually enhanced. The signal strength pseudo-color imaging of point cloud in four wavelengths can represent spatial changes in leaves in different physiological states.

2.3.2. System Evaluation Based on Point Cloud Classification

The MWFL system expands the detection capability of the physiology and growth status through the LIF mechanism for traditional LiDAR. The four added bands multiply the amount of information contained in the system data compared with the single-wavelength LiDAR. The improvement of vegetation-recognition ability via the increased four-wavelength fluorescence must be quantitatively evaluated. In this study, the point cloud with multichannel fluorescence properties was analyzed by classifying the different conditions of the leaves. The point cloud classification analysis included the classification of data within four channels and the comparative classification of the spatial parameter and four channels with that.

Support vector machine (SVM), which is a popular machine learning method, has been widely applied for data classification and regression [43,44]. SVM has certain advantages, such as robustness and demanding small sample size of remote sensing data for training [45]. Such a method is adaptive for classifying the spatial and spectral feature data of the system. This method was used for point cloud classification, to demonstrate the effectiveness of the 3D fluorescence data of the system for vegetation detection.

The classification for system data is for Scene 2 because the data of Scene 1 are the representation of fluorescence detection in a planar form on the MWFL system. The single-, double-, and four-wavelength spectral data from Scene 2 were used as input eigenvalues of the model classifier for classification and analysis. The normal vector is a commonly used parameter and is related to the spatial structure in vegetation detection [46]. The normal vectors of the point cloud were used to indicate the recognition ability of the single-wavelength LiDAR. The classification results of the fluorescence data of four wavelengths with normal vectors were compared. In the training process of SVM classification, due to the difference in the sample sizes of ground categories, the training samples were selected within a category in turn. The classification selected 2-fold cross-validation—that is, 50% training and 50% testing—and SVM kernel function choose the linear.

Moreover, the overall classification accuracy of the point cloud results can be affected by the imbalance of the sample size of each category [47]. The Kappa coefficient [48] was also used as a parameter to evaluate the overall classification in combination with the classification accuracy.

3. Results

For 3D imaging of fluorescence, the space point cloud is formed by reflective ranging of the ultraviolet light and the rotary step values of the scanning. The point cloud data include the distance values obtained by TOF method and the step values. In the experiments of Scene 1 and Scene 2, the detection distance is about 3.5 m, and the point–point distance is about 3.5 mm. The space size of Scene 1 is 0.12 m × 0.22 m, and Scene 2 is 0.25 m × 0.28 m.

Figure 5 shows the 3D imaging result of the fluorescence point cloud of Scene 1. The results of point cloud imaging showed the characterization of the vegetation's physiological states by the four-channel fluorescence signal, given that Scene 1 landed the leaves on the blackboard.

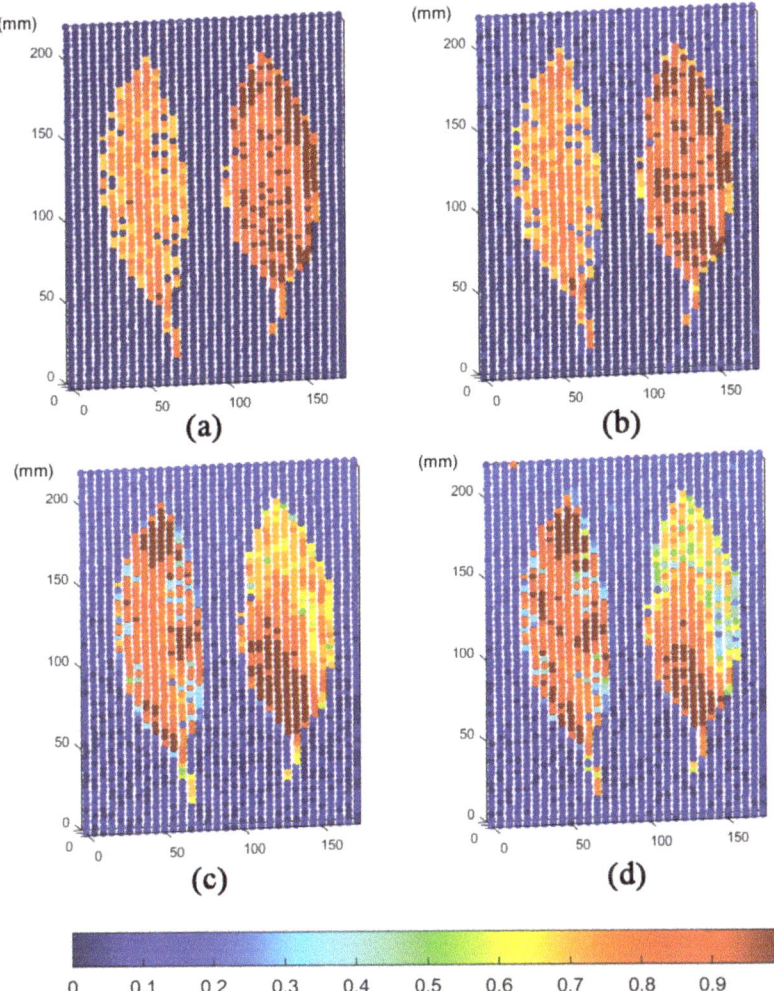

Figure 5. Single-wavelength fluorescence 3D imaging based on scene 1 scanning experiment of the MWFL system: (**a**) 460 nm, (**b**) 525 nm, (**c**) 685 nm, and (**d**) 740 nm wavelength signal intensity pseudo-color 3D imaging. The intensity values were subjected to histogram equalization (HE) and normalization.

The imaging results of Figure 5 demonstrate that the leaf and non-leaf backgrounds are displayed in the form of point cloud and show significant differences. In the fresh green, yellow, and brown areas of leaves, the four-wavelength spectral point cloud imaging presents visual features that match themselves. However, the left and right leaves in Scene 1 exhibit significant differences in these four characteristic wavelengths. The leaf on the left represents the green state of the vegetation. The intensity values at 460 and 525 nm wavelengths are significantly lower than the yellow and brown-leaf regions on the right. The brown area of the right leaf exhibits an extremely high intensity at 460 and 525 nm wavelengths due to the increased degree of the yellowing of the leaves. Fluorescence intensity values at 685 and 740 nm are correlated in most regions of point cloud. The green area of the left leaf and the yellow area of the right leaf are stronger than the brown area because the chlorophyll content in the latter was exhausted. The tip portion of the upper side of the green leaf on the left has similar intensity values, at 685 and 740 nm, to the high intensity of the yellow region. The realistic picture of Scene 1 shows that the green color in this area was declining. The actual colors between the two areas are similar. The high-intensity values of most green areas might be due to the decrease of water content or the nonlinear relationship between chlorophyll fluorescence intensity and chlorophyll content. In the portion where the yellow and brown areas on the right side of the right leaf are bordered, that is, the position of point B at wavelength selection in Section 2.1.1, the fluorescence intensity at 685 nm is slightly stronger than that at 740 nm. Such an outcome is consistent with the test result of the continuous spectrum on point B. This finding indicates that a change buffer distribution of the internal physiological state occurs between the yellow and brown areas of the leaf. The spatial distribution change was revealed by fluorescence imaging of the MWFL system.

Experiments on the spatial distribution of Scene 2 based on MWFL system are performed (Figure 6a). The spatial 3D geometry distribution of the potted vegetation in Figure 4 is presented by the MWFL system. The manual labels were given as real categories of Scene 2, in preparation for classification (Figure 6b). In comparison with Figure 4, Scene 2 was divided into four categories: flowerpot, yellow leaves, withered leaves, and fresh green leaves (Figure 6b).

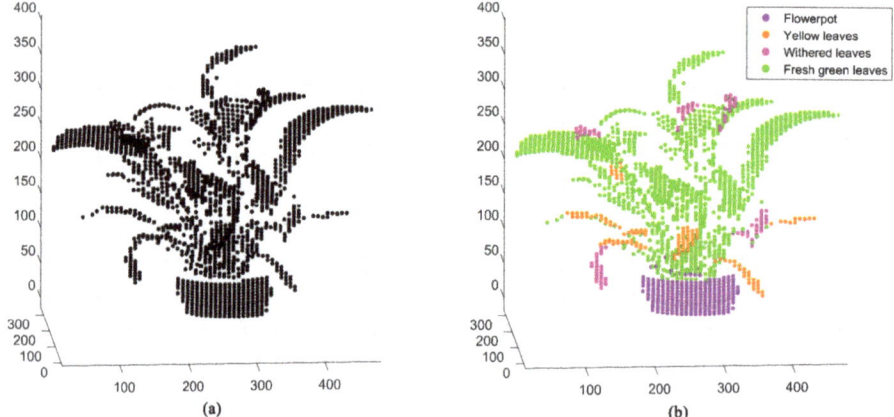

Figure 6. Scene 2 scanning experiment 3D point cloud based on MWFL system. (**a**) Three-dimensional distribution of spatial point cloud; (**b**) The ground truth categories given for classification.

Figure 7 displays the four-wavelength 3D spectral intensity imaging results of Scene 2. Four single-wavelength signal intensity pseudo-color 3D imaging interpretations are presented. In the point cloud imaging of Scene 2, the imaging results of the four channels show an excellent spatial distribution and spectral detection capability. The flowerpot exhibited low values in the four-wavelength 3D fluorescence intensity imaging. Green leaves exhibit high intensity at 685 and 740 nm wavelengths. Such leaves also present weak fluorescence signals at 460 and 525 nm wavelengths. Meanwhile, the withered leaves in this scene show high intensity at 460 and 525 nm wavelengths. The signals at 685 and 740 nm wavelengths are barely high. However, the yellow leaves exhibited low intensity in four channels. After that, we classified and analyzed the fluorescence point cloud data obtained by the MWFL system. This step was conducted to quantitatively estimate and describe the recognition ability of vegetation in Scene 2.

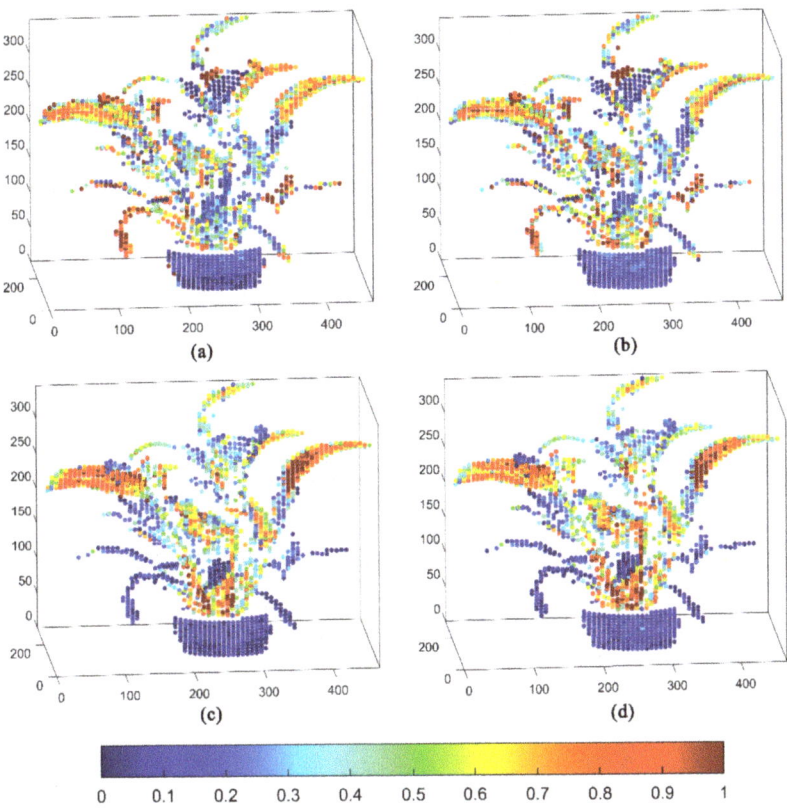

Figure 7. Single-wavelength fluorescence 3D imaging based on the Scene 2 scanning experiment of the MWFL system: (**a**) 460 nm, (**b**) 525 nm, (**c**) 685 nm, and (**d**) 740 nm wavelength signal intensity pseudo-color 3D imaging. The intensity values were subjected to HE and normalization.

4. Discussion

The four single-wavelength spectral signal intensity imaging on the potted vegetation of Scene 2 demonstrated the spatial and spectral combined imaging potential of the MWFL system. The spatial variation distribution of the leaf spectrum in Scene 2 showed the continuous change of the spectrum in space to a certain extent.

Due to the influence of the spatial-distribution conditions, the fluorescence intensity can be affected by factors such as the distance and angle of the system observation [49]. In addition, the laser echo has the mixtures of vegetation and background targets during the scanning experiment. These factors all affect the expression of fluorescence in 3D point cloud. Figure 8 shows the original signal-intensity distribution of different ground categories in Scene 2. As shown in Figure 8a, it is clear that green leaves exhibit high chlorophyll fluorescence intensity, and flowerpots, as non-plant targets, have almost no signal in these two wavelengths. Yellow leaves in Scene 2 have almost no chlorophyll fluorescence, with some fluorescence emission in 460 nm wavelength. Compared with yellow leaves, withered leaves have a significantly enhanced fluorescence emission at 460 nm. From Figure 8b, the correlation between the intensity of two chlorophyll fluorescence bands 685 and 740 nm in Scene 2 is high. However, the difference between these two bands can be reflected in the imaging of Scene 1. The distribution of the intensity of the features in Scene 2 in the channel shows the separability of the data, which means the possibility of the classification.

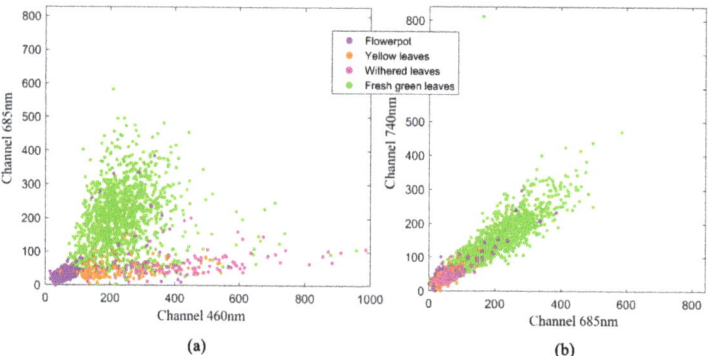

Figure 8. Scatter distribution of the channels' intensity of the ground categories in scanning experiment of Scene 2: (**a**) Channels 460 nm vs. 685 nm; (**b**) Channels 685 nm vs. 740 nm.

The effectiveness of the data obtained by the four added channels, especially the fluorescence data, was evaluated by point cloud classification. SVM acted as a classifier to distinguish the categories of Scene 2. We presented the classification results of two single-wavelengths (460 and 685 nm), double-wavelength combination (460 nm + 685 nm), and four wavelengths as input eigenvalues (Figure 9). The classification accuracies of the four single-wavelength were basically similar. Therefore, two of four single-wavelength were representatively displayed as minimum and maximum classification accuracies. The results of the double-wavelength combination classification were equal; hence, only one combination was selected. Table 2 shows the confusion matrix of Figure 9a–d.

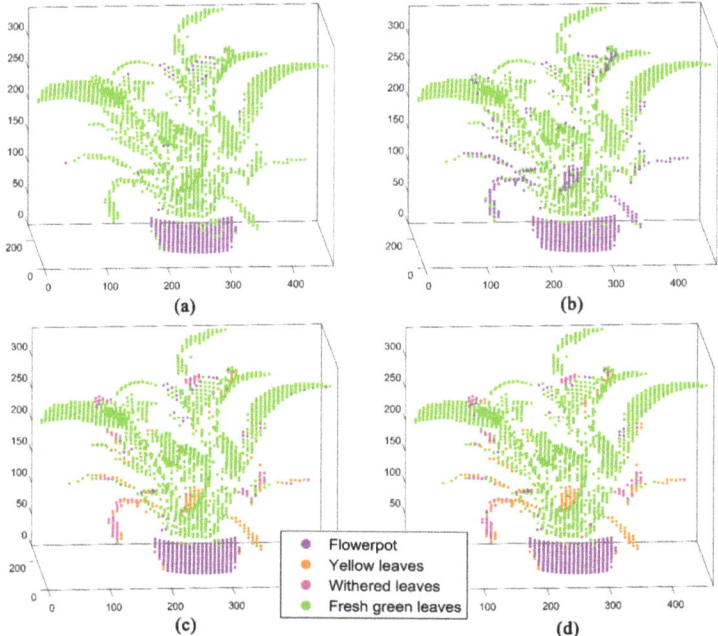

Figure 9. Different wavelength spectral data classification result graphs based on the MWFL system scanning in Scene 2. (**a**) Single-wavelength (460 nm); (**b**) Single-wavelength (685 nm); (**c**) Double-wavelength (460 nm + 685 nm); (**d**) Four-wavelength intensity data for classification.

Table 2. Confusion matrices of different wavelengths' spectrum data classification results for Scene 2 (corresponding to Figure 9).

Ground Truth		Predicted Class				Producer Accuracy
		Flowerpot	Withered Leaves	Yellow Leaves	Fresh Green Leaves	
(a) 460 nm	Flowerpot	108	0	0	140	0.44
	Withered leaves	16	0	0	100	0
	Yellow leaves	4	0	0	129	0
	Fresh green leaves	132	0	0	1147	0.90
	User accuracy	0.42	0	0	0.76	
	Overall accuracy (%): 70.7%					
	Kappa coefficient: 0.17					
(b) 685 nm	Flowerpot	235	0	0	13	0.95
	Withered leaves	30	5	0	81	0.04
	Yellow leaves	1	1	0	131	0
	Fresh green leaves	159	6	0	1114	0.87
	User accuracy	0.55	0.42	0	0.83	
	Overall accuracy (%): 76.2%					
	Kappa coefficient: 0.43					
(c) 460 nm + 685 nm	Flowerpot	236	2	3	7	0.95
	Withered leaves	51	13	6	46	0.11
	Yellow leaves	5	12	18	98	0.14
	Fresh green leaves	55	22	25	1177	0.92
	User accuracy	0.68	0.27	0.35	0.89	
	Overall accuracy (%): 81.3%					
	Kappa coefficient: 0.56					
(d) Four wavelengths	Flowerpot	240	0	4	4	0.97
	Withered leaves	7	57	19	33	0.49
	Yellow leaves	0	13	69	51	0.52
	Fresh green leaves	24	20	23	1212	0.95
	User accuracy	0.89	0.63	0.60	0.93	
	Overall accuracy (%): 88.9%					
	Kappa coefficient: 0.75					

Figure 9 and Table 2 demonstrate that the species-recognition accuracy is gradually increasing from single to double to four wavelengths. Such accuracy was limited in the case where only single-wavelength data were applied. The overall accuracies of the single-wavelength classifications are 70.7% and 76.2%. Such a result is attributed to the simple category division, and the number of fresh green leaves account for a large proportion, thereby resulting in a high classification. However, the confusion matrices demonstrate that the single-wavelength data have almost no ability to distinguish between yellow and withered leaves of the vegetation. The Kappa coefficients of the two single-wavelength classifications also illustrate that. The classification to the kappa values of 0.17 and 0.43 were not ideal. If the double-wavelength data were used for classification, then the recognition ability would be significantly improved. The classification accuracy of this case is 81.3%, and the Kappa coefficient increased to 0.56, reflecting an improvement compared with those of single-wavelength data. The result of the four-wavelength classification reveals that the classification accuracy reaches 88.9%, and the Kappa coefficient increases to 0.75. Such a finding indicates that the classification results (see Figure 9d) are consistent. The application of four-wavelength data further improves the identification capability of vegetation physiological states. This finding illustrates the necessity for four wavelengths to detect vegetation fluorescence. From the classification of the flowerpot, the system also has a certain degree of detection ability for the background objects during vegetation-detection fieldwork.

The normal vector was used as the representative parameter of the spatial structural state to be classified (Figure 10a). The normal vectors were computed by searching the neighbor points of the single point on the basis of the KNN algorithm and calculating the vertical pointing of the fitted plane. The normal vectors and four-wavelength signal values were used together for classification (Figure 10b). Table 3 shows the confusion matrix diagrams of Figure 10a,b.

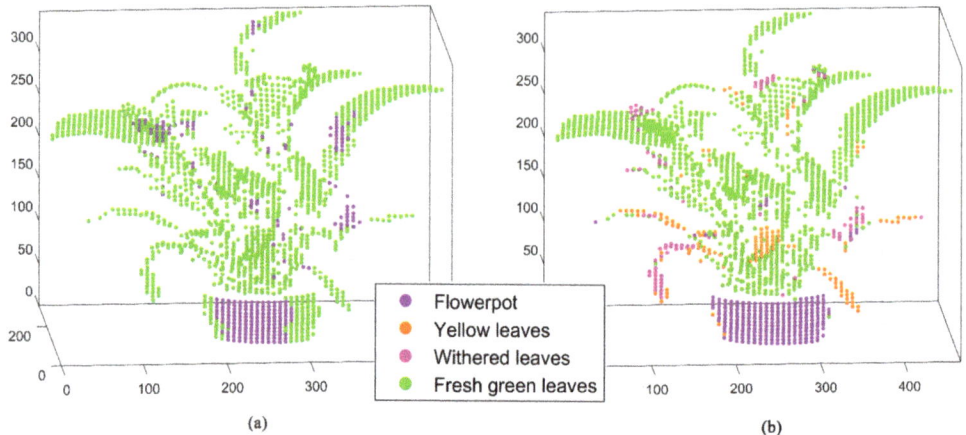

(a)   (b)

Figure 10. Normal vectors and normal vectors with four-channel spectral classification results (Scene 2): (**a**) Normal vectors data and (**b**) Normal vectors with four-channel spectral data for classification.

Table 3. Confusion matrices of the normal vectors and normal vectors with four-channel spectral classification results for Scene 2 (corresponding to Figure 10).

Ground Truth		Predicted Class				Producer Accuracy
		Flowerpot	Withered Leaves	Yellow Leaves	Fresh Green Leaves	
(a) Normal vectors	Flowerpot	107	0	0	141	0.43
	Withered leaves	8	7	6	95	0.06
	Yellow leaves	3	0	9	121	0.07
	Fresh green leaves	48	0	0	1231	0.96
User accuracy		0.64	1.0	0.60	0.78	
Overall accuracy (%): 76.2%						
Kappa coefficient: 0.29						
(b) Normal vectors + four wavelengths	Flowerpot	244	0	0	4	0.98
	Withered leaves	4	73	13	26	0.63
	Yellow leaves	3	14	91	25	0.68
	Fresh green leaves	6	14	24	1235	0.97
User accuracy		0.95	0.72	0.71	0.96	
Overall accuracy (%): 92.5%						
Kappa coefficient: 0.84						

The results in Figure 10 and Table 3 illustrate that the spatial parameter normal vectors present a very unsatisfactory classification outcome for the complex structure of vegetation (classification accuracy 76.2% and Kappa coefficient 0.29). In terms of the point cloud of Scene 2, the spatial shapes and angles of vegetation leaves vary, and certain leaves have few points. Figure 10a demonstrates that the normal vectors can distinguish a part of the flowerpot and most fresh green leaves, which are also related to a large number of the fresh green leaves' points. The normal vectors for yellow and withered leaves are completely indistinguishable. However, the ability of fluorescence to indicate the physiological state are exerted when the normal vectors and the four-wavelength spectral data are combined as multi-eigenvalues (Figure 9b). The classification accuracy improves to 92.5% with the significant enhancement of the producer, user, and overall accuracies. The kappa coefficient increases to 0.84, which is a promotion relative to the four-wavelength classification.

From the results of classification, the detection of vegetation fluorescence was improved to a higher level compared to the space detection capability of the single-wavelength LiDAR. Therefore, the four-wavelength signals detected by the MWFL system can effectively improve the recognition ability of different growth states of vegetation through the LIF mechanism. Such a mechanism effectively works with spatial parameters, which single-wavelength LiDAR possesses. The coupled detection of these two mechanisms has great potential for the remote-sensing field.

5. Conclusions

The proposed MWFL system expands the fluorescence characteristic generated by the LIF in four wavelengths, on the basis of the ranging LiDAR. We believe that the four-wavelength detectors added to the system could represent the internal components of the vegetation. The data form of the MWFL system can be coupled with the 3D spatial structural state and the physiological state information of vegetation monitoring through data organization. The combination of two mechanisms enhances the ability to identify and monitor vegetation targets. The significance of 3D fluorescence imaging of vegetation is that it not only expresses the growth status from the outer space, but also expresses the stress status of the internal physiological status. For different types of vegetation, in addition to the different spatial-expansion states of external growth, the internal biochemical content also varies greatly. The ability of fluorescence features to qualitatively and quantitatively indicate vegetation has improved the capability of LiDAR monitoring. This monitoring method is of great benefit to forestry development and precision agriculture.

At present, the signals of the 460 and 525 nm wavelengths have a relatively high correlation. The necessity for designing these two bands may be performed by the quantitative monitoring of the vegetation in the future.

Our analysis reflects the effectiveness of the 3D fluorescence imaging monitoring on the basis of the MWFL system for vegetation remote sensing. The technical upgrades and performance optimization of the system are required if there are platform operations and large space–time scale applications. In the MWFL system, the radiation correction of distance and angular polarization are beneficial for the quantitative monitoring of the surface vegetation.

Author Contributions: Conceptualization, S.S. and W.G.; data curation, J.S.; investigation, X.Z., S.S. and B.C.; project administration, W.G.; Validation, J.Y., B.C., K.G. and B.C.; writing—original draft, X.Z.; writing—review and editing, J.S. All authors have read and agreed to the published version of the manuscript.

Funding: This research was funded by the National Key R&D Program of China (Grant No. 2018YFB0504500), the National Natural Science Foundation of China (Grant No. 41971307, 41601360), the Natural Science Foundation of Hubei Province (Grant No. 2019CFB532), and the Wuhan Morning Light Plan of Youth Science and Technology (2017050304010308).

Conflicts of Interest: The authors declare no conflict of interest.

References

1. Anderegg, W.R.L.; Schwalm, C.; Biondi, F.; Camarero, J.J.; Koch, G.; Litvak, M.; Ogle, K.; Shaw, J.D.; Shevliakova, E.; Williams, A.P. Pervasive drought legacies in forest ecosystems and their implications for carbon cycle models. *Science* **2015**, *349*, 528–532. [CrossRef]
2. Bohnert, H.J.; Jensen, R.G. Strategies for engineering water-stress tolerance in plants. *Trends Biotechnol.* **1996**, *14*, 89–97. [CrossRef]
3. Croft, H.; Chen, J.M.; Zhang, Y.; Simic, A. Modelling leaf chlorophyll content in broadleaf and needle leaf canopies from ground, CASI, Landsat TM 5 and MERIS reflectance data. *Remote Sens. Environ.* **2013**, *133*, 128–140. [CrossRef]
4. Dalponte, M.; Orka, H.O.; Gobakken, T.; Gianelle, D.; Naesset, E. Tree species classification in boreal forests with hyperspectral data. *IEEE T. Geosci. Remote Sens.* **2013**, *51*, 2632–2645. [CrossRef]

5. Wu, C.; Niu, Z.; Tang, Q.; Huang, W. Predicting vegetation water content in wheat using normalized difference water indices derived from ground measurements. *J. Plant Res.* **2009**, *122*, 317–326. [CrossRef]

6. Yang, J.; Gong, W.; Shi, S.; Du, L.; Sun, J.; Song, S. The effective of different excitation wavelengths on the identification of plant species based on Fluorescence LIDAR. *Int. Arch. Photogramm. Remote Sens. Spat. Inf. Sci.* **2016**, *41*, 147–150. [CrossRef]

7. Van Leeuwen, M.; Nieuwenhuis, M. Retrieval of forest structural parameters using LiDAR remote sensing. *Eur. J. Forest Res.* **2010**, *129*, 749–770. [CrossRef]

8. Ren, X.; Altmann, Y.; Tobin, R.; Mccarthy, A.; Mclaughlin, S.; Buller, G.S. Wavelength-time coding for multispectral 3D imaging using single-photon LiDAR. *Opt. Express* **2018**, *26*, 30146. [CrossRef]

9. Moges, S.M.; Raun, W.R.; Mullen, R.W.; Freeman, K.W.; Johnson, G.V.; Solie, J.B. Evaluation of green, red, and near infrared bands for predicting winter wheat biomass, nitrogen uptake, and final grain yield. *J. Plant Nutr.* **2005**, *27*, 1431–1441. [CrossRef]

10. Duniec, J.T.; Thorne, S.W. Environmental effects on fluorescence quantum efficiencies and lifetimes: A semiclassical approach. *J. Phys. C Solid State Phys.* **1979**, *12*, 4109–4117. [CrossRef]

11. Guanter, L.; Zhang, Y.; Jung, M.; Joiner, J.; Voigt, M.; Berry, J.A.; Frankenberg, C.; Huete, A.R.; Zarco-Tejada, P.; Lee, J.E.; et al. Global and time-resolved monitoring of crop photosynthesis with chlorophyll fluorescence. *Proc. Natl. Acad. Sci. USA* **2014**, *111*, E1327–E1333. [CrossRef] [PubMed]

12. Plascyk, J.A.; Gabriel, F.C. The fraunhofer line discriminator MKII-an airborne instrument for precise and standardized ecological luminescence measurement. *IEEE T. Instrum. Meas.* **1975**, *24*, 306–313. [CrossRef]

13. Alonso, L.; Gomez-Chova, L.; Vila-Frances, J.; Amoros-Lopez, J.; Guanter, L.; Calpe, J.; Moreno, J. Improved fraunhofer line discrimination method for vegetation fluorescence quantification. *IEEE Geosci. Remote Sens. Lett.* **2008**, *5*, 620–624. [CrossRef]

14. Mazzoni, M.; Falorni, P.; Del Bianco, S. Sun-induced leaf fluorescence retrieval in the O2-B atmospheric absorption band. *Opt. Express* **2008**, *16*, 7014–7022. [CrossRef]

15. Tremblay, N.; Wang, Z.; Cerovic, Z.G. Sensing crop nitrogen status with fluorescence indicators. A review. *Agron. Sustain. Dev.* **2012**, *32*, 451–464. [CrossRef]

16. Chappelle, E.W.; Wood, F.M.; McMurtrey, J.E.; Newcomb, W.W. Laser-induced fluorescence of green plants 1: A technique for the remote detection of plant stress and species differentiation. *Appl. Opt.* **1984**, *23*, 134. [CrossRef]

17. Chappelle, E.W.; Wood, F.M.; Wayne Newcomb, W.; McMurtrey, J.E. Laser-induced fluorescence of green plants 3: LIF spectral signatures of five major plant types. *Appl. Opt.* **1985**, *24*, 74. [CrossRef]

18. Lichtenthaler, H.K.; Buschmann, C.; Rinderle, U.; Schmuck, G. Application of chlorophyll fluorescence in ecophysiology. *Radiat. Environ. Bioph.* **1986**, *25*, 297–308. [CrossRef]

19. Subhash, N.; Wenzel, O.; Lichtenthaler, H.K. Changes in blue-green and chlorophyll fluorescence emission and fluorescence ratios during senescence of tobacco plants. *Remote Sens. Environ.* **1999**, *69*, 215–223. [CrossRef]

20. Yang, J.; Sun, J.; Du, L.; Chen, B.; Zhang, Z.; Shi, S.; Gong, W. Effect of fluorescence characteristics and different algorithms on the estimation of leaf nitrogen content based on laser-induced fluorescence lidar in paddy rice. *Opt. Express* **2017**, *25*, 3743. [CrossRef]

21. Leufen, G.; Noga, G.; Hunsche, M. Fluorescence indices for the proximal sensing of powdery mildew, nitrogen supply and water deficit in sugar beet leaves. *Agriculture* **2014**, *4*, 58–78. [CrossRef]

22. Apostol, S.; Viau, A.A.; Tremblay, N. A comparison of multiwavelength laser-induced fluorescence parameters for the remote sensing of nitrogen stress in field-cultivated corn. *Can. J. Remote Sens.* **2007**, *33*, 150. [CrossRef]

23. Kharcheva, A.V. Fluorescence intensities ratio F685/F740 for maple leaves during seasonal color changes and with fungal infection. In *Saratov Fall Meeting 2013: Optical Technologies in Biophysics and Medicine XV and Laser Physics and Photonics XV*; SPIE: Bellingham, WA, USA, 2014; Volume 9031, p. 90310S.

24. Yang, J.; Song, S.; Du, L.; Shi, S.; Gong, W.; Sun, J.; Chen, B. Analyzing the effect of fluorescence characteristics on leaf nitrogen concentration estimation. *Remote Sens.* **2018**, *10*, 1402. [CrossRef]

25. Sun, J.; Shi, S.; Yang, J.; Chen, B.; Gong, W.; Du, L.; Mao, F.; Song, S. Estimating leaf chlorophyll status using hyperspectral lidar measurements by PROSPECT model inversion. *Remote Sens. Environ.* **2018**, *212*, 1–7. [CrossRef]

26. Babichenko, S.; Dudelzak, A.; Poryvkina, L. Laser remote sensing of coastal and terrestrial pollution by FLS-LIDAR. *EARSeL eProc.* **2004**, *3*, 1–7.

27. Lennon, M.; Babichenko, S.; Thomas, N.; Mariette, V.; Mercier, G.E.G.; Lisin, A. Detection and mapping of oil slicks in the sea by combined use of hyperspectral imagery and laser induced fluorescence. *EARSeL eProc.* **2006**, *5*, 120–128.

28. Ohm, K.; Reuter, R.; Stolze, M.; Willkomm, R. Shipboard oceanographic fluorescence lidar development and evaluation based on measurements in Antarctic waters. *EARSeL Adv. Remote Sens.* **1997**, *5*, 104–113.

29. Langsdorf, G.; Buschmann, C.; Sowinska, M.; Babani, F.; Mokry, M.; Timmermann, F.; Lichtenthaler, H.K. Multicolour Fluorescence imaging of sugar beet leaves with different nitrogen status by flash lamp UV-excitation. *Photosynthetica* **2000**, *38*, 539–551. [CrossRef]

30. Kim, M.S.; McMurtrey, J.E.; Mulchi, C.L.; Daughtry, C.S.T.; Chappelle, E.W.; Chen, Y. Steady-state multispectral fluorescence imaging system for plant leaves. *Appl. Opt.* **2001**, *40*, 157. [CrossRef]

31. Cadet, É.; Samson, G. Detection and discrimination of nutrient deficiencies in sunflower by blue-green and chlorophyll-a fluorescence imaging. *J. Plant Nutr.* **2011**, *34*, 2114. [CrossRef]

32. Chappelle, E.W.; McMurtrey, J.E.; Kim, M.S. Identification of the pigment responsible for the blue fluorescence band in the laser induced fluorescence (LIF) spectra of green plants, and the potential use of this band in remotely estimating rates of photosynthesis. *Remote Sens. Environ.* **1991**, *36*, 213–218. [CrossRef]

33. Hak, R.; Lichtenthaler, H.K.; Rinderle, U. Decrease of the chlorophyll fluorescence ratio F690/F730 during greening and development of leaves. *Radiat. Environ. Biophys.* **1990**, *29*, 329–336. [CrossRef] [PubMed]

34. Saito, Y.; Kanoh, M.; Hatake, K.; Kawahara, T.D.; Nomura, A. Investigation of laser-induced fluorescence of several natural leaves for application to lidar vegetation monitoring. *Appl. Opt.* **1998**, *37*, 431. [CrossRef] [PubMed]

35. Lichtenthaler, H.; Miehé, J. Fluorescence imaging as a diagnostic tool for plant stress. *Trends Plant Sci.* **1997**, *2*, 316–320. [CrossRef]

36. Kim, M.S.; Lefcourt, A.M.; Chen, Y.R. Multispectral laser-induced fluorescence imaging system for large biological samples. *Appl. Opt.* **2003**, *42*, 3927–3934. [CrossRef] [PubMed]

37. Pérez-Bueno, M.L.; Pineda, M.; Cabeza, F.M.; Barón, M. Multicolor Fluorescence imaging as a candidate for disease detection in plant phenotyping. *Front. Plant Sci.* **2016**, *7*, 1790. [CrossRef] [PubMed]

38. Wang, X.; Duan, Z.; Brydegaard, M.; Svanberg, S.; Zhao, G. Drone-based area scanning of vegetation fluorescence height profiles using a miniaturized hyperspectral lidar system. *Appl. Phys. B* **2018**, *124*, 207. [CrossRef]

39. Svanberg, S. Fluorescence lidar monitoring of vegetation status. *Phys. Scr.* **1995**, *1995*, 79. [CrossRef]

40. Lee, J.B.; Woodyatt, A.S.; Berman, M. Enhancement of high spectral resolution remote-sensing data by a noise-adjusted principal components transform. *IEEE T. Geosci. Remote* **1990**, *28*, 295–304. [CrossRef]

41. Ali, M.; Clausi, D. Using the Canny Edge Detector for Feature Extraction and Enhancement of Remote Sensing Images. In *IGARSS 2001. Scanning the Present and Resolving the Future, Proceedings of the IEEE 2001 International Geoscience and Remote Sensing Symposium, Sydney, Australia, 9–13 July 2001*; IEEE: Piscataway, NJ, USA, 2001; Volume 5, pp. 2298–2300.

42. Stark, J.A. Adaptive image contrast enhancement using generalizations of histogram equalization. *IEEE Trans. Image Process.* **2000**, *9*, 889–896. [CrossRef]

43. Tarabalka, Y.; Fauvel, M.; Chanussot, J.; Benediktsson, J.A. SVM- and MRF-based method for accurate classification of hyperspectral images. *IEEE Geosci. Remote Sens. Lett.* **2010**, *7*, 736–740. [CrossRef]

44. Rumpf, T.; Römer, C.; Weis, M.; Sökefeld, M.; Gerhards, R.; Plümer, L. Sequential support vector machine classification for small-grain weed species discrimination with special regard to cirsium arvense and galium aparine. *Comput. Electron. Agric.* **2012**, *80*, 89–96. [CrossRef]

45. Mountrakis, G.; Im, J.; Ogole, C. Support vector machines in remote sensing: A review. *ISPRS J. Photogramm.* **2011**, *66*, 247–259. [CrossRef]

46. Secord, J.; Zakhor, A. Tree detection in LiDAR data. In Proceedings of the 2006 IEEE Southwest Symposium on Image Analysis and Interpretation, Denver, CO, USA, 26–28 March 2006.

47. Nguyen, G.H.; Bouzerdoum, A.; Phung, S.L. Learning pattern classification tasks with imbalanced data Sets. *Pattern Recogn.* **2009**, 193–208.

48. Viera, A.J.; Garrett, J.M. Understanding interobserver agreement: The kappa statistic. *Fam. Med.* **2005**, *37*, 360–363.

49. Yang, J.; Cheng, Y.; Du, L.; Gong, W.; Shi, S.; Sun, J.; Chen, B. Analyzing the effect of the incidence angle on chlorophyll fluorescence intensity based on laser-induced fluorescence lidar. *Opt. Express* **2019**, *27*, 12541. [CrossRef]

sensors

Review

Current Development and Applications of Super-Resolution Ultrasound Imaging

Qiyang Chen [1,2,†], Hyeju Song [3,†], Jaesok Yu [3,4,*] and Kang Kim [1,2,5,6,7,*]

1 Department of Bioengineering, School of Engineering, University of Pittsburgh, Pittsburgh, PA 15261, USA; qic41@pitt.edu
2 Center for Ultrasound Molecular Imaging and Therapeutics, Department of Medicine, School of Medicine, University of Pittsburgh, Pittsburgh, PA 15261, USA
3 Department of Robotics Engineering, Daegu Gyeongbuk Institute of Science & Technology (DGIST), Daegu 42988, Korea; hyeju@dgist.ac.kr
4 DGIST Robotics Research Center, Daegu Gyeongbuk Institute of Science & Technology (DGIST), Daegu 42988, Korea
5 Division of Cardiology, Department of Medicine, School of Medicine, University of Pittsburgh, Pittsburgh, PA 15261, USA
6 McGowan Institute of Regenerative Medicine, University of Pittsburgh, Pittsburgh, PA 15219, USA
7 Department of Mechanical Engineering and Materials Science, School of Engineering, University of Pittsburgh, Pittsburgh, PA 15261, USA
* Correspondence: jaesok.yu@dgist.ac.kr (J.Y.); kangkim@pitt.edu (K.K.)
† These authors contributed equally to this work.

Citation: Chen, Q.; Song, H.; Yu, J.; Kim, K. Current Development and Applications of Super-Resolution Ultrasound Imaging. *Sensors* **2021**, 21, 2417. https://doi.org/10.3390/s21072417

Academic Editors: Changho Lee and Changhan Yoon

Received: 6 February 2021
Accepted: 24 March 2021
Published: 1 April 2021

Publisher's Note: MDPI stays neutral with regard to jurisdictional claims in published maps and institutional affiliations.

Abstract: Abnormal changes of the microvasculature are reported to be key evidence of the development of several critical diseases, including cancer, progressive kidney disease, and atherosclerotic plaque. Super-resolution ultrasound imaging is an emerging technology that can identify the microvasculature noninvasively, with unprecedented spatial resolution beyond the acoustic diffraction limit. Therefore, it is a promising approach for diagnosing and monitoring the development of diseases. In this review, we introduce current super-resolution ultrasound imaging approaches and their preclinical applications on different animals and disease models. Future directions and challenges to overcome for clinical translations are also discussed.

Keywords: super-resolution; ultrasound imaging; deep learning; clinical applications

1. Introduction

Abnormal alterations, including the development, degeneration, and regeneration, of the microvasculature, are reported to be associated with several critical diseases, such as tumor development [1–3], progressive kidney disease [4–7], and the development of atherosclerotic plaque [8–10]. Therefore, changes of microvasculature would serve as a useful index for diagnostics and prognostics of such diseases. Several imaging modalities, including microcomputed tomography (micro-CT) [11,12], optical coherence tomography (OCT) [13,14], and magnetic resonance imaging (MRI) [15–19], have been employed successfully in preclinical studies to image the changes of microvasculature inside target organs. Although these imaging methods achieved a high spatial resolution, they have their own their own limitations. Micro-CT is limited by hazardous radiation and contrast agents, while OCT suffers relatively poor imaging depth. As for MRI, imaging systems are bulky and costly, which hinder widespread or repeated applications. Ultrasound imaging that has the advantage of safety, noninvasiveness, portability, affordability, and ease of use, has been explored as a potential approach for imaging microvessels. Conventional noninvasive ultrasound imaging methods for imaging vessels mainly include Doppler ultrasound imaging [20] and contrast-enhanced ultrasound (CEU) imaging [21–25]. However, neither of these techniques provides sufficient spatial resolution for assessing microvessels,

323

mainly due to the acoustic diffraction limit, which is half of the wavelength of the operation ultrasound frequency. Therefore, an ultrasound imaging technique that can achieve spatial resolution beyond the acoustic diffraction limit would be encouraging for broad use in clinics for diseases that are associated with abnormal alterations in the microvasculature.

In 2006, optical super-resolution imaging techniques, including fluorescence photoactivated localization microscopy [26], photoactivated localization microscopy (PALM) [27], and stochastic optical reconstruction microscopy (STORM) [28], were first introduced. The basic idea of super-resolution imaging is to localize the centroid of each randomly blinking fluorescence source based on the system point spread function (PSF). The location information of each blinking fluorophore was stacked up over a substantial sequential dataset that was captured by a fast camera to form an image spatially resolved in subwavelength resolution. By this approach, a spatial resolution down to tens of nanometers was achieved. Inspired by optical super-resolution imaging techniques, super-resolution ultrasound (SRU) imaging was introduced [29–33] for noninvasive imaging of microvasculature by using ultrasound contrast agents that travel through vascular network to replace the role of fluorophores in optical super-resolution imaging. While spatial resolution was sacrificed compared to optical super-resolution, due to the limit of ultrasound operating frequency, the ultrasound approach achieved a larger imaging depth. The technology components of SRU imaging mainly consist of ultrafast, ultrasound imaging [34], a state-of-the-art clutter filter [35] that extracts microbubble signals, and novel microbubble localization algorithms [29–33,36,37] that pinpoint the original locations of the microbubbles. The developed SRU imaging technologies have been successfully tested in preclinical studies with different animal and disease models, which demonstrated great potential for future clinical applications. Some representative conventional technical approaches of SRU, and their in vivo applications, were well described and summarized in a recent review paper [38]. In addition to these conventional SRU approaches, a deep learning approach has recently been adopted for SRU imaging [39–42]. In this review, we introduce the different SRU imaging approaches, especially the deep learning approach, and summarize current preclinical studies in different disease models that have been successfully performed. The vision for future clinical applications, and the major challenges for SRU imaging to overcome, are also discussed.

2. General Technical Components of Super-Resolution Ultrasound Imaging

Figure 1a illustrates the overall block diagram of SRU imaging. The blinking fluorescence sources in the photoactivated localization microscopy can be replaced with microbubbles, used for the contrast agent of ultrasound in SRU imaging, as the spatial locations of microbubbles in the bloodstream are stochastically changed. The following localization technique is a method to find the centroid of the single microbubble signal, localizing each point source with subwavelength precision. Throughout this process, spatial resolution can be improved by up to one-tenth of the wavelength, theoretically [29]. Note that the acoustic response to the point source can be estimated to the PSF of the imaging system. Therefore, the extraction of a single microbubble signal from the original image data is an essential key component for implementing SRU imaging. The first trial for decluttering is a subtraction between neighboring frames to remove the stationary tissue component and maintain moving microbubble signals [32]. Other clutter filtering techniques used in Doppler imaging are also studied for decluttering purpose. Researchers show that the combination of the singular value decomposition-based adaptive clutter filter and a large number of the spatiotemporal image sets, with the ultrafast imaging, would outperform traditional infinite impulse response (IIR) filter-based clutter filtering techniques [35]. This method decomposes the large-sized elongated skinny matrix into the spatial and temporal basis vector matrixes and a diagonal eigenvalue matrix-weighting factor. The combination of these decomposed vectors represents several components of the images, such as stationary tissue, slow-moving tissue, fast-moving particle-microbubbles, and randomly varied value noise. Therefore, microbubbles signals could be exclusively extracted from the images with an adequately selected rank of vector-matrix combinations. The signals, other than the selected ranks, are then removed to maintain only the valid microbubble signals.

The next core component is a method seeking the point source location. Each microbubble location can be precisely localized in subwavelength resolution by fitting with the predetermined PSF of the imaging system [29]. This method achieves the spatial resolution up to 10 microns ($\approx\lambda/10$, with a custom-made 128-element linear array transducer centered at 15 MHz), which is further beyond the acoustic diffraction limit, as shown in Figure 1c [29]. Note that the imaging system's PSF is assumed as the fixed two-dimensional Gaussian function determined by the transmit wave characteristic. However, this approach requires a huge number of the dataset, 75,000 frames (150 s), for a single super-resolved image in this study. Long data acquisition is the main drawback for clinical applications, except imaging the brain, which can be possibly fixed in position during the scan period due to the motion artifact. In the following studies, therefore, several groups suggest techniques in efforts to improve temporal resolution. These include Super-Resolution Optical Fluctuation Imaging (SOFI)-based [43] and deconvolution-based [36] SRU imaging technologies to broaden the clinical applications to other organs and diseases, as shown in Figure 2 [36].

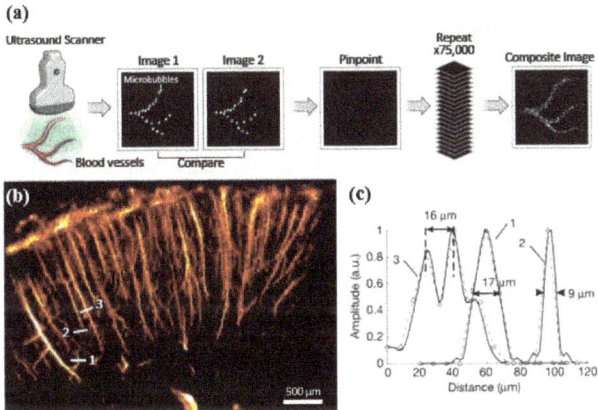

Figure 1. (**a**) An illustration of the concept of SRU imaging (reprinted with permission from Ref. [44]. Copyright 2015 Springer Nature). (**b**) The reconstructed super-resolved brain microvasculature with the resolution of $\lambda/10$ [29]. (**c**) Interpolated profiles along the marked lines (reprinted with permission from Ref. [29]. Copyright 2015 Springer Nature).

SOFI-based super-resolution imaging uses a relatively high concentration of microbubbles, while the traditional super-resolution method utilizes a diluted concentration for better separation of microbubbles [43]. Bar-Zion et al. [43] suggest a parametric model of the contrast-enhanced ultrasound signal of microbubbles to quantify the volume cell, instead of counting the number of microbubbles. High order statistics calculations could improve spatial resolution by 60% at the 4th moment, as shown in Figure 2a, using a L15-4 linear array transducer. They successfully demonstrate their super-resolution imaging approach using a rabbit kidney tumor model with only 150 frames of data, which allows for a 500-times faster scan time than the prior method. The SOFI-based method that utilized high order statistical computations [43] achieved higher temporal resolution than the previous ultrasound localization microscopy [29]. However, spatial resolution and the signal-to-noise ratio were compromised because the dynamic range of image intensity increased as the higher-order statistics were used. Some researchers suggest using a nonlocal means (NLM) denoising filter on the spatiotemporal domain to remove noise from the background signal, while preserving the signal from flowing microbubbles [37]. It should be noted that use of the spatial domain filter to eliminate noise is, in general, challenging, as shown in the previous study, as the amplitude of the background noise looks very similar to signals from microbubbles. They then applied bipartite graph-based microbubble tracking, with persistence control for enhanced microbubble signal quality and tracking fidelity. The localized microbubbles located in each frame could be paired with, and followed by, microbubbles at adjacent frames. When using a 128-element linear array transducer, centered at 8 MHz, vessels as small as 57 μm at depth of 2 cm were reconstructed. Moreover, microvessels 76 μm apart were distinguished in a rabbit kidney in vivo.

(a) SOFI-based super-resolution ultrasound imaging

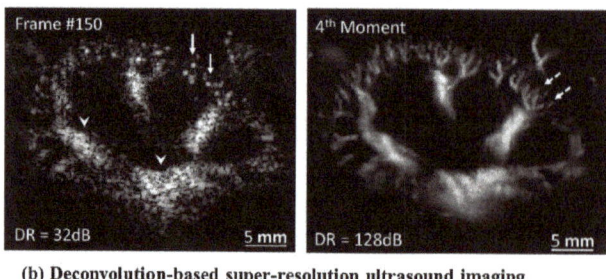

(b) Deconvolution-based super-resolution ultrasound imaging

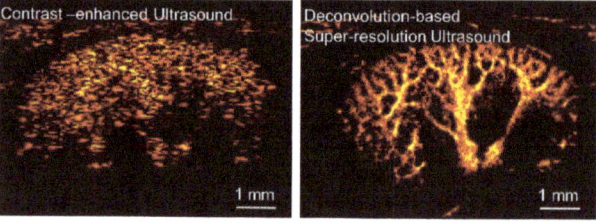

Figure 2. (a) SOFI-based super-resolution ultrasound imaging—left panel shows a frame of B-mode; right panel shows the reconstructed image. (Reprinted with permission from Ref. [43]. Copyright 2017 IEEE). **(b)** Deconvolution-based super-resolution ultrasound imaging—left panel shows a frame of B-mode; right panel shows the reconstructed image. (Reprinted with permission from Ref. [45]. Copyright 2020 Elsevier).

Another trial to improve temporal resolution is the employment of deconvolution and spatiotemporal-interframe-correlation (STIC) data acquisition techniques [36]. Regardless of the local density of the microbubbles, the deconvolution approach can localize each microbubble location from the clumped microbubble signal. It therefore enables the utilization of all acquired frames. Note that clumped microbubble signals have to be discarded in other approaches, resulting in a long scan time. Therefore, deconvolution-based SRU imaging only uses 300 image frames to reconstruct a single super-resolved ultrasound image while maintaining a spatial resolution of 41 μm; that is, 1/5 of the wavelength with a 128-element linear array transducer centered at 7.7 MHz. The calculation complexity of the deconvolution method [36] is lower than the above methods while it utilizes the iteration procedure. However, many physiological events are still faster than the data acquisition speed with a deconvolution of 0.6 s. Researchers implemented the STIC data acquisition technique, that was used for the 3D fetal cardiography, to capture rapid physiological events [46]. STIC algorithms allow for a realignment of sequentially acquired image data based on the reference signal, such as the cardiac pulsation, to make use of more frames that are synchronized. Figure 2b shows typical SRU images in an in vivo acute kidney injury in a mouse model, demonstrating the clinical feasibility for kidney applications [37,45]. Furthermore, the implemented SRU imaging technologies were further validated by comparison with micro-CT images in the following study [45].

One main challenge of super-resolution ultrasound imaging technology is maintaining the spatial and temporal resolution at the same time. Ultrasound localization microscopy achieved the spatial resolution of 1/10 of the wavelength beyond the acoustic diffraction limit. However, this method scarifies the temporal resolution, as the huge dataset, ~75,000 frames, corresponding to data acquisition time of 150 s, are required to track every flowing individual microbubble [29]. The SOFI-based method achieved a higher temporal resolution of ~150 frames, and a data acquisition time of 0.3 s-, with the time-dependent statistics of the microbubbles [43]. It also increased the spatial resolution around a factor of $\sqrt{2}$, which provides further fine spatial resolution beyond the acoustic diffraction limit. The deconvolution-based super-resolution ultrasound imaging method offers compromised spatial and temporal resolutions between the ultrasound localized microscopy and the SOFI approach. The spatial resolution of 1/5 of the wavelength and the temporal resolution of 0.6 s (~300 frames) were achieved [36]. Depending on their advantages, these methods could be applied for different applications. The overall performances of the representative SRU imaging technologies are compared in Table 1.

Table 1. The overall performances of the representative SRU imaging technologies. * All technologies offer spatial resolution beyond the acoustic diffraction limit.

	Localization [32]	SOFI [43]	Deconvolution [36]
Spatial Resolution *	High (9~17 μm)	Low (227.3 ± 9.0 μm)	Middle (41 μm)
Numbers of Frames for Reconstruction	75,000 frames	150 frames	300 frames
Temporal Resolution	150 s @ 500 Hz	0.3 s @ 500 Hz	0.6 s @ 500 Hz
Microbubble Concentration	Low (Diluted, 2×10^8 MBs/mL, Bolus Injection of 1.5 mL)	High (1.2×10^{10} MBs/mL, Bolus Injection of 0.5 mL)	High (1.2×10^{10} MBs/mL, Bolus Injection of 0.2 mL)
Application	Brain	Kidney	Atherosclerosis, Kidney

3. Deep Learning-Based Super-Resolution Ultrasound Imaging

A single super-resolved image can be only reconstructed with a sufficient number of frames of the localized microbubbles for a sufficient signal-to-noise ratio. This large-scale data acquisition results in a relatively long scan time, which may introduce potential motion artifacts. The low consistency of tissue caused by motion could decrease localization accuracy. Thus, a practical limitation of SRU imaging for clinical translation is the trade-off between data acquisition time and localization accuracy. The deep learning-based approach has demonstrated promising achievements, both in temporal accuracy and in reconstruction accuracy, when using a relatively high-concentration microbubble injection.

The deep learning-based ultrasound localization microscopy (Deep-ULM) is the first trial to let artificial intelligence separate individual microbubble signals from the dense microbubble cloud signal [39]. An increased concentration of microbubbles would reduce overall data acquisition time. The Deep-ULM, inspired by the deep learning network for super-resolution stochastic optical-resolution microscopy (Deep-STORM), adopts a network based on the fully convolutional U-net, performing the nonlinear end-to-end mapping between low-resolution input frames to high-resolution outputs, as shown in Figure 3 [39,47–49]. For the synthetic training dataset, randomly located microbubble positions were generated first. The diameters of the microbubbles were also randomly determined, making them similar to the actual microbubble signals. Then, the convolution between the simulated microbubbles and the point spread function can work as a synthetic low-resolution ultrasound image. Then, the simulated ultrasound images were paired with the actual locations of the microbubbles for training through the network. The encoder extracted the dense and aggregated features from low-resolution images. In the decoding layers, the features extracted in the encoding layers were upsampled and deconvolved to align the value for high-resolution image reconstruction. After the training, the trained mapping process reconstructs the low-resolution ultrasound B-mode image to a high-resolution image through feature extraction and upsampling. Besides, the Deep-ULM reduces the computational complexity with the GPU acceleration, allowing it to resolve 1250 high-resolution patches of 128 × 128 pixels within a second. Therefore, the model-based approach has great benefit and potential for implementing the real-time imaging system.

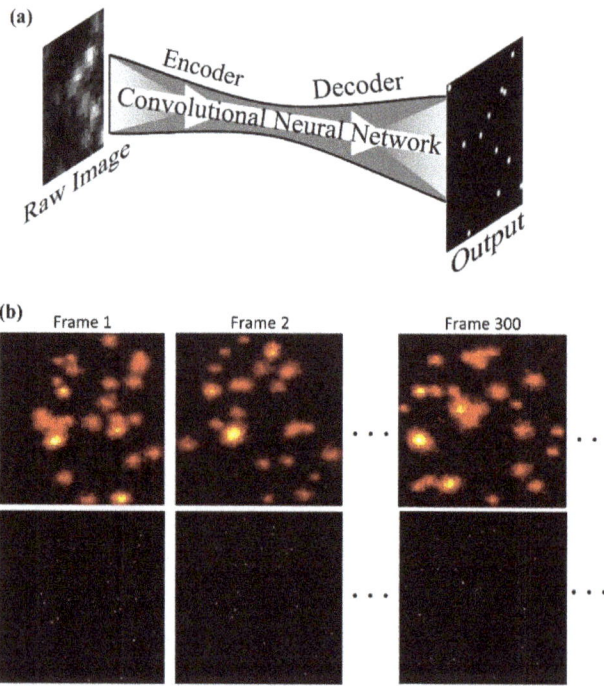

Figure 3. Super-resolution ultrasound localization microscopy through deep learning. (**a**) U-net based architecture for Deep-ULM [39], mSPCN-ULM [40], and a multiple-targets detecting network [47]. (Reprinted with permission from Ref. [47]. Copyright 2018 OSA) (**b**) An example of deep learning-based localization techniques—translation from the low-resolution image (top panel) to high-resolution frames (bottom panel) by the deep learning network [40,49]. (Reprinted with permission from Ref. [40]. Copyright 2020 IEEE).

Several modified or alternative network structures have been studied for a faster data processing time in the following research. For example, the convolutional neural network (CNN) was suggested for identifying individual scatters using high concentration microbubbles [50]. This network has a similar structure to U-net and is composed of an encoder–decoder with pooling and un-pooling, but without skip connections. Unlike the previous deep learning-based ULM method, the ground truth data were acquired after the radio frequency (RF) dataset was simulated. The binary confidence maps obtained through the simulated RF dataset can generate the ground truth data. In addition, the RF signals were not beamformed with the delay-and-sum algorithm, but delayed and sampled to have the same number of confidence maps along the axial direction. Considering that the training dataset is a key factor for trained network performance, it is worth more than passing attention to the improvement of deep learning networks based on a precisely tailored synthetic dataset.

Recently, modified subpixel convolutional neural network (mSPCN) architecture with residual blocks has been suggested for optimization without exhausted parameter tuning and fast data processing speed [40–42]. The subpixel convolutional neural network architecture in the mSPCN-ULM decodes features with a trained upscaling filter and reduces computational complexity. Moreover, residual learning allows an increase in the network depth without losing gradient, and improves training accuracy without a parameter tuning process, as shown in Figure 4. The mSPCN-ULM showed increased temporal resolution using a higher microbubble concentration compared with the above deep learning techniques. The training process was conducted using the synthetic data created in the same way as the previous Deep-ULM method presented. However, further performance degradation is expected in vivo cases since the discrepancy between the synthetic training dataset and the real in vivo data would become more extensive due to several artifacts and nonlinear responses.

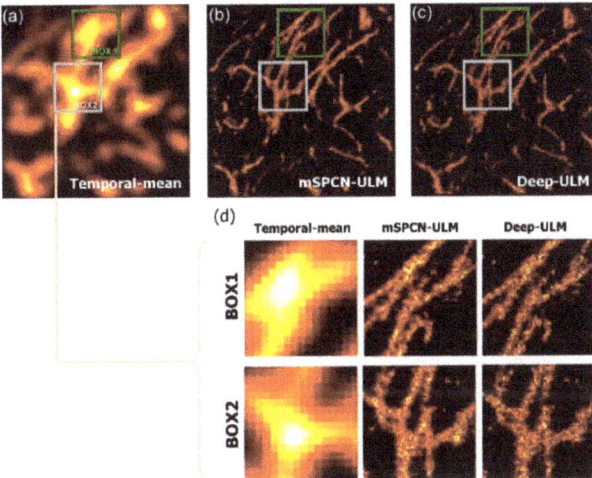

Figure 4. Comparison between mSPCN-ULM [40] and Deep-ULM [39]. (**a**) The temporal–mean image was obtained by averaging all the US images. (**b**) Super-resolution images obtained by the mSPCN-ULM. (**c**) Deep-ULM. (**d**) The magnified sections of (**a**–**c**) [40]. (Reprinted with permission from Ref. [40]. Copyright 2020 IEEE).

Deep learning technology is also applied in another applications, such as the decluttering process [51]. The extraction of small microbubble signals from only a few pixels in the image with noise is challenging. Researchers have employed a 3D convolutional neural network (3D-CNN) to solve this problem. This network has been known as an optimal network for human action recognition in airport surveillance video sequences, which is similar to flowing microbubble detection in the B-mode image sequence (2D + 1D). The proposed method's performance is comparable to the use of singular value decomposition (SVD) filtering in conventional SRU imaging sequences with a lightweight computational burden.

So far, deep learning networks trained with the noised synthetic data showed a resolution comparable to existing SRU imaging. However, the number of in vivo studies with deep learning are limited so far. Therefore, further investigations, especially in vivo evaluations, of deep learning applications should follow. Below, Table 2 shows an overall comparison of the representative deep learning-based technologies currently used in SRU imaging.

Table 2. An overall comparison of the representative deep learning-based technologies currently used in SRU imaging. * Only in vivo data were used for training. The dosage of injected microbubble is 2.5×10^7 MBs in 60 μL saline.

	Deep-ULM [39]	CNN Based Network for Multiple Target Detection [50]	mSPCN-ULM [40]	Deep 3D CNN for Spatiotemporal Filtering [51]
Target	Localization from the dense microbubbles	Localization from the dense microbubbles	Localization from the dense microbubbles	Microbubble extraction
Microbubble Concentration of Synthetic Data for Training	High (~2.6 MBs/mm²)	High (~2.44 MBs/mm²)	Very high (~6.4 MBs/mm²)	N/A *
Network Type	U-net	Convolutional neural network	Modified subpixel convolutional neural network	3-D convolutional neural network
Training Dataset	Synthetic data and unique data generated for each iteration	10,240 synthetic data	10,000 synthetic data	9000 frames acquired from five subjects
Spatial Resolution	~30 μm	27~46 μm	24~28 μm	25 μm
Applied Activation Function	Leaky rectified linear unit (ReLU)	Leaky ReLU	ReLU	ReLU

4. Current Biomedical Applications of Super-Resolution Ultrasound Imaging

With unprecedented spatial resolution and practically reasonable temporal resolution achieved, SRU could be a promising diagnostic tool for diseases associated with abnormal vascular alterations. It also has the potential to be a preferred approach for monitoring disease progression and therapeutic efficacy due to its noninvasiveness, low cost, safety, and widespread accessibility. In this section, some representative applications of SRU in preclinical studies on different organs and disease models, and a very limited first-in-human use, are introduced and discussed.

4.1. Cancer

Cancer is the second leading cause of death in the world [52]. Early detection of malignant lesions can greatly increase the chances of successful treatment [53]. One of the early changes that can differentiate cancer from normal tissues is malignant angiogenesis, which has been recognized as an important biomarker for cancer diagnostics [2,54]. The features of the microvascular network associated with malignant tumors, including density, branching, size, and inhomogeneity, have been observed to be abnormal compared to that of healthy tissue [3,55–58]. In past decades, superharmonic contrast ultrasound imaging, also known as acoustic angiography, has been utilized to visualizing the microvasculature and detect the morphology abnormalities associated with tumor-induced angiogenesis in vivo [59–64]. However, the performance of this imaging technique suffered mainly from the limit of the spatial resolution constrained by the acoustic diffraction limit of the operating ultrasound frequency. An SRU that can overcome this limitation has been explored to detect microvascular changes at much higher resolution and sensitivity, both at an early stage and during tumor progression.

For demonstrating the proof-of-concept and further improving SRU technology, animal tumor models have been adopted in several in vivo studies [43,65,66]. In 2016, Lin et al. successfully imaged the subcutaneous fibrosarcoma tumors implanted in a rat in vivo, with a ten-fold resolution improvement compared to conventional ultrasound imaging by using SRU. This study demonstrated the imaging capability of SRU and the potential of characterizing a tumor-associated microvascular angiogenesis [65]. In the following study, their group evaluated the sensitivity of SRU imaging on the same rat tumor model using microbubbles of different sizes, and showed the sensitivity improvement by using larger microbubbles. For the purpose of shortening the scan time, Bar-Zion et al. proposed an SRU imaging technique with a methodology that was used in super-resolution optical fluctuation imaging (SOFI). This technology was tested by imaging the vasculature around and inside the hind-limb intramuscular VX-2 tumor, and the improvement in temporal resolution was presented [43].

After demonstrating the proof-of-concept in tumor microvasculature imaging, preclinical studies have been conducted to examine the capability of SRU for tumor diagnosis. Lin et al. performed SRU imaging in three dimensions on tumor-bearing rats implanted with subcutaneous fibrosarcoma and compared the microvascular features with the healthy rats [67]. An L11-5 linear probe (Verasonics Inc., Redmond, WA, USA) was mounted to a motorized precision motion stage synchronized with the imaging system to perform the 3D scan. The reconstructed microvascular images by SRU showed a greatly improved spatial resolution compared to the traditional acoustic angiography. As shown in Figure 5a from the study, vessels in the tumor-bearing tissues had a higher tortuosity compared to the control, which implied tumor-associated microvascular angiogenesis. The results demonstrated the potential of differentiating diseased and healthy tissues by evaluating vascular structure using SRU imaging. With the help of the fine details of the vasculature network provided by SRU technology, the capability of SRU for discriminating different tumor types was also shown by the study of Opacic et al. [68]. The fine vascular networks in tumors with different vascular phenotypes were reconstructed by motion model SRU imaging (Figure 5b). Functional parameters, including relative blood volume (rBV), blood flow direction, blood flow velocity, distances to vessels, distances, and velocities, were able to be derived by SRU imaging, and utilized to differentiate different tumor types with the verification of histology successfully.

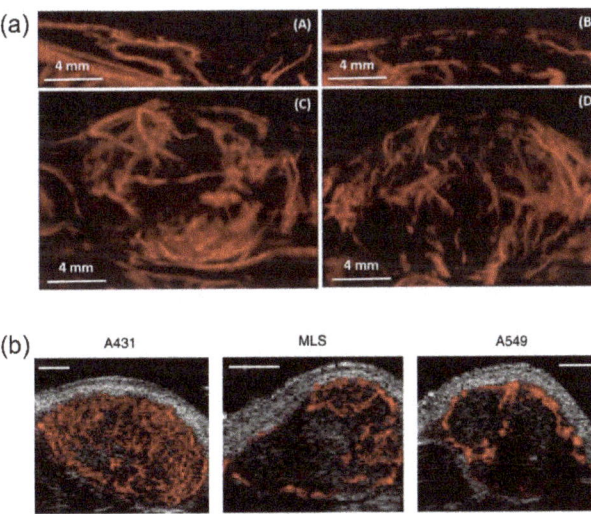

Figure 5. Representative SRU images of microvasculature in tumor-bearing tissues. (**a**) Maximum intensity projections of 3D SRU imaging on healthy rats (upper panel), and tumor-bearing rats (lower panel). (Reprinted with permission from Ref. [67]. Copyright 2017 Ivyspring International Publisher) (**b**) SRU imaging of tumors with different vascular phenotypes. (Reprinted with permission from Ref. [68]. Copyright 2018 Springer Nature).

SRU imaging was further evaluated on specific cancer types, which is the pathway towards clinical translation. Breast cancer, which is the most common type of cancer in women, with the second highest mortality rate [69], is also one focused area for the applications of SRU. Ghost et al. applied SRU imaging to longitudinally monitor changes in the tumor microvascular network of triple-negative breast cancer-bearing mice in response to the treatment [70]. The vessel-to-tissue ratio of the tumor tissue was found decreased progressively after the tumor-targeted therapeutic (Figure 6a), which was consistent with the immunohistological findings. This study suggested the potential of in vivo SRU imaging for monitoring early tumor response to drug treatment. Clinical pilot studies of SRU imaging on patients with breast cancer was further conducted by the Schmitz group [68,71]. Motion model SRU imaging was performed on patients with breast cancer after treatment with first, second, and third cycles of neoadjuvant chemotherapy. By SRU imaging, improved spatial resolution and functional information, including flow velocities, could be derived (Figure 6b) [71], which outperform conventional CEU imaging. The increase in rBV of the tumor tissue and the decrease in tumor size were found after the treatment [68]. The studies could be a scheme for further extended clinical studies and to promote future clinical applications.

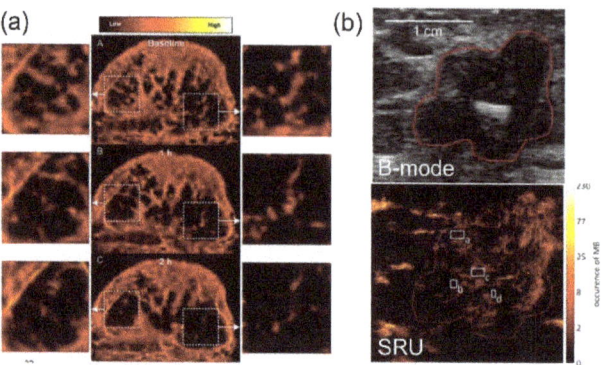

Figure 6. Representative SRU images of microvasculature from (**a**) breast cancer-bearing mice in response to the treatment (Reprinted with permission from Ref. [70]. Copyright 2017 IEEE), and (**b**) the patient with triple-negative breast carcinoma. (Reprinted with permission from Ref. [71]. Copyright 2019 IEEE).

In general, as one of the key features of a tumor is a dense microvasculature network, the tumor model would serve as a good candidate for demonstrating the imaging capability of SRU and for validating the technical improvements with new approaches for SRU. For the potential applications of SRU imaging on cancer, the studies mentioned above show that, with the fine structure of the microvascular networks reconstructed by SRU, several functional parameters can be accurately derived to help diagnose malignant tumors or differentiate different tumor types. Several studies have been performed, specifically on breast cancer, and reported promising results. Experiments on human subjects were also initiated. Extended clinical studies in the near future are expected for the clinical translation of this technology.

4.2. Kidney

Chronic kidney disease (CKD), which has a high incidence rate among adults [72,73], is typically induced by several risk factors, including diabetes, high blood pressure, heart disease, an episode of acute kidney injury, etc. [72,74]. One mechanism for the progression of CKD is the degradation of the renal microvasculature and perfusion impairment [75,76]. Therefore, the detection of renal microvascular changes would be of great importance for the early diagnosis and monitoring of CKD. However, diagnostic tools that enable noninvasive diagnostics and monitoring of renal microvascular alterations during progressive kidney disease are still lacking. Conventional ultrasound imaging techniques, including contrast-enhanced ultrasound imaging [21,22] and Doppler ultrasound imaging [20], have been explored to evaluate microvascular changes during the disease's progression. However, the spatial resolution is not ideal, mainly due to its insufficient sensitivity and the acoustic diffraction limit [77]. The emerging SRU imaging technique would be a promising approach to overcoming these barriers.

Several studies have already been successfully conducted on animal kidneys in vivo to show the capability, as well as the technical improvements, of imaging the renal microvasculature by SRU [37,43,78,79]. Foiret et al. depicted the microvascular structure and characterized the vessels with a flow rate below 2 mm/s of rat kidney, with Contrast Pulse Sequencing (CPS) mode using a 6.9 MHz ultrasound probe (CL15-7, Phillips ATL, MA, USA) [78]. Song et al. proposed the spatiotemporal NLM denoising method, together with the bipartite graph microbubble pairing and tracking method, and showed improved performance of SRU on rabbit kidney [37]. With an operating frequency of 8 MHz and mechanical index of 0.4, a single renal microvessel as small as 57 μm was identified, and microvessels that were 76 μm apart were clearly separated. Their group further developed the Kalman filter-based SRU method and presented a robust measurement of the renal microvascular flow with reduced MB events in the rabbit kidney [79]. Figure 7a,b shows the representative SRU images of the renal vascular network in rat and rabbit kidneys from the studies mentioned above.

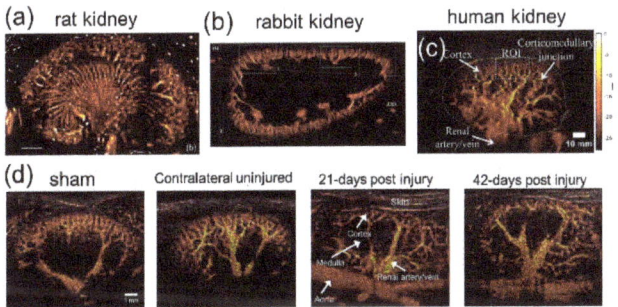

Figure 7. Representative SRU images of microvasculature from rat kidney (Reprinted with permission from [78], Copyright 2017 Springer Nature) (**a**); rabbit kidney using Kalman filter-based method (Reprinted with permission from Ref. [79]. Copyright 2020 IEEE) (**b**); healthy human kidney (**c**); mouse kidney in group of sham, contralateral uninjured, 21-days after ischemia–reperfusion injury, and 42-days post injury (Reprinted with permission from Ref. [45]. Copyright 2020 Elsevier) (**d**).

On the way towards the clinical application of SRU imaging on kidneys, more affirmative data from studies on clinically relevant animal models are required. Yang et al. performed SRU imaging on an acute ischemic–reperfused rat kidney and a normal rat kidney to investigate the in vivo feasibility of evaluating microvascular changes during progressive kidney disease. The results showed that the blood flow speed in the injured rat kidney (<10 mm/s) was much lower than that in the healthy kidney (~30 mm/s) [80]. Similar results was achieved by Andersen et al., suggesting that blood flow in the renal microvasculature was measured to be slower after ischemia and reperfusion by SRU imaging [81]. Studies that demonstrated the feasibility of SRU for identifying microvascular alterations during the disease progression, with a larger group of animals and histological verifications, were performed by Chen et al. [45]. In the study, SRU imaging was performed in vivo on mouse kidneys of four different groups (n = 5), including control, kidneys post 21 days of ischemia–reperfusion injury, and kidneys post 42 days of injury (Figure 7d), followed by the histological analysis with a CD31 stain. The results showed that SRU imaging was able to identify renal microvessels as small as 32 μm (<1/3 λ at the frequency of 15 MHz) in vivo and allow for quantification of the changes in kidney morphology and vasculature, including size, rBV, vessel density, and tortuosity, during the progression of the kidney injury. Changes in renal vascular density in the corticomedullary area were validated by a CD31 stain, and a relatively strong correlation was found between SRU and histological measurement. While the former two studies focused more on the change in flow velocity, the latter only examined the features derived from structural information. Future studies that investigate both structural and flow information on the groups of animals in disease models is highly sought.

The next step towards the future translation of SRU imaging on kidneys would be studies on human subjects, including healthy and CKD patients. Studies have been initiated and some preliminary results have already been presented at conferences (Figure 7c) [82]. Motion artifact is one of the major issues to address before successful future translations of kidney SRU imaging, since the breathing motion affects the locations of the organs in the abdomen significantly. In the animal studies, most of the groups applied block matching algorithms on the envelope of B-mode data to estimate and correct the translational breathing motion in lateral and axial directions [37,45,79–81]. Foiret et al. utilized linear optimization to better correct both translational and rotational motion [78]. However, out-of-plane motion remains a problem for 2D kidney imaging by using the 1D array transducer. Moreover, breathing motion during human kidney imaging can be more critical compared to experiments on anesthetized animals. A short scan time might offer a practical solution so that an SRU imaging session can be completed with minimized motion artifacts while a human subject holds their breath. In addition, further improvements that enhance MB signals in depth, in clinical abdominal imaging conditions and extended experiments on a larger group of human subjects, are required for future clinical translations of SRU imaging.

4.3. Other Applications

SRU was also applied to other organs or animal models, such as the brain, the femoral artery with atherosclerotic plaque (AP), etc.

The pathological process of the small vessels in the brain has been recognized as a contributor to cognitive impairment and dementia [83–85]. Therefore, an imaging tool that can resolve small vessels in the brain would be beneficial for the diagnostics and therapeutics of such neurological diseases. Errico et al. initiated a study of brain microvasculature imaging using an SRU technique in 2015 (Figure 1c) [29]. Rat brain microvasculature was imaged with a 15 MHz ultrasound probe through the thinned skull. Vessels as small as 9 µm ($1/10\ \lambda$) were resolved and the in-plane blood flow profile was achieved, although the scan time was quite long (150 s). Recently, Huang et al. proposed a method that separates spatially overlapping MB events into subpopulations based on spatiotemporal differences in flow dynamics, and successfully visualized chicken embryo brain vasculature with a shortened scan time (~17s) (Figure 8a) [86]. Compared to the kidney imaging, a relatively long scan time could be acceptable for brain imaging, for which the physiologic motions are less pronounced if the ultrasound transducer is fixed to the head. For the future clinical translation of SRU in brain imaging, attenuation and aberration from the skull that significantly degrades imaging performance remains a big challenge which requires further investigation.

Another potential application of SRU is to monitor the development of atherosclerotic plaque (AP) and predict AP rupture by imaging vasa vasorum (VV) near major vessels. It has been reported that abnormal proliferation of VV and the infiltration into the AP core is key evidence of AP progression and vulnerability [10,87–89]. Due to the tiny size of VV, it is challenging to imaging VV in vivo. In a pilot study, Yu et al. successfully identified VV in a rabbit AP in vivo, with the spatial resolution of 45 µm, by the deconvolution-based SRU imaging technique (Figure 8b) [36]. In the follow-up study by the group, the abnormal proliferation of VV near the rabbit femoral artery that was identified by SRU was further validated with subsequent histology and ex vivo microcomputed tomography (µCT), histopathology, and morphology [90]. The experiment protocols and results would encourage extended preclinical studies with larger group of animals and potential human studies in the future.

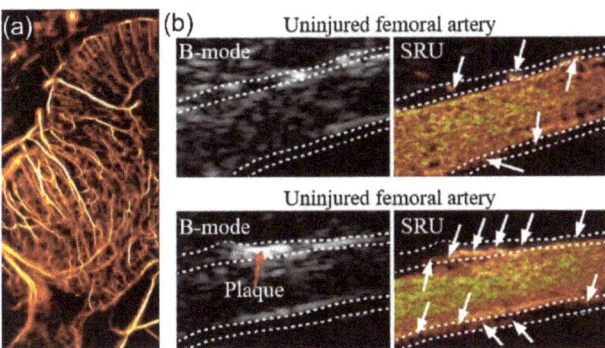

Figure 8. Representative SRU images of microvasculature in chicken embryo brain (Reprinted with permission from Ref. [86]. Copyright 2020 Springer Nature) (**a**), and VV in uninjured (upper panel) and injured (lower panel) rabbit femoral arteries (Reprinted with permission from Ref. [90]. Copyright 2020 IEEE) (**b**).

Besides the in vivo studies mentioned above, promising results of microvascular imaging by SRU were also achieved in vivo in organs, including rabbit lymph nodes [91], rabbit eyeballs [92], mouse liver [93], and human tibialis anterior muscles [94], which may contribute to broader applications of SRU imaging.

5. Limitations and Future Directions

As we discussed above, SRU imaging has been continuously improved and applied in vivo for different applications in animal models and very limited human uses. However, there are still several major limitations that hinder the eventual clinical translations of this novel technology.

One of the major limitations is the relatively long scan time. To reconstruct a single SRU image, the accumulation of a large-scale microbubble backscatter of signals is required, which results in a long acquisition time. A typical SRU algorithm that localizes the center of the spatially isolated MBs requires a scan time of several minutes to collect the necessary data [29]. Since SRU imaging is highly sensitive to motion artifacts, the image quality will be notably degraded due to physiological or externally induced motion, which is inevitable during a long scan time. Considering the freehand scan that would be practiced in future clinical applications, the long scan time would be a big barrier. Harput et al. [94] suggested a two-stage motion correction algorithm which calculates the combination of affine and nonrigid image registrations through the motion estimation from the B-mode image, and corrects the motion artifacts of CEU image. The rigid motion artifact correction using a phase-correlation technique was also suggested to remove the blurring caused by subwavelength motions [95]. Although motion correction can help mitigate the motion artifacts [94,95], it still cannot perfectly solve the problem of motion in real cases, which combines rigid motion, nonrigid motion, and out-of-plane motions. Some algorithms that can shorten the scan time have been developed [36,86,92,96], but the spatial resolution is more or less compromised compared to the algorithm that localizes the isolated MBs. The large-scale data will also require a high computation cost for image reconstruction. Utilization of a GPU for data processing could be a solution to reduce the computation load and to possibly realize SRU image processing in real-time in the future. In some studies of SRU imaging using a matrix array, a GPU has already been successfully applied to significantly shorten the computation time for large-scale data set [97–99].

Another limitation is that image quality is highly dependent on MB concentration and distribution in the blood vessels [77,100]. A low concentration in the vessels will lead to a long scan time and low signal contrast, while a high concentration will degrade localization accuracy and, thus, the spatial resolution of the image. Since MBs are systematically administrated to the blood vessels, it would be difficult to control the real MB concentration in the target area in the clinical practice. Moreover, the dosage may need to be adjusted for different applications in order to have appropriate MB concentration in vessels of the target organs. Deep learning approaches would be a promising direction to overcome this challenge while further evaluations are needed. Moreover, in order to make the technology for broader application in clinics, with no possible safety concerns for some populations, including pregnant women or subjects potentially allergic to microbubbles etc., it is ideal to develop a contrast-agent-free super-resolution. However, no fully developed approaches with ideal performance have been reported so far, except some pioneering initiative works presented in recent conferences [101].

Currently, most in vivo clinical studies of super-resolution ultrasound imaging techniques are conducted on a 2D B-scan with a 1-D array ultrasound transducer. However, 2D cross sectional imaging has limited resolution in the elevational direction, determined by the relatively large elevational beamwidth. The 3D SRU imaging techniques using 2D array arranged 1024 elements in 32×32 matrix showed a promising in vitro study result, with the subwavelength resolution in both lateral and elevational directions [97]. The use of a fully sampled 2D matrix array, however, is suffering from the heavy computational complexities of handling huge volumetric data. Several approaches, including (1) an FPGA–GPU structure-based ultrasound system [102], (2) frequency domain beamforming [103], and (3) novel transducer configurations such as the sparse array [98] and the row-column array [99], are suggested to overcome such limitations. An FPGA–GPU structure-based ultrasound system could manage the huge data size because it benefits from both the FPGA, for high-speed data transfer, and the GPU for processing [102]. Another study performs the beamforming of 3D volumetric imaging in the frequency domain to reduce computational complexity [103]. In addition to these approaches, the sparse array [98]-based volumetric super-resolution ultrasound imaging technique utilizes half of the channels compared to the fully sampled array and achieves a comparable resolution [104]. Further research on sparse array-based super-resolution ultrasound imaging should continue to deal with the grating lobe [99] and the fastidious optimization process [104] of the sparse array [105]. The other approach, the row–column array-based super-resolution imaging method [99], reduces the number of connections from N^2 to 2N in the $N \times N$ 2D array by utilizing two orthogonal arrays [104]. Although the resolution of the row–column array method is comparable, edge artifacts caused by the long element should be suppressed with mechanical apodization [106].

Besides, for abdominal, transcranial, and brain applications in humans, the tradeoff between spatial resolution and penetration depth needs to be considered; the imaging depth will be significantly larger in human subjects. It will be more challenging to detect microbubbles in the microvessels at deep depth due to acoustic attenuation. Techniques that can enhance the transmitted energy without destroying MBs, such as coded excitation [107–110], may be one potential solution. Utilization of microbubbles with larger sizes may also help enhance signals, as reported in the previous study [66].

6. Conclusions

In the past decade, SRU imaging technology that can achieve microvascular images with spatial resolution beyond the acoustic diffraction limit has been developed and continuously improved. With the help of unprecedented spatial resolution and reasonable temporal resolution, this technology could significantly enhance the diagnosis and monitoring of the diseases that are associated with abnormal changes of microvasculature. A number of preclinical studies have already demonstrated the feasibility in vivo on different models, including tumors, kidneys, brain imaging, etc. Overall, while some challenges exist for future clinical translation, SRU imaging technology still holds a great potential for broad clinical applications with a high impact.

Author Contributions: Q.C. conceived the structure of this manuscript, drafted, and edited the manuscript. H.S. drafted and edited the manuscript. J.Y. drafted, reviewed, edited, and supervised the writing of the manuscript. K.K. reviewed, edited, and supervised the writing of the manuscript. All authors have read and agreed to the published version of the manuscript.

Funding: This work was supported by the National Research Foundation of Korea (NRF) grant funded by the Korea government (MSIT) (No. 2020R1C1C1009488, NRF-2018R1A5A1025511) and by the Korea Medical Device Development Fund grant funded by the Korea government (the Ministry of Science and ICT, the Ministry of Trade, Industry and Energy, the Ministry of Health & Welfare, Republic of Korea, the Ministry of Food and Drug Safety) (Project Number: 202011C01).

Institutional Review Board Statement: Not applicable.

Informed Consent Statement: Not applicable.

Data Availability Statement: Not applicable.

Conflicts of Interest: The authors declare no conflict of interest.

References

1. Kerbel, R.S. Tumor angiogenesis. *N. Engl. J. Med.* **2008**, *358*, 2039–2049. [CrossRef]
2. Hanahan, D.; Weinberg, R.A. The hallmarks of cancer. *Cell* **2000**, *100*, 57–70. [CrossRef]
3. Bullitt, E.; Lin, N.U.; Ewend, M.G.; Zeng, D.; Winer, E.P.; Carey, L.A.; Smith, J.K. Tumor therapeutic response and vessel tortuosity: Preliminary report in metastatic breast cancer. In *Lecture Notes in Computer Science*; Springer: Berlin/Heidelberg, Germany, 2006; Volume 4191, pp. 561–568.
4. Basile, D.P.; Bonventre, J.V.; Mehta, R.; Nangaku, M.; Unwin, R.; Rosner, M.H.; Kellum, J.A.; Ronco, C.; ADQI XIII Work Group. Progression after AKI: Understanding Maladaptive Repair Processes to Predict and Identify Therapeutic Treatments. *J. Am. Soc. Nephrol.* **2016**, *27*, 687–697. [CrossRef] [PubMed]
5. Hörbelt, M.; Lee, S.-Y.; Mang, H.E.; Knipe, N.L.; Sado, Y.; Kribben, A.; Sutton, T.A. Acute and chronic microvascular alterations in a mouse model of ischemic acute kidney injury. *Am. J. Physiol. Physiol.* **2007**, *293*, F688–F695. [CrossRef] [PubMed]
6. Molitoris, B.A. Therapeutic translation in acute kidney injury: The epithelial/endothelial axis. *J. Clin. Investig.* **2014**, *124*, 2355–2363. [CrossRef] [PubMed]
7. Kramann, R.; Tanaka, M.; Humphreys, B.D. Fluorescence Microangiography for Quantitative Assessment of Peritubular Capillary Changes after AKI in Mice. *J. Am. Soc. Nephrol.* **2014**, *25*, 1924–1931. [CrossRef]
8. Ritman, E.L.; Lerman, A. The Dynamic Vasa Vasorum. *Cardiovasc. Res.* **2007**, *75*, 649–658. [CrossRef]
9. Xu, J.; Lu, X.; Shi, G.-P. Vasa vasorum in atherosclerosis and clinical significance. *Int. J. Mol. Sci.* **2015**, *16*, 11574–11608. [CrossRef] [PubMed]
10. Moreno, P.R.; Purushothaman, K.R.; Fuster, V.; Echeverri, D.; Truszczynska, H.; Sharma, S.K.; Badimon, J.J.; O'Connor, W.N. Plaque Neovascularization Is Increased in Ruptured Atherosclerotic Lesions of Human Aorta. *Circulation* **2004**, *110*, 2032–2038. [CrossRef]
11. Hyafil, F.; Cornily, J.-C.; Feig, J.E.; Gordon, R.; Vucic, E.; Amirbekian, V.; Fisher, E.A.; Fuster, V.; Feldman, L.J.; Fayad, Z.A. Noninvasive detection of macrophages using a nanoparticulate contrast agent for computed tomography. *Nat. Med.* **2007**, *13*, 636–641. [CrossRef]
12. Sadeghi, M.M.; Glover, D.K.; Lanza, G.M.; Fayad, Z.A.; Johnson, L.L. Imaging Atherosclerosis and Vulnerable Plaque. *J. Nucl. Med.* **2010**, *51*, 51S–65S. [CrossRef]
13. Kubo, T.; Imanishi, T.; Takarada, S.; Kuroi, A.; Ueno, S.; Yamano, T.; Tanimoto, T.; Matsuo, Y.; Masho, T.; Kitabata, H.; et al. Assessment of Culprit Lesion Morphology in Acute Myocardial Infarction. *J. Am. Coll. Cardiol.* **2007**, *50*, 933–939. [CrossRef]
14. Jang, I.-K.; Tearney, G.J.; MacNeill, B.; Takano, M.; Moselewski, F.; Iftima, N.; Shishkov, M.; Houser, S.; Aretz, H.T.; Halpern, E.F.; et al. In Vivo Characterization of Coronary Atherosclerotic Plaque by Use of Optical Coherence Tomography. *Circulation* **2005**, *111*, 1551–1555. [CrossRef]
15. Winter, P.M.; Morawski, A.M.; Caruthers, S.D.; Fuhrhop, R.W.; Zhang, H.; Williams, T.A.; Allen, J.S.; Lacy, E.K.; Robertson, J.D.; Lanza, G.M.; et al. Molecular Imaging of Angiogenesis in Early-Stage Atherosclerosis with $\alpha_v\beta_3$-Integrin–Targeted Nanoparticles. *Circulation* **2003**, *108*, 2270–2274. [CrossRef] [PubMed]
16. Cai, K.; Caruthers, S.D.; Huang, W.; Williams, T.A.; Zhang, H.; Wickline, S.A.; Lanza, G.M.; Winter, P.M. MR molecular imaging of aortic angiogenesis. *JACC. Cardiovasc. Imaging* **2010**, *3*, 824–832. [CrossRef] [PubMed]
17. Kerwin, W.; Hooker, A.; Spilker, M.; Vicini, P.; Ferguson, M.; Hatsukami, T.; Yuan, C. Quantitative Magnetic Resonance Imaging Analysis of Neovasculature Volume in Carotid Atherosclerotic Plaque. *Circulation* **2003**, *107*, 851–856. [CrossRef] [PubMed]
18. Prowle, J.R.; Molan, M.P.; Hornsey, E.; Bellomo, R. Measurement of renal blood flow by phase-contrast magnetic resonance imaging during septic acute kidney injury. *Crit. Care Med.* **2012**, *40*, 1768–1776. [CrossRef] [PubMed]
19. Inoue, T.; Kozawa, E.; Okada, H.; Inukai, K.; Watanabe, S.; Kikuta, T.; Watanabe, Y.; Takenaka, T.; Katayama, S.; Tanaka, J.; et al. Noninvasive evaluation of kidney hypoxia and fibrosis using magnetic resonance imaging. *J. Am. Soc. Nephrol.* **2011**, *22*, 1429–1434. [CrossRef] [PubMed]

20. Faubel, S.; Patel, N.U.; Lockhart, M.E.; Cadnapaphornchai, M.A. Renal relevant radiology: Use of ultrasonography in patients with AKI. *Clin. J. Am. Soc. Nephrol.* **2014**, *9*, 382–394. [CrossRef]
21. Cao, W.; Cui, S.; Yang, L.; Wu, C.; Liu, J.; Yang, F.; Liu, Y.; Bin, J.; Hou, F.F. Contrast-Enhanced Ultrasound for Assessing Renal Perfusion Impairment and Predicting Acute Kidney Injury to Chronic Kidney Disease Progression. *Antioxid. Redox Signal.* **2017**, *27*, 1397–1411. [CrossRef]
22. Hull, T.D.; Agarwal, A.; Hoyt, K. New Ultrasound Techniques Promise Further Advances in AKI and CKD. *J. Am. Soc. Nephrol.* **2017**, *28*, 3452–3460. [CrossRef] [PubMed]
23. Staub, D.; Schinkel, A.F.L.; Coll, B.; Coli, S.; van der Steen, A.F.W.; Reed, J.D.; Krueger, C.; Thomenius, K.E.; Adam, D.; Sijbrands, E.J.; et al. Contrast-enhanced ultrasound imaging of the vasa vasorum: From early atherosclerosis to the identification of unstable plaques. *JACC. Cardiovasc. Imaging* **2010**, *3*, 761–771. [CrossRef] [PubMed]
24. Moguillansky, D.; Leng, X.; Carson, A.; Lavery, L.; Schwartz, A.; Chen, X.; Villanueva, F.S. Quantification of plaque neovascularization using contrast ultrasound: A histologic validation. *Eur. Heart J.* **2011**, *32*, 646–653. [CrossRef]
25. Magnoni, M.; Coli, S.; Marrocco-Trischitta, M.M.; Melisurgo, G.; De Dominicis, D.; Cianflone, D.; Chiesa, R.; Feinstein, S.B.; Maseri, A. Contrast-enhanced ultrasound imaging of periadventitial vasa vasorum in human carotid arteries. *Eur. J. Echocardiogr.* **2008**, *10*, 260–264. [CrossRef]
26. Hess, S.T.; Girirajan, T.P.K.; Mason, M.D. Ultra-high resolution imaging by fluorescence photoactivation localization microscopy. *Biophys. J.* **2006**, *91*, 4258–4272. [CrossRef]
27. Betzig, E.; Patterson, G.H.; Sougrat, R.; Lindwasser, O.W.; Olenych, S.; Bonifacino, J.S.; Davidson, M.W.; Lippincott-Schwartz, J.; Hess, H.F. Imaging intracellular fluorescent proteins at nanometer resolution. *Science* **2006**, *313*, 1642–1645. [CrossRef] [PubMed]
28. Rust, M.J.; Bates, M.; Zhuang, X. Sub-diffraction-limit imaging by stochastic optical reconstruction microscopy (STORM). *Nat. Methods* **2006**, *3*, 793–796. [CrossRef]
29. Errico, C.; Pierre, J.; Pezet, S.; Desailly, Y.; Lenkei, Z.; Couture, O.; Tanter, M. Ultrafast ultrasound localization microscopy for deep super-resolution vascular imaging. *Nature* **2015**, *527*, 499–502. [CrossRef]
30. O'Reilly, M.A.; Hynynen, K. A super-resolution ultrasound method for brain vascular mapping. *Med. Phys.* **2013**, *40*. [CrossRef]
31. Viessmann, O.M.; Eckersley, R.J.; Christensen-Jeffries, K.; Tang, M.X.; Dunsby, C. Acoustic super-resolution with ultrasound and microbubbles. *Phys. Med. Biol.* **2013**, *58*, 6447–6458. [CrossRef]
32. Desailly, Y.; Couture, O.; Fink, M.; Tanter, M. Sono-activated ultrasound localization microscopy. *Appl. Phys. Lett.* **2013**, *103*, 174107. [CrossRef]
33. Christensen-Jeffries, K.; Browning, R.J.; Tang, M.-X.; Dunsby, C.; Eckersley, R.J. In vivo acoustic super-resolution and super-resolved velocity mapping using microbubbles. *IEEE Trans. Med. Imaging* **2015**, *34*, 433–440. [CrossRef] [PubMed]
34. Tanter, M.; Fink, M. Ultrafast imaging in biomedical ultrasound. *IEEE Trans. Ultrason. Ferroelectr. Freq. Control* **2014**, *61*, 102–119. [CrossRef] [PubMed]
35. Demené, C.; Deffieux, T.; Pernot, M.; Osmanski, B.F.; Biran, V.; Gennisson, J.L.; Sieu, L.A.; Bergel, A.; Franqui, S.; Correas, J.M.; et al. Spatiotemporal Clutter Filtering of Ultrafast Ultrasound Data Highly Increases Doppler and fUltrasound Sensitivity. *IEEE Trans. Med. Imaging* **2015**, *34*, 2271–2285. [CrossRef]
36. Yu, J.; Lavery, L.; Kim, K. Super-resolution ultrasound imaging method for microvasculature in vivo with a high temporal accuracy. *Sci. Rep.* **2018**, *8*, 13918. [CrossRef]
37. Song, P.; Trzasko, J.D.; Manduca, A.; Huang, R.; Kadirvel, R.; Kallmes, D.F.; Chen, S. Improved Super-Resolution Ultrasound Microvessel Imaging with Spatiotemporal Nonlocal Means Filtering and Bipartite Graph-Based Microbubble Tracking. *IEEE Trans. Ultrason. Ferroelectr. Freq. Control* **2018**, *65*, 149–167. [CrossRef] [PubMed]
38. Christensen-Jeffries, K.; Couture, O.; Dayton, P.A.; Eldar, Y.C.; Hynynen, K.; Kiessling, F.; O'Reilly, M.; Pinton, G.F.; Schmitz, G.; Tang, M.X.; et al. Super-resolution Ultrasound Imaging. *Ultrasound Med. Biol.* **2020**, *46*, 865–891. [CrossRef]
39. Van Sloun, R.J.G.; Solomon, O.; Bruce, M.; Khaing, Z.Z.; Eldar, Y.C.; Mischi, M. Deep Learning for Super-resolution Vascular Ultrasound Imaging. In Proceedings of the International Conference on Acoustics, Speech and Signal Processing, Brighton, UK, 12–17 May 2019; pp. 1055–1059.
40. Liu, X.; Zhou, T.; Lu, M.; Yang, Y.; He, Q.; Luo, J. Deep Learning for Ultrasound Localization Microscopy. *IEEE Trans. Med. Imaging* **2020**, *39*, 3064–3078. [CrossRef]
41. Shi, W.; Caballero, J.; Huszár, F.; Totz, J.; Aitken, A.P.; Bishop, R.; Rueckert, D.; Wang, Z. *Real-Time Single Image and Video Super-Resolution Using an Efficient Sub-Pixel Convolutional Neural Network*; IEEE Computer Society: Washington, DC, USA; pp. 1874–1883.
42. He, K.; Zhang, X.; Ren, S.; Sun, J. Deep residual learning for image recognition. In Proceedings of the IEEE Conference on Computer Vision and Pattern Recognition, Las Vegas, NV, USA, 26 June–1 July 2016; IEEE Computer Society: Washington, DC, USA, 2016; pp. 770–778.
43. Bar-zion, A.; Tremblay-darveau, C.; Solomon, O.; Adam, D.; Eldar, Y.C. Fast VascularUltrasound Imaging with Enhanced Spatial Resolution and Background Rejection. *IEEE Trans. Med. Imaging* **2017**, *36*, 169–180. [CrossRef]
44. Cox, B.; Beard, P. Super-resolution ultrasound. *Nature* **2015**, *527*, 451–452. [CrossRef] [PubMed]
45. Chen, Q.; Yu, J.; Rush, B.M.; Stocker, S.D.; Tan, R.J.; Kim, K. Ultrasound super-resolution imaging provides a noninvasive assessment of renal microvasculature changes during mouse acute kidney injury. *Kidney Int.* **2020**, *98*, 355–365. [CrossRef] [PubMed]

46. Ionescu, C. The benefits of 3D-4D fetal echocardiography. *Maedica* **2010**, *5*, 45–50.
47. Nehme, E.; Weiss, L.E.; Michaeli, T.; Shechtman, Y. Deep-STORM: Super-resolution single-molecule microscopy by deep learning. *Optica* **2018**, *5*, 458–464. [CrossRef]
48. Van Sloun, R.J.G.; Solomon, O.; Bruce, M.; Khaing, Z.Z.; Wijkstra, H.; Eldar, Y.C.; Mischi, M. Super-resolution Ultrasound Localization Microscopy through Deep Learning. *IEEE Trans. Med Imaging* **2021**. [CrossRef]
49. Ronneberger, O.; Fischer, P.; Brox, T. U-net: Convolutional networks for biomedical image segmentation. In *Lecture Notes in Computer Science*; Springer: Berlin/Heidelberg, Germany, 2015; Volume 9351, pp. 234–241.
50. Youn, J.; Ommen, M.L.; Stuart, M.B.; Thomsen, E.V.; Larsen, N.B.; Jensen, J.A. Detection and Localization of Ultrasound Scatterers Using Convolutional Neural Networks. *IEEE Trans. Med. Imaging* **2020**, *39*, 3855–3867. [CrossRef] [PubMed]
51. Brown, K.G.; Ghosh, D.; Hoyt, K. Deep Learning of Spatiotemporal Filtering for Fast Super-Resolution Ultrasound Imaging. *IEEE Trans. Ultrason. Ferroelectr. Freq. Control* **2020**, *67*, 1820–1829. [CrossRef]
52. Cancer. Available online: http://www.who.int/en/news-room/fact-sheets/detail/cancer (accessed on 30 September 2020).
53. WHO | Early Detection of Cancer. Available online: https://www.who.int/cancer/detection/en/ (accessed on 30 September 2020).
54. Folkman, J. Tumor angiogenesis. *Adv. Cancer Res.* **1974**, *19*, 331–358. [CrossRef]
55. Augustin, H.G. Commentary on folkman: How is blood vessel growth regulated in normal and neoplastic tissue? *Cancer Res.* **2016**, *76*, 2854–2856. [CrossRef]
56. Ruoslahti, E. Specialization of tumour vasculature. *Nat. Rev. Cancer* **2002**, *2*, 83–90. [CrossRef]
57. Ehling, J.; Theek, B.; Gremse, F.; Baetke, S.; Möckel, D.; Maynard, J.; Ricketts, S.A.; Grüll, H.; Neeman, M.; Knuechel, R.; et al. Micro-CT imaging of tumor angiogenesis: Quantitative measures describing micromorphology and vascularization. *Am. J. Pathol.* **2014**, *184*, 431–441. [CrossRef] [PubMed]
58. Chang, M.P.; Jin, M.G.; Hyun, J.L.; Kim, M.A.; Kim, H.C.; Kwang, G.K.; Chang, H.L.; Im, J.G. FN13762 murine breast cancer: Region-by-region correlation of first-pass perfusion CT indexes with histologic vascular parameters. *Radiology* **2009**, *251*, 721–730. [CrossRef]
59. Gessner, R.C.; Frederick, C.B.; Foster, F.S.; Dayton, P.A. Acoustic angiography: A new imaging modality for assessing microvasculature architecture. *Int. J. Biomed. Imaging* **2013**, *2013*. [CrossRef]
60. Gessner, R.C.; Aylward, S.R.; Dayton, P.A. Mapping microvasculature with acoustic angiography yields quantifiable differences between healthy and tumor-bearing tissue volumes in a rodent model. *Radiology* **2012**, *264*, 733–740. [CrossRef]
61. Shelton, S.E.; Lee, Y.Z.; Lee, M.; Cherin, E.; Foster, F.S.; Aylward, S.R.; Dayton, P.A. Quantification of microvascular tortuosity during tumor evolution using acoustic angiography. *Ultrasound Med. Biol.* **2015**, *41*, 1896–1904. [CrossRef]
62. Shelton, S.E.; Lindsey, B.D.; Tsuruta, J.K.; Foster, F.S.; Dayton, P.A. Molecular Acoustic Angiography: A New Technique for High-resolution Superharmonic Ultrasound Molecular Imaging. *Ultrasound Med. Biol.* **2016**, *42*, 769–781. [CrossRef] [PubMed]
63. Rao, S.R.; Shelton, S.E.; Dayton, P.A. The "Fingerprint" of Cancer Extends Beyond Solid Tumor Boundaries: Assessment with a Novel Ultrasound Imaging Approach. *IEEE Trans. Biomed. Eng.* **2016**, *63*, 1082–1086. [CrossRef] [PubMed]
64. Lindsey, B.D.; Shelton, S.E.; Foster, F.S.; Dayton, P.A. Assessment of Molecular Acoustic Angiography for Combined Microvascular and Molecular Imaging in Preclinical Tumor Models. *Mol. Imaging Biol.* **2017**, *19*, 194–202. [CrossRef] [PubMed]
65. Lin, F.; Rojas, J.D.; Dayton, P.A. Super resolution contrast ultrasound imaging: Analysis of imaging resolution and application to imaging tumor angiogenesis. In Proceedings of the IEEE International Ultrasonics Symposium (IUS), Tours, France, 18–21 September 2016.
66. Lin, F.; Tsuruta, J.K.; Rojas, J.D.; Dayton, P.A. Optimizing Sensitivity of Ultrasound Contrast-Enhanced Super-Resolution Imaging by Tailoring Size Distribution of Microbubble Contrast Agent. *Ultrasound Med. Biol.* **2017**, *43*, 2488–2493. [CrossRef]
67. Lin, F.; Shelton, S.E.; Espíndola, D.; Rojas, J.D.; Pinton, G.; Dayton, P.A. 3-D ultrasound localization microscopy for identifying microvascular morphology features of tumor angiogenesis at a resolution beyond the diffraction limit of conventional ultrasound. *Theranostics* **2017**, *7*, 196–204. [CrossRef]
68. Opacic, T.; Dencks, S.; Theek, B.; Piepenbrock, M.; Ackermann, D.; Rix, A.; Lammers, T.; Stickeler, E.; Delorme, S.; Schmitz, G.; et al. Motion model ultrasound localization microscopy for preclinical and clinical multiparametric tumor characterization. *Nat. Commun.* **2018**, *9*, 1527. [CrossRef]
69. Chacón, R.D.; Costanzo, M.V. Triple-negative breast cancer. *Breast Cancer Res.* **2010**, *12*, S3. [CrossRef] [PubMed]
70. Ghosh, D.; Xiong, F.; Sirsi, S.R.; Mattrey, R.; Brekken, R.; Kim, J.W.; Hoyt, K. Monitoring early tumor response to vascular targeted therapy using super-resolution ultrasound imaging. In Proceedings of the IEEE International Ultrasonics Symposium (IUS), Washington, DC, USA, 6–9 September 2017; pp. 1–4. [CrossRef]
71. Dencks, S.; Piepenbrock, M.; Opacic, T.; Krauspe, B.; Stickeler, E.; Kiessling, F.; Schmitz, G. Clinical Pilot Application of Super-Resolution US Imaging in Breast Cancer. *IEEE Trans. Ultrason. Ferroelectr. Freq. Control* **2019**, *66*, 517–526. [CrossRef] [PubMed]
72. National Institute of Diabetes and Digestive and Kidney Diseases. What Is Chronic Kidney Disease? | NIDDK. Available online: https://www.niddk.nih.gov/health-information/kidney-disease/chronic-kidney-disease-ckd/what-is-chronic-kidney-disease (accessed on 12 October 2020).
73. Division of Diabetes Translation; National Center for Chronic Disease Prevention and Health Promotion; Centers for Disease Control and Prevention. *National Chronic Kidney Disease Fact Sheet 2017*; Centers for Disease Control and Prevention: Atlanta, GA, USA, 2017.

74. Heung, M.; Chawla, L.S. Acute Kidney Injury: Gateway to Chronic Kidney Disease. *Nephron Clin. Pract.* **2014**, *127*, 30–34. [CrossRef] [PubMed]
75. Basile, D.P.; Donohoe, D.; Roethe, K.; Osborn, J.L. Renal ischemic injury results in permanent damage to peritubular capillaries and influences long-term function. *Am. J. Physiol. Physiol.* **2001**, *281*, F887–F899. [CrossRef]
76. Tsuruoka, K.; Yasuda, T.; Koitabashi, K.; Yazawa, M.; Shimazaki, M.; Sakurada, T.; Shirai, S.; Shibagaki, Y.; Kimura, K.; Tsujimoto, F. Evaluation of renal microcirculation by contrast-enhanced ultrasound with sonazoid^{TM} as a contrast agent: Comparison between normal subjects and patients with chronic kidney disease. *Int. Heart J.* **2010**, *51*, 176–182. [CrossRef]
77. Couture, O.; Hingot, V.; Heiles, B.; Muleki-Seya, P.; Tanter, M. Ultrasound Localization Microscopy and Super-Resolution: A State of the Art. *IEEE Trans. Ultrason. Ferroelectr. Freq. Control* **2018**, *65*, 1304–1320. [CrossRef]
78. Foiret, J.; Zhang, H.; Ilovitsh, T.; Mahakian, L.; Tam, S.; Ferrara, K.W. Ultrasound localization microscopy to image and assess microvasculature in a rat kidney. *Sci. Rep.* **2017**, *7*, 13662. [CrossRef]
79. Tang, S.; Song, P.; Trzasko, J.D.; Lowerison, M.; Huang, C.; Gong, P.; Lok, U.W.; Manduca, A.; Chen, S. Kalman Filter-Based Microbubble Tracking for Robust Super-Resolution Ultrasound Microvessel Imaging. *IEEE Trans. Ultrason. Ferroelectr. Freq. Control* **2020**, *67*, 1738–1751. [CrossRef]
80. Yang, Y.; He, Q.; Zhang, H.; Qiu, L.; Qian, L.; Lee, F.-F.; Liu, Z.; Luo, J. Assessment of Diabetic Kidney Disease Using Ultrasound Localization Microscopy: An In Vivo Feasibility Study in Rats. In Proceedings of the 2018 IEEE International Ultrasonics Symposium, Kobe, Japan, 22–25 October 2018; pp. 1–4. [CrossRef]
81. Andersen, S.B.; Hoyos, C.A.V.; Taghavi, I.; Gran, F.; Hansen, K.L.; Sorensen, C.M.; Jensen, J.A.; Nielsen, M.B. Super-Resolution Ultrasound Imaging of Rat Kidneys before and after Ischemia-Reperfusion. In Proceedings of the IEEE International Ultrasonics Symposium (IUS), Glasgow, UK, 6–9 October 2019.
82. Chen, Q.; Kumar, A.; Tan, R.J.; Kim, K. Ultrasound Super-Resolution Imaging Algorithm for a Curved Array Transducer for Human Kidney Imaging. In Proceedings of the 2019 IEEE International Ultrasonics Symposium (IUS), Glasgow, UK, 6–9 October 2019; p. 1.
83. Snyder, H.M.; Corriveau, R.A.; Craft, S.; Faber, J.E.; Greenberg, S.M.; Knopman, D.; Lamb, B.T.; Montine, T.J.; Nedergaard, M.; Schaffer, C.B.; et al. Vascular contributions to cognitive impairment and dementia including Alzheimer's disease. *Alzheimer's Dement.* **2015**, *11*, 710–717. [CrossRef] [PubMed]
84. Pantoni, L. Cerebral small vessel disease: From pathogenesis and clinical characteristics to therapeutic challenges. *Lancet Neurol.* **2010**, *9*, 689–701. [CrossRef]
85. Wardlaw, J.M.; Smith, E.E.; Biessels, G.J.; Cordonnier, C.; Fazekas, F.; Frayne, R.; Lindley, R.I.; O'Brien, J.T.; Barkhof, F.; Benavente, O.R.; et al. Neuroimaging standards for research into small vessel disease and its contribution to ageing and neurodegeneration. *Lancet Neurol.* **2013**, *12*, 822–838. [CrossRef]
86. Huang, C.; Lowerison, M.R.; Trzasko, J.D.; Manduca, A.; Bresler, Y.; Tang, S.; Gong, P.; Lok, U.W.; Song, P.; Chen, S. Short Acquisition Time Super-Resolution Ultrasound Microvessel Imaging via Microbubble Separation. *Sci. Rep.* **2020**, *10*, 6007. [CrossRef]
87. Kolodgie, F.D.; Gold, H.K.; Burke, A.P.; Fowler, D.R.; Kruth, H.S.; Weber, D.K.; Farb, A.; Guerrero, L.J.; Hayase, M.; Kutys, R.; et al. Intraplaque hemorrhage and progression of coronary atheroma. *N. Engl. J. Med.* **2003**, *349*, 2316–2325. [CrossRef] [PubMed]
88. Virmani, R.; Kolodgie, F.D.; Burke, A.P.; Finn, A.V.; Gold, H.K.; Tulenko, T.N.; Wrenn, S.P.; Narula, J. Atherosclerotic Plaque Progression and Vulnerability to Rupture. *Arterioscler. Thromb. Vasc. Biol.* **2005**, *25*, 2054–2061. [CrossRef] [PubMed]
89. Gössl, M.; Versari, D.; Hildebrandt, H.A.; Bajanowski, T.; Sangiorgi, G.; Erbel, R.; Ritman, E.L.; Lerman, L.O.; Lerman, A. Segmental heterogeneity of vasa vasorum neovascularization in human coronary atherosclerosis. *JACC. Cardiovasc. Imaging* **2010**, *3*, 32–40. [CrossRef]
90. Chen, Q.; Yu, J.; Lukashova, L.; Latoche, J.D.; Zhu, J.; Lavery, L.; Verdelis, K.; Anderson, C.J.; Kim, K. Validation of Ultrasound Super-Resolution Imaging of Vasa Vasorum in Rabbit Atherosclerotic Plaques. *IEEE Trans. Ultrason. Ferroelectr. Freq. Control* **2020**, *67*, 1725–1729. [CrossRef] [PubMed]
91. Zhu, J.; Rowland, E.M.; Harput, S.; Riemer, K.; Leow, C.H.; Clark, B.; Cox, K.; Lim, A.; Christensen-Jeffries, K.; Zhang, G.; et al. 3D Super-Resolution US Imaging of Rabbit Lymph Node Vasculature in Vivo by Using Microbubbles. *Radiology* **2019**, *291*, 642–650. [CrossRef]
92. Qian, X.; Kang, H.; Li, R.; Lu, G.; Du, Z.; Shung, K.K.; Humayun, M.S.; Zhou, Q. In vivo Visualization of Eye Vasculature using Super-resolution Ultrasound Microvessel Imaging. *IEEE Trans. Biomed. Eng.* **2020**. [CrossRef]
93. Hao, Y.; Wang, Q.; Yang, Y.; Liu, Z.; He, Q.; Wei, L.; Luo, J. Non-rigid Motion Correction for Ultrasound Localization Microscopy of the Liver in vivo. In Proceedings of the IEEE International Ultrasonics Symposium (IUS), Glasgow, UK, 6–9 October 2019; pp. 1–4. [CrossRef]
94. Harput, S.; Christensen-Jeffries, K.; Brown, J.; Li, Y.; Williams, K.J.; Davies, A.H.; Eckersley, R.J.; Dunsby, C.; Tang, M.-X. Two-Stage Motion Correction for Super-Resolution Ultrasound Imaging in Human Lower Limb. *IEEE Trans. Ultrason. Ferroelectr. Freq. Control* **2018**, *65*, 803–814. [CrossRef]
95. Hingot, V.; Errico, C.; Tanter, M.; Couture, O. Subwavelength motion-correction for ultrafast ultrasound localization microscopy. *Ultrasonics* **2017**, *77*, 17–21. [CrossRef]
96. Bar-Zion, A.; Solomon, O.; Tremblay-Darveau, C.; Adam, D.; Eldar, Y.C. Sushi: Sparsity-based ultrasound super-resolution hemodynamic imaging. *IEEE Trans. Ultrason. Ferroelectr. Freq. Control* **2018**, *65*, 2365–2380. [CrossRef] [PubMed]

97. Heiles, B.; Correia, M.; Hingot, V.; Pernot, M.; Provost, J.; Tanter, M.; Couture, O. Ultrafast 3D Ultrasound Localization Microscopy Using a 32 × 32 Matrix Array. *IEEE Trans. Med. Imaging* **2019**, *38*, 2005–2015. [CrossRef] [PubMed]

98. Harput, S.; Tortoli, P.; Eckersley, R.J.; Dunsby, C.; Tang, M.X.; Christensen-Jeffries, K.; Ramalli, A.; Brown, J.; Zhu, J.; Zhang, G.; et al. 3-D Super-Resolution Ultrasound Imaging with a 2-D Sparse Array. *IEEE Trans. Ultrason. Ferroelectr. Freq. Control* **2020**, *67*, 269–277. [CrossRef]

99. Jensen, J.A.; Tomov, B.G.; Ommen, M.L.; Øygard, S.H.; Schou, M.; Sams, T.; Stuart, M.B.; Beers, C.; Thomsen, E.V.; Larsen, N.B. Three-Dimensional Super-Resolution Imaging Using a Row-Column Array. *IEEE Trans. Ultrason. Ferroelectr. Freq. Control* **2020**, *67*, 538–546. [CrossRef]

100. Brown, J.; Christensen-Jeffries, K.; Harput, S.; Zhang, G.; Zhu, J.; Dunsby, C.; Tang, M.X.; Eckersley, R.J. Investigation of Microbubble Detection Methods for Super-Resolution Imaging of Microvasculature. *IEEE Trans. Ultrason. Ferroelectr. Freq. Control* **2019**, *66*, 676–691. [CrossRef]

101. Bar-Zion, A.; Solomon, O.; Maresca, D.; Shapiro, M.G.; Eldar, Y.C. Towards Vascular Ultrasound Super-Resolution without Contrast Agents. In Proceedings of the 2019 IEEE International Ultrasonics Symposium (IUS), Glasgow, UK, 6–9 October 2019; p. 1.

102. Boni, E.; Yu, A.C.H.; Freear, S.; Jensen, J.A.; Tortoli, P. Ultrasound open platforms for next-generation imaging technique development. *IEEE Trans. Ultrason. Ferroelectr. Freq. Control* **2018**, *65*, 1078–1092. [CrossRef]

103. Burshtein, A.; Birk, M.; Chernyakova, T.; Eilam, A.; Kempinski, A.; Eldar, Y.C. Sub-Nyquist Sampling and Fourier Domain Beamforming in Volumetric Ultrasound Imaging. *IEEE Trans. Ultrason. Ferroelectr. Freq. Control* **2016**, *63*, 703–716. [CrossRef] [PubMed]

104. Yu, J.; Yoon, H.; Khalifa, Y.M.; Emelianov, S.Y. Design of a Volumetric Imaging Sequence Using a Vantage-256 Ultrasound Research Platform Multiplexed with a 1024-Element Fully Sampled Matrix Array. *IEEE Trans. Ultrason. Ferroelectr. Freq. Control* **2020**, *67*, 248–257. [CrossRef]

105. Yoon, H.; Song, T.K. Sparse rectangular and spiral array designs for 3D medical ultrasound imaging. *Sensors* **2020**, *20*, 173. [CrossRef]

106. Rasmussen, M.F.; Christiansen, T.L.; Thomsen, E.V.; Jensen, J.A. 3-D imaging using row-column-addressed arrays with integrated apodization -Part i: Apodization design and line element beamforming. *IEEE Trans. Ultrason. Ferroelectr. Freq. Control* **2015**, *62*, 947–958. [CrossRef]

107. O'Donnell, M. Coded excitation system for improving the penetration of real-time phased-array imaging systems. *IEEE Trans. Ultrason. Ferroelectr. Freq. Control* **1992**, *39*, 341–351. [CrossRef] [PubMed]

108. Nowicki, A.; Secomski, W.; Trots, I.; Litniewski, J. Extending penetration depth using coded ultrasonography. *Bull. Pol. Acad. Sci.* **2004**, *52*, 215–220.

109. Principles, S.I.B.; Shen, J.; Member, S.; Ebbini, E.S. A New Coded-Excitation Ultrasound Imaging. *Ultrason. Ferroelectr. Freq. Control IEEE Trans.* **1996**, *43*, 131–140.

110. Chiao, R.Y.; Hao, X. Coded excitation for diagnostic ultrasound: A system developer's perspective. *IEEE Trans. Ultrason. Ferroelectr. Freq. Control* **2005**, *52*, 160–170. [CrossRef] [PubMed]

MDPI

St. Alban-Anlage 66

4052 Basel

Switzerland

Tel. +41 61 683 77 34

Fax +41 61 302 89 18

www.mdpi.com

Sensors Editorial Office

E-mail: sensors@mdpi.com

www.mdpi.com/journal/sensors